Molecular
Epidemiology

Molecular Epidemiology

Principles and Practices

Edited by

Paul A. Schulte

Centers for Disease Control and Prevention
National Institute for Occupational Safety and Health
Robert A. Taft Laboratories
Cincinnati, Ohio

Frederica P. Perera

Department of Epidemiology
Columbia University
New York, New York

Academic Press, Inc.

Harcourt Brace & Company

San Diego   New York   Boston   London   Sydney   Tokyo   Toronto

Academic Press, Inc.
1250 Sixth Avenue, San Diego, California 92101-4311

United Kingdom Edition published by
Academic Press Limited
24–28 Oval Road, London NW1 7DX

Library of Congress Cataloging-in-Publication Data

Molecular epidemiology : principles and practices / edited by Paul A.
    Schulte, Frederica Perera.
        p.   cm.
     ISBN 0-12-632345-3
     1. Molecular epidemiology.   2. Biochemical markers.   I. Schulte,
    Paul A.   II. Perera, Frederica P.
     [DNLM: 1. Biological Markers.   2. Epidemiologic Factors.
    3. Molecular Biology.   QH 506 M71925]
    RA652.5.M65   1993
    614.4–dc20
    DNLM/DLC
    for Library of Congress   92-49193
                                                                 CIP

PRINTED IN THE UNITED STATES OF AMERICA
    93 94 95 96 97   MP   9 8 7 6 5 4 3 2 1

Contents

Part II PRACTICAL APPLICATIONS

11 Carcinogenesis

Frederica P. Perera and Regina Santella

12 Infectious Disease

Lee H. Harrison and Diane E. Griffin

13 Cardiovascular Disease

Paul A. Schulte, Nathaniel Rothman, and Melissa A. Austin

22   Epilogue

Nathaniel Rothman

Contributors

Numbers in parentheses indicate the pages on which the authors' contributions begin.

Nabih R. Asal (469), Department of Biostatistics and Epidemiology, University of Oklahoma Health Sciences Center, Oklahoma City, Oklahoma 73104

Melissa A. Austin (341), Department of Epidemiology, School of Public Health and Community Medicine, University of Washington, Seattle, Washington 98185

Janice S. Dorman (365), Department of Epidemiology, Graduate School of Public Health, University of Pittsburgh, Pittsburgh, Pennsylvania 15261

Pierre O. Droz (137), Institute of Occupational Health Sciences, University of Lausanne, 1005 Lausanne, Switzerland

Lawrence J. Fine (547), Division of Surveillance Hazard Evaluations, and Field Studies, National Institute for Occupational Safety and Health, Centers for Disease Control and Prevention, Robert A. Taft Laboratories, Cincinnati, Ohio 45226

Vincent F. Garry (497), Laboratory of Environmental Medicine and Pathology, University of Minnesota, Minneapolis, Minnesota 55414

Diane E. Griffin (301), Departments of Neurology and Medicine, Johns Hopkins Hospital, Baltimore, Maryland 21205

Jack D. Griffith (497), Health Effects Research Laboratory, Environmental Protection Agency, Research Triangle Park, North Carolina 27711

Elaine W. Gunter (217), National Center for Environmental Health and Injury Control, Centers for Disease Control and Prevention, Atlanta, Georgia 30333

Prabodh K. Gupta (443), Department of Pathology and Laboratory Medicine, The University of Pennsylvania Medical Center, Philadelphia, Pennsylvania 19104

Lee H. Harrison (301), Departments of International Health and Medicine, Johns Hopkins School of Hygiene and Public Health and Medicine, The Johns Hopkins University, Baltimore, Maryland 21205

Dale Hattis (251), Center for Technology Policy Industrial Development, Massachusetts Institute of Technology, Cambridge, Massachusetts 02139

George P. Hemstreet III (469), Department of Urology, University of Oklahoma, Oklahoma City, Oklahoma 73104

Gerry M. Henningsen (547), Environmental Protection Agency, Denver, Colorado 80202

Vicki Stover Hertzberg (199), Division of Biostatistics, Department of Environmental Health, University of Cincinnati, Cincinnati, Ohio 45267

Robert E. Hurst (45, 469), Department of Urology, University of Oklahoma Health Sciences Center, Oklahoma City, Oklahoma 73104

Muin J. Khoury (365), Division of Birth Defects and Developmental Disabilities, Center for Environmental Health and Injury Control, Centers for Disease Control and Prevention, Atlanta, Georgia 30333

Grace Kawas Lemasters (385), Department of Environmental Health, College of Medicine, University of Cincinnati, Cincinnati, Ohio 45207

J. Patrick Mastin (547), Immunochemistry Research Section, Applied Biology Branch, Division of Biomedical and Behavioral Science, National Institute for Occupational Safety and Health, Cincinnati, Ohio 45226

James L. Mulshine (443), Biomarkers and Prevention Research Branch, Division of Cancer Prevention and Control, National Cancer Institute, National Institutes of Health, Bethesda, Maryland 20814

Frederica P. Perera (79, 277), School of Public Health, Columbia University, New York, New York 10032

Norman J. Pressman (443), Cell Systems International, Inc., Rockland, Delaware 19732

Jian Yu Rao (45), Departments of Urology, Biochemistry, and Molecular Biology and Environmental Health Sciences, Health Sciences Center, University of Oklahoma, Oklahoma City, Oklahoma 73104

Nathaniel Rothman (159, 341, 565), Occupational Studies Section, National Cancer Institute, Bethesda, Maryland 20892

Estelle Russek-Cohen (199), Department of Animal Sciences, University of Maryland, College Park, Maryland 20742

Regina Santella (277), School of Public Health, Columbia University, New York, New York 10032

David Schottenfeld (159), Department of Epidemiology, School of Public Health, University of Michigan, Ann Arbor, Michigan 48109

Paul A. Schulte (3, 79, 109, 159, 235, 341, 385, 407), National Institute for Occupational Safety and Health, Centers for Disease Control and Prevention, Robert A. Taft Laboratories, Cincinnati, Ohio 45226

Ken Silver (251), Center for Technology Policy Industrial Development, Massachusetts Institute of Technology, Cambridge, Massachusetts 02139

Carlo H. Tamburro (517), Departments of Medicine and Pharmacology and Toxicology, School of Medicine, University of Louisville, Louisville, Kentucky 40292

Melvyn S. Tockman (443), Department of Environmental Health Sciences, Johns Hopkins School of Hygiene and Public Health, Johns Hopkins University, Baltimore, Maryland 21205

Paolo Vineis (109), Dipartimento di Scienze Biomediche, a Oncologia Umana, Universita di Turino, 10126 Turino, Italy

Robert F. Vogt, Jr. (109, 407), Center for Environmental Health, Centers for Disease Control and Prevention, Atlanta, Georgia 30333

Deborah M. Winn (217), Division of Health Interview Statistics, National Center for Health Statistics, Hyattsville, Maryland 29782

John L. Wong (517), Departments of Chemistry and Pharmacology and Toxicology, University of Louisville, Louisville, Kentucky 40292

Preface

Epidemiologists, now and in the future, will be asked to use increasingly powerful biologic markers of exposure, disease, or susceptibility. These markers promise to reduce misclassification and detect disease earlier in its natural history by identifying biological changes on increasingly smaller scales, eventually examining individual molecular perturbations. Hence, "molecular epidemiology" describes opportunities for inclusion of current and future generations of biomarkers in epidemiologic research.

Molecular epidemiology utilizes the same paradigm as traditional epidemiology. However, it presents the opportunity to use a new resolving power in the assessment of exposure–disease relationships. The resolving power, involving a continuum of events between exposure and disease, can provide new approaches to research, prevention, and intervention.

"Biomarkers" is a contemporary term that describes the current generation of biologic indicators. This term might seem like a new label for well-known phenomena, but it signals a range of new issues and practical questions that will confront epidemiologists and other scientists as they attempt to use biologic markers. Some of these issues include determining the pharmacokinetic relationship between a marker and exposure; characterizing a marker as validated at the laboratory and at the population level; epidemiologic issues surrounding the collection and banking of specimens; and ethical issues in communicating group findings to individuals. While certain aspects of these questions have been addressed elsewhere, there is no current text that addresses all of them using examples of current biologic markers.

Other questions that face researchers include: Should metabolic molecular phenotypes and genotypes be used as confounders or as effect modifiers? What is the gold standard for validating markers of exposure and effect? And, how should biologic marker data be analyzed and entered into statistical models?

The promise of biologic markers to enhance epidemiologic research will only be realized when biological markers are routinely utilized. Biologic markers are identified and developed in the laboratory, but, before they can become useful epidemiologic tools, they need to be validated and field tested.

Taking markers to the field, however, will require attention to issues of validity, sampling, variability, range of normal, background prevalence, specimen collection, study design, and analysis. This process requires communication between laboratory scientists, epidemiologists, statisticians, industrial hygienists, and other public health professionals. The basis of that communication should be the need to understand the requirements of each discipline with regard to biologic markers.

The goal of this text is to provide a compendium of the epidemiologic issues and methods pertinent to biologic markers. For the laboratory scientists, this text is designed to enumerate the considerations necessary for valid field research. For the epidemiologists and other field scientists, this text should also provide a resource on the salient and subtle features of biologic markers. The book should serve as a primer for both laboratory and field scientists who will shape the emerging field of molecular epidemiology.

To the extent possible, we will identify epidemiologic methodologies that will transcend current specific biologic markers and apply to any emerging markers. To be successful in that endeavor, we need to build on our knowledge of those markers currently in use.

In this book, we have tried to provide a broad view of the principles and practices that might be useful in performing molecular epidemiologic research. To some, "molecular epidemiology" is an exciting phrase that conveys the potential for incorporating biologic markers, especially ones depicting events at the genetic or molecular level, into epidemiology. For others, "molecular epidemiology" evokes skepticism that arises from the belief that there is nothing new in molecular epidemiology or even from the view that such reductionist approaches are antithetical to public health. Much of the reason for this divergence of views is due to the fact that the body of knowledge and practices that encompass molecular epidemiology has not been assembled or presented in one place or in a unified fashion. We have tried to remedy this and provide a text that attempts to describe what comprises the hybrid discipline of molecular epidemiology. We don't actually argue that it is a discipline, and in fact, it may not be. But for heuristic purposes, we describe it as if it were a field.

We hope we are not being audacious when we attempt to summarize such a diverse set of efforts, investigations, and practices under a single heading, but it appears there is a common theme that is now running through most biomedical and public health research, namely, that new scientific understanding and technological capabilities are allowing us to "see" how the processes of life work at the most basic levels. This influences how we define disease, discern its types, and identify precursor conditions. It also allows us to measure what interacts with the molecules of life and what distorts them. We are entering an era where we no longer have to guess as much about what is or has happened to a person with regard to exposures, susceptibility, or risk. We can use these new potential capabilities to supplement the long-tested, observational, and analytical methods of epidemiology to answer questions about the causes and mechanisms of disease.

This book has been written for audiences from different disciplines and it tries to erect guideposts toward the conduct of effective research using biologic markers in human populations. It is not intended to teach epidemiology to laboratory scientists or molecular biology to epidemiologists. It introduces laboratory scientists to a framework for using and considering their work in conjunction with a long-established approach for studying diseases in populations. It introduces epidemiologists to the range of practices, techniques, and considerations necessary for applying biologic markers to epidemiologic research.

To those already practicing molecular epidemiology, we hope it will serve as a compendium of approaches, strategies, issues, and guidelines that will allow their thinking to coalesce and move forward. We asked the contributing authors to think globally, that is, not only in terms of which markers are currently being used or are on the horizon, but also what issues in methodology transcend any particular marker that might become obsolete, and what practices are most important in performing field studies in their area of research. We attempted to weave the theme through all the presentations. To help in that regard, we asked the authors to use the categories of "exposure," "disease or outcome," and "susceptibility" in which to frame their discussions.

The first part of the book addresses general principles for molecular epidemiology. In Chapter 1, we define the terms, discuss what molecular epidemiology actually is or can be, and trace the historic contributions to it. Chapter 2 describes the basic molecular biologic information that is necessary for understanding the field and much of the discussion throughout the book. Chapters 3 and 4 address those factors which influence whether the biologic markers that are used in molecular epidemiology are valid and reliable, and how this can be approached and substantiated. Chapter 5 addresses how exposure to xenobiotics can be assessed through pharmacokinetics and pharmacodynamic modeling. In Chapter 6, the range of epidemiologic study designs and structural questions that can form a framework for any molecular epidemiologic study are considered. Chapter 7 addresses statistical questions in using biologic markers. Chapter 8 addresses the collection and storage of biologic specimens, which is a crucial requirement of successful molecular epidemiologic research. Chapter 9 opens the door to the thorny issues involved in the interpretation and communication of molecular epidemiologic information, which has a broad range of potential impacts. In Chapter 10, the use of biologic markers and molecular epidemiology to enhance the field of risk assessment is discussed.

The second half of the book traces the practice of molecular epidemiology, or at least the use of biologic markers, in the different disease categories. Authors were asked to provide a status report on the extent to which biologic markers have been used in epidemiologic or other human studies and what markers are on the horizon. Chapters 11, 12, 13, and 14 cover four disease categories—carcinogenesis, infectious, cardiovascular, and genetic—that have been fertile ground from which molecular epidemiology has grown. Chapters 15, 16, and 17 on reproductive, immunologic, and pulmonary diseases represent areas rich in biologic markers. Reproductive epidemiology, in the last decade, has shown an increase in the use of biologic markers. Immunologic research, essentially a marker-based discipline and long a partner with infectious disease epidemiology, now is poised to use molecular epidemiology to assess immunotoxicologic endpoints. Pulmonary research, which has a history of using physiologic markers, is now moving to the incorporation of molecular and genetic markers.

Chapter 18 pertains to urologic diseases, clearly a laboratory for assessing the continuum of biologic markers. Chapters 19, 20, and 21 on neurologic, liver, and musculoskeletal diseases represent hard-to-get-at organ systems for the purpose of assessing biologic markers. The interesting and creative attempts in these areas are discussed.

Finally, we asked our colleague, Nat Rothman, to muse a bit about the potrayal of molecular epidemiology and the forces that push and pull at the field (Chapter 22).

Molecular epidemiology is an interdisciplinary science. We have attempted to capture that spirit and have tried to further it. We hope the book will be a useful resource for those interested in understanding, pursuing, and advancing the practice of molecular epidemiology.

Many people have contributed to the writing, assembly, review, and editing of this book. We are grateful to all the contributing authors and to the following people for helpful discussions, review, or support: Aaron Blair, Diane Brenner, Neil Caporaso, Richard Carlson, Claffertine Cheeks, Michael Dohn, Vera Drake, Lawrence Fine, Gwen Finegan, Marilyn Fingerhut, Joan Friedland, Chris Gersic, Barbara Grajewski, Marie Haring-Sweeney, Richard Hayes, Ellen Heineman, Gerry Henningsen, Robert Herrick, Reggie Kuhns, Joan Levine, Kathryn Linenger, Jack Mayer, Alma McLemore, Medical Art Company, Juanita Nelson, Robert Oster, Suzzana Park, Ken Radak, Robert Rinsky, Nat Rothman, Alberto Salvan, Regina Santella, Mark Schiffman, Susan Schmidt, Joseph Selby, Velda Smiley, Lee Sonke, Kyle Steenland, Patricia Stewart, Anne Stirnkorb, Lynn St. Clair, Davis Stroop, Glenn Talaska, Deliang Tang, Martha Waters, Timothy Wilcosky, and Leslie Yee.

Finally, we are grateful to Charlotte Brabants, Gayle Early, and the staff at Academic Press for their enthusiasm and support.

Part I

GENERAL PRINCIPLES

1

A Conceptual and Historical Framework for Molecular Epidemiology

Paul A. Schulte

We are in the era of molecular research. Between 1970 and 1990,[1] the number of medical journals with the word "molecular" in the title grew from 31 to 90, signaling that the understanding of biologic phenomena has proceeded to the molecular level. This evolution resulted from advances in molecular biology, genetics, analytical chemistry, and other basic sciences. It is now possible to detect smaller amounts of analytes and contaminants and smaller biological changes, as well as to identify mechanisms at the cellular and molecular levels. Progress in the molecular approach to biology and medicine has stimulated and excited both the public and researchers, who now believe these advances can be applied to the study, prevention, and control of health risks faced by human populations. The term "molecular epidemiology" may be used to describe such an approach: the incorporation of molecular, cellular, and other biologic measurements into epidemiologic research.

The use of molecular markers represents a quantum leap in the evolution of epidemiologic ideas. Epidemiology has evolved through development and inclusion of many advances such as the systematic collection and analysis of vital statistics; delineation of the triad of agent, host, and vector (applied in infectious and chronic diseases); refined exposure assessments such as dietary questionnaires and job exposure matrices; clearly delineated study designs (longitudinal and case-control); and heightened computational and statistical capabilities (maximum likelihood estimators, logistic and Poisson regression). To this list now must be added technologically powerful measures of biologic variables, that is, biologic markers indicating events at the physio-

---

[1] Comparison of National Library of Medicine journal listings for 1970 and 1990.

logic, cellular, subcellular, and molecular levels. Molecular epidemiology is the use of these biologic markers in epidemiologic research. Although use of biologic markers is not new to epidemiology, the current generation of markers enhances past approaches. The use of validated biologic markers can contribute the following opportunities and capabilities to epidemiologic research:

1. delineation of a continuum of events between an exposure and a resultant disease;
2. identification of exposures to smaller amounts of xenobiotics and enhanced dose reconstruction;
3. identification of events earlier in the natural history of clinical diseases and on a smaller scale;
4. reduction of misclassification of dependent and independent variables;
5. indication of mechanisms by which an exposure and a disease are related;
6. better accounting for variability and effect modification; and
7. enhanced individual and group risk assessments.

Collectively, these capabilities provide additional tools for the epidemiologist studying questions on the etiology, prevention, and control of disease (Fig. 1.1).

Molecular epidemiology is a natural confluence of powerful developments in basic biomedical sciences and the field-tested methods of epidemiology. Although "molecular epidemiology" can be viewed as an evolutionary step in epidemiology, a supplemental set of tools, or even a separate discipline, it generally does not represent a shift in the basic paradigm of epidemiology. This new approach allows for more accurate comparisons among groups, further clarification of mechanisms, and more specialized assessment of individual risk functions, all of which have been established in historical epidemiology. Molecular epidemiology does not have all of the characteristics of a distinct discipline or even of a branch of epidemiology. Rather, it is better seen as a diverse range of approaches and techniques that can supplement the field of epidemiology and boost the field to a new level of opportunity and capability.

Capabilities of Molecular Epidemiology

Delineation of a Continuum of Events between Exposure and Disease

Figure 1.2 shows the evolution of the concept of a continuum useful in molecular epidemiologic research [Perera and Weinstein, 1982; Gann *et al.*, 1985; National Research Council (NRC) 1987; Hulka and Wilcosky, 1988;

SEQUENCE LEADING TO NEOPLASIA   INTERVENTION STRATEGIES

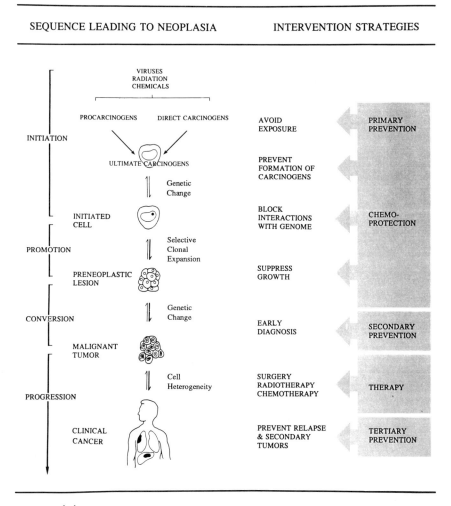

FIGURE *1.1*   Intervention strategies that can utilize molecular epidemiologic methods: The example of cancer. Molecular epidemiologic approaches can be used to identify etiologic factors and to assess effectiveness of chemoprevention, secondary and tertiary prevention, and therapy. (Reprinted with permission of Thomas W. Kensler, 1992.)

Schulte, 1989]. A rich level of detail can supplement previous categorical models that linked exposure and disease. The concept of a continuum of events between exposure and disease provides the opportunity to insure that epidemiologic research has a biologic basis for hypotheses and provides the analyses to test these ideas. New opportunities for classical and hybrid epidemiologic study designs can be applied to this continuum (see Chapter 6).

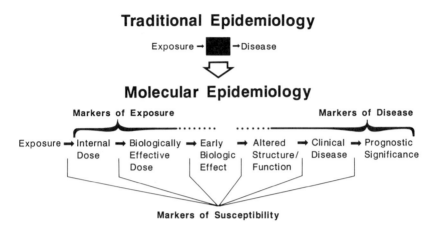

FIGURE 1.2   Evolution of the detailed continuum for molecular epidemiologic research.

Identification of Exposures to Smaller Amounts of Xenobiotics and Enhanced Dose Reconstruction

The powerful tools of molecular biology, analytical chemistry, and related disciplines now allow exposure determinations on the order of 1 part in 10^{18} or 10^{21} (Abdel-Baky and Giese, 1991; Wild and Montesano, 1991). This ability to identify small amounts of a xenobiotic makes more active consideration of "background" levels of target xenobiotics in nominally nonexposed subjects important when designing studies and assessing covariates and confounding factors.

A key capability of molecular epidemiology is assessing past exposures and reconstructing doses received from past exposures by using biologic measurements on samples taken from small groups of subjects (Ehrenberg *et al.*, 1974; Calleman *et al.*, 1978; Thilly, 1982; Wogan and Gorelick, 1989; Groopman *et al.*, 1991; Mendelsohn, 1991; Taylor *et al.*, 1992). This procedure is termed "biologic dosimetry." Biologic dosimetry can complement traditional methods of dose reconstruction by using personal dosimeters to measure ambient exposure, by estimating body burdens through sampling fat, urine, or other materials, or by detecting adducts, gene mutations, chromosome aberrations, or other relevant markers (Mendelsohn, 1991).

Identification of Events Earlier in the Natural History

As shown in Figure 1.2, when a continuum or part of a continuum between an exposure and a disease is identified and understood, it is possible to focus on preclinical rather than clinical events. Thus, asymptomatic individuals who are at increased risk of manifesting clinical disease can be identified (Hatch and Friedman-Jimenez, 1991). Some examples of indicators include decrease in CD4 lymphocytes in HIV-infected persons (Detels *et al.*, 1987), expression of p300 in bladder cells in people at risk of bladder cancer (Rao *et al.,* 1991), elevated levels of lipoprotein Lp(a) in persons at risk for cardiovascular disease (Murai *et al.,* 1986), and various sperm parameters in individuals at risk of reduced fertility (Bostofte *et al.,* 1988).

The ability to identify prodromal events expands the pool of potential "cases" for epidemiologic studies (Hattis, 1988). It permits studies of interventions that can have impact on the group being studied as well as on the individuals to whom the results can be generalized (Cullen, 1989; Greenwald *et al.,* 1990).

Reduction of Misclassification of Variables

Misclassification of exposure and disease variables is a major weakness of epidemiologic studies (Rothman, 1986; Hogue and Brewster, 1988). Better classification of exposure than that achieved using historical characteristics and measurements may be accomplished by assessing markers of internal and biologically effective doses (Hogue and Brewster, 1988; Hulka and Wilcosky, 1988; Landrigan, 1988; Schulte, 1989). More homogeneous disease groupings can be defined using markers of effect such as specific mutations indicative of exposure (mutational spectra) (Shields and Harris, 1991). The validity and precision of point estimates may be increased as misclassifications are reduced.

Indication of Mechanisms

Delineating a continuum of events between exposure and disease provides opportunities for insight into the mechanism of action (Vogelstein *et al.,* 1988). Much epidemiologic research has been based on theorization about mechanisms, or at least some prior speculation that exposure and outcome are related. Molecular epidemiologic approaches facilitate testing the association between mechanistic events in a defined continuum (Ehrenberg, 1974; Harris *et al.,* 1987; Hatch and Stein, 1987; Perera, 1987b; Schulte, 1989; Harris C. C., 1991; Hatch and Friedman-Jimenez, 1991). Knowledge of the mechanism can guide future research and intervention applications.

Accounting for Variability and Effect Modification

Perhaps one of the greatest contributions of molecular epidemiology is the ability to discern the role of host factors, particularly genetic factors, in accounting for variation in response (Omenn, 1982; Cartwright *et al.,* 1982; Rajput-Williams *et al.,* 1988; Kuller, 1991; Shields and Harris, 1991; Wetmur *et al.,* 1991; van Noord, 1992). Why similarly exposed people do not get the same diseases is a target question for molecular epidemiology. In most disease systems, susceptibility markers are being identified and evaluated. These markers can be incorporated into epidemiologic models as effect modifiers (Hulka *et al.,* 1990).

Enhanced Individual and Group Risk Assessments

The use of epidemiologic data to provide individual and group risk assessments is well established (Paul, 1930; Truett *et al.*, 1967). For example, individual risk functions have played a strong role in cardiovascular disease research and control (Truett *et al.*, 1967), in pulmonary and occupational medicine (Ingram and McFadden, 1981), in infectious disease control (Paul, 1930), and in genetic epidemiology and counseling (Ahearn and Hochberg, 1988). Molecular epidemiology can enhance individual and group risk assessments by providing more person-specific information, allowing extrapolation of risk from one group to another, from animal species to humans, and from groups to individuals (Perera, 1987a; Harris *et al.*, 1987; Alavanja *et al.*, 1987; Harris, 1991; Shields and Harris, 1991). The "parallelogram approach" used in genetic toxicology is a model for animal-to-human extrapolation (Sobels, 1982). A marker appropriate to both species (animal and human) that can be related to exposure–disease relationships in the animal can serve as the basis for predicting effects in exposed humans. Similarly, extrapolation from group to group, group to individual, or individual to group follows the same general model (see examples in Figures 1.3 and 1.4). Identification of a detailed continuum of events between an exposure and disease, coupled with covariates of the event variables in multivariate models,

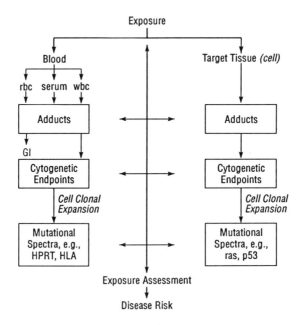

FIGURE 1.3 Exposure assessment paradigm for linking carcinogen–macromolecular adducts, cytogenetic aberrations and mutational spectra. GI, gastrointestinal tract. (Reprinted with permission from C. C. Harris, 1991, copyright CRC Press, Inc.)

permits the calculation of individual risk functions. (See Truett *et al.*, 1967, for an example using serum lipid biomarkers and cardiovascular disease risk functions.) Molecular markers can heighten the specificity of these functions and allow reduced confidence intervals around estimates. Not only is it now possible to say that a middle-aged man with heart disease and a cholesterol level above 240 mg/dl will have a one-in-five chance of dying from a heart attack within 10 years; it may soon be possible to indicate which man that will be (Begley *et al.*, 1991).

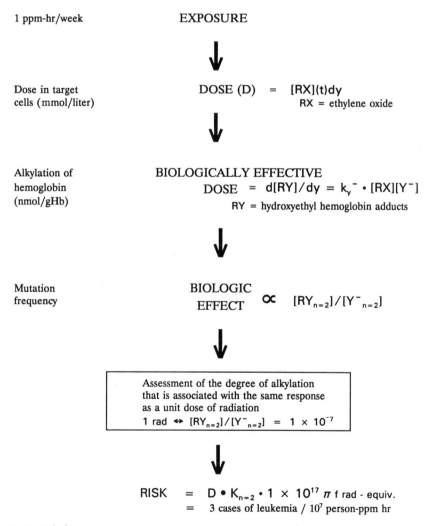

FIGURE 1.4 Example of risk assessment of workers exposed to ethylene oxide at a Swedish plant. (Adapted from Osterman-Golkar and Bergmark, 1988, and Ehrenberg *et al.*, 1980.)

Some readers may view the term "molecular epidemiology" as an oxy-moron. Epidemiology is the study of health effects in groups of people. "Molecular" and "cellular" indicate assessment of the individual at the component level. Epidemiology relies on observation and inference of associations between variables. Molecular and cellular sciences use experimental proof of cause and effect. Despite these different scopes, scales, and approaches, molecular sciences and epidemiology are compatible, if not inevitably linked (Kuller, 1991). There is a historical basis for such a link, as well as a conceptual and epistemological framework for epidemiologic research that incorporates biologic measurements of human processes (Paul, 1958; Paul and White, 1973; Kilbourne, 1979; Feinstein, 1991).

In a sense, molecular epidemiology is a signpost that flags the need to incorporate an understanding of biologic phenomena at the physiologic, cellular, and molecular levels into epidemiologic research (Fig. 1.5). Epidemiologists long have used biologic markers (e.g., antibody titers, serum lipids, blood lead). However, in the past when high "exposures" and single outcomes were more prevalent and frequent, epidemiologists argued that knowledge of associations was more useful than understanding the mechanisms, since prevention through control of exposures was often feasible even in the absence of understanding cellular processes. (Snow, 1855; Maclure and MacMahon, 1980; Hatch and Stein, 1987). Previous success in public health led to the identification of the major single or primary cause of diseases. Today, exposures are often smaller and mixed; understanding mechanisms could be more important in determining appropriate intervention strategies. The health conditions of interest today are multicausal; to investigate them requires a wide array of disciplines.

If molecular epidemiology has characteristics of a field or speciality, they are its hybrid interdisciplinary qualities. This new specialty requires attention to new organizational and educational structures and adherence to principles and practices derived from both its molecular biology and epidemiology roots. Molecular epidemiology is not a fundamental departure from the past, but an evolutionary step. In this chapter, I trace the conceptual and historical development of the "field" of molecular epidemiology.

The goal of molecular epidemiology should be to supplement and integrate, not to replace, existing methods. Molecular epidemiology is also a heuristic term used to describe an enhanced capability of epidemiology to understand disease in terms of the interaction of environment and heredity. Although this practice too has been part of the epidemiologic tradition, it generally has been confined to the specialty area of "genetic epidemiology."

Researchers easily can become arrogant in the face of the potential that molecular epidemiology offers. Believing that merely making molecular measurements leads to enhanced understanding of biologic phenomena is tempting. However, this is not always true and can be misleading, because such a reductionist approach fails to pay attention to the social and cultural char-

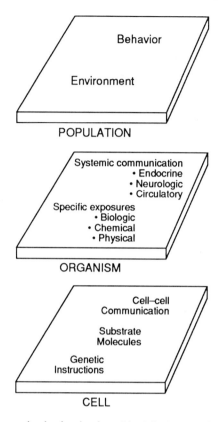

FIGURE 1.5   Component levels of molecular epidemiologic research: The example of cancer. Biologic measurements of individuals and their components, as independent and dependent variables, have a long history of use in epidemiology. The current generation of measurements can be made at the molecular level. (Figure courtesy of John D. Potter.)

acteristics of human populations and to the impact that this type of research has at the population level. Ultimately, research at both the "micro" and the "macro" level is necessary.

Strengths and Limitations of Observational Epidemiology

Epidemiology is the study of the distribution and determinants of health-related states and events in populations and the application of the results of this study to control health problems (Last, 1988). The focus of epidemiology is the group rather than the individual; understanding is gained through inferences drawn from observations within and among groups. Causation is inferred rather than proved. The strengths and weaknesses of epidemiology derive from the field's primary goal: identification and control of the causes of human diseases. The strength of epidemiology is that it does not require extrapolation from other species or laboratory experiments. The weakness of the field stems from ethical constraints imposed on human experimentation and, thus, the requirement to study human disease in its natural state. The epidemiologic study of "free living" humans is more problematic than the study of controlled animals. Factors such as history, geography, social characteristics, and status are exceedingly powerful predictors of the health of human populations. Whereas laboratory animals are usually genetically homogeneous and live in controlled conditions, humans are genetically heterogeneous and live in diverse conditions. In short, because of the subject, all major methods of epidemiology (except for clinical trials) are essentially observational and nonexperimental. Drawing inferences about causation from observational studies is considerably more difficult than drawing them from experiments that use random samples and controls [National Academy of Sciences (NAS), 1991b].

Much of epidemiology has been essentially ecologic (Hogue and Brewster, 1988; Hulka and Wilcosky, 1988). Often an individual is assigned the characteristics of a group; categorical descriptors are used to assess risk factors and health status. These descriptors range from dichotomous characterizations (exposed or not; diseased or not) to more quantitative representations of categories (high, medium, or low exposure). Analytic epidemiology uses measurements made on each subject or applies some specific criterion to classify subjects. In many instances, these measurements are surrogates, for example, measures of ambient air exposure for dose or measures of symptoms in lieu of underlying disease. In general, in epidemiology the group is the unit of comparison.

The epidemiologic approach has been highly successful and is the cornerstone of public health. From John Snow's (1855) assessment and control of the cholera epidemic in London to the understanding of the role of environmental factors in cancer and cardiovascular disease, epidemiology has contributed to finding causes and remedies. Despite these contributions, epidemiology has been limited by its fundamental conceptual and technical characteristics. Ecologic characterization of variables has hindered the epidemiologic approach in addressing the interdependence of multiple agents and variations in susceptibility that lead to disease. Epidemiology is also limited in its ability to classify individuals into exposure categories (misclassification), to identify mechanisms of action, and to detect disease at a time in its natural history when intervention in the study population or in other preclinical populations would be most effective (Higginson 1977b; Hogue and Brewster, 1988; Hulka and Wilcosky, 1988).

Emerging abilities to assess exposure and disease at the cellular and molecular levels are promising supplements to traditional epidemiology. Instead of characterizing groups solely by geographic location, job title, or questionnaire-derived history, it is possible to measure dose by collecting biologic specimens and assessing xenobiotic interactions with biologic sites and molecules (Schulte 1987, 1989; Rogan, 1988; Hoffman *et al.*, 1991). At the other end of the spectrum, instead of making comparisons of cases of frank disease with controls, it will be possible to make assessments based on preclinical events such as abnormal DNA content (Hemstreet *et al.*, 1988) or oncogene alteration (Shields and Harris, 1991), once these end points are established as predictive of clinical disease (Rogan, 1988). Additionally, molecular methods make it possible to distinguish subtypes of clinical disease that have potentially different etiologies (Hollstein *et al.*, 1991; Taylor *et al.*, 1992).

A long-observed weakness of epidemiology is its limited ability to address host factors that contribute to variable responses (Rose, 1990). Why similarly exposed people do not all acquire the same disease is a difficult question. Certainly, gross categories of host factors, such as age, race, and sex, are controlled routinely in epidemiologic studies. However, genetic factors generally have been considered to a lesser extent than environmental factors. Evaluation of gene–environment interactions has been minimal. "In the past epidemiologists tended to favor a 'black box' approach to their work: they have measured inputs from external agents and from susceptibility factors and disease outcomes; but they have not been concerned with explaining how the two come to be related" (Rose, 1990).

Molecular epidemiology is not just a term that describes adding new techniques to epidemiology. Rather, it represents an opportunity to use new resolving powers to develop theories of disease causation that acknowledge complex interactions involved in the health–disease process. Considering disease at the molecular level without confronting the other events, such as genetic differences and competing biochemical processes, that occur at the molecular level is not sufficient. The question addressed in this volume is how to integrate these molecular biologic capabilities—measurements made in individuals—into a science that uses comparisons of groups to find causes of disease and opportunities for health protection.

Molecular Epidemiology—The Use of Biologic Markers in Epidemiologic Research

Definitions

Events in the Continuum between Exposure and Disease

A functional definition of molecular epidemiology is the use of biologic markers or biologic measurements in epidemiologic research. Biologic markers (or biomarkers) generally include biochemical, molecular, genetic, immunologic, or physiologic signals of events in biologic systems (NRC, 1987). The events represented can be depicted as parts of a continuum between a causal initiating event (sometimes an exposure to a xenobiotic substance) and resultant disease. By definition, a continuum is a whole, no part of which can be distinguished from neighboring parts except by arbitrary divisions (The American Heritage Dictionary, 1976). Thus, it is important to remember that the various heuristic components of the continuum shown in Fig. 1.2 are arbitrary.

The proposed continuum between exposure and disease has been described in a number of reports (Perera and Weinstein, 1982; Hatch and Stein, 1987; NRC, 1987; Perera 1987a,b; Schulte, 1989) and is shown in Figure 1.2. Between *exposure* (E) in the environment and the development of *clinical disease* (CD), four generic component classes of biologic markers have been identified: the *internal dose* (ID), the *biologically effective dose* (BED), *early biologic effects* (EBE), and *altered structure and function* (ASF). Clinical disease can be represented not only by biologic markers for the current disease but also by markers for *prognostic significance* (PS). Each marker represents an event in the continuum. The relationships among the markers are influenced by various factors (such as genetic or other host characteristics) that reflect susceptibility to any of the events in the continuum. These indicators for susceptibility also can be represented by markers.

Definition of all the marker events has been elaborated elsewhere (NRC, 1987; Hulka and Wilcosky, 1988) but is summarized briefly here. The continuum between cigarette smoking and lung cancer serves as illustration. ID is the amount of a xenobiotic substance or its metabolites found in a biologic medium (e.g., serum cotinine as an indicator of nicotine). The BED is the amount of that xenobiotic material that interacts with critical subcellular, cellular, and tissue targets, or with an established surrogate tissue (e.g., DNA adducts in peripheral lymphocytes). The BED represents the integration of exposure and effect modification by the host. A marker EBE (e.g., sister chromatid exchange) represents an event correlated with, and possibly predictive of, health impairment. Altered structure or function (e.g., abnormal sputum cytology) and DNA hyperploidy are precursor biologic changes that are more closely related to the development of disease. Markers of CD (e.g., tumor-associated antigen) and of PS (e.g., tumor markers such as CA-125) show the presence or future development of disease, respectively. Markers of susceptibility are indicators of increased (or decreased) risk for any component in the continuum [e.g., extensive debrisoquine metabolizers are at 4- to 6-fold increased risk for lung cancer (Ayesh *et al.*, 1984)].

Relationship between a Marker and the Event It Marks

When considering how biologic markers can be used in epidemiologic research, it is useful to reflect on the nature of the relationship between the marker and the event it marks (Lucier and Thompson, 1987). The semantics of describing a marker in this regard are confusing. Does the marker represent an event, is it an event itself, is it a correlate of the event, or is it a predictor of the event? The answers to these questions may affect who is sampled, how and when they are sampled, and what confounders or effect modifiers are considered. For example, the HPRT gene mutation may be used as a surrogate for other target genes (such as mutated p53 gene) that may be involved in the development of cancer; a hemoglobin adduct, although highly correlated with recent exposures to aklylating agents such as ethylene oxide, may not represent exposures that occur years prior to the collection of a blood specimen, since the dosimetric capacity of the hemoglobin is related to the four-month life of the erythrocyte. Thus, a biologic marker often refers to the use made of a piece of biologic information rather than to a specific type of information (Henderson *et al.*, 1989).

Is a marker different from a test or an assay? Often, biologic markers and tests or assays for a marker are considered the same because, without the assay, the marker cannot be demonstrated. Strictly speaking, they are different and care should be taken not to gloss over the differences. Analytically, a test or assay is said to be valid if it performs "truthfully" in the presence or absence of a marker. The attributes of a test are not necessarily those of the marker. The marker's attributes may pertain to its nature and natural history. A marker may exist although no assay is sensitive enough to detect it. The measurements of a marker involves differentiation of a signal above background noise. A test that has a large signal-to-noise ratio is considered a good test. The signal-to-noise ratio is not generally a set point, but a curve showing the ratio under different conditions. Ultimately, characteristics of the test conditions, the marker, or both may determine the signal-to-noise ratio. These issues are discussed further in Chapters 3 and 4.

Organizational Aspects

Molecular epidemiologic studies require interdisciplinary collaboration between population and field scientists (such as epidemiologists, statisticians, industrial hygienists, exposure assessors, and clinicians) and laboratory scientists from disciplines such as molecular biology, genetics, immunology, biochemistry, pathology, and clinical and analytical chemistry. Interdisciplinary collaboration is not new to epidemiology, but the level and extent of diversity of disciplines that molecular epidemiology will require is unprecedented.

Collaboration requires attention to the underlying assumptions, paradigms, and languages of various disciplines as well as to issues of the institutional context of research (Stein and Jessop, 1988). Every discipline uses assumptions and paradigms to approach a research question. Often these conventions are so fundamental and integrated that investigators may not be conscious of them and, hence, rarely recognize them or make them explicit when interacting with members of other disciplines (Stein and Jessop, 1988). Problems in collaborative research can occur when these fundamentals are not shared. Epidemiologists generally speak in terms of groups and risks to groups. Laboratory scientists tend to focus on individuals or components of an individual. Different disciplines may use common terms in very different ways. Words such as "sensitivity," "small," "valid," "normal," "bias," and "epidemiology" may have different meanings in various disciplines. Too often, the differences may be subtle, for example, understanding of the term "sensitivity." To the laboratory scientist, it refers to the extent to which an assay is capable of detecting a particular marker at some concentration. To an epidemiologist, the same word means the extent to which people with a marker respond positively on a test.

In addition to conceptual and language barriers, interdisciplinary collaboration often may have administrative barriers (Feinstein, 1991). If a study involves disciplines in two or more academic or organizational departments, several questions might arise that must be answered clearly. Which department is in charge? On whose books is the project accounted? To which journals are papers submitted? Who can access which data? Who is the first author? What is the priority for additional assays on specimens? Many administrative difficulties can be avoided if good communication is established from the outset. Effective collaborative work requires mutual commitment to the value of the work, respect for what each participant can offer, intellectual flexibility, tact, patience, and persistence (Stein and Jessop, 1988).

Protocols for molecular epidemiologic studies often differ from traditional epidemiology, not only in the interdisciplinary nature of the research, but also in the need or desire to maintain flexibility to amend the protocol or study to include discovery of new markers. This variability is best handled by anticipating such possibilities, by including references to such changes in informed consent documents, and by processing and storing samples appropriately.

Crucial to the use of biologic markers in human populations is the requirement that biologic markers be validated adequately before their deployment in hypothesis testing studies (Rogan, 1988). The term "validated" not only has different meanings for the laboratory and population scientists, but also leads to different approaches to validation (see Chapters 3 and 4). Therefore, consistent guidelines for validating markers at the laboratory and the field levels are essential. Laboratory validation requires assessing the adequacy and accuracy of assays. Field validation requires determining the population variability and predictive value of the assay in terms of the exposure or disease of interest.

An organizational model for conducting molecular epidemiologic studies is shown in Figure 1.6. Critical in these types of studies are the need for specialists in study design and for subject selection, exposure assessment, and marker assay development to be integrated in all aspects of the work. The amount of data generated is generally much greater than in a classic epidemiologic study because, in addition to questionnaires and recorded data common to both, data are accumulated from biologic specimen analy-

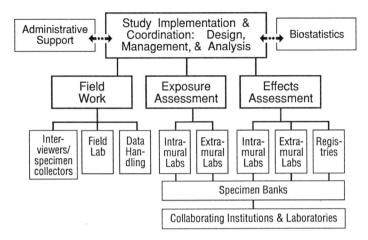

FIGURE 1.6   Project teams required for conducting interdisciplinary molecular epidemiologic studies. (Adapted from Everson, 1989.)

sis. For example, a study of 30 mortuary science students that assessed cytogenetic effects and DNA repair caused by formaldehyde exposure required 1800 slides and generated more than 3600 data points (Suruda *et al.*, 1992). Some of these data came from questionnaires and some from biologic measurements.

The organizational model in Figure 1.6 pertains to studies for validating or using biologic markers. Validation studies require a similar effort as field studies but contain certain differences that may involve animal or *in vitro* research. Validation often requires large, prospective population-based studies that are costly, time consuming, and resource intensive. Prior to such studies, small-scale pilot studies should be made. Pilot studies can provide estimates of assay variability and interindividual variability that are necessary for power calculations. The field of molecular epidemiology has been characterized by pilot studies in highly exposed or high-risk groups or in patient series. Based on the findings of these studies, larger studies in general groups are considered. (See Chapter 6 for a typology of molecular epidemiologic studies.)

Many problems are attendant to developing a marker assay for use in epidemiologic studies. One problem is the difficulty of transforming research tools and research laboratories into productive facilities capable of handling large numbers of specimens in a cost-effective and timely manner. Other challenges include testing for dose response, marker persistence, correlation with other markers, inter- and intraperson variation, and correlation with clinical responses.

As new and more reliable assays are developed, scientists will be pressured to use them to replace previous assays. When a marker assay is new, measurements may differ from those in previous assays for the same marker.

Although it may be reasonable to wait until laboratory techniques and estimates of variability display consistency, it is not reasonable or feasible to wait until a technique is so standard that no refinement of it is likely to occur. Pressure to use the new technique also promotes assessment of comparability of old and new results and, ultimately, making adjustments or corrections. A good balance must be achieved between old and new approaches. In some cases, employing both techniques in a study may be useful.

Molecular epidemiology also requires paying attention to the collection, handling, and storage of biologic specimens (see Chapter 8). Specimens must be stored to be usable for analyses not yet developed in addition to currently known analyses. The actual collection of biologic specimens also may require attention to timing, so specimens are collected when the influence of factors such as therapeutic treatments will be minimal. For example, a study of the role of oncogenes and colon cancer will be more informative if specimens are collected before DNA-altering chemotherapy or radiation is administered. Other "new" issues also arise, such as how to address the difficulty of obtaining biologic specimens from controls. For example, in a case-control study of colon cancer, how can usable colon specimens be obtained from controls?

Use of markers of susceptibility also may have a great impact on the organization and conduct of molecular epidemiologic research. Biologic monitoring data derived from molecular epidemiologic studies are likely to be sought by a diversity of societal groups ranging from insurers and employers to potential mates. The possibility that individual markers can predict future characteristics and outcomes makes them a target of interest. The use of susceptibility markers will be required increasingly in molecular epidemiologic studies that assess exposure or effects at the DNA level, since effect modification or confounding caused by inherited or other nonstudy factors can influence associations under study. The analysis of biologic marker data may create new obligations for researchers that were not present in traditional epidemiology (Schulte, 1991). What is the responsibility for follow-up of "abnormal" results, how will individuals with these results be treated, and what will be done if other pathologic conditions are observed (see Chapter 9)?

The use of biologic markers of effect that permit focus on and detection of preclinical and extremely early disease also raises questions about the true preventability and treatability of these conditions (Yach, 1990). At present, insufficient information is available to determine whether particular levels of a certain biologic marker reflect normal ranges of that marker in a person over time or whether they reflect early stages of a preventable disease (Yach, 1990). Researchers currently face this challenge.

Finally, from an organizational viewpoint, epidemiologists may potentially be polarized into two worlds: one of molecular epidemiologists who emphasize the molecular and genetic causes of disease and the other of social epidemiologists who stress the role of social, psychological, and economic factors in health. Neither approach alone will satisfactorily address the health issues of the current era. History has shown that complete reliance on reductionist approaches is antithetical to public health, yet failure to use the powerful tools available also will not safeguard public health. A synthesis of the two approaches is needed that can address the entire scope of health issues (Yach, 1990).

Historic Contributions to Molecular Epidemiology

If molecular epidemiology can be described as the use of biologic markers and measurements in epidemiologic research, then the historic contributions to this approach come from those disciplines that have made advances in relating biologic measurements to health and disease. The principles and practices discussed in this volume and other publications on the topic (Hulka *et al.*, 1990; Gledhill and Mauro, 1990; Garner *et al.*, 1991; Groopman and Skipper, 1991) build on a rich and diverse history of biologic measurements. Many of the current troubling issues facing molecular epidemiology were encountered when the tools leading to molecular epidemiology were developed. Although not entirely exhaustive, the list of some disciplines that have contributed to molecular epidemiology include bacteriology, immunology, and infectious disease epidemiology; pathology and clinical chemistry; carcinogenesis and oncology; occupational medicine and toxicology; cardiovascular epidemiology; genetics, molecular biology, and genetic epidemiology; and traditional epidemiology and biostatistics. In these disciplines, the techniques and building blocks of molecular epidemiology have their history.

Bacteriology, Immunology, and Infectious Disease Epidemiology

The first consideration of cellular biomarkers in medical research appeared in the study of infectious disease. Even before the development of the microscope, Italian physician Girolamo Fracastoro wrote in 1546 that the "seeds" or germs of contagious diseases were carried from person to person (cited in Clendening, 1942). This speculation later was confirmed, after the development of the microscope by Van Leeuwenhoek (1632–1723) and the conclusion by Schwann in 1839 that the cell was the fundamental unit of living matter (cited in Venzmer, 1968). Subsequently, Pasteur, Koch, Gram, Von Pettenkofer, and others in the late 1800s began to isolate specific organisms responsible for disease. The detection of bacteria in biologic specimens indicated exposure or disease (depending on the study). Immunology was built on the use of biologic markers indicative of exposure, effect, or susceptibility. In contrast to bacteriology, in which bacteria were "markers for themselves," immunologists used indirect markers indicative of infection, for example, white cell counts and antibody titers. In the 1920s, epidemiologists found a way to use these biomarkers to address the difficult problem of measuring degrees of susceptibility and resistance to disease in the human host (Paul and White, 1973). Before this time, resistance and susceptibility were identified by asking the patient or family whether he had experienced various common contagious diseases. However, with advances such as the Schick and tuberculin skin tests, a method was found that made use of a biologic marker to assess exposure and susceptibility in individuals or groups.

Frost's (1928) work in New York City and Baltimore was among the earliest to use biomarkers to identify age-specific immunity patterns. The use of serologic tests, such as the Wassermann test for syphilis in segments of urban populations in Baltimore in the 1920s, signaled the beginning of one of the earliest precursors to molecular epidemiology, "seroepidemiology" (Williams, 1920). A classic example was the work of Aycock and Kramer (1930), which showed a rural–urban difference in both diphtheria and poliomyelitis using skin and serum biomarkers, leading to the conclusion that both infections were spread by close human contact. The term "serological epidemiology" is attributed to John Paul in 1935, who was instrumental in the development and use of serum surveys in epidemiology (White, 1973). Paul identified issues and applications of seriologic epidemiology that foreshadowed what accompanies the current problems and issues in molecular epidemiology. These issues include attention to selecting an appropriate sample group in a population; determination of sample size; storage and handling of specimens; and development of longitudinal studies.

Other contributions to molecular epidemiology are assays that take advantage of immunologic binding and specificity. The introduction of the radioimmunoassay (Yalow and Berson, 1960) is considered one of the most important advances in biologic measurement (Chard, 1990). The immunoassay is a major tool of molecular epidemiology. Much of the work detecting the interaction of xenobiotics and macromolecules such as DNA and RNA involves immunoassay methods. This type of assessment has become known as "molecular dosimetry" (Perera, 1987a) or "biological dosimetry" (Mendelsohn, 1991). The approach seeks to focus on the biologically effective dose by measuring the bound xenobiotics.

The term "molecular epidemiology" may have been used first in modern infectious disease research. In the early 1970s, the term appeared in a number of papers, Kilbourne (1973) published *The Molecular Epidemiology of Influenza*. Subsequently, occasional papers were published in the infectious disease literature; among the earliest were Pereira *et al.* (1976), Kilbourne (1979), Pappenheimer and Murphy (1983), and Follett (1984).

Kilbourne described the approach as involving molecular determinants of epidemiologic events and used influenza to illustrate the approach. (See Figure 1.7 for the level of detail and specificity the molecular approach provided.) The rationale for such an approach also has been described by Balayeva in relation to rickettsiae: "Conventional serological and biological tech-

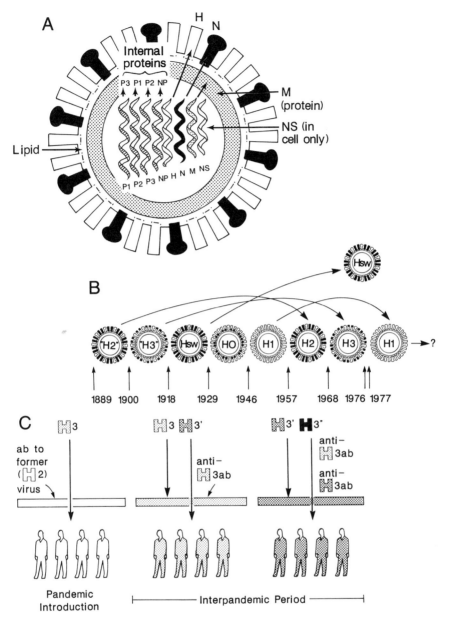

FIGURE 1.7 Molecular determinants of epidemiologic events. (With permission from Kilbourne, 1979.) (A) Schematic of influenza virion, indicating the 8 viral genes and their polypeptide products. Relative sizes of the RNA genes are only approximate; structural details of their relationship to one another and to the internal proteins P1–P3 and NP are not known. H, hemagglutinin glycoprotein; N, neuraminidase glycoprotein. (B) Influenza pandemics of the past century. Schematic of periods of prevalence of major hemagglutinin subtypes of influenza A viruses. Year of introduction of "new" subtypes indicated by arrows. H0, H1, etc., refer to hemagglutinin subtype of the viruses without reference to their neuraminidase antigens. Subtypes Hsw, H0, and H1 are more closely related antigenically than they are to H2 or H3. Identification of viral subtypes prior to 1933 is inferential on the basis of studies of human antibodies. (C) Selection of antigenic mutants are as a function of population antibody. New pandemic viral subtype H3 transcends barrier of antibody to unrelated previously prevalent virus H2 and readily infects the population. When H3 infects a critical percentage of the population its survival is impeded, and antigenically changed mutant H3′, and later H3″, have survival advantage (minor and antigenic variation or antigenic "drift").

niques for characterizing rickettsiae do not seem to be capable of providing answers to all the ecological and epidemiological questions surrounding these organisms. New techniques for identification and epidemiological assessment are required" (Balayeva, 1989).

Pathology and Clinical Chemistry

A century ago, Virchow (1821–1902) foreshadowed what contemporary molecular epidemiology could be by relating clinical disease to cellular pathology (Virchow, 1858). Subsequently, understanding of the cellular basis for disease grew and cellular changes were characterized according to whether they indicated homeostasis, normal variations, or pathogenesis.

Many of the basic principles for the practice of molecular epidemiology are derived from the discipline of clinical chemistry, which comprises the development and performance of specific chemical analyses for diagnostic purposes. Clinical chemistry cannot be sharply separated from the related sciences of pathology, hematology, immunology, and bacteriology (Richterich, 1969). Modern clinical chemistry can be traced back to 1850, when measurements of serum electrolytes were used to distinguish cholera patients from healthy individuals during an epidemic (Schmidt, 1850). However, at that time these measurements were not considered to be of any pathogenic, diagnostic, or therapeutic value. More than 50 years elapsed before electrolyte measurements were considered significant in the field of pediatrics (Richterich, 1969). As clinical chemistry evolved, the fundamental principles of "range of normal," individual variation, sensitivity, and specificity of a test were developed. These principles are part of the foundation of molecular epidemiology. To this end, the classic work of Vecchio (1966), *Predictive value of a single diagnostic test in unselected populations,* and that of Galen and Gambino (1975), *Beyond Normality: The Predictive Value and Efficiency of Medical Diagnoses,* are significant because they provide a quantitative approach to assessing validity. Interestingly, the development of the work of Galen and Gambino, two clinical pathologists, was influenced by the epidemiologist Mervyn Susser.[2] Susser presented the concepts of sensitivity, specificity, and predictive value; Vecchio, and Galen and Gambino applied these concepts to clinical laboratory data.

Quantitative methods in pathology have a long history that contributes directly to molecular epidemiology. In the late 1800s and early 1900s, the many hypotheses of cancer causation relied on identifying quantitative changes in chromosomes in cancer cells (Baak and Tosi, 1991). The beginning of modern cytometry often is dated to the early 1930s when Caspersson and colleagues showed that proteins and nucleic acids in epithelial tumor cells differed from those in normal cells (see Caspersson and Zech, 1972).

---

[2] In 1972, Galen attended an epidemiology seminar led by Professor Mervyn Susser, who described the terms "sensitivity," "specific," and "predictive value."

Automation and quantitative analysis of DNA, cell surface receptors, and other markers is now a major tool in many branches of medicine (Anderson *et al.,* 1984; Koss, 1987).

Carcinogenesis and Oncology

Environmental Carcinogenesis

The study of environmental carcinogenesis, whether caused by chemical, biologic, or physical agents, has contributed some of the broad concepts that form the foundation of molecular epidemiology. The understanding of cancer as a multistage process (Berenblum and Shubik, 1949) and the efforts to find tumor markers provided the background for the concept of a continuum of events between a xenobiotic exposure and a resultant disease. The two-stage and multistage models of chemical carcinogenesis were based on the work of Rous and Kidd (1941) and expanded by Berenblum and colleagues (1949). These models provided a framework within which to explain the temporal continuum between a chemical exposure and development of a cancer. Such models, coupled with an increased understanding of the natural history of cancer, presage current thinking about cancer biomarkers.

In 1958, Foulds described the natural history of cancer by suggesting that the biologic properties and behaviors of neoplastic cells during progression are determined by numerous "unit characteristics" such as growth rate, histologic type, responsiveness to hormones, and invasiveness. Beginning in the late 1960s, Ehrenberg (1974), Calleman *et al.* (1978), Ostermann-Golkar (1983), and colleagues have contributed to an understanding of the components of the continuum between exposure to electrophilic carcinogens and cancer. These investigators have demonstrated how the dose of genotoxic agents can be assessed by measuring the products of their covalent bonds to hemoglobin and DNA, a process known as molecular or biologic dosimetry. Subsequently, the researchers compared the alkylation effect of a dose of a genotoxic chemical to the known alkylation effect of radiation and using these "rad equivalents," performed risk assessments for leukemia in workers exposed to those chemicals.

Other insights into the continuum of events between exposure and disease came from the field of radiation carcinogenesis. In this area, indicators of dose and markers of early biologic effects were used to predict health outcomes (Beebe *et al.,* 1978). The concepts of "internal dose" and "biologically effective dose") were developed in the area of radiation carcinogenesis (Gledhill and Mauro, 1990). The assessment of biologic effects, specifically mutations, in animal and human populations exposed to radiation used biologic markers of metabolic phenotypes, karyotypic differences, and gene rearrangements. The term "mutational epidemiology" was applied to these uses of biomarkers, first with radiation and later with mutagenic chemicals (Miller, 1983).

More recent history of molecular epidemiology can be traced to efforts in the 1970s and 1980s to address environmental carcinogensis. Wynder and Reddy (1974) are credited with using the term "metabolic epidemiology" to describe the incorporation of measurements of bile acids and cholesterol into epidemiologic studies to assess the role of diet in colon cancer. Later, the term "biochemical epidemiology" was used to describe testing of the molecular hypotheses at the clinical and population levels (Harris *et al.,* 1985). For example, whereas it was once common to identify individuals at high risk of cancer due to lifestyle or occupation, the susceptible individual now may be identified at the molecular level for some cancers. Outside the realm of infectious disease, the use of the term "molecular epidemiology" has been ascribed to Higginson (1977) in his paper, "The role of the pathologist in environmental research and public health." However, seminal thinking in this area was found in the work of Lower (Lower *et al.,* 1979; and Lower, 1982) and Perera and Weinstein (1982). Lower's work addressed the interaction between exposure to a carcinogen and metabolic phenotypes. He described an interdisciplinary approach to molecular epidemiologic research. The work of Perera and Weinstein (1982) demonstrated in detail the potential for the use of biologic markers of dose to provide a more accurate appraisal of exposure than ecologic descriptors can. Perera and Weinstein (1982) first used the term "molecular cancer epidemiology" to describe this approach. They discussed carcinogen–DNA adducts measured by highly sensitive antibodies as examples of promising markers in studies of environmental carcinogenesis. They concluded that adduct formation ultimately might be useful in identifying individuals at high risk of cancer.

Tumor Markers

One of the clearest historical precursors of the use of biologic markers was the quest to discover cancer at an early and intervenable stage. In the early 1930s, Zondek (1942) considered using human chorionic gonadotropin in body fluids to diagnose tumors of gestational and germ cell origin. The term "markers," as in "tumor markers," has been used widely in oncology and cancer research. The search for a tumor marker or test that would detect all or most forms of cancer was based on the supposition that neoplasia in all its many forms results in a unique change, either in a component of body fluids or in a host response phenomenon (Bagshawe, 1983). No such marker has been found. However, numerous markers are used routinely as part of the process to diagnose or confirm a diagnosis for specific cancer types. These include, for example, carcinoembryonic antigen (CEA) for gastrointestinal tumors, serum acid phosphatase for carcinoma of the prostate, 5-hydroxy indoleacetic acid in the urine for carcinoid tumors, α-fetoprotein for liver cancer, and thyrocalcitonin for modular carcinoma of the thyroid (Ghosh and Rob, 1987; Klee and Go, 1987). The introduction in 1975 of monoclonal antibody technology revolutionized serologic and biochemical analysis by allowing the development of sensitive and specific probes of human cancer (Wright and Cox, 1987). Tumor markers have been considered for use in screening, diagnosing, staging, and monitoring treatment and recurrence of disease and in determining the efficacy of specific forms of therapy. From an epidemiologic viewpoint, most of the studies of tumor markers have focused on efforts to validate whether a marker indicates a diagnosis of cancer and whether the predictive value of a positive test was sufficient to render a cost-effective method for screening the general population.

A great lesson for molecular epidemiologists can be learned from the history of the quest for tumor markers. As one reviewer noted, ". . . it is not uncommon for scientists to tumble on some phenomenon that appears to distinguish patients with cancer from other people and before long the scientist is liable to be carried along on a wave of blind conviction and messianic fervour" (Bagshawe, 1983). The lesson is that some of the papers that report on a "cancer test" provide inadequate technical information; early studies may have included researchers not blinded to the ultimate disease status of the subject or a host of other factors that can fool an investigator. Critical evaluation of the initial milieu in which a marker was identified and developed is important.

The history of tumor marker research also provides excellent examples of past attempts to validate a marker and use it in a screening program. Some of the best examples of early tumor markers have been developed in the fields of cancer cytology and cytogenetics. Since the conceptualization of the cell theory by Schwann (1839), researchers have been on a quest for the basic cellular markers of disease. Cellular changes have been the primary focus of study for many diseases, particularly cancer. The very definition of cancer is based on characteristics indicative of loss of cellular growth control. As refinements in technology progressed, subcellular markers such as chromosomal changes that could be linked to human cancers were found. In 1972, the 9,22 translocation was identified in chronic myeloid leukemia (Rowley, 1973). To date, at least 70 recurring translocations have been detected in human malignant cells (Rowley, 1990). Using chromosome aberrations as markers, genes relevant to the process of malignant transformations have been located. Ultimately, it has been feasible to measure DNA bases, which are key to the carcinogenic process. Measurement of the DNA characteristics in cells build on the work of Feulgen and Rossenbeck (1924), who described the stain that subsequently proved specific for double-stranded DNA (Koss, 1990). In the 1930s, Caspersson and colleagues began measuring the fluorescent intensity of DNA and RNA in cells (Koss, 1990; Baak and Tosi, 1991). This work was part of the foundation of quantitative methods in pathology that foreshadowed the measurement of biologic changes as variables in epidemiology.

The use of Papanicolaou cytology as a marker of preclinical cervical cancer demonstrates how a good marker can lead to effective intervention when a disease is in a treatable state. The work of Papanicolaou and Traut (1943) in staining and classifying cells led to this important marker of cancer diagnosis. The history of the Pap test provides a lesson in how a new biomarker may not be used in a timely fashion. Approximately 27 years lapsed between the development and the adoption of the Pap test. Greenwald (Greenwald *et al.*, 1990) identified three main reasons for this delay: (1) failure to recognize and pursue the potential of cervical cytology, (2) lack of clinical trials to verify the test's effectiveness, and (3) failure to use screening in those population groups at highest risk.

Mutational Spectra

The location and type of mutations in a specific sequence of nucleotides define a mutational spectrum (Hollstein *et al.*, 1991). The ability to determine mutational spectra has been demonstrated in genetic toxicology and cancer research and may prove to be a major component to molecular epidemiology. The discovery of chemical-induced chromosome mutations first occurred in studies of plants (Oehlkers, 1943; Basler, 1987). More detailed analysis of mutations occurred when Benzer (1961) first demonstrated the sequence specificity of spontaneous and induced mutations in the *rII* locus of the bacteriophage T4. Two key observations were made: (1) For a given agent, several sites in the gene displayed higher mutation frequencies than others. (2) The distribution of mutations as a function of base-pair position was different for spontaneous mutations and for mutations induced by a wide variety of chemicals (Thilly *et al.*, 1982).

From this and other research we have learned that spontaneous and induced mutations are not distributed randomly with respect to type and position in the genome. Various researchers, including Albertini (1982) in the 1970s and early 1980s, developed assays on T lymphocytes as practical approaches to monitoring human mutagenicity. Thilly *et al.* (1982) demonstrated an approach to distinguishing between spontaneous and induced mutations using specific forward mutation assays, each of which detects only missense mutations at a small number of base pairs in the human genome. In the late 1970s and early 1980s, the term "mutational epidemiology" began to be used to describe surveillance and monitoring of individuals exposed to known and suspected mutagens (Hook, 1982). The phrase also encompassed etiological epidemiologic studies (Miller, 1983).

In 1991, Hollstein *et al.* focused on the pattern of base substitution mutations in the *p53* gene observed in human cancers. Different malignancies showed different patterns of mutations. By comparing the mutation spectra at the same locus, for a variety of tumors of different proposed etiology, it may be possible to distinguish the etiological contributions of exogenous and endogenous factors to human carcinogenesis. (See Figure 1.8 for an illustration of specific mutations for various cancers.) For example, for liver tumors

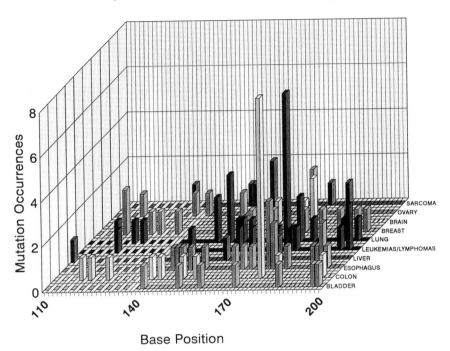

FIGURE 1.8 Localization of p53 base substitutions in human cancers. Only a section (codons 110–200) of p53 is shown to illustrate this mutational spectra. Adapted from Hollstein *et al.*, 1991. In many instances, analysis of the p53 gene in tumors was limited to exons 5 through 8, corresponding to residues 126 through 306. Five of the 280 mutations were found outside the 200-codon stretch presented here, four of which resulted in chain-terminating codons.

in persons from geographic areas in which aflatoxin B is a cancer risk factor, most mutations of the *p53* gene are at one nucleotide pair (G to T) of codon 249 (Bressac *et al.*, 1991; Hsu, 1991), suggesting a major role for aflatoxin B. The findings in mitogenesis experiments, in which aflatoxin B was administered to rodents and found to induce G to T transversions, suggest an ability to distinguish cancer caused by aflatoxin exposure and that caused by other factors (Muench *et al.*, 1983 McMahon, *et al.*, 1990). Chemical or physical agents are believed to produce a spectrum of mutational events that may be specific for each exposure, creating a mutational fingerprint (See Figure 1.9).

Occupational Medicine and Toxicology

The study of occupational disease and toxicology is another historical root of molecular epidemiology. Since the early part of the twentieth century, occupational exposures have been confirmed by biologic monitoring (Hamil-

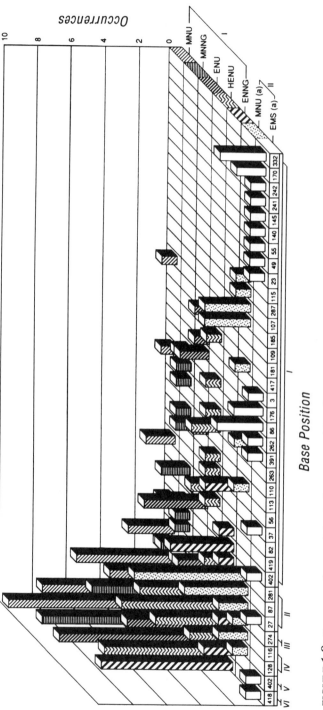

FIGURE 1.9 Frequency of mutated sites in the *gpt* gene of *E. coli* exposed to selected alkylating agents. EMS, ethyl methanesulfonate; MNU, *N*-methyl-*N*-nitrosourea; ENNG, *N*-ethyl-*N'*-nitro-*N*-nitrosoguanidine; HENU, 1-(2-hydroxyethyl)-1-nitrosourea; ENU, *N*-ethyl-*N*-nitrosourea; MNNG, *N*-methyl-*N'*-nitro-*N*-nitrosoguanidine. Roman numerals represent either base positions or chemical types that clustered in a multivariate analysis. This figure represents data collected by Benigni *et al.*, (1992). (Reprinted with permission.)

ton, 1943). Biologic monitoring of markers of exposure was advocated in the 1950s as an essential element of industrial hygiene programs (Elkins, 1954). The forerunner of the continuum of biologic markers shown in Figure 1.2 came from the work of Zielhuis (1978), Lauwerys (1983), Hernberg and Aitio (1987), and others. Four conditions for meaningful biologic monitoring have been identified from their work:

1. The substance and/or its metabolites are present in some tissue or body fluid suitable for sampling.
2. Valid practical and analytical methods are available.
3. The measurement is correct.
4. The result can be interpreted in terms of health risk or exposure (Tola and Hernberg, 1981).

That external exposure data is a poor approximation of individual dose is of particular concern in occupational epidemiology. If an individual dose is the integration of exposures from various routes and sources, both known and unknown, there may not be a high correlation of dose to a measure of external exposure. This problem was addressed by a World Health Organization study group (1975), which found that "an important conceptual advance in the assessment of toxic hazards is the increasing understanding that a good biological exposure test need not correlate with measurements of workroom air." Such correlations often are hampered by several factors, including exposures via nonrespiratory routes, additional nonoccupational exposures that accumulate in the person, use of respirators, and interindividual variations with respect to personal hygiene, smoking, metabolic processes, and physical work demands. This very discrepancy makes biologic monitoring a useful supplement to measurements of workroom air as an indicator of individual risk (Droz *et al.*, 1991).

A major historic parallel for molecular epidemiology is the testing of pulmonary function in worker populations exposed to airborne pollutants (Ingram and McFadden, 1981). This study exemplifies an early use of physiologic markers to establish dose–response relationships and predict risk. The challenge to assess the variability, sensitivity, accuracy, and precision of such tests serves as a lesson for those attempting to use molecular markers as surrogates for outcome and to assess individual risk.

Cardiovascular Epidemiology

Cardiovascular epidemiology is another historic precursor of molecular epidemiology. Biomarkers have played a significant role in the epidemiologic research of cardiovascular disease. The physiologic marker blood pressure has been assessed widely and, when elevated, has been found in epidemiologic studies to be a risk factor for cardiovascular disease. At the biochemical level, cholesterol and lipoproteins are included among the numerous risk factors for atherosclerosis. Studies of familial hyperlipidemias represent early examples of the use of biologic markers in assessment of gene–environment interactions. Research using these serum or plasma markers [e.g., the Framingham Heart Study (Kannel *et al.*, 1976), the Tecumseh Study (Epstein *et al.*, 1970), and other studies (Lipid Research Clinics Program (LRCP), 1984)] provided early lessons on the need for standardizing laboratory techniques for handling biomarkers. This need was most evident in the Lipid Research Clinics Program to determine the prevalence of various levels of lipids and lipoproteins in 11 North American populations. Triglycerides and cholesterol levels were determined for more than 75,000 individuals (LRCP, 1984). Serum cholesterol and other lipids have been used in epidemiologic studies as markers of exposure (to dietary fat), as markers of effect (risk factor for atherosclerosis), and as markers of susceptibility (LDL receptor defect) in genetic epidemiologic studies.

Cardiovascular disease research also provides a model for individual risk characterization, which has been described as a major advance of molecular epidemiology (Shields and Harris, 1991). Truett and colleagues (1967) described the calculations of individual risk functions based on multivariable analysis of covariates (including biologic markers and demographic and behavioral factors).

Genetics, Molecular Biology, and Genetic Epidemiology

The science of genetics was built on the concept that offspring phenotype is a biomarker of events at the gene level. This field provides good examples of how science uses observation and inference at the macro level. Visualization of cells—and eventually chromosomes—facilitated definition of mitosis and meiosis, so by the 1890s enough evidence was available to establish a chromosomal theory of inheritance (Mayr, 1982). In the twentieth century, progress in technique has allowed precise determination of chromosomal number; development of methods to detect how chromosomes function; recognition of Q-, G-, R-, T-, and C-bands of stained chromosomes (Caspersson and Zech, 1972); and understanding of chromatin structure (Jeppesen and Bower, 1989).

Genetics and molecular biology overlap as research extends to the level of the gene. Even before it was "seen," the gene was hypothesized, albeit among numerous competing hypotheses, as the unit of inheritance (see Mayr, 1982, for review), based on observations at the macro rather than the micro level. However, a more complete understanding evolved from research on the biochemistry of living processes.

Genetics and molecular biology rely in part on an understanding of the information macromolecules: proteins and nucleic acids. The central dogma of molecular biology is that information flows from DNA to RNA to proteins (Figure 1.10). This dogma also describes the flow of genetic information be-

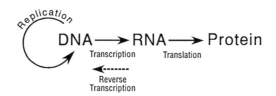

FIGURE *1.10* Information flow in molecular biology. Reprinted with permission from Maxon and Daugherty, 1985.

tween generations of somatic cells (mitosis) and germinal cells (meiosis). However, even this view is subject to change, as evidenced by the finding that RNA, once thought of as a passive carrier of genetic information, can sometimes function as an enzyme as well (Woldrop, 1992).

The sampling frame for molecular epidemiology is likely to be composed largely of the nucleotide sequences that constitute DNA (Figure 1.11). In 1883, Roux recognized that the basic process of transmission of genetic information was the division of the cell nucleus into two identical halves. Subsequently, however, the crucial event was recognized as the actual doubling of the genetic material, followed by its segregation into two daughter cells. In 1958, replication of DNA was demonstrated by Meselson and Stahl to be semi-conservative, that is, each daughter double helix receives one complete parental stand of DNA (Meselson and Stahl, 1958). This single strand serves as a template on which a new strand is synthesized from free nucleotides. The significance of this discovery is that the genetic information in a cell was shown to be reproducible indefinitely and—barring mutations—precisely.

Genetic analysis at the macro and micro levels has provided an opportunity to quantify variability in populations. "Variation" has been the topic that sparked creative thinking in biology as well as a controversial and perplexing one for molecular epidemiology (and promises to continue to be so). However, Mayr (1982) observed that Western thinking for 2000 years after Plato was dominated by "essentialism," or the belief that there were a limited number of immutable essences. Much of epidemiologic (and, for that matter, biomedical scientific) thinking to date has been essentialist. For example, the early statistics used in public health by Graunt (1620–1674) and Quetelet (1796–1874) attempted to calculate true values to overcome the confusing effects of variation. Quetelet, (see Mailley, 1875), who attempted to use a mathematical function of height and weight as a biomarker, hoped to calculate characteristics of the average person. To him and like thinkers, variations were nothing but "errors" around mean values. In the nineteenth century, new ways to view nature began to spread; a concept called "population thinking" developed that stresses the uniqueness of everything in the organic world. To population thinkers, there is no "typical" individual and mean values are abstractions (Mayr, 1982).

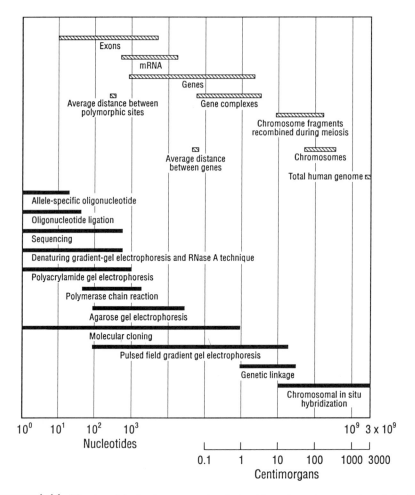

FIGURE 1.11 The size of the human genome in relationship to the various genetic and physical methods that are being applied to obtaining a map. Hatched lines, Regions of informational units. Solid lines, Techniques used in mapping. (Modified from Landegren *et al.,* 1988 in Weatherall, 1991.)

With the development of electrophoretic and other analytical techniques, researchers have discovered far more variability in proteins and their underlying genetic information than population geneticists and evolutionary biologists expected (Janetos, 1988). Similarly, with the introduction of recombinant DNA technology, the genomes of prokaryotes and eukaryotes have been found to be far more diverse and plastic than originally thought (Janetos, 1988).

Another lesson from genetics has been the perspective on the relative roles of heredity and environment in causing or exacerbating disease. Despite a long-standing "nature–nurture" debate, most scientists support the concept that phenotypic variation generally results from both environmental and hereditary influences. "Genetic epidemiology" has been particularly useful in disentangling the effects of heredity and environment. Biologic markers have been used in genetic epidemiology at the macro level (an individual trait, such as eye color) and at the micro level (e.g., HLA antigens) in twin studies, adoption studies, path analysis, analysis of cultural transmission of disease risk factors, and studies of association between specific genotypes and diseases. Since the early 1950s, the terms "pharmacogenetics" (Vogel, 1959), "ecogenetics" (Brewer, 1971), and "occupational ecogenetics" (Mulvihill, 1986) have been used to describe fields studying the interaction of genetic factors and exogenous agents. By 1975, a report by the National Academy of Sciences (1975) listed 92 human genetic disorders believed to predispose the affected individuals to the toxic effects of pollutants. DNA polymorphisms have been used to assess genetic predisposition to various diseases, such as cardiovascular disease, cancer, and diabetes, that may put persons with exposure to xenobiotics at increased risk (Cartwright *et al.*, 1982; Rajput-Williams *et al.*, 1988; Trucco and Dorman, 1989; Caporaso *et al.*, 1990; Weatherall, 1991).

Since the time of Mendel, genetics and genetic epidemiology have used the concept of "genetic markers." A genetic marker is a variant allele that is used to label a biologic process or structure throughout the course of an experiment or epidemiologic study (Suzuki *et al.*, 1986). That Mendel used the genes for pea shape, pea color, and so on was largely irrelevant; he used them as markers to trace the hereditary processes of segregation and assortment. DNA technology has provided a method for obtaining a large number of new genetic markers, which facilitates studying heritable differences in DNA sequences. Unlike classical expressed markers, DNA polymorphisms can be detected whether or not a given sequence encodes a protein (Conneally *et al.*, 1984).

Ultimately, for epidemiologic purposes, the technical developments for amplifying small samples of DNA (the polymerase chain reaction; Saiki 1985) and for determining the sequences of genes will be considered landmark events. These approaches allow the use of small amounts of sample collected from contemporary subjects or from stored or archived specimens. When applied to a properly designed study, the polymerase chain reaction (PCR) techniques can allow powerful comparisons among various groups. Moreover, a high resolution map of the entire human genome could be developed using restriction fragment length polymorphisms (described in Botsein *et al.*, 1980).

Molecular biology, which has undergone a meteoric rise since the discovery of the structure of DNA (Watson and Crick, 1953), has been the source of powerful tools in genetics that will be employed in molecular epidemiology. The discoveries of molecular biology have simplified and unified biology, even as they have shown the immense diversity. These discoveries are a driving force behind molecular epidemiology, since they promise to permit more accurate classification of subjects, more detailed appraisal of mechanisms, and earlier opportunities for prevention and control.

Epidemiologic and Statistical Context

A number of authors have discussed how "molecular epidemiology" fits into the historic framework of epidemiology (Perera and Weinstein, 1982; Hatch and Stein, 1987; Schulte, 1987; 1989; Vandenbroucke, 1988; Hulka *et al.,* 1990; Loomis and Wing, 1990; Yach, 1990). Vandenbroucke (1988) believes the use of molecular biology to identify individual steps in disease causation is reminiscent of the nineteenth century conflict between the miasma and contagion theories of disease. For the miasmists, the competing idea was the germ theory; for the epidemiologists of today, it is molecular biology.

Loomis and Wing (1990) see the contest between the miasma and contagion theories as growing from a tension as old as Western medicine between the often unglamorous work of the hygienist and the heroic capabilities of the healer. The miasma and contagion theories share a derivation from a global scientific paradigm that has been called "Cartesian reductionism." As Loomis and Wing caution:

> In epidemiology this paradigm equates causal inquiry with discovering the inherent dose–response relationships between agents and diseases through rigorous hypothesis testing. The weakness of causal theories arising from the Cartesian paradigm is their failure to recognize that the action of interdependent parts (agents, aspects of the environment, and individuals) is not an immutable, historical characteristic of those parts but is, instead, dependent on the properties of the whole system in which they operate. (Loomis and Wing, 1990)

Data on biologic markers often have been used as topical examples in the evolution of statistical epidemiologic ideas. Generally, use of the markers was not the driving force behind an idea, but a case in point. For example, Woolf's paper (1955), "On estimating the relation between blood group and disease," presented the first opportunity to assess the common odds ratio estimation for a set of two-by-two tables and a test for heterogeneity of the odds ratio across tables (Greenland, 1987).

Measurements of lipids, lipoproteins, and blood pressure were among the variables used in heart disease research that forced researchers to introduce concepts of multivariate analysis to overcome the "thinness of data" encountered in ordinary stratified analysis (Greenland, 1987). Similarly, when Gordon (1974) addressed the hazards of using multivariate analysis, such as the assumption of a linear combination of variables, biomarkers were used as examples. They also were used to illustrate the issue of combining categorical and continuous variables.

Many of the issues about validity and reproducibility of biological markers came from work on nutritional epidemiology (Keys *et al.*, 1965; Willet *et al.*, 1983; Romieu *et al.*, 1990; Bates, *et al.*, 1991). The primary efforts have involved validation of biochemical markers as predictors of dietary intake and the use of biologic markers to validate food frequency questionnaires.

A significant contribution to the statistical handling of biologic marker data was the seminal work of Finney (1952), *Statistical Method in Biological Assay*. Finney traced the history of the biologic assay dating from the late nineteenth century. His focus encompassed Erlich's investigations into the standardization of diphtheria antitoxin; developments in pharmacology, endocrinology, and plant pathology; and the works by Coward (1947) and Gaddum (1948) that directed attention to the statistical considerations. Key in Finney's work is the elucidation of concepts of the statistical validity of a bioassay and his recommendation pertinent to molecular epidemiology:

> Unless the assayist himself knows enough about the statistical logic and structure of programs, he may feel compelled to use methods suited to his experiments or may overlook indications that anomalies or unsuspected errors make a particular set of data misleading. The great expansion in some types of assay (notably radioimmunoassays) that are essential to patient care surely renders neglect of quality in statistical analysis as blameworthy as neglect of proper maintenance of equipment or negligence in identifying and measuring samples. (Finney, 1952)

Much of what is portrayed currently as molecular epidemiology appears not to have heeded this lesson. Such studies often fail to involve discussion of study design, statistical analysis, sources of bias, misclassification, or errors. Considering changes found in a gene to be indicative of the "truth" about a phase in the natural history of disease may be acceptable as part of a pilot process in molecular epidemiology, but will not support a firm foundation for inference, comparison, or generalization without attention to statistical considerations.

Many of the issues that will face practitioners of molecular epidemiology will be similar to those identified by John Paul (1958) in his publication, *Clinical Epidemiology* and include issues pertaining to host susceptibility, understanding the natural history of disease, interacting with the subject, and collecting biologic specimens. Clinical epidemiology is a forerunner of molecular epidemiology in these ways.

The recent history of molecular epidemiology was built on work done in the 1980s and early 1990s in molecular dosimetry (reviewed in Gledhill and Mauro, 1990; Garner *et al.*, 1991; Groopman and Skipper, 1991), and in pilot efforts to incorporate biologic markers into epidemiologic research. These efforts were mostly in the area of environmental health research. Papers and monographs (Perera and Weinstein, 1982; Tannenbaum and Skipper, 1984; Harris *et al.*, 1985; Wogan and Gorelick, 1985; Schulte, 1987; Hulka and Wilcosky, 1988) in the 1980s formed the systematic body of literature on methods and techniques of this nascent field. The publication by Hulka *et al.* (1990), *Biological Markers in Epidemiology*, was the first epidemiology-oriented text in the field. This work, although mostly focused on cancer-related biomarkers, described many of the basic issues that, in retrospect, will be deemed characteristic of this approach, including when to use a biologic marker in epidemiologic research, properties of the marker, sample size, and control of confounding, analysis, and interpretation of marker. Other related publications (Bertazzi and Duca, 1987; Hatch and Stein, 1987; Perera *et al.*, 1987, 1990; Schulte, 1987, 1989; Vineis *et al.*, 1990; NRC, 1991a, b) also provide guidance on methodologic issues.

Conclusion

Currently, molecular epidemiology is an evolving set of techniques rather than a well-conceived and practiced discipline. The "field" may become more rigorous by application of lessons learned from other disciplines that have a history of obtaining and using biologic markers. Biomedical science can be characterized by a process of increasingly refined biologic measurements and their use in etiologic, preventive, and therapeutic research. I have described this history briefly, with particular reference to predecessor disciplines that contribute to a body of knowledge that could be useful to molecular epidemiology. This is a rich history that reveals how many of the concerns about molecular epidemiology have been confronted previously. Researchers can learn from this history and use it as a foundation to establish good principles in molecular epidemiology.

References

Abdel-Baky, S., and Giese, R. W. (1991). Gas chromatography electron capture negative-ion mass spectrometry at the zeptomole level. *Anal. Chem.* **63**, 2986–2989.

Ahearn, J. M., and Hochberg, M. C. (1988). Epidemiology and genetics of ankylosing spondylitis. *J. Rheumatol.* **15** (**Suppl. 6**), 22–28.

Alavanja, M., Aron, J., Brown, C., and Chandler, J. (1987). Cancer risk-assessment models: Anticipated contributions from biochemical epidemiology. *J. Natl. Cancer Inst.* **4**, 633–643.

Albertini, R. J., Castle, K. L., Borcherding, W. R. (1982). T-cell cloning to detect the mutant 6-thioquanine-resistant lymphocytes present in peripheral blood. *Proc. Nat. Acad. Sci. U.S.A.* **79**, 6617–6621.

Anderson, N. L., Hoffman, J. P., Gemmell, A., and Taylor, J. (1984). Global approaches to quantitative analysis of gene-expression patterns observed by use of two-dimensional gel electrophoresis. *Clin. Chem.* **30**, 2031–2036.

Aycock, W. L., and Kramer, S. D. (1930). Immunity to poliomyelitis in normal individuals in urban and rural communities as indicated by the neutralization test. *J. Prev. Med.* **4**, 189–200.

Ayesh, R., Idle, J. R., Richie, J. C., Crothers, M. J., and Hetzel, M. R. (1984). Metabolic oxidation phenotypes as markers for susceptibility to lung cancer. *Nature* **312**, 169–170.

Baak, J. P. A., and Tosi, P. (1991). Historical background of quantitative pathology. *In* "Manual of Quantitative Pathology in Cancer Diagnosis and Prognosis" (J. P. A. Baak, ed.), pp. 3–5. Springer-Verlag, Berlin.

Bagshawe, K. D. (1983). Tumour markers—Where do we go from here? Br. J. Cancer **48**, 167–175.

Balayeva, N. (1989). Approaches to the molecular epidemiology of rickettsioses. *Eur. J. Epidemiol.* **5**, 414–419.

Basler, A. (1987). Scientific justification of testing chromosome mutations and regulatory requirements for the assessment of mutagenicity. *In* "Cytogenetics" (G. Obe and A. Basler, eds.), pp. 379–393. Springer-Verlag, Berlin.

Bates, C. J., Thurnham, D. I., Bingham, S. A., Margetts, B. M., and Nelson, M. (1991). Biochemical markers of nutrient intake. *In* "Design Concepts in Nutritional Epidemiology" (B. M. Margetts and M. Nelson eds.), pp. 192–265. Oxford University Press, New York.

Beebe, G. W., Kato, H., and Land, C. E. (1978). Studies of the mortality of A-bomb survivors. 6. Mortality and radiation dose 1950–74. *Radiat. Res.* **75**, 138–201.

Begley, S., Glick, D., and Wilson, L. (1991). Not in the stars but in our genes. *Newsweek* October 21, 56–57.

Benigni, R., Palombo, F., and Dogliotti, E. (1992). Multivariate statistical analysis of mutational spectra of alkylating agents. *Mutation Research* **267**, 77–88.

Benzer, S. (1961). On the topography of the genetic fine structure. *Proc. Natl. Acad. Sci. U.S.A.* **47**, 403–415.

Berenblum, I., and Shubik, P. (1949). An experimental study of the initiating state of carcinogenesis and re-evaluation of the somatic cell mutation theory of cancer. *Brit. J. Cancer* **3**, 109–118.

Bertazzi, P. A., and Duca, P. S. (1987). The use of biological indicators in the surveillance of groups of individuals. *In* "Occupational and Environmental Chemical Hazards: Cellular and Biochemical Indices for Monitoring Toxicity" (V. Foa, E. A. Emmett, M. Maroni, and A. Colombi, eds.), pp. 61–68. Ellis Horwood, Chichester, England.

Bostofte, E., Bagger, P., Michael, A., and Stakemann, G. (1988). Fertility prognosis for infertile men: Results of follow-up study of semen analysis from two different populations evaluated by the Cox regression model. *Urology–Andrology* **54**, 1100–1106.

Botstein, D., White, R. L., Skolnick, M., and Davis, R. W. (1980). Construction of genetic linkage map in man using restriction fragment length polymorphisms. *Am. J. Hum. Genet.* **32**, 314–331.

Bressac, B., New, M., Wunds, J., and Ozturk, M. (1991). Selective G to T mutations of p53 gene in hepatocellular carcinoma from Southern Africa. *Nature* (London) **350**, 429–431.

Brewer, J. C. (1971). Annotations: Human ecology, and expanding role for the human geneticist. *Am. J. Hum. Genet.* **23**, 92–94.

Bull, R. J. (1989). "Decision Model for the Development of Biomarkers of Exposure." U.S. Environmental Protection Agency. EPA 600/x-89/163. U.S. Government Printing Office, Washington, D.C.

Calleman, C. J., Ehrenberg, L., Jansson, B. Ostermann-Golkar, S., Segerback, D., Svansson, K., and Wachtmeister, C. A. (1978). Monitoring and risk assessment by means of alkyl groups in hemoglobin in persons occupationally exposed to ethylene oxide. *J. Environ. Pathol. Toxicol.* **2**, 427–442.

Caporaso, N. E., Tucker, M. A., Hoover, R. N., Hayes, R. B., Pickle, N. W., Issaq, H. J., Muschik, G. M., Green-Gallo, L., Burvys, P., Arsnor, S., Resau, J. H., Trump, B. F., Tollerud, D., Weston, A., and Harris, C. C. (1990). Lung cancer and the debrisoquine metabolic phenotype. *J. Natl. Cancer Inst.* **82**, 1264–1272.

Cartwright, R. A., Glashen, R. W., Rodgers, H. J., Ahmad, R. A., Barham-Hall, T., and Higgins, E. (1982). The role of N-acetyltransferase phenotypes in bladder carcinogensis. *Lancet* ii, 842–896.

Caspersson, T., and Zech L. (eds.) (1972). "Chromosome Identification." Nobel Foundation, Stockholm.

Chard, T. (1990). "An Introduction to Radioimmunoassay and Related Techniques." Elsevier Science, Amsterdam.

Clendening, L. (1942). "Source Book of Medical History," pp. 106–121. Dovery Publications, New York.

Conneally P. M., Wallace, M. R., Gusella, J. F., and Wexler, N. S. (1984). Huntington Disease: estimation of heterozygote status using linked genetic markers. *Genet. Epid.* 1, 81–88.

Coward, K. H. (1947). "The Biological Standardisation of the Vitamins," (2d Ed.) Baillieve, Tindal, & Cox, London.

Cullen, M. R. (1989). The role of clinical investigations in biological markers research. *Environ. Res.* 50, 1–10.

Detels, R., Visscher, B. R., Fahey, J. L., Sever, J. L., Gravell, M., Madden, D. L., Schwartz, K., Dudley, J. P., English, D. A., Powers, H., Clark, V. A., and Gottlieb, M. S. (1987). Predictors of clinical AIDS in young homosexual men in a high-risk population. *Int. J. Epidemiol.* 16, 271–276.

Droz, P. O., Berode, M., and Wu, M. M. (1991). Evaluation of concomitant biological and air monitoring results. *Appl. Occ. Env. Hyg.* 6, 465–474.

Ehrenberg, L. (1974). Genetic toxicity of environmental chemicals. *Acta Biol. Ingosl. Genetika* 6, 367–398.

Ehrenberg, L., Hallstrom, T., Ostermann-Golkar, S., and Wennberg, I. (1974). Evaluation of genetic risks of alkylating agents: Tissue doses in the mouse from air contaminated with ethylene oxide. *Mutat. Res.* 24, 83–103.

Elkins, H. B. (1954). Analyses of biological materials as indices of exposure to organic solvents. *Arch. Ind. Hyg. Occ. Med.* 9, 212–220.

Epstein, F. H., Napier, J. A., Block, W. D., Hayner, N. S., Higgins, M. P., Johnson, B. C., Keller, J. B., Mitzner, H. L., Montoyo, H. J., Ostrandor, L. D., and Ullman, B. M. (1970). The Tecumseh study: Design, progress, and prospectives. *Arch. Environ. Health* 21, 402–407.

Everson, R. B. (1989). Conceptual models for biochemical epidemiology. *Proc. Am. Assoc. Cancer Res. Ann. Mg.* 30, 678–679.

Feinstein, A. R. (1991). Scientific paradigms and ethical problems in epidemiologic research. *J. Clin. Epidemiol.* 44 (Suppl. 1), 1195–1235.

Feulgen, R., Rossenbeck, H. (1924). Mikrosopisch—chemischer Nachweis einer Nukleinsäure vom Typus der Thymonukleinsäure und die daraut beruhende elektive Färbung von Zellkernen in mikroscopische Präparaten. *Z. Physiol. Chem.* 135, 203–224.

Finney, D. J. (1952). "Statistical Method in Biological Assay." Charles Griffin, London.

Follett, E. A. C., Sanders, R. C., Beards, G. M., Hundley, F., and Desselberger, U. (1984). Molecular epidemiology of human rotaviruses. *J. Hyg. Camb.* 92, 209–222.

Foulds, L. (1958). The natural history of cancer. *J. Chron. Dis.* 8, 2–37.

Frost, W. U. (1928). Infection, immunity, and disease in the epidemiology of diphtheria. *J. Prev. Med.* 2, 305.

Gaddum, J. H. (1948). "Pharmacology." Oxford University Press, London.

Galen, R. S., and Gambino, S. R. (1975). "Beyond Normality: The Predictive Value and Efficiency of Medical Diagnoses." John Wiley and Sons, New York.

Gann, P. H., Davis, D. L., and Perera, F. (1985). "Biologic markers in environmental epidemiology." Paper presented at the 5th workshop of the Scientific Group on Methodologies for the Safety Evaluation of Chemicals (SGOMSEC), sponsored by the World Health Organization, 12–16 August, Mexico City.

Garner, R. C., Farmer, P. B., Steel, G. T., and Wright, A. S. (eds.) (1991). "Human Carcinogen Exposure. Biomonitoring and Risk Assessment." IRL Press, Oxford.

Ghosh, B. C., and Rob, C. G. (1987). Tumor markers. *In* "Tumor Markers and Associated Antigens" (B. C. Ghosh and L. Ghosh, eds.), pp. 1–10. McGraw-Hill, New York.

Gledhill, B. L., and Mauro, F. (eds.) (1990). "New Horizons in Biological Dosimetry." Wiley-Liss, New York.

Gordon, T. (1974). Hazards in the use of logistic function with reference to data from prospective cardiovascular studies. *J. Chron. Dis.* **27**, 97–102.

Graunt, J. (1662). "Natural and Political Observations Made upon the Bills of Mortality." Republished by The Johns Hopkins University Press, Baltimore, 1939.

Greenland, S. (ed.) (1987). "Evolution of Epidemiologic Ideas." Epidemiology Resources, Chestnut Hill.

Greenwald, P., Cullen, J. W., and Weed, D. (1990). Introduction: Cancer prevention and controls. *Sem. Oncol.* **17**, 383–390.

Groopman, J. D., and Skipper, P. L. (1991). "Molecular Dosimetry and Human Cancer: Analytical, Epidemiological, and Social Considerations." CRC Press, Boca Raton, Florida.

Groopman, J. D., Sabbioni, G., and Wild, C. P. (1991). Molecular dosimetry of human aflatoxin exposures. *In* "Molecular Dosimetry and Human Cancer: Analytical, Epidemiological, and Social Considerations" (J. D. Groopman and P. L. Skipper, eds.), pp. 303–324. CRC Press, Boca Raton, Florida.

Hamilton, A. (1943). "Exploring the Dangerous Trades." Little Brown, Boston.

Harris, A. L. (1991) Telling changes of base. *Nature (London)* **350**, 377–378.

Harris, C. C. (1991). Molecular epidemiology: Overview of biochemical and molecular basis. *In* "Molecular Dosimetry and Human Cancer: Analytical, Epidemiological and Social Considerations." (J. D. Groopman and P. L. Skipper, eds.), pp. 15–26. CRC Press, Boca Raton, Florida.

Harris, C. C., Vahakangas, K., Autrup, H., Trivers, G. E., Shamsuddin, A. K. M., Trump, B. F., Bowman, B. M., and Mann, D. L. (1985). Biochemical and molecular epidemiology of human cancer risk. *In* "The Pathologist and the Environment." (D. Scarpelli and J. Craighead, eds.) International Academy of Pathology Monograph No. 26, pp. 140–167.

Harris, C. C., Weston, A., Willey, J. C., Trivers, G. E., and Mann, D. L. (1987). Biochemical and molecular epidemiology of human cancer: Indicators of carcinogen exposure, DNA damage, and genetic predisposition. *Environ. Health Perspect.* **75**, 109–119.

Hatch, M. C., and Friedman-Jimenez, G. (1991). Using reproductive effect marker to observe subclinical events, reduce misclassification, and explore mechanism. *Env. Health Perspect.* **90**, 255–259.

Hatch, M. C., and Stein, Z. A. (1987). The role of epidemiology in assessing chemical induced disease. *In* "Mechanisms of Cell Injury: Implications for Human Health" (B. A. Fowler, ed.), pp. 303–314. John Wiley and Sons, New York.

Hattis, D. (1988). The use of biological markers in risk assessment. *Stat. Sci.* **3**, 358–366.

Hemstreet, G. P., Schulte, P. A., Ringen, K., Stringer, W., and Altekruse, E. B. (1988). DNA hyperploidy as a marker for biological response to bladder carcinogen exposure. *Int. J. Cancer* **42**, 817–820.

Henderson, R. F., Bechtold, W. E., Bond, J. A., and Sun, J. D. (1989). The use of biological markers in toxicology. *Crit. Rev. Tox.* **20**, 65–82.

Hernberg, S. (1987). Report of Session 1. *In* "Occupational and Environmental Chemical Hazards: Cellular and Biochemical Indices for Monitoring Toxicity" (V. Foa, E. A. Emmett, M. Maroni, and A. Colombi, eds.), pp. 121–124. Ellis Horwood, Chichester, England.

Hernberg, S. and Aitio, A. (1987). Validation of biological monitoring tests. *In* "Occupational and Environmental Chemical Hazards: Cellular and Biochemical Indices for Monitoring Toxicity" (V. Foa, E. A. Emmett, M. Maroni, and A. Colombi, eds.), pp. 41–49. Ellis Horwood, Chichester, England.

Higginson, J. (1977a). The role of the pathologist in environmental medicine and public health. *Am. J. Pathol.* **86**, 460–484.

Higginson, J. (1977b). Changing concepts in cancer prevention: Limitations and implications for future research in environmental carcinogenesis. *Cancer Res.* **48**, 1381–1389.

Hoffman, D., Hayley, N. J., Djordjevic, K. D., Brunnemann, K. D., and Hecht, S. S. (1991). Biomarkers to exposure to tobacco products. In "Human Carcinogen Exposure" (R. C.

Garner, P. B. Farmer, G. T. Steel, and A. S. Wright, eds.), pp. 275–284. Oxford: New York.

Hogue, J. R., and Brewster, M. A. (1988). Developmental risks: Epidemiologic advances in health assessment. *In* "Epidemiology and Health Risk Assessment" (L. Gordis, ed.), pp. 61–81. Oxford University Press, New York.

Hollstein, M., Sidransky, D., Vogelstein, B., and Harris, C. C. (1991). p53 mutations in human cancers. *Science* **253**, 49–53.

Hook, E. (1982). Perspectives in mutation epidemiology: Epidemiologic and design aspects of studies of somatic chromosome breakage and sister chromatid exchange. *Mutat. Res.* **99**, 373–383.

Hsu, I. C., Metcalf, R. A., Sun, T., Welsh, J. A., Wang, N. J., and Harris, C. C. (1991). Mutational hotspot in the p53 hepatocellular carcinomas. *Nature* (London) **350**, 427–428.

Hulka, B. S., and Wilcosky, T. (1988). Biological markers in epidemiologic research. *Arch. Environ. Health* **43**, 83–89.

Hulka, B. S., Wilcosky, T. C., and Griffith, J. D. (1990). "Biological Markers in Epidemiology." Oxford University Press, New York.

Ingram, R. H., Jr., and McFadden, E. R., Jr. (1981). Physiologic measurements providing enhanced sensitivity in detecting early effects of inhalants. *In* "Occupational Lung Diseases: Research Approaches and Methods" (H. Weill and M. Turner-Warwick, eds.), pp. 87–98. Marcel Dekker, New York.

Janetos, A. C. (1988). Biological variability. *In* "Variation in Susceptibility to Inhaled Pollutants" (J. D. Brain, B. D. Beck, A. J. Warren, and R. A. Shaikh, eds.), pp. 9–29. The Johns Hopkins University Press, Baltimore.

Jeppesen, P., and Bower, D. J. (1989). Towards understanding the structure of *eukaryotic* chromosomes. *In* "Cytogenetics" (G. Obe and A. Basler, eds.), pp. 1–29. Springer-Verlag, Berlin.

Kannel, W. B., McGee, D., and Gordon, T. (1976). A general cardiovascular risk profile: The Framingham Study. *Am. J. Cardiol.* **38**, 46–51.

Keys, A., Anderson, J. T., and Grande, F. (1965). Serum cholesterol response to changes in the diet. IV. Particular saturated fatty acids in the diet. *Metabolism* **14**, 376–387.

Kilbourne, E. D. (1973). The molecular epidemiology of influenza. *J. Infect. Dis.* **127**, 478–487.

Kilbourne, E. (1979). "Molecular Epidemiology: Influenza as Archetype." *The Harvey Lectures* **73**, 225–228.

Klee, G. G., and Go, V. L. W. (1987). Carcinoembryonic antigen and its role in clinical practice. *In* "Tumor Markers and Tumor Associated Antigens" (B. C. Ghosh and L. Ghosh, eds.), pp. 22–43. McGraw-Hill, New York.

Koss, L. G. (1987). Automated cytology and histology. A historical perspective. *Anal. Quant. Cytol. Histol.* **9**, 369–374.

Koss, L. G. (1990). The future of cytology. *Acta Cytologica* **34**, 1–9.

Kuller, L. (1991). Epidemiology is the study of "epidemics" and their prevention. *Am. J. Epidemiol.* **134**, 1051–1056.

Landrigan, P. J. (1988). Relation of body burden measures to ambient measures. *In* "Epidemiology and Health Risk Assessment" (L. Gordis, ed.), pp. 139–147. Oxford University Press, New York.

Last, J. M. (ed.) (1988). "A Dictionary of Epidemiology." 2nd Edition, Oxford University Press, New York.

Lauwerys, R. R. (1983). "Industrial Chemical Exposure: Guidelines for Biological Monitoring." Biomedical Publications, Davis, California.

Lipid Research Clinics Program (1974). "Manual of Laboratory Operations. Vol. 1. Lipid and Lipoprotein Analysis," 2d Ed. U.S. Government Printing Office, Washington, D.C.

Lipid Research Clinics Program (1984). The Lipid Research Clinics Coronary Primary Prevention Trial Results. I. Reduction in incidence of coronary heart disease. *J. Am. Med. Assoc.* **251**, 351–364.

Loomis, D., and Wing, S. (1990). Is molecular epidemiology a germ theory for the end of the twentieth century? *Int. J. Epidemiol.* **19**, 1–3.

Lower, G. M. (1982). Concepts in causality: chemically induced human urinary bladder cancer. *Cancer* **49**, 1056–1066.

Lower, G. M., Nilsson, T., Nelson, C. E., Wolf, H., Gamsley, T. E., and Bryan, G. T. (1979). N-acetyltransferase and risk in urinary bladder cancer; approaches in molecular epidemiology. Preliminary results in Sweden. *Environ. Health Perspect.* **29**, 71–79.

Lucier, G. W., and Thompson, C. L. (1987). Issues in biochemical applications to risk assessment: When can lymphocytes be used as surogate markers. *Environ. Health Perspect.* **76**, 187–191.

Maclure, K. M., and MacMahon, B. (1980). An epidemiologic perspective of environmental carcinogenesis. *Epidemiol. Rev.* **2**, 19–48.

Mailley, E. (1875). Essai sur la vie et les ouverage de L.A.J. Quetelet. F. Hayes, Brussells.

McMahon, G., Davis, E., Huber, L. J., Kim, Y., and Nognn, G. (1990). Characterization of C-Ki-*ras* and N-*ras* oncogenes in aflatoxin B1-induced rat liver tumors. *Proc. Natl. Acad. Sci. U.S.A.* **87**, 1104–1108.

Maxon, L. R., and Daugherty C. H. (1985). "Genetics." William C. Brown, Dubuque, Iowa.

Mayr, E. (1982). "The Growth of Biological Thought." Harvard University Press, Cambridge.

Mendelsohn, M. L. (1991). An introduction to biological dosimetry. *In* "New Horizons in Biological Dosimetry" (B. L. Gledhill and F. Mauro, eds.), pp. 1–9. Wiley-Liss, New York.

Meselson, M., and Stahl, F. (1958). The replication of DNA in *Escherichia coli*. *Proc. Nat. Acad. Sci., U.S.A.* **44**, 671–682.

Miller, J. R. (1983). Perspectives in mutation epidemiology. 4. General principles and considerations. *Mutat. Res.* **114**, 425–447.

Muench, K. F., Misra, R. P., and Humayun, M. Z. (1983). Sequence specificity in aflatoxin B1-DNA interactions. *Proc. Natl. Acad. Sci. USA* **80**, 6–10.

Mulvihill, J. J. (1986). Occupational ecogenetics: Gene–environmental interactions in the workplace. *J. Occup. Med.* **28**, 1093–1095.

Murai, A., Miyahara, T., Fujimoto, N., Matsuda, M., and Kameyama, M. (1986). Lp[a] lipoprotein as a risk factor for coronary heart disease and cerebral infarction. *Atherosclerosis* **59**, 199–204.

National Academy of Sciences. (1975). Special risks due to inborn errors of metabolism. In "Principles for Evaluating Chemicals in the Environment." p. 331. National Academy Press, Washington, D.C.

National Academy of Sciences (1991a). "Human Exposure Assessment for Airborne Pollutants. Advances and Opportunities." National Academy Press, Washington, D.C.

National Academy of Sciences (1991b). "Environmental Epidemiology." National Academy Press, Washington, D.C.

National Research Council (1987). Biological markers in environmental health research. *Environ. Health Perspect.* **74**, 1–191.

Oehlkers, F. (1943). Die Auslösung von Chromosomenmutationen in der Meiose durch eine Wirkung von Chemikalien. *Z. Ind. Abst. Verbundsl.* **81**, 313–341.

Omenn, G. S. (1982). Predictive identification of hypersusceptible individuals. *J. Occup. Med.* **24**, 369–374.

Ostermann-Golkar, S. (1983). Tissue doses in man: Implications in risk assessment. *In* "Developments in the Science and Practice of Toxicology" (A. W. Hayes, R. C. Schnell, and T. S. Miya, eds.), pp. 289–298. Elsevier Science, Amsterdam.

Papanicolaou, G. N., and Traut, H. F. (1943). Diagnosis of uterine cancer by the vaginal smear. Commonwealth Fund, New York.

Pappenheimer, A. M., Jr., and Murphy, J. R. (1983). Studies on the molecular epidemiology of diphtheria. *Lancet* October 22, 923–926.

Paul, J. R. (1958). "Clinical Epidemiology." University of Chicago Press, Chicago.

Paul, J. R., and White, C. (1973). "Serological Epidemiology." Academic Press, New York.

Pereira, L., Cassai, E., Honess, R. W., Roizman, B., Terni, M., and Nahmias, A. (1976). Variability in the structural polypeptides of herpes simplex virus 1 strains: Potential application in molecular epidemiology. *Infect. Immun.* **13**, 211–220.

Perera, F. P. (1987a). The potential usefulness of biological markers in risk assessment. *Environ. Health Perspect.* **76**, 141–145.

Perera, F. P. (1987b) Molecular cancer epidemiology: A new tool in cancer prevention. *J. Natl. Cancer Inst.* **78**, 887–898.

Perera, F. P., and Weinstein, I. B. (1982). Molecular epidemiology and carcinogen–DNA adduct detection: New approaches to studies of human cancer causation. *J. Chron. Dis.* **35**, 581–600.

Perera, F. P., Schulte, P. A., Santella, R. M., and Brenner, D. (1990). DNA adducts and related markers in assessing the risk of complex mixtures. *In* "Genetic Toxicology of Complex Mixtures" (M. D. Waters, F. B. Daniel, J. Lewtas, M. M. Moore, and S. Nesnow, eds.), pp. 271–290. Plenum Press, New York.

Rajput-Williams, T. J., Knott, Wallis, S. C., Sweetman, P., Yarnell, J., Cox, N., Bell, G. I., Miller, N. E., and Scott, J. (1988). Variation of apoliprotein-B gene is associated with obesity, high blood cholesterol levels, and increased risk of coronary heart disease. *Lancet* ii, 1442–1446.

Rao, J. Y., Hemstreet, G. P., Hurst, R. E., Bonner, R. A., Jones, P. L., Kyung, W. M., and Fradet, Y. (1991). "Biochemical Mapping of Bladder Cancer Tumorigenesis" (submitted).

Richterich, R. (1969). "Clinical Chemistry." pp. 3–5. S. Karger, Basel.

Rogan, W. J. (1988). Relation of surrogate measures to measures of exposure. *In* "Epidemiology and Health Risk Assessment" (L. Gordis, ed.), pp. 148–158. Oxford University Press, New York.

Romieu, I., Stampfer, M. J., Stryker, W. S., Hernandez, M., Kaplan, L., Sober, A., Bosner, B., and Willett, W. C. (1990). Food predictors of plasma beta-carotene and alpha-tocopherol: Validation of a food frequency questionnaire. *Am. J. Epidemiol.* **13**, 864–876.

Rose, G. (1990). Preventive cardiology: What lies ahead. *Prev. Med.* **19**, 97–104.

Rothman, K. (1986). "Modern Epidemiology." Little Brown, Boston.

Rous, P., and Kidd, J. G. (1941). Conditional neoplasms and subthreshold neoplastic states. *J. Exp. Med.* **73**, 365–389.

Roux, W. (1883). "Über die Bedeutung der Kerntheilungs Figuren." Engelmann, Leipzig.

Rowley, J. D. (1973). A new consistent chromosomal abnormality in chronic myelogenous leukaemia identified by quinacrine fluorescence and Giemsa staining. *Nature (London)* **243**, 290–293.

Rowley, J. D. (1990). Molecular cytogenetics: Rosetta Stone for understanding cancer. Twenty-ninth GHA Clowes Memorial Award Lecture. *Cancer Res.* **50**, 3816–3825.

Saiki, R. K., Arnheim, N., and Erlich, H. A. (1985). A novel method for detection of polymorphic restriction sites by clearance of oligonucleotide probes: Application to sickle cell anemia. *BioTechnology* **3**, 1008–1012.

Suzuki, D. T., Griffiths, A. J. F., Miller, J. H., and Lewontin, R. C. (1986). "An Introduction to Genetic Analysis." Freeman, New York.

Schmidt, C. (1850). "Characteristics of Epidemic Cholera." Rehyer, Leipzig.

Schulte, P. A. (1987). Methodologic issues in the use of biologic markers in epidemiologic research. *Am. J. Epidemiol.* **126**, 1006–1016.

Schulte, P. A. (1989). A conceptual framework for the validation and use of biologic markers. *Environ. Res.* **48**, 129–144.

Schulte, P. A. (1991). Contributions of biological markers to occupational health. *Am. J. Ind. Med.* **20**, 435–446.

Schwann, T. (1839). "Mikroskopische Untersuchungen über die Übereinstimung in der Struktur und dem Wachstum der Tiere and Pflanzen." Springer-Verlag, Berlin.

Shields, P. G., and Harris C. C. (1991). Molecular epidemiology and the genetics of environmental cancer. *J. Am. Med. Assoc.* **246**, 681–687.

Snow, J. (1855). "On the Mode of Communication of Cholera. (2nd ed.) Churchill, London. Reproduced in "Snow and Cholera." Common-Wealth Fund, New York, 1936. Reprinted by Hafner, New York, 1965.

Sobels, F. H. (1982). The parallelogram: an indirect approach for the assessment of genetic risks from chemical mutagens. *Prog. Mut. Res.* **3**, 323–327.

Soussi, T, Caron de Fromentel C., May, P. (1990). Structural aspects of the p53 protein in relation to gene evolution. *Oncogene* **5**, 945–52.

Stein, R. E. K., and Jessop, D. J. (1988). Thoughts on interdisciplinary research. *J. Clin. Epidemiol.* **41**, 813–815.

Suruda, A., Schulte, P., Boeniger, M., Hayes, R. B., Livingston, G. K., Steenland, K., Stewart, P., Herrick, R., Douthitt, D., and Fingerhut, M. A. (1992). Cytogenetic effects of formaldehyde exposure in students of mortuary science. (*in press*)

Tannenbaum, S. R., and Skipper, P. L. (1984). Biological aspects to the evaluation of risk: Dosimetry of carcinogens in man. *Fund. Appl. Toxicol.* **4**, 5367–5370.

Taylor, J. A., Sandler, D. P., Bloomfield, C. D., Shore, D. L., Ball, E. D., Neubauer, A. R., McIntyre, O. R., and Liu, E. (1992). *Ras* oncogene activation and occupational exposures in acute myeloid leukemia. *J. Natl. Cancer Inst.* **84**, 1626–1632.

The American Heritage Dictionary (1976). (W. Morris, ed.), p. 289. Houghton Mifflin, Boston.

Thilly, W. G., Phaik-Mooi, L., and Skopek, T. R. (1982). Potential of mutational spectra for diagnosing the cause of genetic change in human cell populations. *Banbury Report* **13**, 453–465.

Thompson, S. G. (1983). A method of analysis of laboratory data in an epidemiological study where time trends are present. *Stat. Med.* **2**, 147–153.

Tola, S., and Hernberg, S. (1981). Strategies of biological monitoring. *In* "Recent Advances in Occupational Health" (J. C. McDonald, ed.), pp. 199–209. Churchill Livingstone, Edinburgh.

Trucco, M., and Dorman, J. S. (1989). Immunogenetics of insulin-dependent diabetes mellitus in humans. *CRC Crit. Rev. Immunol.* **9**, 201–245.

Truett, J., Cornfield, J., and Kannel, W. (1967). A multivariate analysis of coronary heart disease in Framingham. *J. Chronic Dis.* **20**, 511–524.

Vandenbroucke, J. P. (1988). Is 'the causes of cancer' a miasma theory for the end of the twentieth century? *Int. J. Epidemiol.* **17**, 708–709.

van Noord, P. A. H. (1992). "Selenium and Human Cancer Risk: Nail Keratin as a Tool in Metabolic Epidemiology." Thesis Publishers, Amsterdam.

Vecchio, T. J. (1966). Predictive value of a single diagnostic test in unselected populations. *N. Engl. J. Med.* **274**, 1171–1173.

Venzmer, G. R. (1968). "Five Thousand Years of Medicine." Taplinger, New York.

Vineis, P., and Caporaso, N. (1988). Applications of biochemical epidemiology in the study of human carcinogenesis. *Tumori* **74**, 19–26.

Vineis, P., Faggiano, F., and Terracini, B. (1990). Biochemical epidemiology: Uses in the study of human carcinogenesis. *Terat. Carcin. Meta.* **10**, 231–237.

Virchow, R. (1858). [English translation 1971]. "Cellular Pathology." Dover, New York.

Vogel, F. (1959). Modern problems in human genetics. *Ergeb. Inn. Med. Kinderheilkd.* **12**, 52–125.

Vogelstein, B., Fearon, E. R., Hamilton, S. R., Kern, S. E., Preisinger, A. C., Leppert, M., Nakamura Y., White, R., Smits, A. M. M., and Bos, J. L. (1988). Genetic alterations during colorectal tumor development. *N. Eng. J. Med.* **319**, 525–532.

Waldrop, M. M. (1992). Finding RNA makes proteins give 'RNA world' a big boost. *Science* **256**, 1396–1397.

Watson, J. D., and Crick, F. H. (1953). Molecular structure of nucleic acids in a structure for deoxyribose nucleic acid. *Nature (London)* **171**, 737–740.

Weatherall, D. J. (1991). "The New Genetics and Clinical Practice" 3d Ed. Oxford University Press, Oxford.

Wetmur, J. G., Lehnert, G., and Desnick, R. J. (1991). The delta-aminolevulinate dehydratase polymorphism: Higher blood lead levels in lead workers and environmental exposed children with the 1-2 and 2-2 isozymes. *Environ. Res.* **56**, 109–119.

White, C. (1973). Preface. *In* "Serological Epidemiology" (J. R. Paul and C. White, eds.) Academic Press, New York.

Wild, C. P., and Montesano, R. (1991). Detection of alkylated DNA adducts in human tissues. *In* "Molecular Dosimetry and Human Cancer" (J. D. Groopman and P. L. Skipper, eds.), pp. 263–280. CRC Press, Boca Raton, Florida.

Willett, W. C., Stampfer, M. C., Underwood, B. A., Speizer, F. E., Rosner, B., and Hennekins, C. H. (1983). Validation of a dietary questionnaire with plasma carotenoid and alpha-tocopherol. *Am. J. Clin. Nutr.* **38**, 631–639.

Williams, J. W. (1920). The value of the Wasserman reaction in obstetrics based upon the study of 4547 consecutive cases. *Johns Hopkins Hosp. Bull.* **31**, 335–342.

Wogan, G. N., and Gorelick, N. J. (1985). Chemical and biochemical dosimetry of exposure to genotoxic chemicals. *Environ. Health Perspect.* **62**, 5–18.

Woolf, B. (1955). On estimating the relation between blood group and disease. *Ann. Hum. Genet.* **19**, 251–253.

World Health Organization (1975). "Early Detection of Health Impairment in Occupational Exposure to Health Hazards. (WHO Technical Report No. 571). World Health Organization, Geneva.

Wright, G. L., Jr., and Cox, A. D. (1987). Monoclonal antibodies to human tumor antigens. *In* "Morphological Tumor Markers" (G. Seifert, ed.), pp. 1–18. Springer-Verlag, New York.

Wynder, E. C., and Reddy, B. S. (1974). Metabolic epidemiology of colorectal cancer. *Cancer* **34**, 801–806.

Yach, D. (1990). Biological markers: Broadening or narrowing the scope of epidemiology. *J. Clin. Epidemiol.* **43**, 309–310.

Yallow, R. S., and Berson, S. A. (1960). Immunoassay of endogenous plasma insulin in man. *J. Clin. Invest.* **39**, 1157–1175.

Zielhuis, R. L. (1978). Biological monitoring. *Scand. J. Work Environ. Health* **4**, 1–18.

Zondek, B. (1942). The importance of increased production and excretion of gonadotrophic hormone for diagnosis of hydatidiform mole. *J. Obstet. Gynecol. Br. Emp.* **49**, 397.

2

Molecular Biology in Epidemiology

Robert E. Hurst
and Jian Yu Rao

Introduction

The revolution in molecular biology began with perhaps one of the finest, most concise papers ever written: Crick and Watson's seminal paper (1953) in which they demonstrated the double-stranded nature of DNA and how it was produced by base pairing. In a classical understatement, they stated, "It has not escaped our attention that the specific pairing [of bases in DNA] we have postulated immediately suggests a possible copying mechanism for the genetic material." The 1960s saw the realization of results that flowed from this simple statement in the unraveling of the genetic code and the elucidation of the mechanisms of DNA transcription into messenger RNA (mRNA) and mRNA translation into proteins. In the 1970s, recombinant DNA technology and techniques to rapidly sequence nucleic acids emerged. Recombinant DNA technology allowed individual genes to be isolated and replicated; rapid sequencing techniques made it practical to sequence stretches of up to a few thousand base pairs with relatively little effort. The explosive radiation of this basic knowledge into almost every area of biology occurred in the 1980s, providing an understanding at the molecular level of the mechanisms and regulation of cell growth, the response to drugs and toxic chemicals, carcinogenesis, progression and metastasis, as well as many other biologic processes.

In this chapter, we provide an overview of the major concepts of molecular biology, presenting a brief outline of the major techniques and their strengths, weaknesses, and practical limitations. We present how these concepts can be used to map the origins of human disease at a molecular level and outline a scenario for integrating molecular biologic monitoring into

Molecular Epidemiology: Principles and Practices

45

field epidemiologic studies. Of necessity, molecular biology is presented simply; our presentation assumes some knowledge on the part of the reader.

Basic Principles of Molecular Biology

The use of molecular biologic data in epidemiologic research requires a familiarity with how genes are organized and how they function. DNA constitutes the genetic material of the cell. Other than some rare methylated bases, DNA is composed of only four bases, the pyrimidines thymine (T) and cyto-

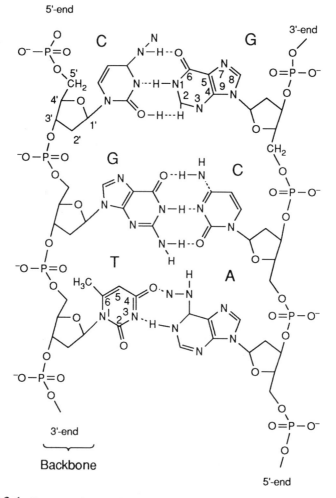

FIGURE 2.1   Structure of 3 complementary base pairs showing numbering systems for bases (unprimed) and deoxyribose (primed).

sine (C) and the purines adenine (A) and guanine (G). A segment of three base pairs is shown in Figure 2.1. The strands are arranged in the familiar double helix. The backbone consists of alternating sugars and phosphates; the bases are attached to the sugars at carbon 1. The numbering systems of sugars and bases are indicated. The strands have a polarity, indicated by the designations 5' and 3', as defined by the numbering system of the sugars. The strands are also complementary in structure because of the hydrogen bonding scheme in which A always bonds with T (or the closely related base uracil, U, found in RNA) and G always bonds with C. Thus, either strand can serve as the template for synthesis of the other. Note that A–T pairs have two hydrogen bonds whereas G–C pairs have three. The extra bond confers extra stability on G–C pairs. Information is stored in a hierarchy of structure. The lowest level is the individual nucleotide. A *codon* consists of 3 nucleotides and specifies a single amino acid. A *gene* is the functional unit that specifies a protein. A *chromosome* contains several thousand genes and is the smallest replicating unit. The *genome* is the complete set of genetic information that an organism contains.

The structure of a generic stretch of mammalian DNA is illustrated in Figure 2.2. This figure presents a solitary coding gene and a gene family separated by long stretches of noncoding DNA. In general, less than 30% of the total genome is ever transcribed into RNA, and only a fraction of that is ever translated into protein. At least 20 kilobase pairs (kb; 1000 base pairs) separate the solitary coding gene from other expressed elements. In general, many of these stretches consist of repetitious sequences that usually constitute over 70% of the total genome. The structure of the repetitious sequences can repeat on scales of every 5–10 base pairs (bp), every 100–300 bp, or even every 5000–6000 bp. Usually, the repetitious structures are not exact copies of each other; repeats of high to moderate *sequence homology* (similarity in sequence) are found. Of particular interest are *Alu* sequences, found in mammals, that repeat on a scale of 150–300 bp. These sequences have a species specificity and can be used to mark genes that are introduced into cells of

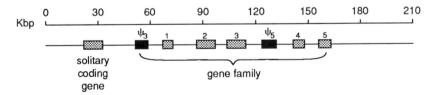

FIGURE 2.2   Structure of a large segment of mammalian DNA, including a gene family. Boxes indicate genes, which are sequences that are transcribed into RNA. Shown are a solitary coding gene and a gene family such as the globin family or keratin family. Untranscribed pseudogenes (ψ) are also shown. A more detailed structure of a coding gene and RNA processing is shown in Fig. 2.3.

another species. *Alu* sequences are common in introns and may play a part
in mRNA splicing. Also present in these noncoding regions are sequences
that bind to various proteins and regulate the transcription of genes. These
control regions can range upstream (in the 5′ direction) up to several thousand bp from the start of the gene.

Downstream from the solitary coding gene in the figure is a gene family.
Such gene families are common and code for proteins that are structurally
related and whose expression is coordinated in some way. Rarely, however,
are all the members expressed at the same time in the same cell. The globin
genes, for example, are expressed at different times during embryogenesis.
The keratins are expressed in different cell types; only a few of the over 20
genes are expressed in any one cell. Gene families also may contain pseudogenes, which are homologous in structure to the other members of the gene
family but are not expressed. Gene duplication, with or without recombination of parts of other genes, is apparently the most common mechanism by
which new proteins are produced. Unequal recombination, in which one
strand receives both copies of a gene, is the mechanism for such duplication.
How the expression of gene families is regulated and coordinated is an important unanswered question.

Figure 2.3 illustrates the structure of a gene on a finer scale and the transcription of its sequence into RNA together, in addition to the further processing that the RNA undergoes to produce mRNA used as the template for
protein synthesis. The eukaryotic gene sequence almost invariably consists of
sequences that are expressed (exons), that is, encode protein, and intervening
sequences (introns) that do not encode protein. Regulatory sequences are
found upstream of the initiation site and, less commonly, downstream. One
regulatory sequence that is found very commonly at -20 to -30 bp is a
"TATA box," so called because it contains the sequence TATA or TATAAT.
The TATA box orients RNA polymerase II, the enzyme responsible for transcription of most protein genes; most eukaryotic genes seem to have such a
sequence. In general, promoter elements are found in the regions between
-500 and -100 and -70 and -30 bp, where 0 bp is the start of the transcribed region. These regulatory sequences bind regulatory proteins that enhance or inhibit the binding of RNA polymerase and constitute the mechanism by which the expression of a gene can be regulated or coordinated with
the expression of other genes. Another sequence that commonly is found in
these regions is CAAT, which whimsically has been named "CAAT box."

The first step in transcription is production of a "primary transcript"
RNA by RNA polymerase II. This transcript is complementary to the template DNA sequence, ranges from the initiation sequence on the DNA to a
termination sequence, and includes both exons and introns.

Transcription starts with the binding of RNA polymerase II. This binding generaly requires a transcription factor. Transcription factors are proteins that recognize the region of DNA to be transcribed and enable poly-

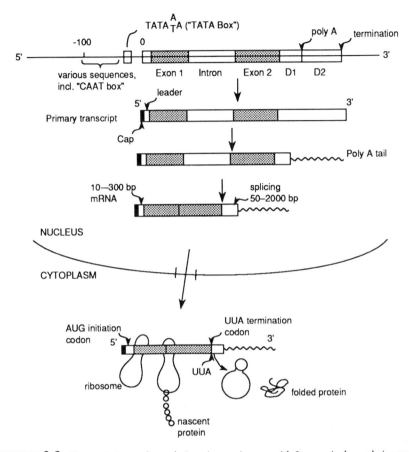

FIGURE 2.3   Transcription and translation. A generic gene with 2 exons is shown being transcribed to the RNA primary transcript, which is, in turn, processed to the mature mRNA. The mRNA is exported to the cytoplasm, where several ribosomes can bind sequentially and translate the nucleic acid sequences to protein.

merase binding. The RNA polymerase proceeds in a $3' \rightarrow 5'$ direction along one strand of the DNA and synthesizes the complementary RNA sequence, including exons, introns, and untranslated sequences on both the $5'$ and $3'$ end. This molecule is referred to as a "primary transcript." A convention exists concerning the sense and complementarity of DNA and transcribed mRNA. The two strands of DNA are labeled "sense" and "antisense." Transcription is considered to proceed from the antisense strand in a $3' \rightarrow 5'$ direction, which is $5' \rightarrow 3'$ on the sense strand. Thus, the sense strand of the DNA and the RNA transcribed from the antisense strand have the same direction and sequence, except U in RNA substitutes for T in DNA.

The cellular nucleus contains a large variety of primary transcripts, known collectively as "heterogeneous nuclear RNA" (hnRNA). The first step in RNA processing is to cap the 5' end by adding a guanylate nucleotide to the 5' end of the chain in a unique 5' → 5' arrangement. This G is then methylated. The RNA no longer has a free 5'-OH group. The next step in processing consists of removing and destroying sequences at the 3' end that extend past the actual protein-coding sequence. A long repetitive sequence of adenylate nucleotides is then added to the 3' end of the RNA. This "poly(A) tail" is approximately 250 nucleotides long in mammals. After this reaction, a complex series occurs during which the intron sequences are clipped out and the exon sequences are spliced together very precisely. The product of this reaction is the mature mRNA, which is exported to the cytoplasm where it is translated into protein by specialized organelles called ribosomes. Most proteins are further modified by one or more of several possible posttranslational modification steps, including proteolytic cleavage, addition of carbohydrate or lipid moieties, and modification of amino acids.

TABLE 2.1   **Terms Used in Molecular Biology**

anneal	To incubate DNA at a temperature a few degrees below its melting point to allow complementary sequences to hybridize.
cDNA	Complementary DNA is derived from RNA by reverse transcriptase. If prepared from mRNA, the usual template for cDNA, it differs from genomic DNA in not containing introns.
clone	A genetic unit (cell, plasmid, vector) derived from a single progenitor.
codon	A three-base sequence that codes for a single amino acid.
denaturing gel	A gel for electrophoresis in which the molecules are not in their native conformation (single-stranded DNA, random coil proteins).
epitope	The specific portion of a protein recognized by an antibody.
exon	A section of DNA that is expressed in a protein sequence.
gene	The functional unit of genetic inheritance and a sequence of DNA that codes for a single polypeptide.
genome	The complete genetic information of an organization.
genomic DNA	DNA derived from the genome (vs cDNA).
hnRNA	Heterogeneous nuclear RNA. The primary transcript.
hybridoma	A hybrid cell derived from fusing an immortal cancer cell (myeloma cell) with an antibody-producing B-cell.
immunogen	A protein used to immunize an animal to produce antibodies.
intron	Intervening sequence between coding regions of DNA.
library	A collection of plasmids containing recombinant DNA fragments obtained from genomic or cDNA.
ladder gel	The gel resulting from DNA sequencing; so called because the result resembles a ladder.
mRNA	Messenger RNA. RNA that is used as a template to code for protein synthesis.
melt	To denature (dissociate into two single strands) double-stranded DNA by heating.

continued

The reader will have noted by now that molecular biology has a complex and unique specialized language that may not be familiar to everyone. Table 2.1 lists definitions for most of the important terms of molecular biology.

Gene expression occurs within a cell that is part of a tissue, an organ, and an organism; such cells are not autonomous. The growth and differentiation of cells, and hence the expression of genes, is regulated closely by endocrine and paracrine mechanisms. Endocrine regulation is defined as regulation by a chemical signal generated by other cells distantly located from the target cell, whereas paracrine regulation is regulation by chemical signals generated by adjacent cells. In general, most tissues consist of overlying cells of epithelial origin, frequently containing glandular cells, and a stroma, or connective tissue layer or membrane, that underlies the epithelium. Organs are composed of collections of tissues whose actions are coordinated. Cells communicate in the form of peptide signals that bind to highly specific receptors on the surface of the cell. Such peptides and receptors exist in large num-

TABLE 2.1   **Terms Used in Molecular Biology (continued)**

northern blot	A technique for RNA analysis in which RNAs are separated by electrophoresis, transferred ("blotted") to a carrier, and visualized by hybridizing to radioactive probes.
oligonucleotide	A short (<20) sequence of bases.
oncogene	A gene that codes for a protein involved in cell growth that unless properly regulated will result in inappropriate growth.
plasmid	A genetic element, usually circular, derived from bacteria and used to carry inserted genetic elements (see vector).
recombinant DNA	The techniques for recombining DNA from disparate sources into a single molecule.
restriction nuclease	An endonuclease that cleaves double-stranded DNA at precise sequences.
restriction fragment length polymorphism (RFLP)	A difference in the sequences of DNA from the two chromosomal copies such that cleavage by restriction nucleases produces different sized fragments from each copy.
restriction site	A DNA sequence recognized by a restriction nuclease.
sequence homology	The similarity in sequence between different nucleic acids. Homology is usually expressed as a percentage.
stringency	Solution conditions and temperature for hybridization that regulate the degree of homology required for hybridization.
Southern blot	A technique of DNA analysis in which DNA fragments are separated by electrophoresis, transferred ("blotted") to a carrier, and visualized by hybridizing to radioactive probes.
tumor suppressor gene	A regulatory gene that prevents inappropriate cell growth.
vector	A genetic element, usually a plasmid or virus, used to carry exogenous DNA inserts for replication and manipulation.
western blot	A technique of protein analysis in which proteins are separated by electrophoresis, transferred ("blotted") to a carrier, and visualized using an antibody against the protein and immunochemical visualization.

bers, giving the organism the ability to regulate the large number of organs and cell types independently. In turn, these receptors communicate with the interior of the cell, where they interface with a limited number of intracellular systems regulating growth, differentiation, and function. The generalized mechanism is that the binding of a peptide signal initiates a specific program of protein expression within the cell; the signal and receptor confer tissue specificity. At the molecular level, this mechanism is controlled by the various initiator and inhibitor sequences on the DNA and the proteins that bind to them.

Molecular biology has developed truly amazing techniques to probe these systems for genes of interest from a large collection of possible genes. The mammalian genome has been estimated to consist of 100,000 genes; some 5,000–10,000 normally are expressed in any given cell type. Much of molecular biology is finding a "needle in the haystack," where the "needle" is a gene sequence or protein of interest in a "haystack" of other sequences or proteins. Molecular epidemiology seeks to apply these techniques to understanding the causes of disease and to developing markers that can chart the progress of disease. These techniques can advance infectious disease epidemiology by defining and detecting infectious organisms with accuracy and sensitivity not possible with conventional serologic techniques. In addition, molecular techniques used to examine the mRNAs in a given cell or tissue directly identify the genes that are expressed and how that expression may change with exposure to infectious or toxic agents. Also these procedures elucidate how other agents may modify these responses. Finally, using monoclonal antibodies and other protein chemistry techniques, the levels of specific protein markers that ultimately govern the behavior of cells and define their phenotypes can be measured with high precision.

Core Techniques of Molecular Biology

This section reviews the general principles and some applications to epidemiology of the major methods of molecular biology. Specific experimental details can be found in several excellent laboratory texts cited at the end of this chapter.

Restriction Endonucleases

The DNA of even a small chromosome is enormous, consisting of millions of base pairs. Such a large molecule cannot be manipulated conveniently. One of the first breakthroughs of molecular biology was the identification of *restriction endonucleases,* which are enzymes found in bacteria that cleave double-stranded DNA at precise sequences that are often palindromic. A palindrome reads the same backward and forward. The classic word palindrome is "Madam I'm Adam." A palindromic DNA sequence is CGATCG because its complementary sequence GCTAGC reads the same from right to left (5' to 3'). These DNA restriction enzymes serve as bacterial defenses against foreign viral DNA, since their own DNA is protected by methylation of one or more bases in the *restriction site,* or sequence of bases defining the specificity of the restriction enzyme. Restriction endonucleases are named for the organism of origin; a roman numeral indicates the order of discovery. Thus, *Eco*RI is the first enzyme isolated from the R strain of *Escherichia coli; Hpa*II is the second enzyme to be isolated from *Haemophilus aphrophilus.* Restriction endonucleases allow precise cleavage of DNA into fragments of reproducible size that can be separated by electrophoresis or other techniques. The size of the fragments produced generally is a function of the number of bases in the restriction site; the larger the site, the less frequent it is, and the fewer there are in a genome. "Four cutters," which detect tetrameric sequences, usually cleave genomic DNA into small fragments, whereas "eight cutters" usually produce gene-sized fragments because of their lower frequency. Interestingly, a site recognized by the eight cutter *Not*I frequently is found at or near the beginning of many mammalian genes and rarely anywhere else. Restriction endonucleases can leave "blunt" ends, as illustrated for *Bal*I in Figure 2.4, or "sticky" or "overhanging" ends, as illustrated for *Eco*RI. As discussed later, these properties are useful in recombinant DNA technology for ligating DNA fragments into vectors that can be amplified or multiplied.

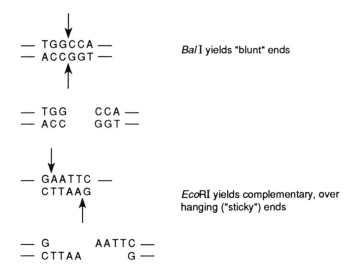

FIGURE 2.4 Actions of restriction nucleases. *Bal*I yields blunt ends, *Eco*RI yields "sticky" ends.

Hybridization

A fundamental property of nucleic acid is base complementarity or base pairing, the formation of hydrogen bonds that allows one strand of a nucleic acid to specify a complementary strand. Adenine (A) always pairs with thymine (T) or uracil (U) and guanine (G) always pairs with cytosine (C). This formation of complementary sequences stabilized by hydrogen bonds is the major principle by which molecular biology finds the "needle in the haystack." Hybridization is one of the cornerstones of molecular biology. A probe consisting of a segment of known sequence is isolated or synthesized and is used to "fish out" the complementary sequence from a large sample of heterogeneous material.

Although the principle of hybridization is simply the recognition of complementary sequences, the technology depends on an understanding of the conditions that stabilize the DNA double helix and how these conditions are manipulated to disrupt, or *melt,* the DNA into two single strands or to allow the double-stranded helix to form again, or *anneal.* The first step in any hybridization experiment is to convert double-stranded DNA into single-stranded DNA to allow the probe to bind to complementary sequences. Failure to properly meet the conditions required for this process will lead to incorrect results.

The melting of a double-stranded helix is a sudden event that occurs over a narrow temperature range, as illustrated in Figure 2.5. As the temperature is raised, a point is reached at which the thermal energy begins to overcome the bond energy of the hydrogen bonds that stabilize the helix, so it begins to come apart. Because the stabilization of the helix is cooperative, that is, the formation of one base pair increases the probability that a neighboring base

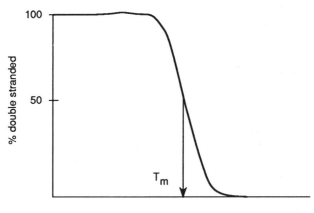

FIGURE 2.5   Melting and annealing of DNA.

pair will form, the transition to single strands occurs suddenly. The melting temperature, or T_m, represents the temperature at which half the bonds are broken. DNAs have different T_ms as determined by the percentage of G–C pairs, among other factors. Because G–C pairs form three hydrogen bonds and A–T pairs only two, the former are more stable; therefore, T_m increases with GC content.

After dissociation of the DNA into its constituent single strands, the probe is added and double strands are allowed to reform. This process is called *annealing* and depends on several factors, including the concentration of DNA, the time allowed for annealing, the temperature, and the concentration of salts in the solution. The latter two variables are referred to collectively as *stringency*. The process of annealing is statistical and depends on successive formation and breaking of hydrogen bonds, always leading to successive minimization of potential energy, which corresponds to formation of the maximum number of hydrogen bonds. If the annealing temperature is too low, insufficient thermal energy will be available to support this process. The binding that occurs tends to be nonspecific. Rapid cooling on ice tends to suppress hydrogen bond formation and preserves single strands. Annealing at too high a temperature will melt the strands continuously.

Manipulating stringency allows the experimenter to control the degree of allowable mismatch between the probe and its target sequence. Stringent conditions are those of relatively high temperature and low salt concentrations which favor helix formation only between exactly complementary sequences. By carefully maintaining the temperature just under T_m, it is sometimes possible to distinguish single base pair mutations in small DNA fragments. Under lower stringency, some mismatches are allowed. Less stringent conditions can be used to probe for homologous genes or to use probes from another species in conserved genes. Thus, the entire family of hemoglobin or *ras* genes, for example, can be detected using a probe under lower stringency, that is exactly complementary to one member of the family. Likewise, a readily available mouse probe sometimes may be substituted for an unavailable human probe. The concentrations of reactants and the time allowed for reaction are also important in hybridization; longer times lead to more complete hybridization. Mathematical models, which are beyond the scope of this chapter, have been derived to describe hybridization. These factors are discussed at much greater depth in laboratory manuals.

The simplest use of hybridization is the dot or slot blot. In this procedure, a drop of single-stranded DNA-containing solution is spotted onto a nitrocellulose or nylon membrane and immobilized by one of several techniques. The membrane is then "prehybridized" with sheared, heterologous DNA (e.g. a salmon sperm DNA) or RNA (e.g., tRNA) to reduce nonspecific binding of radioactive probe. The radioactive probe is then added and allowed to hybridize, usually overnight. The membrane is washed under conditions of appropriate stringency until no further radioactivity is detectable in the wash solution. The radioactivity remaining on the membrane is bound specifically. The blot is developed by laying a piece of X-ray film over the membrane and placing the "sandwich" in the freezer. Disintegrations of the radioactive tracer will expose the film. Freezer temperatures slow the reactions in the film emulsion, increasing resolution. Thus, the presence of a particular sequence and an estimate of the amount can be obtained from the density of the spot.

Electrophoresis

Another important technique of molecular biology is electrophoresis, in which proteins or nucleic acids are separated by size and charge. In practice, all electrophoretic techniques are carried out using a supporting gel of controlled pore size. This gel minimizes diffusion, which tends to degrade resolution. Usually gels are prepared from polyacrylamide or agarose. Polyacrylamide is used to separate smaller molecules, whereas agarose is used for larger ones. Although the theory of electrophoresis shows that the charge actually governs migration of molecules in an electric field, most separations are by size and are governed by size. This sieving occurs because the holes in the gel are approximately the size of the molecules being separated; larger molecules are held up and migrate more slowly than smaller ones. Moreover, with DNA, each nucleotide has one charge, so all DNA molecules have very nearly the same charge density, or ratio of charge to molecular weight. For proteins this is not true, since the charge of proteins is highly dependent on its amino acid composition. However, by adding the charged detergent sodium dodecyl sulfate (SDS), which binds strongly and in large numbers to proteins, the native charge of the molecules is overwhelmed by the large numbers of SDS molecules, leading to a separation of proteins mainly by size as well. The size of the pores in the gel is regulated by the composition of the gel. Polyacrylamide gels are formed by adding a chemical cross-linking agent to acrylamide; the higher the concentration of cross-linker, the tighter the gel. Porosity of agarose gels is controlled by the amount of agarose added to the gel. The gel can be *denaturing* or *nondenaturing,* depending on whether agents such as urea or SDS are added. High concentrations of urea disrupt hydrogen bonds, thereby dissociating double-stranded DNA into single strands. For proteins, a sulfhydryl reagent such as mercaptoethanol or dithiothreitol is added to reduce secondary disulfide bonds that help maintain the native conformations of many proteins, allowing the SDS to denature the protein completely.

Polyacrylamide gels are prepared by pouring the liquid gel between two cleaned glass plates separated by a spacer. "Combs," devices used to form wells into which to inject sample at the top of the gel, are placed in the gel and the gel is allowed to polymerize. Gels can be operated in the vertical position, which is usual for DNA sequencing gels or protein gels, or in the horizontal position, which is usual for DNA separations. The sample usually is dissolved in a loading buffer that contains a "tracking dye," which migrates more rapidly than the sample and is used to track the progress of the separation. The loading buffer solution also is more dense than the running buffer surrounding the gel, preventing the sample from floating away when it is applied with a small pipet directly into the well. The separation is started by applying a voltage ranging from several tens of volts for a small gel to 1000 or more volts for a large gel. Nucleic acids and proteins in SDS gels, both being negatively charged, migrate toward the positive electrode (anode). After an appropriate time, the electrophoresis is stopped and the DNA or protein is visualized by one of several techniques.

The simplest technique for visualizing DNA is to stain the gel with ethidium bromide, a dye that binds tightly to double-stranded DNA and fluoresces brightly under ultraviolet light. The ethidium bromide can be added prior to electrophoresis or the entire gel can be bathed in a solution containing the dye; the unbound dye is removed by washing. Blotting techniques can be used also.

The *Southern blot,* named for Edward Southern who developed the technique in 1975, is used to visualize DNA sequences; an analogous technique, the *Northern blot,* is used to visualize RNA sequences. In the typical Southern blot, the gel is laid atop a piece of nylon or nitrocellulose membrane. Several techniques are available for eluting the DNA from the gel, including electroelution, which uses an electric field, or simple capillary action generated by pieces of filter paper overlaid by paper towels, causing a flow of buffer through the gel onto the membrane. Transfer of fragments larger than 5 kb becomes progressively poorer with increasing size. Southern blotting is used to identify the number and sizes of DNA fragments that contain a particular sequence. The technique is used, for example, to identify fragments containing a particular sequence in *restriction fragment length polymorphism* (RFLP) analysis (see subsequent text) and to analyze complementary DNA (cDNA) (see subsequent text).

The Northern blot, so named because it detects the nucleic acid "opposite" that detected by the Southern blot, is an important technique for studying gene expression. The total RNA fraction is isolated from a cell sample or tissue, as described subsequently, and $10-20$ μg are loaded into a lane on a 1% agarose gel containing 2.2 M formaldehyde to destroy secondary structure in the RNA. After electrophoresis, the RNA is transferred to a membrane and probed. The bands are used to identify the size of the primary transcript as well as alternative splicings of the primary transcript into different messages. Mammalian messenger RNA is frequently assembled by differential splicing, leading to different sized messages. The Northern blot also is used to determine whether a particular tissue or cell expresses a certain gene and to estimate the level of expression.

An analogous technique exists for proteins. Proteins are separated by

electrophoresis in a denaturing polyacrylamide gel containing SDS and a reducing agent to eliminate disulfide bonds. These treatments produce a random coil to which a large number of SDS molecules have bound. The bound
SDS suppresses the native charge of the protein and the complex migrates
according to size. Proteins of known molecular weight, available as a mixture
commercially, are used as controls to estimate the size of a new protein. The
protein bands can be stained with silver stain, which has a very high sensitivity to all proteins, with Coomassie Blue, which tends to visualize primarily
glycoproteins (proteins with attached carbohydrate groups), or with other
stains. Specific proteins also can be visualized by blotting the protein to a
nylon or nitrocellulose membrane and incubating with an antibody against
the specific protein. The bound antibody is detected by a second antibody
against the first. The second antibody is labeled with a reporter group, such
as horseradish peroxidase or another enzyme, and, after incubation with the
appropriate substrate, produces a colored band. This technique is called a
Western blot.

Isolation of DNA and RNA

In most of the techniques outlined here, nucleic acid samples of adequate
purity are essential to obtaining accurate and reproducible results. DNA is,
in general, much more stable and more easily isolated than RNA, which
is degraded easily by ubiquitous RNases. An RNase-free laboratory area
should be prepared for work with RNA in which all containers and pipet
tips are autoclaved, gloves are always worn, and solutions are treated with
diethylpyrocarbonate (DEPC), an agent that destroys RNases. The major
method for purification of DNA is the phenol-chloroform extraction. Highly
purified phenol is mixed with the sample under conditions that promote dissociation of proteins from the nucleic acid. Chloroform, which tends to promote protein denaturation and increase the density of the lower phase, usually is added to improve efficiency. The proteins tend to remain in the lower,
organic phase or at the interphase while the nucleic acid remains in the upper
phase. Centrifugation is used to separate the phases. The upper phase is removed and the nucleic acid is recovered by precipitation with ethanol.

   RNA isolation is complicated by the concomitant release of RNases on
cellular lysis, necessitating the addition of RNase inhibitors. Cells can be
lysed with detergent and the total RNA, including mRNA, ribosomal RNA
(rRNA), transfer RNA (tRNA), and hnRNA, can be isolated using a modified phenol-chloroform method. Methods based on use of a strong detergent
and guanidinium isothiocyanate or chloride, to which a reducing agent such
as mercaptoethanol has been added to further denature RNases, frequently
are used to isolate RNA. The quality of total RNA needs to be checked by
electrophoresis in a denaturing gel (to break up secondary structure). Sharp
bands corresponding to rRNA are usually an indicator that high-quality

RNA has been isolated. Usually the mRNA fraction rather than total RNA is desired. mRNA is isolated by chromatography on oligo(dT)-cellulose or poly(U)-Sepharose, taking advantage of base pairing between the poly(A) tail on the mRNA and the oligo(dT) or poly(U). A single oligo(dT) step usually will remove most tRNA and reduce rRNA by about 50%; a second step generally reduces the level of rRNA to about 10% of the original. RNA isolation techniques are continuously being improved. Commercially available kits include a "one-step" isolation of mRNA involving lysis in guanidinium isothiocyanate and oligo(dT) chromatography without the intermediate step of purification of total RNA, on the use of oligo(dT)-coated magnetic beads for "30 minute" isolation of highly purified mRNA.

Synthesis of cDNA

The conversion of unstable mRNA into stable DNA through the use of reverse transcriptase originally isolated from retroviruses is the first step in many procedures. In essence, an *oligonucleotide* primer (a short sequence of bases, generally less than 20) complementary to a short sequence on the mRNA is used to initiate the reaction. Reverse transcriptase then synthesizes a complementary copy in DNA from deoxyribonucleoside triphosphates in the reaction mixture. Because of its complementary nature, the product is referred to as cDNA. Although the process is conceptually simple, several choices must be made and a few shortcomings must be understood. Reactions must be carried out carefully to maximize the synthesis of full-length cDNA and minimize the synthesis of small fragments and a second strand. Second strand synthesis can occur when the cDNA forms a hairpin loop by doubling back on itself. Two reverse transcriptases are available, one from Moloney murine leukemia virus (MMLV) and the other from avian leukemia virus (ALV). The two enzymes are not identical, responding differently to agents such as pyrophosphate, spermidine, and actinomycin D that are added to increase full-length cDNA synthesis and inhibit hairpin loop formation. The choice of primer is also important. When oligo(dT) is used to prime the reverse transcriptase, it is possible to obtain full-length cDNAs. In general, yields of full-length cDNA are less than 50% relative to the mRNA template, introducing a distinct bias toward the 3′ end of the mRNA. A cDNA more representative of the entire mRNA can be synthesized at the expense of obtaining full-length cDNAs by priming with a random mixture of hexamers. A third alternative is to prime with a specific primer that contains a restriction site, an approach that is useful in cloning.

Polymerase Chain Reaction

Polymerase chain reaction (PCR) techniques have made certain experiments possible on a time scale of hours that otherwise would have taken weeks or months. With PCR, cDNAs in the 300–1000 bp size range can be duplicated explosively to microgram quantities from femtogram quantities and, in theory at least, from a single DNA or RNA molecule. The principle of PCR is illustrated in Figure 2.6. The original template can be single or double stranded. The key to PCR is the selective use of specific oligonucleotide primers to prime synthesis only of sequences complementary to the primers. Two primers are needed, one for each strand of the double helix. The primers

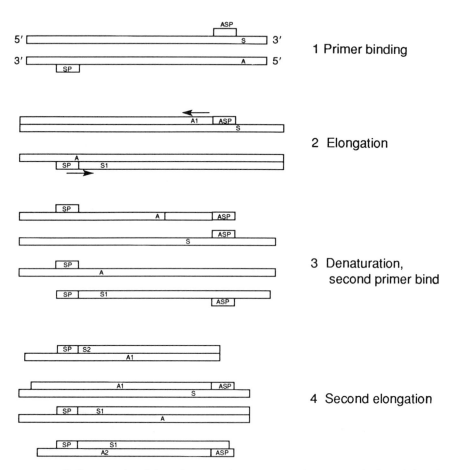

FIGURE 2.6 Principles of the polymerase chain reaction. Chains are melted, cooled, and primers added. The antisense primer (ASP) binds to the sense strand, while the sense primer (SP) binds to the antisense strand. After elongation, 2 complementary chains are produced. The A1 chain is antisense but delimited on the 3′ end by ASP, while the converse is true for the S1 chain. In the second round, the S and A chains again serve as templates for A1 and S1 chains, but the A1 and S1 chains now serve as S2 and A2 chains delimited on *both* ends. S2 and A2 chains then serve as templates for other A2 and S2 chains, which then multiply exponentially.

usually are custom synthesized on automated synthesizers at a cost of a few hundred dollars. The enzyme used in PCR is one of several DNA polymerases isolated from thermophilic bacteria native to hot springs or oceanic vents. Nucleoside triphosphates and magnesium ion also are added to the reaction mixture. If the template is single stranded, the first step consists of synthesis of a complementary second strand by the polymerase. If the template is a double-stranded DNA, this step is skipped. The temperature is increased to above 95°C for 1–2 min to melt the two strands. The reaction mixture is cooled rapidly to near 50°C for 1–2 min to allow the primers to anneal, then raised to 70°C for 30 sec to 10 min to allow the DNA polymerase to synthesize a second complementary strand. The two primers delimit the size of the sequence, as illustrated in Figure 2.6. This program is repeated 20 to 40 times, producing an enormous amplification of the original template. If doubling were perfect each time, this process would produce 1.1×10^{12} copies from each template. In fact, the efficiency is usually considerably less; practical applications usually yield 10^7–10^8 copies.

The primers need not match at the 5' end completely. The ability to introduce sequences at the 5' end of the molecule is one of the strengths of PCR. The experimenter can take advantage of this to, for example, introduce restriction sites than can be used to clone the products. As an example, a primer synthesized to be complementary to 20 bases is synthesized with a CTTAAG sequence at the 5' end. This sequence is not complementary to the target, but is the sequence recognized by *Eco*RI. Because each sequence of the amplified DNA contains one primer sequence, each of the amplified product sequences will contain an *Eco*RI restriction site. Another use is to introduce specific mutations into the product at a specific site. Thus, to introduce an A → G mutation at nucleotide 77 of a particular gene, the primer would incorporate nucleotides 60–80, but G would be substituted for A at position 77. Each of the amplified sequences will now contain a G at position 77 rather than the wild-type A. This approach of *site-directed mutagenesis* can be used to investigate the structure and function of proteins by introducing specific mutations at specific sites.

The particular strength of PCR is that the amplification can be highly selective. For example, the *ras* gene can be amplified selectively in the presence of the remainder of the genome or total RNA isolated from a cell. The amplified *ras* genes can be probed for mutants as a way of determining the ratio of mutant to wildtype genes. Even more exciting, as described in the applications, mutant sequences can be amplified selectively in the presence of normal sequences. PCR amplification also tends to favor relatively short amplified sequences, usually less than 1000 bp. In production of cDNA from mRNA, there usually is a distribution of lengths of cDNAs; the full-length product constitutes 50% or less of the total. There seems to be a limit to the amount of total amplified DNA that can be produced, particularly after 20 cycles or so; shorter sequences seem to have a competitive advantage over longer sequences. Thus, libraries produced from PCR-amplified cDNA derived from mRNA will contain few full-length sequences. However, such libraries are nonetheless useful by producing cDNA probes that can be used later to extract full-length cDNA or to isolate the genomic sequence.

In practice, PCR, like all techniques, has considerable "art" associated with it, as well as several limitations. The selectivity lies almost entirely in the choice of primers. Since many genes share regions of homology, selectivity is rarely as perfect as the experimenter desires. Primers also should be selected carefully to avoid internal complementarity and palindromes, particularly at the 3' end. Primer complementarity produces "primer dimers," often as the main reaction product. Also, contamination from exogenous DNA in the laboratory can be a serious problem. Often, PCRs take place even when no DNA was added to the reaction tube. This potential for contamination requires careful attention to technique. The PCR laboratory should be kept as separate as possible from amplified DNA or DNA plasmid solutions. Separate pipets should be maintained, as should careful attention to control experiments. Nonspecific priming can be a problem also; nucleic acid preparations should be purified carefully by precipitation to remove small fragments that could prime nonspecifically. The melting and annealing temperatures are also important considerations. Sequences high in GC content may not melt completely and sequences high in AT may not anneal completely, or may anneal at a temperature so low it supports nonspecific priming. Unlike many claims, PCR is an exacting technique that must be optimized carefully and carried out with elaborate positive and negative controls to be successful.

Ligase chain reaction (LCR), a modification of the PCR approach, is much better suited to detecting single point mutations. As in PCR, repetitive cycles of denaturation, annealing, and hybridization are used. In contrast to PCR, in which the two primers delimit the fragment to be amplified, LCR uses two primers that completely span the region of interest; if they match the sequence of the template DNA exactly, the two primers will lie flat on the template, leaving only an incomplete sugar–phosphate bond gap between them. A thermostable DNA ligase now joins the exactly matching primers. On the other hand, if they do not match exactly due to, for example, a point mutation or deletion, then DNA ligase will not join the two ends. Repetitive cycling will lead to accumulation of joined oligonucleotide primers if there is an exact match. In the absence of an exact match, no amplification will be seen.

Various techniques for detection have been developed. One of the simplest is biotinylation of one of the primers and addition of a fluorescent or radioactive label to the other. Passing the reaction mixture through oligo(dT) an avidin-coated polymer column will remove the biotinylated primers. If they have been joined to the labeled primers, label will be retained on the column; otherwise it will pass through the column into the effluent. This technique is a very simple screen for point mutations and deletions.

Cloning

The objective in cloning is to isolate and expand a single cDNA obtained from a single mRNA or fragment of genomic DNA and permanently maintain that cDNA in a form that can be duplicated. In general, this problem is approached by making a *library* of cDNAs or genomic fragments, inserting all the cDNAs or genomic fragments in the mixture into vectors. The process is illustrated in Figure 2.7. A *vector* is a DNA molecule, usually circular, that can incorporate the target DNA and be propagated stably in a suitable host cell. Many plasmid DNAs are circular because they have complementary ends that hybridize with each other. Vectors usually are prepared from plasmids or viruses. It is relatively simple to insert foreign sequences, such as promoters or a variety of restriction sites, into plasmids, but viruses generally exhibit a limit to the amount of additional genetic material that can be inserted. The cDNA produced by reverse transcription must be treated further to produce a double-stranded DNA. The RNA that served as the template is degraded partially, leaving the single-stranded cDNA intact. Fragments of

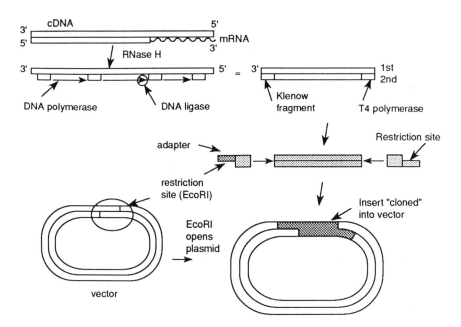

FIGURE 2.7   Cloning. A cDNA is made from mRNA. RNase H cleaves the mRNA, leaving a few fragments to act as primers for DNA polymerase. DNA ligase ligates the junctions, and the Klenow fragment of DNA polymerase and T4-polymerase fill in any unmatched sequences at the ends. Adapters containing restriction sites are added to the blunt ends, making the "insert," which is now ligated into the vector. When these are inserted into bacteria, and a single colony containing a single insert is selected and expanded, the original cDNA is said to be cloned.

the original RNA usually serve as primers for second strand synthesis. The two steps usually are carried out in a single reaction by mixing RNase H to fragment the original RNA, DNA polymerase to synthesize the second strand (using the RNA fragments as primers and the first strand as the template), and DNA ligase to fill in the breaks. The double-stranded cDNA must now have linkers or adapters attached to the ends to produce ends that are complementary to a restriction site, unless the cDNA was synthesized by PCR using primers that contained such sites. Inserts produced by PCR are treated with the appropriate exonuclease to cleave any artifactual adenine groups that are sometimes added upstream of the primers during amplification.

The currently favored route for cloning an insert that does not contain restriction sites is ligation (attachment) of the blunt-ended double-stranded cDNA to appropriate oligonucleotide adaptors using the enzyme T4 ligase. First, the cDNA is blunt-ended, either using the Klenow fragment of DNA polymerase, which fills in any overhangs on the 3′ end, or using T4 polymerase, which fills in 5′ overhanging termini. The vector contains one or more restriction sites that allow it to be cleaved and that will be complementary to a similar restriction site in the insert DNA. After the treatment with the appropriate restriction enzyme, the insert and opened vector both contain complementary single-stranded ends, so they hybridize. The enzyme DNA ligase closes the breaks, creating a stable circular DNA that contains the insert. The vector is introduced into bacteria, a process called transfection. The bacteria grow and duplicate the vector. Unfortunately, the library thus formed is a random collection of fragments with no classification system. It must be screened to identify the clones of interest.

Cloning can be *directional* or *nondirectional,* depending on whether the same or different restriction sites are on the two ends of the insert. If the same restriction site is on both ends, then the orientation of the insert will be random; half the inserts will match (+) and (−) strands of the vector with the insert and half will be in the opposite orientation. Orientation is most significant when expression of the insert is desired.

The screening strategies for the two types of vectors, plasmids and viruses are different. To screen a plasmid, bacteria must be selected that have received the plasmid. This selection is achieved by using plasmids that encode antibiotic resistance; plating the bacteria on a medium that contains the antibiotic selects only those bacteria that received a plasmid. These cells will survive and grow as a clonal colony. Each colony will contain a single cDNA insert.

If the vector contains an initiation site and a promoter, some of the inserts will be in the correct reading frame to produce protein, which can be detected with antibodies. Alternatively, the colonies can be screened by hybridization to a radioactive nucleic acid probe. In either case, the plate containing the colonies is touched to a hybridization filter and the bacteria are lysed. Screening with antibodies is performed similarly to a Western blot, whereas screening with nucleic acids is performed similarly to a Southern blot. When phage vectors are used, they are mixed with a large excess of bacteria. The entire mixture is then plated out. Each bacterium that contains a phage will serve as a focus for further growth and lysis, resulting in clear plaques on an otherwise dense lawn of bacteria. The plate is probed as described using blotting techniques.

When particular clones containing cDNAs of interest are found, they can be grown in almost limitless quantities. The inserts can be manipulated further by, for example, removing the insert with the appropriate restriction nuclease and inserting it into another vector, perhaps one that supports protein synthesis. This manipulation allows production of so-called "recombinant" protein. If placed in a vector active in mammalian cells, the inserted gene can be expressed and translated, a technique that allows gene function to be studied under controlled conditions. Placing an inducible promoter, for example, the androgen- or estrogen-sensitive promoters, in front of the recombinant gene allows it to be turned on and off at will by adding or removing the inducer. Additionally, a strand antisense to the one that is transcribed will bind with the natural sense RNA, thereby selectively stopping the expression of a selected gene.

Sequencing DNA

Sequence analysis using dideoxy chain termination is, by far, the most popular current method because of its relative simplicity. This sequencing technique is dependent on having appropriate polymerase promoter sites built into the cloning vector at or near the insertion site. Usually two sites are built into such vectors, one on each strand. Two popular promoter sites are T7 and Sp6. An appropriate primer complementary to either T7 or Sp6 promoter site is used to initiate synthesis from one strand; another primer is used to initiate synthesis on the other strand, as illustrated in Figure 2.8. Four reactions are set up for each priming reaction; each contains polymerase and all four nucleotides needed for cDNA synthesis. However, each of the reactions also contains a small amount of one radio-labeled dideoxy base analog. A dideoxy nucleotide is capable of being added to the 5' end of a growing chain but, because the sugar does not contain any free OH groups, cannot serve as a substrate for further extension. Thus, the chain is terminated. Because the dideoxy nucleotide represents a small fraction of the total nucleotide added, the statistical probability that any single elongation step will be terminated is small. Therefore, there will be a wide distribution of cDNAs of different lengths, ranging from termination at the first step to, in theory, a full-length cDNA. Next, all four reaction mixtures are subjected to electrophoresis in adjacent lanes in a long *sequencing gel*. The electrophoretic step separates the DNAs by size; autoradiography is used to develop the gel. Because the dideoxy sugars are radiolabeled, a dark band will be noted where

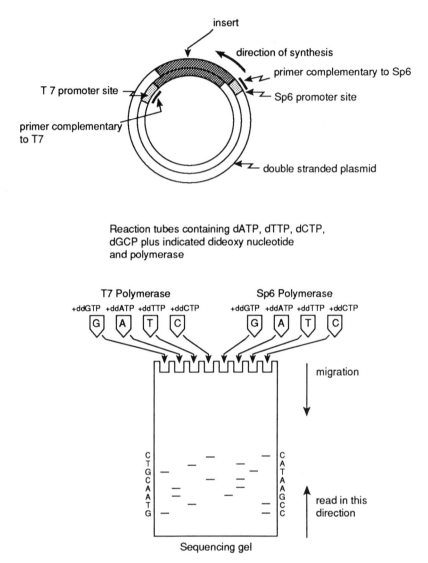

FIGURE 2.8 Sequencing cloned DNA sequences. The sequence from one end is GTAACGTC and GGCTTATG (the complements of CCGAATAC because it is from the complementary strand). If the insert is cloned directionally, then it is possible to determine whether the + or − strand of insert DNA is being sequenced.

a dideoxy sugar was incorporated. The result is a *ladder gel,* so called because of its resemblance to a ladder. A straight edge is laid across all four lanes. Because each reaction contained only a single dideoxy nucleotide, a dark line appears in only one of the four lanes and identifies the base present at that sequence. The ruler is moved up one rung at a time. The sequence is simply read from bottom to top in this manner. The technique is simple, requiring only pipetting of four reaction mixtures, incubation for 1 hr, electrophoresis, and autoradiography. A sequencing reaction can be set up on one day and the sequence read the next. The practical limit of this technology is about 200 bases. However, because primers are used to initiate cDNA synthesis from both ends, a total of about 400 bases can be read. Obtaining complete sequences for longer fragments requires sequencing "nested" or overlapping fragments.

Monoclonal Antibodies

The molecular revolution has affected protein chemistry as well as nucleic acid chemistry. In the usual immune response, multiple B-cell lineages will respond to a single antigen. As the antigenic protein circulates, it encounters B cells that present their antibody on the surface. Some small fraction of the B cells will have some affinity for the protein or *immunogen*. Each such B cell may recognize different portions, or *epitopes*, of the molecule. An *epitope* is a single antigenic determinant or site that selects and induces a specific antibody. Even among those B cells that recognize the same epitope, each one is likely to have a different *affinity* for, or strength of interaction with, the epitope. The resulting stimulated B cells now begin to reproduce rapidly in the spleen or bone marrow and produce antibodies. These polyclonal antibodies represent a multiplicity of antibodies directed at more than one *epitope* and with a range of affinities for the antigen.

In 1975, Köhler and Milstein discovered how to obtain almost limitless quantities of a single antibody. A mouse is immunized with the target protein, preferably in purified form. After the immune response of the mouse has reached its peak, spleen cells are harvested and, in the presence of polyethylene glycol to weaken cell membranes, fused with immortal *myeloma* cells. (Myeloma cells are cancerous lymphocytes capable of continuous immortal growth.) The fused cells are grown in a medium that selects for hybrids so neither the unfused myeloma line nor the spleen cells are able to grow, but the hybrids are able to grow. Of the hybrids, or *hybridomas*, a small fraction will have incorporated the immunoglobulin genes of the spleen cell into the immortal and rapidly replicating myeloma cell. Again we encounter the familiar "needle in a haystack" problem of molecular biology.

After fusion, the hybridomas are distributed in low density into 96-well plates. The supernatants, which contain the antibodies, are screened against the original antigen using an enzyme-linked immunosorbent assay (ELISA), which will detect the desired antibodies. Cells from wells that contain antibody that reacts with the original antigen are cloned. The contents of the well are diluted into another set of 96-well plates so each well contains no more than a single cell. Wells containing cells are tested to insure that the antibodies produced react with the target protein. Each such cell produces a monoclonal antibody against a single epitope of the protein. The hybridomas can be grown in almost limitless quantities, producing large amounts of the monoclonal antibody. A simpler technique has been developed for producing monoclonal antibodies, in which the DNA of spleen cells is isolated and transfected directly into myeloma cells.

Monoclonal antibodies frequently are screened further to identify the epitope against which the antibody reacts and to determine the affinity. In general, 2–10 different clones producing antibody against the antigen of interest will result from one fusion reaction. Some of these antibodies will be against the same epitope, but are likely to differ in affinity. On the other hand, a set of monoclonals against different epitopes can be extraordinarily useful. For example, glycoproteins usually will yield antibodies against both the protein and the carbohydrate portions of the molecule. Also, antibodies against different protein epitopes can be useful in mapping protein structure.

Monoclonal antibodies have revolutionized protein chemistry because of their specificity and their ready availability in reproducible form. It is now sometimes possible to isolate monoclonal antibodies against proteins that have never been isolated and are known only from a gene sequence by synthesizing a peptide from the coding sequence and using the peptide as the immunogen to produce a monoclonal antibody specific to that peptide.

Applications

Restriction Fragment Length Polymorphism Analysis and the Genetics of Kidney Cancer

RFLP analysis is a technique that is being used widely in biology to study gene distribution in populations and to identify certain mutations. Sometimes a mutation occurs that will either create or destroy a restriction site, or a significant deletion is generated within a fragment excised by a restriction nuclease that will alter the size of the fragment. If these changes exist in different forms throughout a population for a single region of DNA, so some individuals are heterozygous for the two forms, the gene is said to display a RFLP. The analysis for RFLPs is illustrated in Figure 2.9. Genomic DNA is digested with restriction enzymes and separated by electrophoresis. The genomic DNA can be derived from tumors, normal tissue, or leukocytes from peripheral blood. The electrophoretic gel is Southern blotted with a radioactive probe to the target DNA region. When a RFLP has been identified, the probe will bind to two sizes of DNA fragments, a "long" form and a "short" form. A large number of RFLPs have been cataloged and probes are available for a small fee from the American Type Culture Collection. The technique is widely used in genetic studies of populations.

Linehan and his group (Anglard, et. al., 1991) have used this approach to investigate allelic loss in renal cell carcinoma. Ten different RFLPs mapped to chromosome 3 were determined in tumor tissue and normal tissue from

Allele

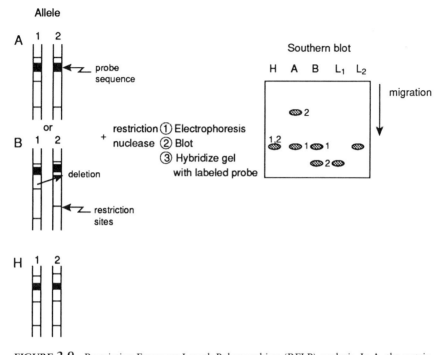

FIGURE 2.9 Restriction Fragment Length Polymorphism (RFLP) analysis. In A, the restriction site in the middle of the DNA segment of allele 2 is lost, either through a deletion or a point mutation within the restriction site. After digestion, electrophoresis, blottings, and hybridizing with the probe sequence, the fragment from allele 2 will be much larger, and hence will have migrated a shorter distance, than the fragment from allele 1. In B, a deletion within the fragment delineated by the two restriction sites in allele 2 will show up as a fragment smaller than that afforded by allele 1. In H, the person is homozygous, and both fragments migrate identically. In L_1 and L_2, alleles 1 and 2, respectively, are lost.

60 patients with renal cell carcinoma. Normal DNA from 58 of the 60 patients was heterozygous at one or more locations. Heterozygosity was identified by the presence of both allelic forms in the same individual. One possible confounding variable in studies of this kind is that tumors frequently contain infiltrating lymphocytes and connective tissue elements that are not altered genetically, either of which can overwhelm the observation of a deletion of one allele in the tumor cells. To account for this possibility the investigators isolated tumor cells from some negative samples; the purified tumor cells showed loss of heterozygosity. If one allele is lost due to complete deletion of the RFLP from one of the two chromosome pairs, the analysis will show the loss of either the long or the short band from the electrophoretogram, as shown in Figure 2.9. Linehan and his collaborators showed loss of heterozygosity at one or more of the 10 RFLP loci tested in 51 patients; loss of heterozygosity in all the loci was seen in 39 patients. The other 12 patients

retained the loci at one end or the other of the large stretch of DNA defined by the short arm of chromosome 3.

The significance of these findings is that they strongly support the existence of a tumor suppressor gene located in the region of the deletions. Loss of this suppressor gene seems to be a primary event in the development of renal cell carcinoma. Interestingly, the investigators also found a constitutional deletion in the same region in von Hippel–Lindau syndrome, a heritable disorder characterized by renal cell carcinoma.

Human Papilloma Virus and Cervical Cancer

Papilloma viruses are small viruses with a DNA genome some 7.9 kb in length that infect keratinocytes. DNA sequence homology represents the main, if not the exclusive tool for classification of these viruses. A papilloma virus is considered to be a new type if it shares less than 50% homology with each of the other known types, as established by hybridization under stringent conditions. On this basis, at least 60 different types have been identified to date. The papilloma viruses are extremely selective about which keratinocytes they infect; HPV-1 and HPV-2 form warts on the plantar and palmar epithelia, respectively, and nowhere else, whereas HPV-16 and HPV-18 infect the genital epithelium, causing genital wars (condyloma), but no other epithelial tissue.

The increase in prevalence of sexually transmitted diseases that has occurred in recent years has provided information that suggests that genital warts may not be entirely benign and that the causative papilloma viruses may play a causal role in human cervical cancer (reviewed by Schlegel, 1990). More than 90% of human cervical carcinomas contain and express HPV; 70% are either HPV-16 or HPV-18. Moreover, the epidemiology of human cervical cancer is similar to that of other sexually transmitted diseases, that is, it is extremely rare in women who are members of celibate religious orders and its prevalence is correlated with the number of sexual partners of both the woman and her primary partner.

These discoveries suggest a mechanism for cervical carcinogenesis by certain types of human papilloma but not by others. The genome of papilloma viruses consist of 9 or 10 genes. The genes are subdivided into 7–8 "early" genes and 2 large "late" genes. The late genes are capsid proteins that encapsulate viral DNA in productive lesions such as warts or condyloma. The early genes are expressed in cervical carcinomas and derived cell lines and represent functions of cell transformation, control of RNA transcription, and viral DNA replication. Of particular interest are the *E6* and *E7* genes. Detailed comparison of the structure of the E6 and E7 proteins from non-oncogenic papilloma viruses such as HPV-6 and HPV-11 and oncogenic forms such as HPV-16 and HPV-18 shows that, although there are large areas of homology, important differences exist as well. The question is whether any of these differences in structure could be associated with differences in function that might explain the oncogenic properties of the latter two types.

Evidence points to the importance of key regulatory genes known generically as *tumor suppressor genes* in maintaining a normal cell phenotype. The *Rb* (retinoblastoma) and *p53* genes have been identified as important members of this family. The observation that the E6 and E7 proteins of oncogenic papilloma viruses bind to the p53 and Rb proteins, respectively, whereas the corresponding proteins from nononcogenic HPV-6 or HPV-11 do not suggest that carcinogenesis may proceed by inactivation of p53 or Rb by binding to the E6 and E7 proteins. Further evidence for this model was obtained in studies in which inducible plasmid vectors that express *antisense E6/E7* mRNA were transfected into cervical carcinoma cells. Antisense RNA is complementary in sequence and binds to normal mRNA. The vectors use a promoter that is active in the presence of dexamethasone and inactive in its absence. This feature allows the expression of the insert, in this case antisense *E6* or *E7,* to be controlled reversibly. When the inserted antisense RNA was expressed actively, the cervical carcinoma cells expressed a normal phenotype in culture, but when the antisense RNA was not expressed they reverted to their usual transformed phenotype.

Diagnosis of HPV infection and its typing is becoming an increasingly important issue in studies of cervical cancer. Until recently, the standard for such typing was the Southern blot. However, the Southern approach suffers from a crucial shortcoming: The tissue must be lysed to extract the DNA, which destroys anatomic structure and mixes normal and malignant tissue together. Of increasing use are one or more techniques of *in situ* hybridization.

The principle of *in situ* hybridization is simple. A tissue section that has been treated to be permeable is treated with a labeled probe that is complementary to the sequence being sought. Most often the probe is labeled radioactively, but new technologies using fluorescent or luminescent probes are being developed. When a radioactive probe is used, the tissue is overlaid with a silver-containing gel. Radioactive disintegrations produce grains of silver metal that can be seen easily with the light microscope at high power. The particular power of *in situ* techniques is that they allow the investigator to view the signal and the morphology of the object that yielded the hybridization signal. Stoler, et. al (1992) used *in situ* hybridization to investigate the distribution of HPV-16 genes through a range of pathologic lesions ranging from dysplasias to squamous carcinoma and adenocarcinoma lesions.

Selective Amplification Techniques *and* ras *Genes*

The *ras* oncogene was one of the first to be discovered. The discovery that this oncogene was found in viral tumors and naturally occurring tumors, as shown by transfection studies, firmly established the oncogene theory as a general theory of carcinogenesis (reviewed by Bos, 1989). The *ras* gene family includes three well-characterized genes that have a high degree of homology and encode proteins of 21,000 molecular weight, referred to generically as p21 proteins. Although the exact role of the p21 proteins is not clear, their function seems to be to modulate the activity of transmembrane receptor–tyrosine kinase complexes. In general, cellular growth is controlled exogenously through the binding of growth factors or growth peptides to external receptors. (The difference between a growth factor and a growth peptide is that a factor has not been identified with a particular chemical structure.) The receptors span the cell membrane; when they are occupied by the peptide, the associated tyrosine kinase on the inner membrane becomes active, initiating a program of intracellular biochemical changes. Mutations in the p21 proteins are theorized to lead to permanent activation of the tyrosine kinase in the absence of growth peptide. The most frequent sites for mutations are codons 12, 13, and 61.

Originally the *ras* genes were identified by transfection into 3T3 mouse fibroblasts. This technique is relatively insensitive. Genomic DNA from naturally occurring tumors was digested with a restriction endonuclease to produce conveniently sized fragments. These fragments were introduced into cultured 3T3 cells. Transfection can be achieved simply by coprecipitating DNA with calcium phosphate. The small crystals are taken up by cells and, in a small percentage, the DNA is incorporated into the cel genome. Foci of transformed cells indicate the presence of a transforming gene, which in this case was identified eventually as the *ras* gene.

The sensitivity of detecting *ras* mutations has been increased markedly by using specific amplification techniques, first with a thermolabile DNA polymerase, later with the more stable polymerases now in use. A nonspecific PCR technique that amplified mutant and wild type sequences equally was used to amplify the region between codons 1–61 of the *ras* gene roughly 10^4 fold in DNA prepared from tumors obtained at thoracotomy (Rodenhuis, et al., 1988). The amplified DNA was probed with oligonucleotides that were complementary to each of the commonly identified mutations under conditions of high stringency. Under such conditions, a single base mismatch would inhibit hybridization. Of interest was the finding that tumors from nonsmokers or those who had quit smoking at least 5 years earlier tended not to contain K-*ras* mutations (9/10 had normal K-*ras*) whereas tumors in smokers tended to show mutated K-*ras* genes (13/32 had mutant K-*ras*).

One of the weaknesses of such an approach is that tumors contain normal tissue and, moreover, represent many steps of selection. Thus, there is no guarantee that finding mutations in one or more genes means that these mutations are causative events. They may represent the effects of cumulative mutation and selection after emergence of the primary tumor. Bert Vogelstein and several colleagues, (1988), investigated the origins of human colon cancer to determine whether *ras* mutations were causative. Colon adenocarcinoma seems to develop in three stages. First, small tubular adenomas form that sometimes develop in the colon and do not progress, in most cases. Occasionally, these simple adenomas progress to a larger, more villous type that frequently contains patches of developing adenocarcinoma. Finally, an invasive adenocarcinoma develops. These investigators concentrated on adenomas, which are early noncancerous premalignant lesions, and early adenocarcinomas that were not likely to have undergone extensive additional genetic rearrangement. Tissue sections were microdissected carefully to isolate only cancer cells. The investigators were interested particularly in separating adenocarcinomatous areas that had developed within adenomas. PCR was used to amplify the *ras* genes from the malignant and premalignant areas selectively. In the smaller tubular adenomas, the incidence of mutated *ras* genes was less than 20% but, in the villous adenomas and the adenocarcinomas, the incidence had risen to about 50%; a large fraction of those lesions contained both adenoma and adenocarcinoma and the mutation was found in both. These results suggested that a *ras* mutation was a very early event associated with the initial progression from tubular to villous adenoma.

A number of techniques have been developed to enhance the sensitivity of the detection of mutated genes. One technique is use of a primer that has a mismatch with the wild-type sequence at the 3' end (illustrated in Figure 2.10A for selective amplification of *ras*). In the wild-type, the mismatch at the 3' end effectively inhibits PCR amplification of the normal sequence, but permits amplification of a specific mutant sequence. A primer selective for each mutant sequence must be included to detect the specific mutation. Another approach involves introducing a restriction endonuclease-sensitive site at the point of the mutation, as illustrated in Figure 2.10B. Sometimes this will occur spontaneously, but more often a slightly mismatched primer must be used. A one- or two-base mismatch will not preclude amplification, as long as it does not occur at the 3' end of the primer. In the illustration, the 6-base sequence encompassing the potential mutation site and the previous codon almost match the sequence for a *Bal*I site in the wild-type. Only the 3 bases on the 5' end of the primer are shown. The mismatch in the second base does not preclude amplification, but all the amplified DNA will contain a TGG sequence rather than the wild-type TCG. The amplification products of the wild-type DNA are sensitive to *Bal*I, but those derived from the mutant DNA are not, since the fourth base from the 5' end is T instead of C. Digestion of the amplified product with *Bal*I will yield RFLPs because mutant sequences will be longer than wild-type sequences. This approach does not require a radioactive probe because staining the gel with ethidium bromide will reveal the different sized fragments.

PCR also can be used to detect gene amplification. Genes in cancers are amplified frequently. The N-*myc* gene frequently is amplified in neuroblastomas, epidermal growth factor receptor is amplified in gliomas, and *neu* is

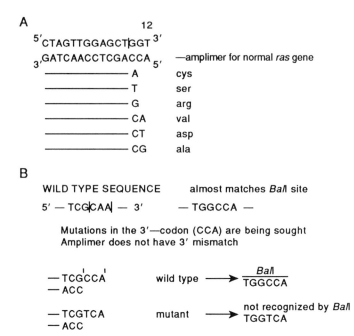

FIGURE 2.10 Increasing the sensitivity of amplification procedures for mutant sequences. A, Selective amplification of *ras*, mutant *ras* sequences by mismatches at the 3' end of the 5'-amplimer (primer). B, Introduction of a restriction nuclease-sensitive site. Any mutation in the marked codon will destroy the *Bal*I site. Mutations in the 3' codon (CCA) are being sought. Amplimer does not have 3' mismatch.

amplified in breast adenocarcinoma. The multidrug resistance gene is amplified in cells that are resistant to chemotherapeutic agents. Quantitation by PCR is not straightforward because of the complex reaction kinetics, but the simplicity of the technique is leading numbers of investigators to publish improvements that allow quantitative PCR.

Northern Blotting and Prostatic Acid Phosphatase

The prostate secretes an acid phosphatase that has been used for over 50 years as a tumor marker in the diagnosis and management of prostate cancer. However, several different forms of acid phosphatase are present in different tissues. Virtually all differentiated cells synthesize lysosomal acid phosphatase. The two forms may not be readily distinguishable and share a 62% homology at the nucleic acid level. A number of enzymatic techniques have been developed to attempt to distinguish the two enzymes reliably without total success. Even some monoclonal antibodies will react with both enzymes. Therefore, supposed prostatic acid phosphatase detected in many organs possibly represents immunologically related forms with common or related epitopes.

Northern blotting was used to determine whether prostatic acid phosphatase was synthesized by tissues or cancers other than prostate (Solin, et al., 1990). The mRNA of prostatic acid phosphatase had been shown earlier to be 3.3 kb in size whereas lysosomal acid phosphatase mRNA was 2.4 kb in size. Using probes specific to the two mRNAs, differentiated prostate cancers were shown to synthesize both prostatic and lysosomal acid phosphatases, as evidenced by bands at 3.3 and 2.4 kb, respectively. However, undifferentiated prostate cancer cells yielded only the 2.4-kb band, confirming the loss of the prostatic acid phosphatase. Wide variations in the amount of prostatic acid phosphatase were seen in cancers and benign prostatic hyperplasia. In addition, when other cancers and other tissues were probed, none were found to produce a 3.3-kb mRNA. Although prostatic acid phosphatase appears to be a gene product specific to differentiated secretory prostatic cells, the amount can vary over wide ranges, and antibodies or enzymatic techniques may not be entirely reliable in measuring the true level of expression.

Mutations in the p53 Suppressor Gene

The *p53* gene is, so far, the most commonly found deletion of a suppressor gene associated with human cancers. The *p53* gene encompasses 16–20 kb of DNA on the short arm of human chromosome 17. The gene contains 11 exons. Mutations of *p53*, including point mutations, allelic loss, rearrangements, or deletions, have been found in most human tumors examined so far, including colon, lung, liver, breast, esophageal, bladder, and stomach cancers, as well as lymphomas and leukemias. At least 300 different mutations have been identified, most in exons 5–8. Cancers from different geographical areas have been suggested to display different mutational spectra, an observation that, if confirmed, is of importance to molecular epidemiology. In common with other suppressor genes, germ line *p53* mutations are associated with development of cancers relatively early in life, whereas somatic deletions are associated with "sporadic" incidence due to environmental exposures.

The variety of mutations has limited hybridization methods. Two approaches are being used; single-strand conformational polymorphism (SSCP) and direct sequencing of amplified segments of the exons most frequently involved in mutations. The SSCP technique is relatively simple. Either genomic DNA or mRNA is used as a template for PCR. The product is then melted, quickly cooled to prevent renaturation, and separated by electrophoresis as single-stranded DNA. The single-stranded DNA usually forms extensive secondary structure, but the presence of a single point mutation in a relatively short fragment (up to 100 bases) usually will alter this secondary structure and, hence, the electrophoretic mobility. Thus, the bands corresponding to the wild-type and the mutant will be distinct. The wild-type band is contributed by normal cells such as infiltrating lymphocytes present in the tumor, from adjacent normal cells, or from tumor cells that do not contain mutated sequences. When a mutation is identified, the DNA can be sequenced fairly readily to establish exactly what kind of mutation has occurred.

Sequencing of *p53* to identify mutations has been proposed by Jones and co-workers (1992) as an approach that can give information about whether an endogenous process or an exogenous agent was responsible for a point mutation. Mammalian DNA often is methylated, which seems to affect regulation of gene expression. The methyl groups usually are added to cytosine–guanine (CpG) doublets. (The designation CpG means the two bases are adjacent to each other on the same chain and not a base pair, which would be designated C–G). However, 5-methylcytosine is somewhat less stable than other bases; approximately 8 deamination events occur each day in the human genome. When deamination occurs, 5-methylcytosine is converted to thymine which, on replication, is likely to result in a mutation; the resulting T–G base pair is likely to be replicated as a T–A base pair on one strand. Of all human germ line mutations associated with human genetic diseases, 30–40% can be associated with this mechanism. Jones reasoned that this kind of mutation was an indicator of an endogenous mutational event and, by extension, other mutations were likely to represent exogenous events. The sequences of *p53* from four body sites—colon, lung, bladder, and liver—were mapped, and showed strikingly different distributions. In colon cancer, 63% of the mutations found were consistent with deamination at methylated CpG. In lung cancer, which is caused predominately by the action of carcinogens in tobacco smoke, a strikingly different pattern emerged; only a small fraction of the mutations were consistent with the deamination mechanism. Bladder cancer, which is known to be associated with smoking (to a lesser degree than lung cancer) and with occupational exposures, shows a distribution intermediate between lung and colon, which may suggest that bladder cancers arise more or less equally from both processes. Hepatocarcinoma, on the other hand, showed a different pattern still; almost all the mutations were clustered at a single site at nucleotide 249 in the codon AGT, as would be predicted from the known chemistry of the reaction of the fungal metabolite and carcinogen aflatoxin with DNA.

Summary

The fusion of molecular biology and epidemiology into molecular epidemiology creates a powerful approach to understanding the origins of disease at the molecular level and to predicting the risk that an individual may carry in his or her genome or the risk that results from a given toxic or carcinogenic exposure. Traditional epidemiology has been concerned with groups of individuals with given group characteristics, and risk has been defined in terms of those group characteristics. Molecular epidemiology offers the possibility of producing much more specific estimates of risk, based on a knowledge of events at the gene level. With this knowledge will come much more specific methods for monitoring the risk of individuals and of groups. Theoretically it may be possible eventually to predict which 5% of smokers will develop lung cancer or to detect very early the signs that, in a given individual, the process of carcinogenesis is proceeding toward an irreversible outcome. However, with this increased ability to predict individual risk come other risks resulting from societal decisions. The essence of insurance is group risk, not individual risk; society soon will have to deal with such difficult issues.

Suggested Further Reading

Textbooks

Alberts, B., Bray, D., Lewis, J., Raff, M., Roberts, K., and J. D. Watson, (1989). Molecular Bioilogy of the Cell," 2d Ed. Garland, New York.

Darnell, J., Lodish, H., and Baltimore, D. (1986). "Molecular Cell Biology." Scientific American Books, New York.

Stryer, L., (1988). "Biochemistry," 3d Ed. Freeman, New York.

Laboratory Manuals

Berger, S. L., and Kimmel, A. R. (eds.) (1987). "Guide to Molecular Cloning Techniques" (Methods in Enzymology, Vol. 152). Academic Press, New York.

Davis, L. G., Dibner, M. D., and Battey, J. F. (1986). "Basic Methods in Molecular Biology." Elsevier, New York.

Sambrook, J., Fritsch, E. F., and Maniatis, T. (1989). "Molecular Cloning, A Laboratory Manual," 2d Ed. Cold Spring Harbor Laboratory Press, Cold Spring Harbor, New York.

References

Anglard, P., Tory, K., Brauch, H., Weiss, G. H., Latif, F., Merino, M. J., Lerman, M. I., Zbar, B. and Linehan, W. M. (1991) "Molecular Analysis of Genetic Changes in the Origin and Development of Renal Cell Carcinoma." Cancer Res., 51, 1071–1077.

Bos, J. L. (1989) "ras Oncogenes in Human Cancer: A Review." Cancer Res. 49, 4682–4689.

Jones, P. A., Rideout, W. M., Shen, J.-C., Spruck, C. H. and Tsai, Y. I. (1992) "Methylation, Mutation and Cancer." BioEssays, 14, 33–36.

Köhler, G. and Milstein, C. (1975). "Continuous Cultures of Fused Cells Secreting Antibody of Predefined Specificity." Nature (London) 256, 495–497.

Rodenhuis, S., Slebos, R. J. C., Boot, A. J. M., Evers, S. G., Mooi, W. J., Wagenaar, S. C., van Bodegom, P. C. and Bos, J. (1988) "Incidence and Possible Clinical Significance of K-ras Oncogene Activation in Adenocarcinoma of the Human Lung." Cancer Res., 48, 5738–5741.

Schlegel, R. (1990) "Papilloma Viruses and Human Cancer." *Seminars in Virology*, **1**, 297–306.

Solin, T., Konturri, M., Pohlman, R. and Vihko, P. (1990). "Gene Expression and Prostate Specificity of Human Prostatic Acid Phosphatase (PAP): Evaluation by RNA Blot Analyses." *Biochim. Biophys. Acta*, **1048**, 72–77.

Southern, E. M. (1975) "Detection of Specific Sequences Among DNA Fragments Separated by Gel Electrophoresis." *J. Mol. Biol.*, **98**, 503–517.

Stoler, M. H., Rhodes, C. R., Whitbeck, A., Wolinsky, S. M., Chow, L. T. and Broker, T. R. (1992) "Human Papilloma Type 16 and 18 Gene Expression in Cervical Neoplasias" *Human Pathol.*, **23**, 117–128.

Vogelstein, B., Fearon, E. R., Hamilton, S. R., Kern, S. E., Preisinger, A. C., Leppert, M., Nakamura, Y., Whyte, R., Smits, A. M. M. and Bos, J. L. (1988) "Genetic Alterations during Colorectal Tumor Development." *N. Engl. J. Med.*, **319**, 525–532.

Watson, J. D. and Crick, F. H. C. (1953) "A Structure for Deoxyribose Nucleic Acid." *Nature (London)* **171**, 737–740.

Note Added in Proof

A number of different methods have been used to quantify the interaction of xenobiotics and DNA, each having particular advantages and disadvantages. Detection of carcinogen-DNA adducts by ^{32}P-postlabelling has been used in many human studies. The method involves hydrolysis of DNA to $3'$-phosphonucleosides and labelling with carrier-free ^{32}P at the $5'$ position. Bulky adducts will have different chromatographic mobility than the unadducted nucleotides and appear as a distinct peak. This procedure has proven useful because of high sensitivity, relatively small DNA requirements, and because it is sensitive to a variety of chemical and naturally occuring carcinogens, several of which can be distinguished by their chromatographic behavior.

Fluorescence methods, which are concentration dependent, have also been used in several cases, but have been limited by relatively great tissue requirements. However, high sensitivity is possible if the molecules are confined to a very small volume, such as the nucleus of a cell. Coupled with specific fluorescence-labelled immunological probes, microscopic photometry might detect as few as 300 adduct molecules per cell.

3

Validation

Paul A. Schulte
and Frederica P. Perera

The contribution of molecular epidemiology to etiologic research, risk assessment, or disease prevention and control depends on the use of valid biologic markers. The use of invalid markers can lead to wrong conclusions and costly programs. Validity is the best approximation of the truth or falsehood of a marker (Cook and Campbell, 1979). Validity is a sense of degree rather than an all-or-none state. To validate the use of biologic measurement as a marker, it is necessary to understand the relationship between the marker and the event or condition of interest. Biologic markers can be validated against exposure, disease, or susceptibility events. To date, most molecular epidemiologic research has involved validating biologic markers (as opposed to using them for etiologic and intervention research or risk assessment). When the validity, reliability, and practicality of a marker have not been demonstrated, pilot studies are useful. Perera (1987), among others, has demonstrated a strategy and approaches for these pilot studies: characterize the marker in known high-dose groups such as chemotherapy patients, proceed to highly exposed groups such as occupational groups, then use less exposed occupational and environmental groups. The goal of these studies is to determine characteristics of markers that must be known prior to their use in large population studies. These characteristics include a dose–response relationship, marker persistence, inter- and intraperson variation, correlation between markers, and correlation with clinical response. For example, Perera *et al.* (1992) studied cancer patients treated with *cis*-platinum (*cis*-DDP)-based chemotherapy and found posttreatment differences in a battery of biologic markers, including increased levels of platinum–protein and platinum–DNA adducts and increased incidence of sister chromatid exchanges, micronuclei, and gene mutation at the glycophorin A locus.

In many, but not all, of the studies (conducted in the 1980s) using biomarkers, particularly genetic and molecular markers, the validity of findings was limited or in question because adequate attention was not given to subject selection, control of confounding, and choice of statistical analyses. Subjects often appeared to be selected with no appreciation of the impact of bias, attention to confounding factors, or attention to sample size, statistical power, or other design features. Granted, many of the early studies were conducted to see if an assay "worked" or to examine how the assay performed under a range of conditions. In most of these cases, investigators had the good sense not to include statistical analyses, since such assessments generally were not appropriate. In other cases, however, studies included practically no discussion of statistical design features or evaluation of the underlying assumptions for the statistical tests employed.

Often in these studies, attention was paid only to one definition of validity of a marker, which has different meanings to laboratory scientists and epidemiologists. To the laboratory scientist, validity generally means the ability of a test to respond in the presence of a marker and to not respond in its absence. To the epidemiologist, validity pertains to predictive value, that is, the probability that a person who has a marker actually experiences the event being indicated. Ultimately, from an epidemiologic viewpoint, a marker will be valid and useful if it reduces misclassification, provides better interpretation of exposure–disease associations, or is useful in prevention or control of disease.

Research involving biologic markers first must determine whether such markers are valid—whether they measure what they are believed to measure—and what they mean with respect to the risk of disease. Only after validation can markers be used effectively for etiologic and disease-control research and in risk assessments. The validation of biologic markers for use in molecular epidemiologic research requires that extensive laboratory work be performed prior to testing in humans. Key to the validation procedure for markers of effect is agreement on what constitutes a "critical effect" (Ashby, 1987; Hernberg, 1987). Exposure to xenobiotic substances results in a range of perturbations to biologic systems. To determine which effects are critical (that is, indicate some aspect of a disease response) and which are merely adaptive, it is necessary to relate critical effects to dose estimates, to determine what factors affect dose, and to define a no-effect level (Ashby, 1987; Hernberg, 1987). A similar but less familiar process is required to validate markers of exposure and markers of susceptibility.

Although homage often is paid to the concept of validity, little attention has been given to its meaning and how to evaluate it (Schulte and Mazzuckelli, 1991). The objective of this chapter is to identify and explore the range of considerations that constitute the concept of validity and to address how validity pertains to the use of biologic markers in epidemiologic risk research and quantitative health risk assessments.

Three broad categories of validity can be distinguished: measurement validity, internal study validity, and external validity (Figure 3.1). Measure-

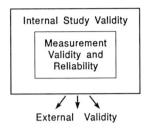

FIGURE 3.1 Three categories of validity.

ment validity has been defined as an expression of the degree to which "... a measurement measures what it purports to measure" (Last, 1988). Internal study validity is the degree to which inferences drawn from a study pertain to the actual subjects in the study (Rothman, 1986). External study validity is the extent to which the findings of a study can be generalized to apply to other populations (Last, 1988).

Biologic markers and the studies that include them need to be shown to have measurement, internal, and external validity before they can be used accurately in etiologic research and quantitative risk assessment. The use of invalid markers can result in nondifferential misclassifications of exposure or outcome, which can lead to underestimation of a true effect (Hogue and Brewster, 1988). Risk assessments based on studies that underestimate a true effect can lead to regulations that provide exposure limits that are thought to be "safe" but, in fact, are not. Conversely, a differential misclassification bias, depending on the direction of the bias, can lead to regulations that provide exposure limits that are either too high or too low. In quantitative risk assessment, the inferences derived from small study groups are generalized to larger populations. The strength of those inferences depends on the methodology of the study, including measurements and other design factors that lead to the results. Invalid measurements, inferences, or generalizations may lead to erroneous risk assessments. In this chapter, the three categories of validity are discussed in terms of how they apply to biologic markers for epidemiologic research and quantitative risk assessment.

Measurement Validity

Biologic measurements are among the principal building blocks of molecular epidemiology. If measurements are invalid, the research and risk assessments constructed from those measurements are likely also to be invalid. Researchers can use three aspects of measurement validity to characterize the markers they intend to employ in a study. Measurement validity reflects the extent to which a marker of a phenomenon has (1) content validity, that is, pertains to the underlying biologic phenomenon; (2) construct validity, that is, correlates with other relevant characteristics of the underlying phenomenon; and (3) criterion validity, that is, predicts some aspect of the underlying phenomenon. In general, these three components of measurement validity are best assessed in terms of the extent or degree to which they apply to the underlying phenomenon, rather than as an all-or-none condition (Nunnaly, 1967). The precision and reliability of measurements of biologic markers also have an effect on their validity and use in research. These issues will be discussed in Chapter 4.

Content Validity

"Content validity is the extent to which a marker represents the underlying biological phenomenon being studied" (Last, 1988). For example, a marker of internal dose will have content validity if it reflects the dose contributed by all routes of exposure. A marker of effect will have content validity if it encompasses the essential characteristics of the disease it represents. In other words, the marker must pertain to the appropriate target organ or the relationship of the marker to the natural history of the disease in question must be unambiguous. For example, a DNA adduct of benzo[*a*]pyrene [B(*a*)P] will have content validity as a marker of exposure in a study of B(*a*)P exposure and lung cancer, since the involvement of DNA in B(*a*)P-induced carcinogenesis is well documented. In contrast, the development of DNA adducts at the N7 position might not have content validity as a marker of biologically effective dose if the O6 methylguanine adduct is shown to be related more clearly to the carcinogenic process. However, the N7 adducts might be reasonably valid markers of the biologically effective dose of B(*a*)P if (1) the production of O6 and N7 adducts is directly proportional, as would be expected if they were produced by the same activated B(*a*)P metabolite; and (2) if relatively little time is allowed for possible differential repair, or the likely effect of differential repair on the measurement is removed during extrapolation of the data to zero time (Schulte and Mazzuckelli, 1991).

A proper assessment of content validity considers the extent to which the marker pertains to the phenomenon of interest (exposure, effect) or the extent to which the marker represents a relevant feature of that phenomenon (Schulte and Mazzuckelli, 1991). For example, to assume that hydroxyethyl histidine adducts of hemoglobin are valid markers of the internal dose of occupational exposure to ethylene oxide would be erroneous. The presumed marker would lack complete content validity, since hydroxyethyl histidine adducts of hemoglobin can result from exposure to other substances that contain ethyl groups. Further, populations with no known exposure to ethylene oxide have been shown to form hydroxyethyl histidine adducts of hemoglobin. Without considering content validity, one might reach erroneous conclusions if only occupational ethylene oxide exposure is assumed to be responsible for the observed adducts. By subtracting the number of adducts attributable to factors other than the exposure under study from the total number of adducts formed, valid measures might be developed. This manipulation requires the evaluation of a nonexposed comparison group and attention to other sources of ethyl groups that generate hemoglobin adducts.

Since content validity is assessed by professional judgment and logical consensus, there are no universally accepted criteria for its determination (Zeller and Carmines, 1980). However, it is possible to strengthen determinations of content validity if judgments are made by a group of experts or specialists with understanding of a particular marker. The focus of such judgments should be the degree to which the marker represents the underlying phenomenon. Establishing content validity is especially difficult in situations in which it is needed most, that is, when there is an incomplete understanding of the underlying characteristics of the exposure–disease process. Content validity should be established before a marker is used in intermediate or large scale human studies.

Construct Validity

Construct validity describes the extent to which a marker corresponds with other relevant characteristics of the underlying phenomenon, that is, the theoretical concepts or constructs concerning the phenomenon under study (Last, 1988). This correspondence is exhibited in part by association of the subject marker with other markers or covariates of the phenomenon. For example, if the characteristics of a phenomenon change with age, a marker with construct validity will change accordingly (Last, 1988). Further, if the marker shows no associations with other variables that are reasonably expected to be linked with the phenomenon under study, then the marker may be of questionable relevance in an epidemiologic study or subsequent risk assessment.

Construct validity is sometimes difficult to distinguish from content validity when describing biologic markers, but should be evaluated whenever general understanding of the underlying phenomenon is not clear. Hence, if a marker is a candidate for inclusion in an epidemiologic study of exposure or outcome and the actual role of the marker in the exposure–outcome continuum has not been established (i.e., its content validity has not been established), it still may be useful as a covariate if it can be shown to have construct validity. For example, if one wants to study the relationship between exposure to a dietary toxicant and some immunologic outcome, a marker of DNA repair could be useful as an independent variable. A marker of DNA repair may not have content validity for a study of immunologic changes after exposure to a dietary toxicant but still might be useful as a marker of biologic age because of its construct validity and, thus, could be used as a covariate in a multivariate model of dietary exposure and immunologic outcome.

Criterion Validity

Criterion validity describes the extent to which a marker correlates with the phenomenon being studied (Last, 1988). The criterion validity of a marker is assessed in terms of its sensitivity, specificity, and predictive value. Griffith *et al.* (1989) have distinguished the terms sensitivity and specificity as they refer to laboratory methods to detect a marker. The terms also are used to describe the ability of a marker to detect an exposure, or to detect or predict an event in a population.

> Laboratory sensitivity to detect a marker refers to the ability of a detection system to respond in the presence of the markers. Population sensitivity, in contrast, is the ratio of numbers of subjects positive for both the marker and the event to the number of subjects with the event.
>
> Laboratory specificity refers to the detection system's ability to fail to respond in the absence of the marker. Population specificity is the ratio of the number of subjects negative for both marker and event to the number of subjects that are negative for the event. (Griffith *et al.*, 1989)

Perhaps the most valuable indicator of whether a marker is valid is the predictive value. Predictive value for a marker of disease is the proportion of people studied with a particular disease among all the people who have the marker. Thus, in terms of whether the marker reflects disease, *predictive value* is

$$\frac{\text{true positives}}{\text{true positives and false positives}}$$

Predictive value can be calculated in terms of those with (predictive value positive) or without (predictive value negative) a marker. The relationship between sensitivity, specificity, and predictive value is shown in Figure 3.2.

Ideally, a marker assay should have high sensitivity and specificity but, in reality, one can increase the value of one aspect at the expense of the other. Thus, it is necessary for the investigator to decide where between a positive and negative test to place the assay result cut off point (Hulka, 1990). The complex relationship between population sensitivity, specificity, and the choice of threshold point is perhaps best depicted through a receiver–operator characteristic (ROC) curve. ROC analysis began during the 1940s as an approach to optimizing signal detection and has become a tool increasingly used to evaluate screening and diagnostic tests and algorithms (Swets *et al.*, 1979; Hanley, 1989; Thompson and Zucchini, 1989). The most common form of the ROC curve depicts the true-positive fraction (TPF) on the vertical axis and the false-positive fraction (FPF) on the horizontal axis (Figure 3.3). The curve is constructed by plotting the TPF and FPF for each decision threshold (or "cut off") value for which data are available, then connecting those points through an appropriate regression methodology (e.g., England, 1988; Tosteson and Begg, 1988). Both linear and probability co-

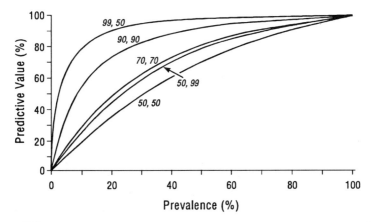

FIGURE 3.2 Relationship between predictive value and prevalence of marker. Numbers on lines represent percentage sensitivity and percentage specificity.

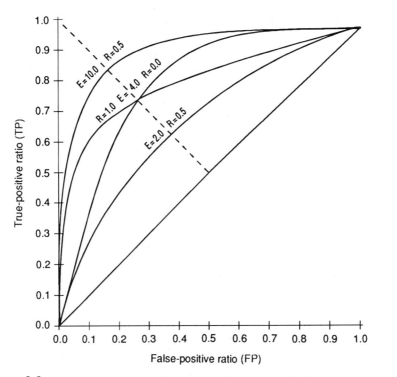

FIGURE 3.3 Data generated from receiver-operator characteristic (ROC) curves can take different shapes depending on sensitivity and specificity at different cut-off points. A general model for fitting these curves includes the parameter E (reflecting the height of the curve) and the parameter R (reflecting the skewness of the curve). (Reprinted with permission from England, 1988.)

ordinates have been used for TPF and FPF axes (Swets *et al.,* 1979; England, 1988). The decision threshold values may be continuous (as with the concentration of a chemical analyte) or categorical (as with radiologic or histopathologic assessment).

ROC curves plotted on linear scales allow a quick visual assessment of assay performance. In general, the more nearly the curve approaches the left corner of the coordinate system, the better the test performance. The point of closest proximity to the left corner represents the optimal decision threshold value, that is, the greatest sensitivity with the least nonspecificity. When comparing multiple assays depicted by ROC curves on the same scale, a simple assessment is provided by noting that the test with the greatest area under the curve provides the best overall combination of sensitivity and specificity (Hanley and McNeil, 1982). However, not all regions of the curve are necessarily of equal interest; more advanced techniques have been applied to comparisons of tests by ROC analysis (McClish, 1987, 1989).

In biomedical science, ROC analyses have been applied extensively to diagnostic imaging, clinical laboratory testing (e.g., Hanequin *et al.,* 1987; Hunink *et al.,* 1990), and the use of biologic markers in psychiatric diagnosis (Somoza and Mossman, 1991). These tests also have been used to address more general questions, such as the gains achieved through replicate analysis (Metz and Shen, 1992) and the efficiency of training neural networks for artificial intelligence (Meistrell, 1990). Of course, like any other method of test evaluation, conventional ROC analysis is entirely dependent on the accuracy of the health end point assessment to obtain true values for sensitivity and specificity. Some investigators have suggested that the use of ROC analysis with correlative information can avoid the need for such independent assessment of health end points (Henkelman *et al.,* 1990); however, great caution is required with such approaches.

Key in the validation and use of biologic markers for studies of human populations is the prevalence of the underlying condition being represented. Figure 3.2 shows how the positive predictive value of a marker is reflective of the underlying prevalence of the marker. Thus, for example, a marker assay that is 90% sensitive and 90% specific will still only have a predictive value of 50% when the prevalence of the underlying event is 10%. Field studies that do not incorporate prevalence considerations in planning are likely not to be able to detect an association, even if one exists.

In general, the sensitivity and positive predictive value of a marker can be increased by studying susceptible populations. For example, as Wilcosky (1992) describes, exposure to acid aerosols/oxidants can exacerbate symptoms of sensitized asthmatics, whereas it will likely generate an inflammatory response in nonasthmatics at lower exposure levels; thus, markers of inflammation will have greater sensitivity in studies of exposed and unexposed asthmatics.

Strategy for Validating Intermediate Markers

The strategy for validating markers that are intermediate between an exogenous exposure and disease involves two steps: (1) selection of candidate markers and (2) identification and quantification of the association between the markers and the disease (Schatzkin *et al.*, 1990). Selection of candidate markers from the large pool of available markers can be accomplished using case series, ecologic studies, and animal or *in vitro* experiments. In some instances, these markers can be used in case-control and cohort study designs (see Chapter 5) to identify an association with disease. This association can be quantified using the attributable proportion (AP) (Cole and MacMahon, 1971; Miettinen, 1974; Schatzkin *et al.*, 1990) or other measures of association such as odds ratio or relative risk. AP is defined as the proportion of cases of disease that is attributable to the marker. The AP can be determined directly from the sensitivity (S) and the relative risk (R), as defined in the formula from the data array in Table 3.1. Schatzkin *et al.* (1990) have developed examples (Table 3.2) of how AP varies with different levels of relative risk and sensitivity. Table 3.2 shows the AP for various levels of S and R. Levels of low S and high R have an AP much less than those with high S and low R.

To assess the validity of an intermediate marker of disease, it is useful to consider the simplified framework in Figure 3.4. In Figure 3.4A, a single marker is shown to be linked causally and, hence, necessary and sufficient for disease. In this case, AP approaches 1.0 and S equals 1.0. This marker would have a large degree of content and criterion validity.

In most instances, more than one pathway is involved; this is shown in Figure 3.4B. Here, neither pathway is necessary for disease but either is sufficient. The content validity may be judged to be relatively high, but the criterion validity is reduced (as evidenced by S, which is less than 1.0). The AP is also less than 1.0. The closer the AP is to 0, the less a marker can be considered a valid surrogate for disease.

TABLE 3.1   Calculation of Attributable Proportion (AP)[a]

	Disease	
Marker	Yes	No
+	A	B
−	C	D

Source: Shatzkin *et al.* (1990).
[a] Sensitivity (S) = A/A + C; relative risk (R) = [A/ (A + B)] / [C/ (C + D)]; using these values, AP = S (1 − 1/R).

TABLE 3.2 Relationship between Attributable Proportion, Relative Risk, and Sensitivity[a]

Sensitivity of intermediate marker	Relative risk for intermediate marker positives								
	2.5	3.5	4.5	5.5	6.5	7.5	8.5	9.5	10.5
0.1	0.03	0.06	0.07	0.08	0.08	0.09	0.09	0.09	0.09
0.3	0.10	0.18	0.21	0.23	0.25	0.25	0.26	0.26	0.27
0.5	0.17	0.30	0.36	0.39	0.41	0.42	0.43	0.44	0.45
0.7	0.23	0.42	0.50	0.54	0.57	0.59	0.61	0.62	0.63
0.9	0.30	0.54	0.64	0.70	0.74	0.76	0.78	0.79	0.81

Source: Adapted from Schatzkin *et al.* (1990).

[a] $[AP = S(1 - 1/R)]$

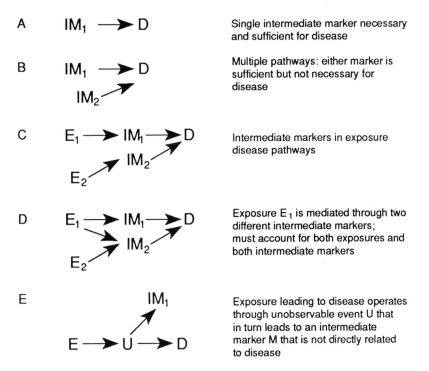

FIGURE 3.4 Framework for considering exposure–marker–disease relationships. (Adapted from Schatzkin *et al.*, 1990.)

In Figure 3.4C, separate exposure–marker–disease pathways are shown. The effects of each exposure are mediated entirely through their intermediate markers. In Figure 3.4D, a different situation is depicted. The effect of exposure E_1 is mediated partially through IM_2. The AP for neither of these markers will be 1.0. The marker that has the highest AP is the one with the best content and criterion validity (Schatzkin *et al.*, 1990).

Another situation is shown in Figure 3.4E. An exposure leading to disease operates through an unobservable event U that, in turn, leads to an intermediate marker that is not directly on the disease pathway. This process can be illustrated, for example, with chromosomal micronuclei, which, although they occur in nonviable cells, reflect genotoxic events similar to ones that may be necessary for carcinogenesis (Schatzkin *et al.*, 1990). In this case, the marker may have some content validity, but not of the greatest degree. The criterion validity of the marker would depend on its relationship to U and the relationship of U to disease. If the latter is not high, the former will be meaningless and the marker cannot be used as a surrogate.

In summary, the quality of epidemiologic studies and risk assessments depends on the quality and validity of measurements. Assessment of validity involves the empirical process of determining that the candidate biomarker is indeed measuring what it intends to measure. To achieve that goal, a range of validation studies should be performed prior to use of the marker in etiologic research or intervention.

A practical example of the employment of the concepts of content, construct, and criterion validity is shown in the scheme in Figure 3.5 and in Table

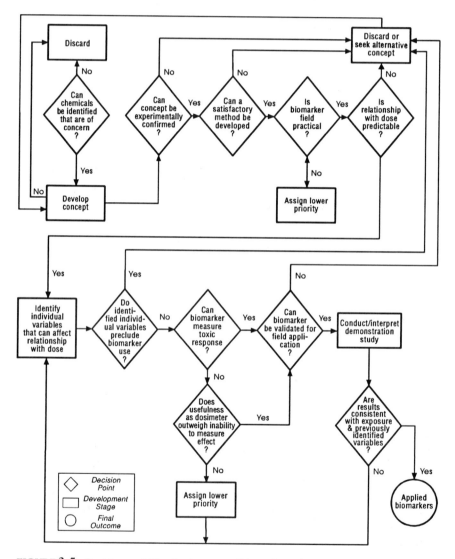

FIGURE 3.5   Decision model for development of biomarkers of exposure. (Bull, 1989; Stevens *et al.*, 1991.)

TABLE 3.3   Steps in the Development of a Biomarker

Step	Action required	Relative importance[a]
1. Chemical selection	Prioritize based on occurrence, significant human exposure, potential for adverse human health effects	C
2. Conceptualization	Identify logical consequence of chemical exposure that might serve as a useful measure of exposure	C
3. Confirmation of concept	Experimentally confirm validity of basic concept	C
4. Develop method of measurement	Identify method for detecting changes in biomarker at doses at or below those producing toxic effects	C
5. Biomarker practical for field?	Develop plausible field methodology and develop sufficient sensitivity of biomarker to monitor existing exposures	L
6. Establish dose–response relationship	Characterize pharmacokinetics and metabolism of chemical (consistent relationship to systematic dose is critical; knowledge of effective dose is limiting)	C,L
7. Identify variables affecting relationship with dose	Establish specificity of response and identify lifestyle, genetic, disease state, therapeutic, or occupational variables that modify the response	C,L
8. Measures toxic effect?	Provides advantage only among biomarkers of equal ability as measures of exposure	N
9. Validation of applicability to humans	Conduct pilot study in small groups of humans with defined exposure gradients to chemical of interest	C
10. Conduct demonstration study	Determine whether variation in response in larger population can be accounted for by known variables	C

[a] C, critical to the application of the biomarker; L, limiting to the application of the biomarker, i.e., places limits on interpretation of results for secondary purposes, e.g., risk assessment; N, nice to have, but not essential to the application of the biomarker (Bull, 1989).

3.3. This example is a decision algorithm developed to assess biologic markers of exposure. An analogous approach could be developed for markers of effect and susceptibility.

Internal Validity

Proper use of biologic markers can enhance the internal validity of epidemiologic studies. The internal validity of a study is the degree to which index and comparison groups are selected and compared so, apart from sampling errors, the observed differences between the dependent variables are attributed only to the hypothesized effect (Last, 1988). Such a result is validity in the estimation of effect, and is dependent on the ability to control bias. Internal study validity has been discussed widely in epidemiology textbooks. In this section, we discuss some issues of internal validity that pertain to the use of biologic markers. Although some of this discussion is specific to markers, other more general issues also merit comment.

Statistical bias is a distortion that may result during evaluation of an association and can occur when subject selection is unequal according to disease or exposure status (Ozonoff and Wartenberg, 1991). When selecting subjects for studies, the investigator must identify factors such as background rates of markers and the range of normal so classification and subject selection are equal for the groups being compared. These issues have been discussed elsewhere (Hulka and Wilcosky, 1988; Schulte, 1989). Bias can lead to misclassification of subjects based on exposure or disease, and may result from failure to adjust for other variables that are also predictive of the disease of interest.

Misclassification

Since the basis of the epidemiologic method is the comparison of rates of disease in exposed versus unexposed populations, a mix of exposed and unexposed or disease and nondisease characteristics (or both) clearly will harm the chances of seeing a difference (Ozonoff and Wartenberg, 1991). This misclassification can be "differential," that is, systematically wrong in one aspect of classifying exposure or disease, or nondifferential, that is, nonsystematic. Differential misclassification of exposure or disease can reduce the validity of a study (Hogue and Brewster, 1988; Hulka and Wilcosky, 1988). Biologic markers that reduce misclassification may enhance study validity. Similarly, biologic markers can contribute to the reduction of nondifferential misclassification. This type of misclassification, which has been considered a lesser threat to validity, can result in bias toward the null value.

The key to valid epidemiologic studies is a strong rationale for selection of the exposure (dose) variables. The choice of exposure variables for individuals exposed to toxic substances can range from anamnestic information gathered by questionnaire to detailed measurement of biologic markers (Rogan, 1988). However, as Rogan (1988) notes, ". . . in the strict sense, any exposure information other than biological effective dose is a surrogate." Thus, the question is, "How closely does the exposure surrogate, used to derive a model, resemble the actual exposure under study?" Valid biologic markers can provide empirical data that is preferable to deductively derived estimates (Rogan, 1988).

For example, Lawrence and Taylor (1985) demonstrated the value of empirical exposure measurements when they were confronted with the problem of assessing historical polychlorinated biphenyl (PCB) exposures among women who manufactured electrical capacitors. The purpose of their investigation was to determine the effects of PCB exposure on reproductive outcomes from 1979 to 1983. Although the investigators did not have actual serum PCB measurements for that period, they did have a complete work history for each subject. The industrial hygiene data allowed classification of each job in terms of a low, medium, or high PCB concentration. The challenge was to choose a surrogate that best approximated the true exposure. The investigators also had sera that had been gathered in 1976 from a sample of workers as part of a general company survey. Using those data, the investigators developed a regression model to estimate the explicit serum PCB concentration as a continuous variable level for each woman during each of her pregnancies between 1979 and 1983. Hence, the serum PCB concentration, derived from a sample of subjects, was used as a biologic marker to construct a more accurate estimate of the true exposure than was available using job-classification data (Lawrence and Taylor, 1985).

Analytical Adjustment for Other Variables

Proper data analysis depends on the choice of the correct mathematical model, especially when multiple variables exist. The strongest models take into account *a priori* hypotheses specific to the topic under study. The incorporation of biologic markers into study designs and mathematical models also implies an understanding of the direction and mechanism of action. For example, in a study of hospital workers exposed to low levels of ethylene oxide (Schulte *et al.*, 1992), the association between hydroxy ethyl hemoglobin adducts and ethylene oxide exposure was assessed using multiple linear regression based on two assumptions: (1) exposure at low doses was linear and (2) the impact of confounding and intervening variables could be evaluated in this approach. Thus, as shown in Table 3.4, the level of hydroxy ethyl adducts showed an exposure–response relationship when group means were adjusted for confounding factors such as age, race, cigarette smoking, and education. Additionally, controlling measurement validity makes it possible partially to control study validity, since measurement errors can produce biased estimates of regression coefficients used in models (Louis, 1988).

Longitudinal studies that employ biologic markers will be used increasingly in molecular epidemiologic studies and quantitative risk assessments. The validity of those study results will depend in part on the analytical approach that is selected (Dwyer *et al.*, 1992). Such studies may involve repeated measures of a continuous random variable; thus, there may be measurement errors that are considered random among persons but are autocorrelated within persons. The use of autoregressive modeling by epidemiologists for the analysis of longitudinal data is increasing, and will be more frequent in studies involving biologic markers. These models permit consideration of the time course of change of a variable (Rosner *et al.*, 1985). Other

TABLE 3.4   Relationship between Ethylene Oxide Exposure and Hemoglobin
Adducts in Hospital Workers

Ethylene oxide exposure (ppm-hr)	Mean[a]	N	Hemoglobin adducts[b] (pmol/mg Hb)
0	0 (0)	8	0.06 (0.02)
>0–32	12.8 (11.0)	32	0.09 (0.01)
>32	105.2 (45.7)	11	0.16 (0.02)
Comparison	0 vs > 0–32		p = 0.27
	0 vs > 32		p = 0.002
	>0–32 vs >32		p = 0.004

[a] Means were adjusted for age, race, cigarette smoking, and education.
[b] Values are means. Standard deviations given in parentheses.

methods for analyzing repetitive measures with a Gaussian error structure
have been reviewed by Louis (1988), who concluded that this area needs
continued statistical, numerical, and interpretive research and development.

External Validity

Application of molecular epidemiologic research often will be in the assessment of health risks to groups not included in the studies. Risk assessment is
an effort to address a condition of incomplete data (Erdreich, 1988) and involves the extrapolation (or generalization) of known exposure–response
data to ill-defined risk situations in target populations. External validity is
the degree to which a study can produce unbiased inferences about those
target populations. For risk assessment, external validity involves the appropriateness of the following extrapolations: within or between populations or
species, from high doses to low doses, and between different organs within a
species. All these efforts can be enhanced using biologic markers common to
each population or species. Allometric assessment of effects in different species can be determined by observing how the same marker varies with similar
exposures. Valid extrapolation requires an understanding of the major events
that can cause such inter- and intraspecies differences. For example, in
chemical carcinogenesis, the following factors appear to play a critical role
in species and organ differences: the overall balance of metabolic activation
and detoxification, the balance of DNA damage and repair, the persistence
of DNA damage, and tumor formation (Slaga, 1988).

Many uncertainties are attendant to extrapolation of data from an epidemiologic study of a smaller group to a large population. The characteristics that make a study internally valid are often barriers to extrapolation. Nevertheless, extrapolation is a current practice in risk assessment. The use of valid biologic markers may allow some evaluation of whether a particular extrapolation is warranted, whether the variability is too extreme, or whether differences in susceptibility have resulted in sensitive subgroups (Erdreich, 1988).

Extrapolation to low doses (or exposures) involves determining (or assuming) the shape of the dose–response curve. In some instances, establishing a dose–response relationship in a risk assessment might be considered a meta-analytic procedure, that is, results from different studies might be combined to provide a larger sample size or a broader range of dose estimates. The validity of this effort can be enhanced if the same markers are used in different studies or if different markers have been shown to be correlated, that is, to have construct validity.

The contribution of macromolecular adducts to low-dose extrapolation has been heralded as a potential improvement to risk assessment. However, the use of biologic markers also can be a source of confusion in risk assessment. Most studies of adducts in humans have not yet demonstrated a clear dose response, perhaps because of the wide variability in human response and the current inability to determine true individual exposures (Perera, 1987, 1988). However, trends with exposure have been observed in occupational settings (Perera et al., 1988). Until the sources of variability can be identified and their impact evaluated, the absence, or faulty characterization, of a dose response will limit the usefulness of this class of biologic markers in risk assessment (Brown, 1988; Motulsky, 1988). A potentially major source of differential susceptibility in dose response is the phenotypic variation of metabolic parameters (Motulsky, 1988). This variation rarely has been considered in epidemiologic studies and risk assessments, partially because such data have not been developed.

The effect of the choice of a dose variable on risk estimates can be severe, especially when the pattern of exposure that the estimates supposedly reflect differs from the predominant pattern experienced by a study cohort (Crump and Allen, 1985). The use of a biologic marker of exposure can help reduce the impact of using an ambiguous dose variable, because it can reflect the true dose more accurately, even in studies in which exposures are observed to have occurred over a wide range. For example, attempts have been made to compare biologically effective doses at high exposures at which tumors are observed to low exposures to determine whether linearity of the carcinogenic effect is a valid assumption. Perera (1987, 1988) has concluded that extensive data on DNA, RNA, and protein binding indicate that macromolecular effects, at the lowest administered doses, generally follow first-order kinetics, that is, the rate of binding in target organs *in vivo* is directly proportional to the administered dose. Since many carcinogens covalently bind to and structurally alter DNA, the adducts that are formed are conceptually valid markers of exposure and, possibly, of effects. Moreover, the ratio of surrogates for DNA adducts, for example, protein adducts, to dose have been shown to be constant over a dose range of 10^{-5} mol/kg to 10^5 mol/kg. However, as Swenberg (1988) asked, ". . . What data-bases are available so that such a molecular dosimetry approach can be validated?" Few carcinogens have been evaluated for which the exposure range is greater than one order of magnitude.

Validation and Selection of Biomarkers

A determination of the validity of a biomarker for molecular epidemiologic studies should be made on the basis of six fundamental criteria. (See Perera, 1987, for review.)

Biologic Relevance

A clear hypothesis or model of exposure–dose or exposure–response relationships must be available that defines the role of the specific marker in relation to other points in the continuum (Perera, 1987). This relationship is content validity. For example, the biologic relevance of carcinogen–DNA adducts as initiating events in carcinogenesis and carboxyhemoglobin levels as indicators of reduced oxygen-bearing capacity of erythrocytes are well established. The meaning of sister chromatid exchanges (SCEs) and H-*ras*-encoded p21 proteins is more ambiguous.

The biologic relevance of markers of exposure is generally easier to establish than that of effect markers. Determining the ability of markers of early biologic effect (such as SCEs or mutated oncogenes) to predict cancers or reproductive risk requires prospective or "nested" case-control studies. Circumstantial evidence could be provided through findings of increased biomarkers in groups (such as workers) with historically high cancer rates. Unfortunately, longitudinal studies have not yet been carried out to assess the predictive ability of effect markers. Indeed, only limited biomarker data are available from cross-sectional studies of workers exposed, in most cases by inhalation, to known occupational carcinogens.

Understanding Pharmacokinetic Aspects

Molecular epidemiology relies on a conceptual model that depicts the exposure–response continuum as a sequence of time series related to xenobiotic exposure, burden, damage, and risk (see Figure 3.6). Analysis of the link between these processes identifies two kinetic conditions that are neces-

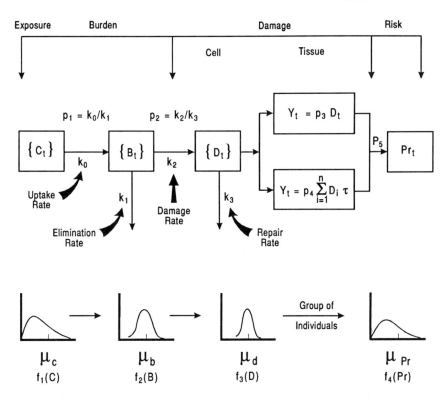

FIGURE 3.6   Conceptual model relating exposure, dose, damage, and risk of disease for an individual worker or uniformly exposed group of workers. The symbols labelled k_0–k_3 represent rate constants for uptake (k_0), elimination (k_1), cellular damage (k_2), and repair (k_3). The proportionality constants labelled p_1–p_5 result from linear transfer functions between succeeding series. The distribution functions shown at the bottom represent the following: exposure [$f_1(C)$], burden [$f_2(B)$], cellular damage [$f_3(D)$], and population risk [$f_4(Pr)$]. (Rappaport, 1991.)

sary for variability of exposure to affect individual risk of chronic disease appreciably (Rappaport, 1991). To illustrate these issues, it is useful to consider occupational exposures to a toxicant. Rappaport (1988, 1991) has described the functional relationships between airborne exposure distributions and subsequent distributions of dose, damage, and risk of disease. Key to understanding these relationships is the realization that environmental exposures vary greatly over time and among individuals. The validity of a marker will depend on its biologic relationship to the process being marked (its content and construct validity) and also on the way in which groups of subjects are sampled. Thus, collection of biologic specimens and attendant exposure assessment is inherently a statistical process. Sampling designs must accommodate the distribution of possible exposures as well as the purpose of the study.

Two considerations are important in addressing kinetic conditions that affect an exposure–disease continuum. First, the variations of exposure from interval to interval must be translated efficiently into burden and damage; that is, there should be no physiologic damping of exposure variability. Second, during the period of intense exposure, the relationship between intense burden and damage must be nonlinear (Rappaport, 1991). Practically speaking, with respect to chronic diseases, the sampling strategy that generally will be useful is one that involves assessment of mean exposure with little regard for short-term fluctuations of exposure. However, if acute effects are the subject of investigation, it may be appropriate to estimate short-term peak exposures (Rappaport, 1991).

Thus, the proper selection and use of biologic markers depends on an understanding of the underlying pharmacokinetics and pharmacodynamics (World Health Organization (WHO), 1986; Andersen, 1987; Smith, 1987). Knowledge of pharmacokinetics is important to determine the frequency of sampling and the tissues or fluids that are most appropriate for study. It also guides the interpretation of the dosimetry and effect data obtained in the target tissue or in a surrogate.

Temporal Relevance

The temporal relationship of markers to external exposure or to disease end points must be clear. Markers have different half-lives in biologic media. Droz *et al.* (1991) have characterized the extent to which exposure history can be represented by a biologic marker. Droz *et al.* (1985) had demonstrated how the measured dose of organic solvents in workers was influenced by the following variables: interday and intraday fluctuation of exposure, physical workload, body build, and metabolism. Individuals classified as having the same exposure still can have a different dose because of the influence of these variables. The timing of marker measurement in relation to exposure influences the ability to detect response; measurements made too early or too late may underestimate the magnitude of the response (Wilcosky, 1992). For example, transient markers such as the influx of neutrophils or eosinophils into the respiratory tract will occur usually during the first 3–7 days of an inflammatory response. Hence, measurement of these markers of inflammation in bronchoalveolar lavage fluid immediately after exposure would underestimate the inflammatory response (Wilcosky, 1993). Whether an exposure marker reflects recent or cumulative exposures, peaks or averages, will depend on the pharmacokinetics of the toxicant and the persistence of the marker in the biologic sample being assayed (which, in large part, is a function of the turnover rate or half-life of the sample, but also is a reflection of repair rates).

An understanding of temporal relevance is essential to developing monitoring strategies and to interpreting results. Most measures of internal dose, for example, reflect recent exposures (ranging, for example, from hours for cotinine to days for benzene in exhaled air). An exception would be a substance that is fat soluble and is stored in adipose tissue. Hemoglobin is a good integrating dosimeter over the 4-month life span of the erythrocyte which, unlike lymphocytes, lacks repair systems. Human serum albumin has a 20- to 25-day half-life. Since it is synthesized in the liver where many carcinogens are metabolically activated, albumin might collect adducts not detected in hemoglobin (Perera, 1987).

The time period reflected by total white blood cell, or lymphocyte, markers is considerably more complex than that reflected by albumin (Perera, 1987; Carrano and Natarajan, 1988). For example, adducts on DNA from lymphocytes can reflect past as well as current exposure, since a subset (T-cells) is very long lived. A review of lymphocyte subpopulations shows that measurements of DNA adducts in these cells will be influenced by the longer lived T cells and will, therefore, reflect exposures that occurred both recently and several decades in the past. T cells make up 60–90% of lymphocytes, which in turn represent 22–28% of peripheral lymphocytes in circulating blood. Thus they constitute a maximum 25% of lymphocytes. The estimated half-life of T cells is 3 years. In contrast, B cells and monocytes constitute 1–2% and 1–7% of circulating lymphocytes, respectively, and have lifetimes ranging from days to weeks. Granulocytes represent the remaining 66–85% of lymphocytes and are short lived (hours to days). Thus, considering only DNA adducts in cells damaged while in circulation, and excluding consideration of adducts in circulating lymphocytes that might result from damage to stem or precursor cells in the bone marrow, when all DNA from a sample of peripheral blood is assayed for DNA adducts, the long-lived T cells will be the major contributors in cases of past discontinued exposure, largely because of their 100–1000 times longer life-span. In addition, lymphocytes have a lower repair activity than do cycling cells; thus, DNA lesions are likely to be more persistent in lymphocytes. In cases of current, recent, or long-term uninterrupted exposure to carcinogens, T cells will contribute less significantly to total adduct measurements. The preponderance of adducts will be measured in the shorter-lived granulocytes, B cells, and monocytes.

In retrospective case-control studies, a permanent marker left decades earlier by the initiating carcinogen would be ideal. Unfortunately, in the case of discontinued past exposures, even the most long-lived markers will be diluted by cell turnover, thereby underestimating the true past or cumulative dose. Only if exposure was continuous and had not changed significantly during the past decades (and only if the disease did not alter metabolism) would current levels of the marker directly represent critical prior exposure. At the very least, however, markers such as adducts reflect individual responsiveness to carcinogenic exposures, provided that metabolism is unchanged by disease. Many other exposure patterns are relevant to case-control and cohort studies (current but interrupted, continuous but of varying magnitude, etc.). Each pattern leads to a different distribution of adducts among lymphocyte populations and, hence, to a varying pattern of persistence.

Understanding "Background" Variability and Confounding Variables

The variability of a biologic marker is a statistical characteristic of groups of subjects or populations (Janetos, 1988). Variability is the result of genetic and environmental factors, separately and interacting. Motulsky (1988) has described the pervasiveness of genetic variation that accounts for human individuality.

> Human physiognomy is unique and no two human beings except identical twins are alike. The involved genes remain unknown. Remarkable genetic individuality also exists for red cell and tissue cell (HLA) groups, in enzymes and proteins. Enzyme variation usually is associated with variable enzyme levels in the normal range, so a person's exact activity level for a given enzyme (i.e., high normal, average, low normal) may be the speed of breakdown of various substances. Protein variation may lead to differential binding of foreign substances. . . . Variability in receptor activity may cause differential metabolism of foreign or endogenous ligands. Various HLA and related markers may lead to differential immunologic reactions that predispose to certain autoimmune diseases. Variability at the DNA level is more striking. Frequent differences occur at the individual nucleotide level (every 500 nucleotides), as do size variations of longer stretches of DNA (minisatellites). Most such DNA variants are phenotypically silent but often can be used as "markers" for closely linked gene loci that specify proteins that have physiologic, biochemical, or immunologic effects (Motulsky, 1988).

This natural variability makes it essential to know the range of biomarker values in a "normal" population. Care must be taken not to be deceived by the extensive variation in biochemical individuality (a "healthy" level for some individuals may indicate a health risk for others) (Schulte, 1987). The range of normal can be quite extensive. For example, it is well known that the cholinesterase level in subjects not exposed to organophosphorus insecticides covers a wide range—a 25% change in the group may overshadow a decrease of 50% in a few subjects (WHO, 1975).

Interindividual and intraindividual variations are important contributors to "noise" or "background" in biomonitoring or epidemiologic studies and should be characterized prior to large-scale application of a particular biomarker. Such data can, however, be generated only through large-scale surveys that employ repeated sampling and efforts to control for confounding variables. Thus, a background study is a significant epidemiologic exercise in itself. With respect to carcinogen–DNA and –protein adducts, significant interindividual and intraindividual variation has been observed with polyaromatic hydrocarbons (PAHs), 4-aminobiphenyl (4–ABP), and nitrosamines (Harris, 1985; Umbenhauer *et al.*, 1985; Bryant *et al.*, 1987; Perera,

1987). SCEs also vary significantly within and among individuals (Carrano and Natarajan, 1988).

As discussed earlier, since biologic markers can be potentially more sensitive than indicators used in conventional epidemiologic methods, there is a greater need to control for confounding or mitigating factors [National Academy of Sciences (NAS), 1991]. Confounding variables in studies using biomarkers may include age, sex, race, cigarette smoking, alcohol consumption, diet, drugs or other environmental exposures, genetic factors, or pre-existing health impairment.

Age, a potential confounding factor of exposure–effect associations, often is addressed in a facile manner. Increasing evidence shows that chronologic age is not an adequate descriptor of biologic age (Ingram, 1988). Biologic age may be more of an influence on marker frequency than chronologic age. However, the definition and measurement of aging is controversial. Efforts to identify and develop biologic markers of aging are underway but are in the early stages of progress (Hart and Turtarro, 1988). Molecular epidemiologic studies may be influenced significantly by the fact that DNA alterations (DNA chemical structure, DNA sequence organization, and gene expression) increase with age (Mullaart *et al.*, 1990). DNA repair also has been shown to correlate well with lifespan in various mammalian species. This correlation has been cited as an indication for the role of DNA repair in the aging process. Any conclusion is complicated by the large interindividual differences in DNA repair. (See Figure 3.7 for a schematic representation of various types of DNA damage).

A more vigorous appraisal of age beyond a chronologic one may be important to conducting effective molecular epidemiologic studies. Biomarkers of aging may be susceptibility markers or may be covariates in exposure–response studies. In designing molecular epidemiologic studies and validating a biologic marker, attention may need to be paid to the comparability of populations in terms of biologic and functional age as well as chronologic age.

Race may be another potentially confounding variable that generally is controlled in epidemiologic studies, but only categorically. Within apparently homogeneous racial categories may exist subcategories distinguished by molecular markers or DNA repair capacities (Weston *et al.*, 1991). For molecular epidemiologic research, classification by race may result in a net loss of information, since such categorization may be misleading and fail to define large categories of interperson variability (Cooper and David, 1986).

Many other potentially confounding factors also need consideration in molecular epidemiologic studies but often are neglected. Most reports on cytogenetic studies have provided no information on smoking and other exposures to carcinogens or mutagens, yet life-style factors (e.g., smoking, diet) and other chemical exposures (e.g., environmental, recreational, medicinal, and drug-related) are potential confounding factors with respect to SCEs

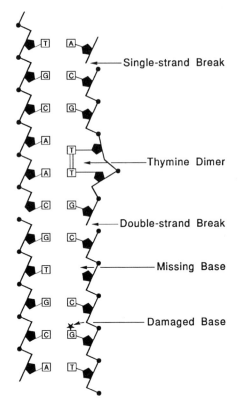

FIGURE 3.7 Schematic of DNA damage. (Reprinted with permission from Baan, 1987. Copyright © 1987 by Springer-Verlag.)

and most other markers. Other potential confounders are host factors (e.g., health and immune status) and exposures that influence the marker of interest. In a study of PAH–DNA adducts in workers, for example, it is necessary to account for all "background" exposures to the chemical and, ideally for factors that could induce or inhibit metabolism and binding of benzo(a)pyrene (BaP). These factors include inducers such as cigarette smoke, charbroiled meat, ethanol, sedatives, and PCBs, as well as inhibitors such as methylxanthines in foods, steroids, solvents, and spray paints. Thus, even pilot studies become fairly complex epidemiologic exercises that require careful interviewing (Perera *et al.*, 1982; NAS, 1991). For example, Figure 3.8 shows the magnitude of a molecular epidemiologic study of 33 mortuary science students evaluated for various biomarkers before and after exposure to formaldehyde during an embalming course. Despite the small number of subjects, many data points were collected (Suruda *et al.*, 1992).

Since most biomarkers are nonspecific, that is, different exposures may

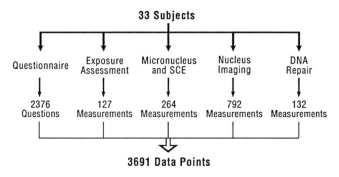

FIGURE 3.8   Magnitude of a molecular epidemiologic study.

cause the same marker response, attention should be paid to the impact of their use in studies. Nonspecific biomarkers may lead to dilution and confounding of exposure–response associations in the presence of other exposures (Weiss and Liff, 1983; Wilcosky, 1993). If two different exposures, E_1 and E_2, cause the same marker response through independent pathways, they increase the overall marker response rate in an additive manner. However, the assessment of the different exposures that cause the same response through independent pathways will be misleading in epidemiologic studies that use relative risk or odds ratio because these measures are based on the assumption of a multiplicative model of association, whereas the two exposures working independently increase overall response in an additive manner. Thus, the relative risk of response of one agent will be influenced by the background response of the other agent. Therefore, the use of risk difference, rather than relative risk, to compare responses will help avoid the problems of dilution from a high background incidence from one exposure (Wilcosky, 1993).

Reproducibility, Sensitivity, Specificity, and Predictive Value of Assays

Assays should be reproducible with limited variability that is attributable to laboratory personnel or assay method (Gann *et al.*, 1985). The same criteria of adequate specificity, sensitivity, and predictive value that apply to the validation of screening methods should be met by biomarkers. Markers of exposure should be "sensitive" and "specific" to toxic exposures, picking up a high percentage of individuals in the exposed group and attributing negative results to a high percentage of unexposed persons. Given interindividual variation, however, not all exposed persons would be expected to be positive. Markers of effect or response should detect a high number of individuals at elevated risk of adverse outcomes. Both types of markers should give a high proportion of "correct" answers.

Summary

Validation of candidate biomarkers for molecular epidemiologic studies is an empirical process that can be approached by producing several different, but convergent, lines of evidence. Key to future efforts in the use of a particular biomarker is the establishment of content, construct, and criterion validity, and attention to factors that can cause confounding and variability in an exposure–response association. To validate a biomarker adequately it is necessary to understand its natural history and temporal relevance to the event being marked. Epidemiologic methods for validation are discussed in Chapter 6.

Acknowledgment

The authors thank Robert F. Vogt, Jr., for the discussion of the ROC curves.

References

Andersen, M. E. (1987). Tissue dosimetry in risk assessment, or what's the problem here anyway? *In* "Pharmacokinetics in Risk Assessment," pp. 8–23. National Academy Press, Washington, D.C.

Ashby, J. (1987). Monitoring human exposure to genotoxic chemicals. *In* "Occupational and Environmental Chemical Hazards: Cellular and Biochemical Indices for Monitoring Toxicity (V. Foa, E. A. Emmett, M. Maroni, A. Colombi, eds.), pp. 411–423. Ellis Horwood, Chichester, England.

Baan, R. A. (1987). DNA damage and cytogenetic endpoint. *In* "Cytogenetics" (G. Obe and A. Basler, eds.), pp. 327–337. Springer-Verlag, Berlin.

Brown, S. L. (1988). Differential susceptibility: Implications for epidemiology, risk assessment, and public policy. *In* "Phenotypic Variation in Populations: Relevance to Risk Assessment" (A. D. Woodhead, M. A. Bender, and R. C. Leonard, eds.), pp. 255–269. Plenum Press, New York.

Bryant, M. S., Skipper P. L., Tannenbaum, S. R., and Maclure, K. M. (1987). Hemoglobin adducts of 4-aminobiphenyl in smokers and nonsmokers. *Cancer Res.* **47**, 602–608.

Bull, R. J. (1989). Decision model for the development of biomarkers of exposure. U.S. Environmental Protection Agency (EPA 600/X-89/163), Las Vegas, Nevada, p. 12.

Carrano, A. V., and Natarajan, A. T. (1988). Considerations for population monitoring using cytogenetic techniques. *Mutat. Res.* **204**, 379–406.

Cole, P., and MacMahon, B. (1971). Attributable risk percent in case-control studies. *Br. J. Prev. Soc. Med.* **25**, 242–244.

Cook, T. D., and Campbell, D. T. (1979). "Quasi-experimentation: Design and Analyses Issues for Field Settings." Rand McNally, Chicago.

Cooper, R., and David, R. (1986). The biological concept of race and its application to public health and epidemiology. *J. Public Health. Policy Law* **11**, 97–115.

Crump, K. S., and Allen, B. C. (1985). Methods for quantitative risk assessment using occupational studies. *Am. Stat.* **39**, 442–450.

Droz, P. O., Berode, M., and Wu, M. M. (1991). Evaluation of concomitant biological and air monitoring results. *Appl. Occup. Environ. Hyg.* **6**, 465–474.

Dwyer, J. H. Feinleib, M., Lippert, P., and Hoffmeister, H. (eds.) (1992). "Statistical Models for Longitudinal Studies of Health." Oxford University Press, New York.

England, W. L. (1988). An exponential model used for optimal threshold selection on ROC curves. *Med. Decis. Making* **8**(2), 120–131.

Erdreich, L. S. (1988). Combining animal and human data: Resolving conflicts, summarizing the evidence. *In* "Epidemiology and Health Risk Assessment" (L. Gordis, ed.), pp. 197–207. Oxford University Press, New York.

Gann, P. H., Davis, D. L. and Perera, F. (1985). "Biological Markers in Environmental Epidemiology: Constraints and Opportunities." Paper presented at the 5th Workshop of the Scientific Group on Methodologies for the Safety and Evaluation of Chemicals, World Health Organization, 12–16 August, Mexico City.

Griffith, J., Duncan, R. C., and Hulka, B. S. (1989). Biochemical and biological markers: Implications for epidemiologic studies. *Arch. Environ. Health* **44**, 375–381.

Hanley, J. A. (1989). Receiver operating characteristic (ROC) methodology: The state of the art. *Crit. Rev. Diagn. Imaging* **29**(3), 307–335.

Hanley, J. A., and McNeil, B. J. (1982). The meaning and use of the area under a receiver operating characteristic (ROC) curve. *Radiology* **143**(1), 29–36.

Hannequin, P., Liehn, J. C., Delisle, M. J., Deltour, G., and Valeyre, J. (1987). ROC analysis in radioimmunoassay: An application to the interpretation of thyroglobulin measurement in the follow-up of thyroid carcinoma. *Eur. J. Nucl. Med.* **13**(4), 203–206.

Harris, C. C. (1985). Future directions in the use of DNA adducts as internal dosimeters for monitoring human exposure to environmental mutagens and carcinogens. *Environ. Health Perspect.* **62**, 185–191.

Hart, R. W., and Turturro, A. (1988). Biomarkers in aging and toxicity. *Environ. Gerontol.* **23**, 241–43.

Henkelman, R. M., Kay, I., and Bronskill, M. J. (1990). Receiver operating characteristic (ROC) analysis without truth. *Med. Decis. Making* **10**(1), 24–29.

Hernberg, S. (1987). Report of Session 1. *In* "Occupational and Environmental Chemical Hazards: Cellular and Biochemical Indices for Monitoring Toxicity." (V. Foa, E. A. Emmett, M. Maroni, A. Columbi, eds.), pp. 121–241. Ellis Horwood, Chichester, England.

Hogue, C. J. R., and Brewster, M. A. (1988). Developmental risks: Epidemiologic advances in health risk assessment. *In* "Epidemiology and Health Risk Assessment" (L. Gordis, ed.), pp. 61–80. Oxford University Press, New York.

Hulka, B. D. (1990). Principles of bladder cancer screening in an intervention trial. *J. Occup. Med.* **32**, 812–816.

Hulka, B. S. and Wilcosky, T. (1988). Biological markers in epidemiologic research. *Arch. Environ. Health* **43**, 83–89.

Hunink, M. G., Richardson, D. K., Doubilet, P. M., and Begg, C B. (1990). Testing for fetal pulmonary maturity: ROC analysis involving covariates, verification bias, and combination testing. *Med. Decis. Making* **10**(3), 201–211.

Ingram, D. K. (1988). Key questions in developing biomarkers of aging. *Environ. Gerontol.* **23**, 429–434.

International Programme on Chemical Safety (1983). Guidelines on studies in environmental epidemiology. *In* "Environmental Health Criteria" vol. 27, pp. 133–136. World Health Organization, Geneva.

Janetos, A. C. (1988). Biological variability. *In* "Variations in Susceptibility to Inhaled Pollutants" (J. D. Brain, B. D. Beck, A. J. Warren, and R. A. Shaikh, eds.), pp. 9–29. Johns Hopkins University Press, Baltimore.

Last, J. M. (ed.) (1988). "A Dictionary of Epidemiology." 2nd Edition, Oxford University Press, New York.

Lawrence, C. E., and Taylor, P. R. (1985). Empirical estimation of exposure in retrospective epidemiologic studies. *In* "Environmental Epidemiology" (F. C. Kopfler and G. F. Craun, eds.), pp. 239–246. Lewis Publishers, Chelsea, Michigan.

Louis, T. A. (1988). General methods for analyzing repeated measure. *Stat. Med.* 7, 39–45.

McClish, D. K. (1987). Comparing the areas under more than two independent ROC curves. *Med. Decis. Making* 7(3), 149–155.

McClish, D. K. (1989). Analyzing a portion of the ROC curve. *Med. Decis. Making* 9(3), 190–195.

Meistrell, M. L. (1990). Evaluation of neural network performance by receiver operating characteristic (ROC) analysis: Examples from the biotechnology domain. *Comput. Meth. Prog. Biomed.* 32(1), 73–80.

Metz, C. E., and Shen, J. H. (1992). Gains in accuracy from replicated readings of diagnostic images: Prediction and assessment in terms of ROC analysis. *Med. Decis. Making* 9(3), 190–195.

Miettinen, O. S. (1974). Proportion of disease caused or prevented by a given exposure, trait, or intervention. *Am. J. Epidemiol.* 99, 325–332.

Motulsky, A. G. (1988). Human genetic individuality and risk assessment. *In* "Phenotypic Variation in Populations: Relevance to Risk Assessment" (A. D. Woodhead, M. A. Bender, and R. C. Leonard, eds.), pp. 7–9. Plenum Press, New York.

Mullaart, E., Lohman, P. H. M., Berends, F., and Vijg, J. C. (1990). DNA damage, metabolism, and aging. *Mutat. Res.* 237, 189–210.

National Academy of Sciences (1991). "Human Exposure Assessment for Airborne Pollutants." National Academy Press, Washington, D.C.

Nunnaly, J. C. (1967). "Psychometric Theory." McGraw-Hill, New York.

Ozonoff, D. M., and Wartenberg, D. (1991). Toxic exposures in a community setting: The epidemiologic approach. *In* "Molecular Dosimetry and Human Cancer: Analytical, Epidemiological, and Social Considerations" (J. D. Groopman and P. L. Skipper, eds.), pp. 77–88. CRC Press, Boca Raton, Florida.

Perera, F. P. (1987). Biological markers in risk assessment. *Environ. Health Perspect.* 76, 141–146.

Perera, F. P. (1988). The significance of DNA and protein adducts in human biomonitoring studies. *Mutat. Res.* 205, 255–269.

Perera, F. P., Poirier, M. C., Yuspa, S. H., Nakayama, J., Jaretzki, A., Curnen, M. M., Knowles, D. M., and Weinstein, I. B. (1982). A pilot project in molecular cancer epidemiology: Determination of benzo[*a*]pyrene–DNA adducts in animal and human tissues by immunoassays. *Carcinogenesis* 3, 1405–1410.

Perera, F. P., Hemminki, K., Young, T. L., Brenner, D., Kelly, G., and Santella, R. M. (1988). Detection of polycyclic aromatic hydrocarbon–DNA adducts in white blood cells of foundry workers. *Cancer Res.* 48, 2288–2291.

Perera, F., Motzer, R. J., Tang, D., Reed, E., Parker, R., Warburton, D., O'Neill, P., Albertini, R., Bigbee, W. L., Jensen, R. H., Santella, R., Tsai, W. Y., Simon-Cereijido, G., Randall, C., and Bosl, G. (1991). Multiple biologic markers in germ cell tumor patients treated with platinum-based chemotherapy. *Cancer Res.* 52, 3558–3565.

Rappaport, S. M. (1988). Biological considerations for designing sampling strategies. *In* "Advances in Air Sampling Strategies" (W. John, ed.), pp. 337–352. Lewis Publishers, Chelsea, Michigan.

Rappaport, S. M. (1991). Assessment of long-term exposures to toxic substances in air. *Ann. Occup. Hyg.* 35, 61–121.

Rogan, W. J. (1988). Relation of surrogate measures to measures of exposure. *In* "Epidemiology and Health Risk Assessment" (L. Gordis, ed.), pp. 148–158. Oxford University Press, New York.

Rosner, B., Munoz, A., Tager, I., Speizer, R., and Weiss, S. (1985). The use of an autoregressive model for the analysis of longitudinal data in epidemiologic studies. *Stat. Med.* **4**, 457–467.

Rothman, K. J. (1986). "Modern Epidemiology." Little, Brown, Boston.

Schatzkin, A., Freedman, L. S., Schiffman, M. H., and Dawsey, S. J. (1990). Validation of intermediate endpoints in cancer research. *J. Natl. Cancer Inst.* **82**, 1746–1752.

Schulte, P. A. (1987). Methodologic issues in the use of biologic markers in epidemiologic research. *Am. J. Epidemiol.* **126**, 1006–1016.

Schulte, P. A. (1989). A conceptual framework for the validation and use of biological markers. *Environ. Res.* **48**, 129–144.

Schulte, P. A., and Mazzuckelli, L. F. (1991). Validation of biological markers for quantitive risk assessment. *Environ. Health Perspect.* **90**, 239–246.

Schulte, P. A., Boeniger, M., Walker J. T., Schober, S. E., Pereira, M. A., Gulati, D. K., Wojciechowski, J. P., Garza, A., Froelich, R., Strauss, G., Halperin, W. E., Herrick, R., Griffith, J. (1992). Biological markers in hospital workers exposed to low levels of ethylene oxide. *Mutat. Res.* **278**, 237–251.

Slaga, T. J. (1988). Interspecies comparisons of tissue DNA damage, repair, fixation, and replication. *Environ. Health Perspect.* **77**, 73–82.

Smith, T. J. (1987). Exposure assessment for occupational epidemiology. *Am. J. Ind. Med.* **12**, 249–268.

Somoza, E., and Mossman, D. (1991). Biological markers and psychiatric diagnosis: Risk–benefit balancing using ROC analysis. *Biol. Psychiatry* **29**(8), 811–826.

Stevens, D. K., Bull, R. J., Nauman, C. H., and Blancato, J. N. (1991). Decision model for biomarkers of exposure. *Regulatory Toxicol. Pharmicol.* **14**, 286–296.

Suruda, A., Schulte, P., Boeniger, M., Hayes, R. B., Livingston, G. H., Steenland, K., Stewart, P., Herrick, R., Douthitt, D., and Fingerhut, M. A. (1993). Cytogenetic effects of formaldehyde exposure in students of mortuary science (submitted).

Swenberg, J. (1988). Banbury Center DNA Adduct Workshop (Commentary). *Mutat. Res.* **203**, 55–68.

Swets, J. A., Pickett, R. M., Whitehead, S. F., Getty, D. J., Schnur, J. A., Swets, J. B., and Freeman, B. A. (1979). Assessment of diagnostic technologies. *Science* **205**, 753–759.

Thompson, M. L., and Zucchini, W. (1989). On the statistical analysis of ROC curves. *Stat. Med.* **8**(10), 1277–1290.

Tosteson, A. N., and Begg, C. B. (1988). A general regression methodology for ROC curve estimation. *Med. Decis. Making* **8**(3), 204–215.

Umbenhauer, D., Wild, C. P., Montesano, R., Saffhill, R., Boyle, J. M., Hahn, K. Y., Thomale, J., Rajewsky, M. I., and Lu, S. H. (1985). O-6 Methyldeoxyguanosine in oesophageal DNA among individuals at high risk of oesophageal cancer. *Int. J. Cancer* **36**, 661–65.

Weiss, N. S., and Liff, J. M. (1983). Accounting for the multicausal nature of disease in the design and analysis of epidemiologic studies. *Am. J. Epidemiol.* **117**, 14–18.

Weston, A., Vineis, P., Caporaso, N. E., Krontiris, T. G., Lonergan, J. A., and Sugimura, H. (1991). Racial variations in the distribution of Ha-ras-1 alleles. *Mol. Carcin.* **4**, 265–268.

Wilcosky, T. C. (1993). Biological markers of intermediate outcomes in studies of indoor air and other complex mixtures, Health Effects Institute research report (*HEI*) (in press).

World Health Organization (1975). Early Detection of Health Impairment in Occupational Exposure to Health Hazards. WHO Technical Report No. 571. World Health Organization, Geneva.

World Health Organization (1986). Principles of toxicokinetic studies. *In* "Environmental Health Criteria" vol. 57, World Health Organization, Geneva.

Zeller, R. A., and Carmines, E. G. (1980). "Measurement in the Social Sciences." Cambridge University Press, Cambridge.

4

Technical Variability in Laboratory Data

Paolo Vineis,
Paul A. Schulte, and
Robert F. Vogt, Jr.

The main focus of this book is the use of laboratory methods to measure biologic markers in epidemiologic studies. The confidence an investigator has in the data, and the ultimate conclusions of a study using biologic markers, depend largely on the consistency of the laboratory analyses and instruments that produce them. Therefore, epidemiologists should be familiar with laboratory quality control procedures for assays that are included in their studies. This chapter presents an overview of assessing reliability in laboratory data and the sources of technical variability.

All measurements are subject to variation from a number of sources. For biologic markers, the main sources are (1) error in the act of measurement; (2) biologic changes from time to time in the same person; and (3) biologic differences among persons. The statistical analysis of biologic laboratory data aims to disentangle the role played by these different sources of variation, trying to separate the "error" attributed to the act of measurement from biologic properties of the observed subjects.

Technical variability is largely a function of instrumentation, reagents, and human error in sample labeling, preparation, and test performance (Stites, 1991). Lack of standardization of procedures and calibrators also can contribute to variability among laboratories (Santella *et al.,* 1988). The contribution of technical variability should be assessed when analyzing the overall variability observed among and within study populations, so the inherent biologic variability within and among individuals can be discerned from true differences among populations. Many analytes, especially hormones and proteins, show cyclical rhythms that can be circadian, monthly, or seasonal in nature. Knowledge of these changes over time is vital to the collection of specimens at appropriate times, to the design for repeat collection, and to the statistical evaluation of the data (Fraser and Harris, 1989). To allow true assessment of biologic variability, assay procedures must be sufficiently robust and free from short-term fluctuations and long-term drift (Bates *et al.,* 1991).

This chapter briefly reviews the general principles of laboratory quality control and assessment of technical variability, and discusses several aspects of epidemiologic interpretation of laboratory results. The chapter concludes with several pragmatic examples that apply these principles, covering a wide range of laboratory methods from comparatively simple techniques employed in routine clinical settings to advanced research techniques that are not yet fully evaluated.

Laboratory Quality Assurance and Control

General Principles of Analytical Variability

Analytical variability may be random, which causes imprecision, or nonrandom, which causes bias. Variability among results generally increases with the analytical interval, that is, the difference in time and location between two assays (Table 4.1). Variation caused by the act of measurement may be attributed to the instrument or the assay or to the observers making the measurement. Seemingly innocuous differences in practices or procedures can contribute to variability. For example, Figures 4.1 and 4.2 and Table 4.2 show the influence of cytopreparatory techniques on measurement of nuclear size in various types of cells.

Imprecision and Bias

Imprecision

Although some random variation accompanies all measurements, undue imprecision generally indicates some inherent weakness in the assay meth-

TABLE *4.1*   Temporal and Locational Sources of Variability in Laboratory Results

Relative degree of variability	Comparison
Minimal ↓ Maximal	1. samples in the same analytical run 2. samples in different analytical runs on the same day 3. samples on different days 4. samples in different laboratories 5. samples with different methods

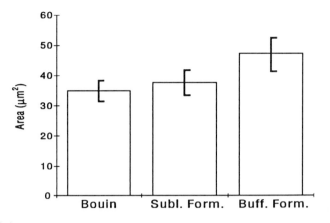

FIGURE *4.1*   Mean and SD of nuclear area of tissue fixed in Bouin's fluid, mercury–formalin (sublimed formalin), and 4% formaldehyde (pH = 7) (buffered formalin). (Reprinted with permission from Baak *et al.*, 1989, in Fleege *et al.*, 1991. Copyright © 1991 by Springer-Verlag.)

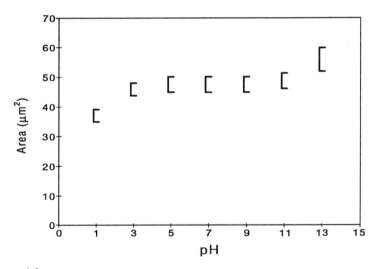

FIGURE *4.2*   Acidity of 4% formaldehyde affects nuclear area, but for 5<pH<9, no significant influences are observed on the mean and SD of the nuclear area of guinea pig liver slices. (Reprinted with permission from Baak *et al.*, 1989, in Fleege *et al.*, 1991. Copyright © 1991 by Springer-Verlag.)

odology, and lowers predictive value by blurring the distinction between true differences in results (Figure 4.3). The smaller the true differences among distributions in the populations tested, the more detrimental is the effect of imprecision on predictive value. As a rule, reliable laboratory tests for continuous variables should have maximal coefficients of variation (CVs) no greater

TABLE 4.2   Influence of Cytopreparatory Techniques on Normal Bladder Cells

Cell processing technique	Nuclear area N (μm^2)	Cellular area C (μm^2)	N/C
Air-drying, May–Grünwald–Giemsa	95	277	0.37
50% ethyl alcohol, Papanicolaou	49	176	0.30
96% ethyl alcohol, Papanicolaou	32	114	0.29
Spray fixative, Papanicolaou	40	131	0.31
No fixation, no staining	60	208	0.34

Source: Modified from Beyer-Boon *et al.*, 1979, in Fleege *et al.*, 1991. Reprinted with permission. Copyright © 1991 by Springer-Verlag.

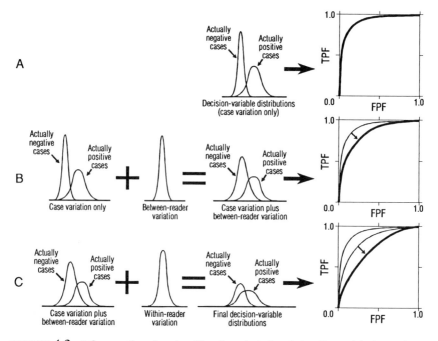

FIGURE 4.3   Influence of random (nonbiased) analytical variation (imprecision) on a hypothetical test performance. (A) Distribution and receiver–operator characteristic (ROC) curve of true values for a decision variable (e.g., a laboratory test result) as a function of the true health end point. Some overlap in test results exists between the actual negative and actual positive cases, but, at the optimal decision variable threshold, about 90% of the true positive cases (TPF) can be identified while including only about 10% false positive cases (FPF). (B) and (C) The addition of analytical imprecision blurs the distinction between test results, creating more apparent overlap between negative and positive cases. The degree of overlap increases with increasing imprecision. The degradation in the efficacy of the test with increasing analytic variability is apparent from the ROC curve (see Chapter 3). At the greatest level of imprecision, the optimal decision variable threshold identifies only about 60% of the true positive cases (TPF) and includes nearly half the false positive cases (FPF). (Reprinted with permission from Metz and Shen, 1992.)

than 10%, that is, the standard deviation of replicates should not exceed 10% of the mean value. The best assays will have low CVs even when results from different laboratories are compared.

For noncontinuous variables, such as histopathologic interpretation, imprecision is the extent to which different readers place the result in different categories. Example 5 discusses this situation in detail.

Bias

Nonrandom variation is generally more serious than random variation, since it can produce the false impression of biologic differences that are actually caused by bias. Much less likely is its ability to mask a true difference through a compensatory bias. A small degree of bias in different methods is common and often is disregarded in clinical assays used for primary medical care. However, bias modifies the sensitivity and specificity of a test and can decrease its cost effectiveness (Figure 4.4). Constant bias definitely can have epidemiologic significance, since it can make the valid comparison of results from different study populations or different study times difficult.

A particularly insidious source of analytical bias is assay drift, in which a laboratory method gives increasingly lower or higher results with time.

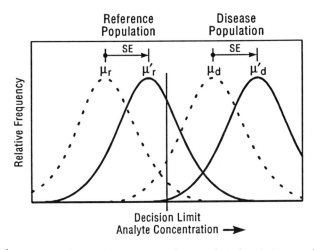

FIGURE **4.4**   Influence of nonrandom (constant bias) analytical variation on a hypothetical test performance. Unbiased assay results are represented by the dotted curves, associated with average values μ_r and μ_d. The vertical line represents the optimized decision limit for the test; results above the limit are considered abnormal. A constant systematic error (SE) shifts all assay results and their respective means to the right (solid curves). The specificity and sensitivity of the test will be altered if the decision threshold based on unbiased data is applied to the biased data. Because the decision threshold was optimized with unbiased data, the biased data will give suboptimal results. Moreover, the bias will suggest a significant difference in the distribution of results, leading to the false impression of biological effects. (Reprinted with permission from Channing-Rodgers, 1987.)

Drift can occur in the same method used in the same laboratory, especially for assays that are not in common use or do not have stable control material. Assay drift can lead to artifactual differences among populations if the specimens are not analyzed at the same time (Figure 4.5). The best insurance against drift is monitoring several control analytes during the course of the

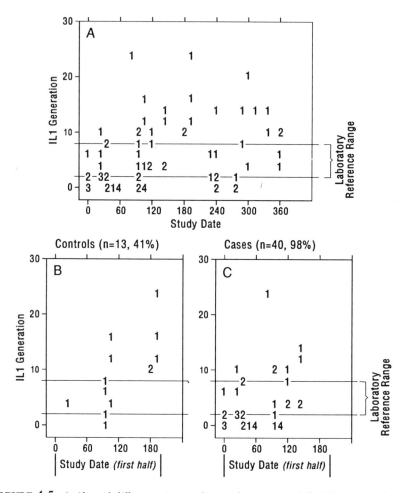

FIGURE 4.5 Artifactual differences in populations due to assay drift. When results for an immune function test (Interleukin-1 generation) were compared between a test and a control group assayed over a 1-year period, the mean value for the test group was significantly lower than that of the control group. However, 98% of the test subjects were assayed in the first half of the study, whereas only 41% of the control subjects were assayed during that time. Further analysis showed that the difference between groups could be accounted for entirely by an upward drift in the assay during the course of the study (A). No significant difference was seen between control (B) and test (C) subjects assayed in the first half of the study. Worth noting is the fact that this artifactual difference due to bias was uncovered by the epidemiologists and not by the laboratory scientists. (Adapted from Daniell, 1991.)

study, randomizing specimen collection and analysis among the different populations in a study, and matching by time elapsed between collection and analysis. Unfortunately, these preventive measures are not always feasible.

Evaluation and Control of Analytical Error

The results of most laboratory assays for continuous variables are obtained using an instrument to measure the strength of a signal created by physical or chemical properties of the analyte. In most cases, the signal strengths are converted into units of size, mass, or functional activity by comparison with the signal strengths obtained from standards or calibrators that have "known" values for the analyte. The relationship between the signal strength and the calibrator value produces the standard curve; results for unknown samples may be obtained by interpolating values of their signal strengths onto the standard curve. The degree of error in the final results depends on imprecision (random variability), which influences all measurements, and several additional factors that may not be random and could cause bias:

1. changes in the samples during storage or preparation for the measurement
2. changes in the way the instrument measures signal strength
3. uncertainty in the relationship that determines the standard curve
4. uncertainty in the true values of the standards
5. differences in the software used to generate results

Good laboratory practice requires a quality assurance/quality control (QA/QC) program that minimizes the error in the final results. Such programs should include:

1. insuring analyte integrity during storage and preparation
2. instrument calibration
3. reviewing mathematical parameters of the standard curve
4. measuring primary or secondary reference material for the analyte
5. multiple measurements on the same sample ("split" samples)
6. measuring samples on which historical results have been obtained within the laboratory
7. measuring samples distributed among different laboratories, that is, surveys and proficiency tests
8. active training programs for the technicians performing the tests, and for clinicians or epidemiologists interpreting them

Some laboratory tests (e.g., histopathologic determinations) are based partially or entirely on human judgment rather than on readings from instruments. Although the same general principles of QA/QC apply in all cases, such tests require approaches that also can assess the contribution of the subjective human component to the overall variability of the results.

Reliability

Reliability encompasses both unsystematic random variation observed from repeated measurements and bias caused by nonrandom variation (Carmines and Zeller, 1979; Massey, 1986; Last, 1988). If he or she has no confidence in the reliability of a marker, an investigator cannot be sure that results obtained in comparative studies are genuine individual or group differences. In the measurement of continuous variables, as with most biologic markers, errors of various kinds are inevitable and true value of the measurement can never be determined (Massey, 1986). If a measure of a biologic marker yields results that differ markedly from one occasion to another, the marker is of little value in epidemiologic research. Thus, reliability of a marker assay must be assessed prior to studies of its validity.

Quantitative indices may help establish the extent of random variation of a biologic marker. These indices can be used to determine whether the reliability of a given measure is sufficient for the purpose being considered. The two most common indices are the standard error of the measurement and the reliability coefficient (Massey, 1986). To assess random errors, multiple measurements are needed to compensate for the fact that the random error in the arithmetic mean of several measurements is likely to be much less than the random error in an individual measurement (Massey, 1986). Multiple measurements should be made at several points. First, multiple analytical measurements of the same biologic specimen (or several biologic specimens from the same individual at the same point in time) must be made to estimate the variability due to random analytical errors. This fraction of the total random variability of a biologic marker is usually minor. Second, multiple measurements of a marker must be made for one individual over time to estimate the intraindividual temporal variability. Third, multiple measurements across different individuals must be made to estimate interindividual variability in the value of the marker. In the second and third cases, random errors are a part of the variability but systematic errors also may contribute to the variability (e.g., circadian cycles in the value of a marker or differences among individuals due to genotype). Most epidemiologic research using biologic markers seldom requires large numbers of individual measurements. Thus, a small number of individuals can be used as a sample of the infinitely larger population to which the distribution refers. The standard error indicates how the mean of that sample is distributed around the mean of the larger population. Hence, the standard error of the mean reflects the reliability of the sample mean as an indicator of the population mean (Massey, 1986). This value may not be as informative as the reliability coefficient for evaluating markers to be used in epidemiologic studies.

The reliability coefficient is technically known as the intraclass coefficient of reliability (Shrout and Fleiss, 1979; Fleiss, 1986) and ranges from 0 to 1. If each measurement is identical, the intraclass coefficient is 1.0. The greater the variation among measurements, the lower the reliability. Fleiss

(1986) has evaluated the impact of unsystematic variation in measurement, described the untoward consequences of unreliability, and recommended how unreliability can be controlled. The untoward consequences described by Fleiss (1986) include the need to increase sample sizes to reduce unreliability, the high rates of misclassification in studies of the association between exposure and disease, and the consequent underestimation of the association between a health measure and the measured extent of exposure to an environmental risk factor. All these factors pertain to studies using biologic markers of exposure or effect. Fleiss (1986) recommends that unreliability be controlled by conducting pilot studies and replicating measurement procedures on each study subject.

In some cases, the measurement of the amount of a marker is not an end in itself, but is used to calculate some other value, thereby propagating measurement errors (Massey, 1986). Since "correct" values from measurements generally are never known, calculations will involve errors. Thus, it is useful to know how errors in individual measurements affect the results of subsequent calculations (Massey, 1986). For example, individual errors in a sum or difference of measurements are added; standard errors are combined with the root sum of squares (Massey, 1986). Acknowledgment of these calculation errors should be included in studies and in subsequent risk assessments. When such errors become significant, appropriate adjustments should be made.

Components of Reliability and Their Assessment

Reliability can be seen as having two components: reproducibility and repeatability. Reproducibility of an assay or a test is a measure of the extent to which the same test result (with an acceptable variation due to random error) will be obtained for the same analysis on the same sample when the test is performed in different laboratories. Repeatability is the measure of the extent to which the same test result will be obtained for the same analysis on the same sample when the test is performed several times in the same laboratory. When measuring reproducibility and repeatability, the concern is not with false positives or negatives to assess how often the test succeeds or fails, but only with agreement or disagreement *between* observers (or laboratories) or *within* the same observer (or laboratory) for the categorical or continuous laboratory result. In contrast with *validity,* there is no standard for comparison with the test results. The methods for measuring reproducibility and repeatability will be described in Example 5.

Repeated Measurements in the Same Individual

Regression to the mean   Originally, the concept of regression to the mean referred to the change of a measurement from one generation to the next. Galton (1886) noted that tall fathers tended to have tall sons but the sons, although taller than average, tended to be less extreme than their fathers. In Galton's example, the variables (x and y) were the heights of fathers and their sons; more frequently, however, the same measurement is carried out on a single individual on two different occasions. Apart from assessing repeatability, regression to the mean occurs when the second measurements are, on average, less extreme than the first. Consider an intervention trial in a population that has been treated with a special diet to lower the level of cholesterol in the blood. If the measurements are repeated in the same individuals after the trial, they may be lower than before, not because the trial is effective but as a consequence of regression to the mean. This effect is particularly important when cholesterol is measured only in a population, than in those individuals who have a value higher than a certain threshold, and subsequently in the same individuals as a remeasurement. The recognition and treatment of regression to the mean in such a situation requires a statistic model, as depicted in Table 4.3.

Regression dilution bias and how to correct it   Regression to the mean can have a diluting effect on an association (Comstock *et al.*, 1992). The concept of regression dilution bias has been illustrated clearly in a paper concerning a study on diastolic blood pressure (DBP) and the risk of coronary heart disease (CHD) (MacMahon *et al.*, 1990). The bias consists in systematic underestimation of the association between CHD and DBP because of regression to the mean in repeated measurements of DBP (Table 4.4).

TABLE 4.3   **Model for Regression to the Mean: Cholesterol Measurements**

$$X_i^T = x_i^T + e_i^T$$

where X represents the value of the variable measured in subject i at time T, x is its "true" value, and e is the random (measurement) error. The regression to the mean (R) is estimated as follows (Davis, 1976).

$$R = \lambda \, (\overline{X} - \mu)$$

where λ is the coefficient of error (the variance of the variable e, expressing random measurement error, divided by the variance of the variable that actually is measured, X), μ is the mean value of the variable at the first measurement in the entire population, and $\overline{X}$ is the mean of the variable X at the second measurement in the selected subgroup.

   For example, a hypothetical trial might target for treatment within a population those individuals with levels of cholesterol higher than 300 mg/dl. Their mean value at second measurement is 315.7, whereas the mean value and variance in the entire population (at first measurement) were, respectively, 217.3 and 2169. The mean and the variance in the entire population are used as estimates of μ and of the variance of X, respectively. The variance of e (measurement error) is 513.8, therefore $\lambda = 513.8/2169 = 0.24$. R will be $0.24 \times (315.7 - 217.3) = 23.6$ mg/dl, the estimated value of the regression to the mean.

TABLE 4.4   **Regression Dilution Bias in the Framingham Study**

Consider the data reported in the following table:
Mean DBP at Baseline (First Measurement) and at Surveys 2 and 4 Years Later in 3776 Men and Women (Framingham Study)

DBP category at baseline (mmHg)	Baseline	2 years later	4 years later
1. <80	70.8	75.7	76.2
2. 80–89	83.6	83.0	83.9
3. 90–99	93.5	91.2	91.3
4. 100–109	103.4	99.2	98.5
5. 100+	116.4	107.3	104.7

(From MacMahon *et al.,* 1990.)

The table clearly shows that subjects who had low blood pressures at first measurement tended to have higher values in the following measurements, whereas the opposite happened with subjects in the highest DBP categories; that is, both groups tended to the mean values in repeated measurements. Since the baseline measurement was used to classify the subjects to estimate the risk of CHD in the follow-up, the consequence of this regression-to-the-mean phenomenon will be an underestimation of pressure values in the lowest categories and an overestimation in the highest categories, entailing a "flattening" of the curve associating the risk of CHD with DBP.

MacMahon *et al.* (1990) have proposed two methods for correcting the regression dilution bias. The first is based on the observation that the difference in mean DBP between the top and bottom categories is about 60% greater for baseline DBP than for DBP remeasured 4 or more years later. Therefore, the authors of the paper corrected the slope of the DBP/disease association, increasing it by 60%. The second procedure consists of multiplying by 1.6 the log odds ratios obtained from a regression analysis of the effect on CHD of different levels of DBP.

Examples of Laboratory Variability for Biologic Markers Used in Epidemiologic Studies

The range of issues pertaining to analytical variability are illustrated in the following examples:

1. CD4 lymphocyte counts in peripheral blood
2. serum immunoglobulin levels
3. interlaboratory variability in assays for DNA adducts
4. serum lipid fraction measurement in correlated exposures
5. variation in the histopathologic determination of cancer
6. genetic analysis by polymerase chain reaction

Example 1: CD4 Lymphocyte Counts in Peripheral Blood

The number of CD4 lymphocytes in the peripheral blood is the key laboratory marker used to assess the status of HIV-infected persons (see Chapter 12). Three different laboratory determinations contribute to determination of the final result:

1. total white blood cell count (WBC)
2. percentage of white cells that are lymphocytes
3. percentage of lymphocytes that are positive for the CD4 surface marker

Total WBC is determined by automated methods that are relatively consistent within and among properly calibrated machines, although some small instrument-specific biases may exist. The percentage of the WBC that is lymphocytes is less well standardized. The reference method is a manual differential count on a dried blood smear, in which human judgment is used to classify lymphocytes and other types of white cells. Manual differential counts tend to be imprecise, since only 100 cells are classified in the usual clinical settings and they may be subject to bias from different analysts (Keopke and Ross, 1985). Automated differential counters are generally more precise since they analyze many more cells. However, the automated counters classify WBC on the basis of physical or chemical properties, and their software is critical in identifying these properties and assigning the appropriate classification. Instrument-specific (as well as software-specific) biases therefore may be present. The percentage of lymphocytes positive for CD4 usually is determined on an entirely different instrument, the fluorescence flow cytometer. The result often involves automated objective and manual subjective contributions, each with considerable dependence on software (Margolick and Vogt, 1992).

An evaluation of results for percentage lymphocytes from an automated cell counter provides a good example of the distinction between random variability and bias (Figure 4.6). When results from fresh blood on the automated counter were compared with results from a 200-cell manual count by an experienced technologist, the differences were small and random; they could be explained readily by inherent counting error in the manual method (unpublished data). However, when the same blood samples were analyzed by the automated counter after overnight storage, they showed a definite bias averaging 20% below the original result. If blood samples older than 20 hr from HIV-infected persons were analyzed by this method and the results applied to calculating the total CD4 counts, they would give the false impression of a less serious illness than would fresh samples analyzed the same way.

Example 2: Serum Immunoglobulin Levels

Serum immunoglobulins (Igs or antibodies) are large protein molecules with three major divisions: IgG, IgA, and IgM (see Chapter 16). They generally are measured by immunochemical reactions that give only an indirect indi-

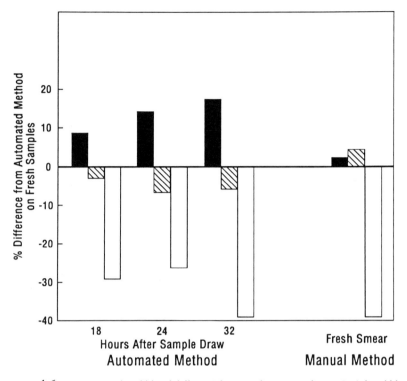

FIGURE 4.6 Bias in peripheral blood differential counts due to sample age. Peripheral blood lymphocyte counts using an automated method were performed on 32 fresh blood samples and again at 18, 24, and 32 hr later. Results from later analyses showed a significant increase in the percentage of lymphocytes (solid) and a concomitant decrease in the percentage of monocytes (open) and neutrophils (hatched). The total leukocyte count did not differ significantly (not shown) and the reference manual method gave comparable results for the lymphocytes and neutrophils as the automated method on fresh tissue.

cator of concentration and must be compared with calibrators to convert to gravimetric (mass) units. An international effort to standardize these measurements was initiated in the late 1970s, when several serum pools were prepared, exchanged between laboratories, and assessed for variability and relative potency (Reimer *et al.*, 1978). Mass units for the Igs and certain other proteins later were assigned to one of these pools, the U.S. National Reference Preparation, and further interlaboratory studies established consensus values with CVs of 9.7%, 9.6%, and 15.4% for IgG, IgA, and IgM, respectively (Reimer *et al.*, 1982).

When the U.S. National Reference Preparation was assayed during an instrument evaluation (study in progress at the U.S. Centers for Disease Control), a significant difference between the results and the 1982 consensus values was apparent (Figure 4.7). Further investigation suggested two reasons for the discrepancies. First, the initial differences in relative potency between

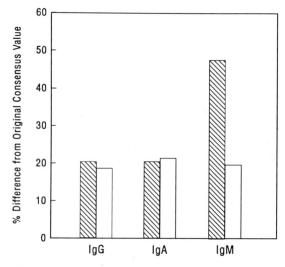

FIGURE 4.7 Differences in serum immunoglobulin levels caused by biased calibration. When a reference preparation was analyzed by a particular method for serum IgG, IgA, and IgM levels, the results were significantly higher than expected. Review of the literature showed that most of the bias for IgG and IgA could be accounted for by a difference in calibrators reported in 1978 but disregarded by the developer of the method. The bias for IgM was much higher than originally reported, suggesting a further effect due to degradation of IgM in the calibrator used to standardize the method. Bias determined in 1992 (hatched); bias measured in 1978 (open).

the pertinent reference pools appear to have been disregarded when the manufacturer of the instrument assigned mass units to its calibrators; this disregard can explain all the discrepancies observed for IgG and IgA. Second, the calibrator used by the manufacturer appears to have degraded in relative potency for IgM. The effects of this degradation were additive with the effects of the original discrepancy (which also showed a lower relative potency), producing a positive bias for current results of over 50%. Since all the quality control material supplied with the system was based on biased calibration, it could not reveal the inconsistency. Only analysis of the primary reference material and comparison with other methods through proficiency testing surveys uncovered the discrepancy.

Example 3: Interlaboratory Variability in Assays for DNA Adducts

Santella *et al.* (1988) were among the first groups to examine the variability in quantitation of human samples from different laboratories involved in molecular epidemiologic research. They examined interlaboratory differences in immunoassay procedures for DNA adducts using antisera elicited against benzo[*a*]pyrene-diol-epoxide-I-modified DNA (BPDE-I-DNA). The assays included competitive enzyme-linked immunosorbent assays (ELISA) with color and fluorescence end point detection and ultrasensitive enzyme radio-immunoassay (USERIA) with a radioactive end point. During the course of the study, two standard *in vitro* BPDE-I-DNA samples modified to high (36 pmol/μg DNA; 1.2 adducts/10^2 nucleotides) and low (4.5 fmol/μg DNA; 1.5 adducts/10^6 nucleotides) levels were used by each laboratory.

All three laboratories used alkaline phosphate conjugate antisera; however, there were differences in type and source of plates, method of plate coating and dilution, source of conjugates, and types of substrate. The comparison showed that the antisera all were elicited against DNA modified to a great extent. Therefore, it was not surprising that the antisera detected adducts in the slightly modified DNA sample with lower efficiency than in the highly modified DNA samples. The discrepancy between highly and slightly modified samples varied between 1.4- and 11.2-fold, depending on the antiserum and the assay.

Example 4: Serum Lipid Fraction Measurement Error
in Correlated Exposures

Investigators have realized that the effect of measurement imprecision in correlated exposures can lead to serious distortion of relative risk estimates from epidemiologic studies (Mosteller and Tukey, 1977; Rosner *et al.*, 1990; Willett, 1990; Phillips and Smith, 1991). This discovery has implications for molecular epidemiologic research. Most molecular epidemiologic studies use continuous data. The problems of estimation due to strong correlations between continuous experiences (collinearity) in multiple logistic models have been studied, but the problem caused by collinearity becomes more acute if exposures are measured imprecisely (Phillips and Smith, 1991).

Using the effects of high density lipoprotein (HDL) cholesterol and triglycerides on the risk of coronary heart disease, Phillips and Smith (1991) assessed the bias in multivariate relative risk estimates due to imprecise measurement of exposures. They found that the degree of bias in relative risk estimates is highly dependent on the amount of measurement imprecision ascribed to exposures. To address the problem, they suggest that epidemiologists seek to prevent measurement imprecisions by improving the design of the studies rather than by trying to correct for the flaw in analyses. Thus, for some exposures, more than one measurement taken over a period of time may be necessary. Epidemiologists should seek study populations in which the correlations between exposures are smaller. These considerations clearly apply to epidemiologic investigations of diet.

Another example of the effect of measurement error on the relationship between a lipid biomarker and coronary heart disease is shown in Table 4.5. The effect of the measurement error also is carried over into assessments of prevention strategies that involve imprecise measurements of biologic markers.

TABLE 4.5 Effect of Measurement Error on the Relationship between Plasma Cholesterol and Fatal Coronary Heart Disease

		As measured	Corrected for measurement error with reliability coefficient of		
			0.75	0.50	0.25
Distribution of plasma cholesterol (mmol/l)					
Mean		5.15	5.15	5.15	5.15
Standard deviation		1.23	1.06	0.87	0.61
10th percentile		3.57	3.78	4.04	4.36
90th percentile		6.72	6.51	6.26	5.94
Dose-response relationship with coronary heart disease					
Overall 18-year CHD mortality (per 1000 men)		74.2	74.2	74.2	74.2
CHD mortality difference (per 1000 men) per mmol/l increase in plasma cholesterol		15.7	20.9	31.4	62.8
18-year CHD mortality (per 1000) in bottom decile		40.0	34.7	25.8	5.8
Population attributable CHD mortality (per 1000)[a]		34.2	39.5	48.4	68.4
Proportion of attributable mortality in top decile[b]		19.4%	19.4%	19.4%	19.4%
"High-risk" strategy 1: Identify and treat all in top decile of true distribution					
Reduction (per 1000) in 18-year CHD mortality assuming reduction of cholesterol with treatment of:	0.5 mmol/l	0.78	1.05	1.57	3.14
	0.75 mmol/l	1.18	1.57	2.36	4.71
	1.0 mmol/l	1.57	2.09	3.14	6.28
"High-risk" strategy 2: Identify and treat all from true top decile in top 25% of measured distribution					
Proportion of true decile identified		(100%)	89.9%	75.0%	62.8%
Reduction (per 1000) in 18-year CHD mortality per 1.0 mmol/l reduction of cholesterol with treatment[c]		1.57	1.88	2.36	3.94
"High-risk" strategy 3: Identify and treat all from true top decile in top 10% of measured distribution					
Proportion of true top decile identified		(100%)	62.9%	46.5%	38.4%
Reduction (per 1000) in 18-year CHD mortality per 1.0 mmol/l reduction of cholesterol with treatment		1.57	1.31	1.46	2.41
"Population-based" strategy: Shift entire distribution of cholesterol levels downward by lifestyle change					
Reduction (per 1000) in 18-year CHD mortality assuming reduction of all cholesterol levels by:	0.05 mmol/l	0.78	1.05	1.57	3.14
	0.1 mmol/l	1.57	2.09	3.14	6.28
	0.25 mmol/l	3.92	5.23	7.85	15.70

Source: Reprinted with permission from Strachan and Rose, 1991. Copyright © 1991 by Pergamon Press, Ltd.

[a] 18-year CHD mortality in the whole population minus 18-year CHD mortality in the lowest decile.

[b] Proportion of population attributable mortality occurring among men in the top decile.

[c] Average reduction among all those offered treatment (taking into account acceptance and compliance).

Willett (1990) has reviewed the methods for correcting epidemiologic measures of association for measurement error. Key in applying such corrections is attention to assumptions regarding the type of error (random or systematic) and the distribution of exposure variables. Statistical methods are available for correction of correlation and regression coefficients and relative risks (using categorical and continuous variables).

Example 5: Variation in the Histopathologic Determination of Cancer

Histopathologic determinations involve the appraisal of specimens by a cytopathologist or pathologist. The subjectivity in this appraisal can lead to variability in repeated determinations. The following hypothetical example illustrates some of the issues involved in use of biologic markers that are determined by methods that involve judgment of the investigator.

Seven laboratories have performed, on 45 subjects, a test that classifies individuals into five categories (normal, inflammatory disease, mild dysplasia, severe dysplasia, cancer; Table 4.6). Each laboratory has repeated the test, after a 2-year delay, on the same specimens. The results are presented in Table 4.6 (number of subjects classified according to the five diagnostic categories). What is the level of interlaboratory and intralaboratory agreement? Consider a simpler format of the same data, that is, only the distinction between cancer and noncancer according to laboratories A and C in the initial set of test results (Table 4.7):

TABLE 4.6

	Laboratory												
	A		B		C		D		E		F		G	
Repetitions	1	2	1	2	1	2	1	2	1	2	1	2	1	2
Normal	9	8	7	6	11	11	4	5	13	9	8	7	8	5
Inflammatory disease	7	13	6	5	14	15	15	15	10	7	8	7	12	12
Mild dysplasia	12	9	17	14	5	7	10	13	6	10	12	13	4	8
Severe dysplasia	3	12	8	14	12	10	11	7	10	16	13	16	10	11
Cancer	4	3	4	4	3	2	3	4	4	3	3	2	4	2
Technically unsatisfactory	0	0	3	2	0	0	2	1	2	0	1	0	7	7

TABLE 4.7

	Laboratory	
	A	C
Noncancer	41	42
Cancer	4	3
Total	45	45

Four cancers have been diagnosed by Laboratory A and three by Laboratory C. However, whether the 3 latter cancers are included in the 4 diagnosed by Laboratory A is not known. In fact, the following are four possibilities, all compatible with Table 4.7.

TABLE 4.7A

Laboratory C	Laboratory A		
	No	Yes	Total
No	38	4	42
Yes	3	0	3
Total	41	4	45

TABLE 4.7B

Laboratory C	Laboratory A		
	No	Yes	Total
No	39	3	42
Yes	2	1	3
Total	41	4	45

TABLE 4.7C

Laboratory C	Laboratory A		
	No	Yes	Total
No	40	2	42
Yes	1	2	3
Total	41	4	45

TABLE 4.7D

Laboratory C	Laboratory A		
	No	Yes	Total
No	41	1	42
Yes	0	3	3
Total	41	4	45

Tables 4.7A–D represent all the possible combinations of the data, given the observed totals in Table 4.7. Table 4.7A corresponds to minimum agreement (none of the cancer cases diagnosed by Laboratory A has been diagnosed by Laboratory C). Table 4.7D corresponds to maximum possible agreement. This kind of representation of the data is the only one that enables the estimation of reproducibility.

A simple way to estimate agreement is to compute the proportion of concordant positive diagnoses:

Table 4.7A   0/45
Table 4.7B   1/45
Table 4.7C   2/45
Table 4.7D   3/45

Another measure is the proportion of total agreement:

Table 4.7A   38/45
Table 4.7B   40/45
Table 4.7C   42/45
Table 4.7D   44/45

The κ Statistic in a Two-by-Two Table

An important limitation of the two methods for computation of agreement is that they do not account for chance. Although in 3 of the 45 subjects the two observers agree on a cancer diagnosis (Table 4.7D), such a result could be obtained by chance rather than because of professional competence. For example, among the 45 cases, 41 might be clearly negative (with no possibility of disagreement), whereas the other 4 could be very difficult to interpret; when deciding about each of the latter, the two laboratories might choose to toss a coin and to diagnose cancer when the result is "heads." In this way, one would observe easily, by chance, three agreements for a positive diagnosis and one disagreement (as in Table 4.7D). A measure that incorporates chance agreement is the kappa statistic,

$$\kappa = (P_o - P_e)/(1 - P_e)$$

where P_o is the observed proportion of concordant diagnoses and P_e is the proportion of concordant diagnoses expected on the basis of chance.

The logic of the test is to compare the observed total agreement (e.g., 44/45 subjects in Table 4.7D) with the agreement expected on the basis of chance alone:

$$42 \times 41/45 = 38$$

rather than 41 and

$$4 \times 3/45 = 0.3$$

rather than 3. The numbers of concordant negative and concordant positive diagnoses to be expected if the two observers agree just by chance, given the marginals reported in Table 4.7, are 38 and 0.3, respectively. Since the test is based on proportions, compute

$$(P_o - P_e) = (44/45 - 38.3/45) = (0.98 - 0.85) = 12.7\%$$

which is the observed agreement in excess over agreement by chance. The value

$$(1 - P_e) = (1 - 38.3/45) = 15\%$$

represents the maximum agreement attainable with the given marginals. Therefore,

$$\kappa = 12.7/15 = 0.85$$

In other words, given the distribution of data reported in Table 4.7D, (38 + 0.3)/45 concordant diagnoses were anticipated as a consequence of chance only. The observed value of 44/45 corresponds to an excess of 12.7% over the expected value:

$$(44/45 - 38.3/45) = 12.7\%$$

However, the maximum possible agreement was not 100% because, given the marginals, the data revealed that cancers were 3 for Laboratory C and 4 for Laboratory A. The total possible agreement was, in fact,

$$(100\% - [38.3/45]) = 15\%$$

The proportion of this maximum value that is represented by the observed excess agreement is 0.85 (12.7/15).

How is this result interpreted? κ is a variable with values that are continuous and range between -1 and $+1$. A value of 0 means that there is no agreement over chance (as if the two observers tossed a coin); a value of -1 means that there is total disagreement (all cancers for Laboratory A are diagnosed as noncancers by Laboratory C and vice versa); a value of $+1$ means that there is total agreement. Intermediate values do not have an obvious interpretation. Rather than interpreting 0.85 as good or excellent agreement, on an absolute scale, the value should be used to make comparisons with other measurements. For example, if two previous surveys estimated κ of 0.99 and 0.93, the conclusion should be that the performance of the two observers who were tested is not as good as that of others in other laboratories.

There are, however, two important ancillary problems. First, with 4 and 3 cancers diagnosed by the two laboratories, there is limited room for a whole range of different values of κ (in fact, as shown before, only four different combinations of the data are possible with the given marginals). Second, κ is influenced strongly by the prevalence of the condition to be identified. Consider another set of data:

TABLE 4.7E

	Laboratory A		
Laboratory C	No	Yes	Total
No	37	2	39
Yes	0	6	6
Total	37	8	45

The difference between Table 4.7D and Table 4.7E is that the prevalence of cancer is twice as high. Now,

$$\kappa = (43/45 - 33/45)/(100 - 33/45) = 22/27 = 81.5\%.$$

If the distribution of the data is

TABLE 4.7F

	Laboratory A		
Laboratory C	No	Yes	Total
No	29	4	33
Yes	0	12	12
Total	29	16	45

that is, the prevalence of the condition is four times higher than in Table 4.7D,

$$\kappa = (41/45 - 25/45)/(100 - 25/45) = 35.5/44.5 = 79.8\%.$$

It is clear, therefore, that an increasing prevalence of the condition is associated with a decreasing value of κ. κ must be used with caution, always specifying the prevalence under consideration. In fact, different surveys are comparable, using the κ statistic, only if the prevalence of the condition is approximately the same.

κ for a Polychotomous Ordinal Variable

When the data are expressed in terms of an ordinal polychotomous variable, as in Table 4.8 (i.e., there are several discrete levels with a natural order, instead of a simple two-by-two table), κ is still a valuable statistic. Natural order is defined as the presence of categories ordered according to a meaningful sequence, such as normal, inflammatory condition, . . . cancer, as in Table 4.8. In other instances, there might be a polychotomous nominal variable, whose values (1, 2, 3, 4, 5) do not mean anything in terms of severity or progression of the disease. One further problem arises here. On many occasions, the difference between one category and an adjacent one is not "constant" and might not even be measurable. In the example reported in Table 4.8, the difference between "normal" and "inflammatory disease" does not have the same meaning, that is, the same clinical significance, as the difference between dysplasia and cancer.

A diagnostic mistake has worse consequences if it involves the second rather than the first pair of adjacent categories. Therefore, "weighted" κ statistic has been suggested as a measure of agreement that includes an adjustment for variable differences between categories. As a system of weights, the square of the derivation of the pair of observations from exact agreement has been proposed. For example, if a subject is classified as normal by Laboratory A and as "inflammatory disease" by Laboratory C, the deviation is 1 and its square is also 1; if a subject is classified as normal by Laboratory A and as cancer by Laboratory C, the deviation is 4 and the square is 16. In other words, this system assigns weights that increase exponentially with increasing deviations from the exact agreement.

The weighted κ is

$$\kappa = 1 - S_{ij}W_{ij}O_{ij}/S_{ij}W_{ij}E_{ij}$$

where S is the sum over cells, O_{ij} is the observed frequency in cell ij, E_{ij} is the expected frequency in cell ij, and W_{ij} is the weight assigned to cell ij. Consider the first measurements performed by Laboratory A and by Laboratory C, reorganized. The diagonal in Table 4.8 represents total agreement between A and C. The greater the distance from the diagonal, the more severe the error. Using the proposed system of weights, the value 1 is assigned to the first line parallel to the diagonal, 4 to the second, 9 to the third, and 16 to the fourth. Weighted κ is, therefore,

$$1 - (9+0+0+0+0+0+1\times1+6+0+0+0+1\times4+6\times1+5+0+0+0$$
$$+1\times4+0+12+0+0+1\times9+0+0+3)/[(9\times11/45)\times0+(9\times14/$$
$$45)\times1+(9\times5/45)\times4+(9\times12/45)\times9+(9\times3/45)\times16\ldots] = 0.61$$

TABLE 4.8[a]

Laboratory C	Laboratory A					
	Normal	I. D.	M. D.	S. D.	Cancer	Totals
Normal	9	0	0	0	0	9
I. D.	1	6	0	0	0	7
M. D.	1	6	5	0	0	12
S. D.	0	1	0	12	0	13
Cancer	0	1	0	0	3	4
Totals	11	14	5	12	3	45

[a] I.D., inflammatory disease; M.D., mild dysplasia; S.D., severe dysplasia.

TABLE *4.9*   Carcinogen–Hemoglobin Adducts

Subject	First laboratory	Second laboratory
1	23.7	24.5
2	15.8	11.2
3	76.0	89.6
4	113.1	99.5
5	125.6	133.0
6	9.8	14.2
7	167.4	144.3
8	23.9	26.5
9	84.4	72.7
10	50.0	43.3

Misuse of κ and Alternative Measures

A few reviews are available on the use and misuse of the κ statistic and on statistical methods for assessing agreement (Walker and Blettner, 1985; Bland and Altman, 1986; Maclure and Willett, 1987; Mazoyer and Mary, 1987; Thompson and Walter, 1988).

The same measures used for reproducibility may be used for repeatability as well. κ should not be used with continuous data (even when grouped); in this case, the alternative measure is the correlation coefficient. For example, Table 4.9 reports a hypothetical series of data concerning two different measurements, done in different laboratories, of the same variable in the same subjects (carcinogen–hemoglobin adducts).

Means are 68.9 for the first laboratory and 65 for the second; the correlation coefficient is 0.98, indicating high reproducibility. The appropriateness of a correlation coefficient is associated with normal distribution of the variables that are compared. A log transformation may be used to linearize nonnormal distributions; alternatively, a nonparametric correlation coefficient (e.g., Spearman) can be used.

When dealing with naturally ordinal data, an alternative to κ is the intraclass correlation coefficient (Snedecor and Cochran, 1978) which is less sensitive to changes in the number of categories (κ tends to decrease with the number of categories).

κ should not be used as a measure of validity, for which sensitivity, specificity, and predictive value are used when dealing with nominal data, and the mean and standard deviation of the difference are used when dealing with continuous data (see Chapter 3).

Example 6: Genetic Analysis by Polymerase Chain Reaction

The epidemiologist who employs highly sensitive assays in field studies needs to be familiar with the potential pitfalls, which can be quite significant.

Error-causing procedures can be introduced at most steps between specimen collection and reporting assay results. Even failure to control or standardize conditions prior to specimen collection, for example, whether the subject fasted, was using a medication, or had a disease, could influence results. The epidemiologist may pass to the laboratory colleague the responsibility of handling the performance and reporting of assays. Although the laboratory scientist should take the lead, the epidemiologist must be acquainted with possible sources of laboratory variation when considering data or comparing results from different laboratories or from the same laboratory over time. The example of the polymerase chain reaction (PCR) might be informative in alerting epidemiologists to the kind of problems that can arise.

PCR has become a widely used research technique in molecular biology because of its exquisite sensitivity and relative ease of application. (See Chapter 2 for a description of PCR.) PCR has been heralded as theoretically capable of amplifying DNA and, hence, identifying any mutations even in 1 cell in 10^6 cells (Lum, 1986). Because the primary extension products synthesized in one cycle can serve as a template in the next, the number of target DNA copies doubles at every cycle; thus, 20 PCR cycles yield a million-fold amplification (McCormick, 1989). This heightened sensitivity also can be the weakness of this technique, because it facilitates the possibility of false positives in assays (Kawasaki and Erlich, 1990; Martin, 1991).

Contamination resulting from "carry-over" from previous PCR reactions can lead to false positive assays. In fact, when performing PCR it is advisable to use separate rooms for initially processing specimens, setting up the assays, running the amplifications, and analyzing the PCR products (Martin, 1991). A panel of "blank" reactions with no template DNA is required to detect potential contamination. Negative and positive controls should be run with each amplification (Martin, 1991).

PCR is being used in a wide variety of applications. Thus, it is not possible to describe a single set of conditions that will guarantee success in all situations (Saiki *et al.,* 1989). The selection of efficient and specific primers remains somewhat empirical. Primers are key in determining the success or failure of amplification reactions. Saiki and colleagues (1989) have proposed the following guidelines that will help in the design of primers:

1. Where possible, select primers with a random base distribution and a GC content similar to that of the fragment being amplified. Try to avoid primers with stretches of polypurines, polypyrimidines, or other unusual sequences.

2. Avoid sequences with significant secondary structure, particularly at the 3′ end of the primer. Computer programs, such as Squiggles or Circles available from the University of Wisconsin, are useful in revealing these structures.

3. Check the primers against each other for complementarity. In par-

ticular, avoid primers with 3' overlaps to reduce the incidence of primer dimers.

Two major problems may occur in PCR assays used in epidemiologic studies (Martin, 1991). First, a lack of sensitivity even with known DNA is possible. This result may seem paradoxical, given the discussion of the sensitivity of the technique. However, the sensitivity can be reduced because of the instability of primer–template binding at the high temperatures necessary for optimal activity of the Taq enzyme. This problem may be addressed by lowering the annealing temperature to 37°C and prolonging the duration of the temperature rise to 72°C (Martin, 1991).

The second important problem encountered with PCR is extensive cross-priming, which can lead to the production of nonspecific products including primer dimers. These responses can be difficult to distinguish from a low level of a specific target sequence. Possible solutions to this problem include increasing the temperature of primer annealing, reducing concentrations of template, primers, enzyme, and dNTP, and using fewer cycles of shorter duration (Martin, 1991).

When sequencing target DNA, another technical artifact is the possibility of the creation of mosaic sequences in the PCR reaction. This artifact occurs when the sequence to be amplified is degraded into very small pieces. The mosaic alleles are created when incompletely extended fragments prime the extension of unintended target sequences, building a compound sequence. It is possible to minimize this outcome by amplifying smaller fragments and reducing the number of PCR cycles (Saiki _et al._, 1988; Paabo _et al._, 1988; King, 1991). This description of technical aspects of the PCR method clarifies how inherent technical problems can influence the performance of a test in population studies.

Conclusion

To summarize, a scientist should perform the following steps to evaluate the technical variability of a test he is proposing for use in populations: (1) assess the coefficient of variation in repeated measurements; (2) assess the reliability of the test, indicating the degree of both random and nonrandom error; (3) assess the two components of reliability, reproducibility and repeatability (e.g., κ statistic, intraclass correlation coefficient); (4) assess the extent of bias (nonrandom error) against random error; (5) with continuous measurements, assess the extent of regression to the mean and its consequences (regression dilution bias); (6) identify temporal changes (e.g., laboratory drift); (7) identify the sources of variability; and (8) correct the measurements for the effects of known sources of variability that are not of biologic significance or of interest to the researcher.

References

Bates, C. J., Thurham, D. I., Bingham, S. A., Margetts, B. M., and Nelson, M. (1991). Biochemical markers of nutrient intake. *In* "Design Concepts in Nutritional Epidemiology" (B. M. Margotts and M. Nelson, eds.), pp. 192–265. Oxford Medical Publications, Oxford.

Bland, J. M., and Altman, D. G. (1986). Statistical methods for measuring agreement between two methods of clinical measurement. *Lancet* 1, 307–310.

Carmines, E. G., and Zeller, R. A. (1979). "Reliability and Validity Assessment." Sage Publications, Beverly Hills.

Channing Rodgers, R. P. (1987). How much quality control is enough? A cost-effectiveness model for clinical laboratory quality control procedures (illustrated by its application to a ligand-assay-based screening program). *Med. Dec. Making* 7, 156–169.

Comstock, G. W., Bush, T. L. and Helzlsouer, K. (1992). Serum retinol, beta-carotene, vitamin E, and selenium as related to subsequent cancer of specific sites. *Am. J. Epidemiol.* 135, 115–121.

Davis, C. E. (1976). The effect of regression to the mean in epidemiological and clinical studies. *Am. J. Epidemiol.* 104, 493–498.

Fleege, J. C., van Diest, P. J., and Baak, J. P. A. (1991). Reliability of quantitative pathological assessments, standards, and quality control. *In* "Manual of Quantitative Pathology in Cancer Diagnosis and Prognosis. (J. P. A. Baak, ed.), pp. 151–161. Springer-Verlag, Berlin.

Fleiss, J. (1986). Statistical factors in early detection of health effects. *In:* "New and Sensitive Indicators of Health Impacts of Environmental Agents" (D. M. Underhill and E. D. Radford, eds.), pp. 9–16. University of Pittsburgh, Pittsburgh, Pennsylvania.

Fraser, C. G., and Harris, E. K. (1989). Generation and application of data on biological variation in clinical chemistry. *Crit. Rev. Clin. Lab. Sci.* 27, 409–437.

Galton, F. (1886). Regression towards mediocrity in hereditary stature. *J. Anthrop. Inst.* 15, 246–263.

Kawasaki, E. and Erlich, H. (1990). Polymerase chain reaction and analysis of cancer cell markers. *J. Natl. Cancer Inst.*, 82, 806–807.

Keopke, J. A., and Ross, D. W. (1985). White blood cell differential: A call for standards. *Blood Cells* 11, 1–10.

King, M. C. (1991). The application of DNA sequencing to a human rights problem. *In* "Molecular Genetic Medicine" (R. Friedman, ed.). Academic Press, San Diego.

Last, J. M. (ed.) (1988). "A Dictionary of Epidemiology." 2nd Ed. Oxford University Press, New York.

Lum, J. B. (1986). Visualization of mRNA transcription of specific genes in human cells and tissues using *in situ* hybridization. *BioTechniques* 4, 32.

Maclure, M., and Willett, W. C. (1987). Misinterpretation and misuse of the kappa statistic. *Am. J. Epidemiol.* 126, 161–169.

MacMahon, S., Peto, R., Cutler, J., Collins, R., Sorlie, P., Neaton, J., Abbott, R., Godwin, J., Dyer, A., and Stamler, J. (1990). Blood pressure, stroke, and coronary heart disease. Part 1. Prolonged differences in blood pressure: Prospective observational studies corrected for the regression dilution bias. Lancet 335; 765–774.

Margolick, J. B., and Vogt, R. F. (1992). Environmental effects on the human immune system and the risk of cancer: Facts and fears in the era of AIDS. *Environ. Carcin. Ecotoxicol. Rev.* C9(2), 155–206.

Martin, W. J. (1991). Polymerase chain reactions: A tool for the modern pathologist. *In* "Molecular Diagnostics in Pathology" (C. M. Fenoglio-Proiser and C. L. Willman, eds.), pp. 21–46. Williams and Wilkins, Baltimore.

Massey, B. S. (1986). "Measures in Science and Engineering." Ellis Horwood, Chichester, England.

Mazoyer, B., and Mary, J. Y. (1987). Kappa as an index of reproducibility: Distribution under the null-hypothesis. *Rév. Épidém. Santé Publ.* **35**, 474–481.

McCormick, F. (1989). The polymerase chain reaction and cancer diagnosis. *Cancer Cells* **1**, 56–61.

Metz, C. E. and Shen, J. H. (1992). Gains in accuracy from replicated readings of diagnostic images: prediction and assessment in terms of ROC analysis. *Med. Dec. Making* **12**, 60–75.

Mosteller, F., and Tukey, J. W. (1977). "Data Analysis and Regression. A Second Course in Statistics." Addison-Wesley, Philippines.

Paabos, S., Gifford, J. A., and Wilson, A. C. (1988). Mitochondrial DNA sequence from a 7000-year-old brain. *Nucleic Acids Res.* **16**, 9775–9787.

Phillips, A. N., and Smith, G. D. (1991). How independent are "independent" effects? Relative risk estimation when correlated exposures are measured imprecisely. *J. Clin. Epidemiol.* **44**, 1223–1231.

Reimer, C. B., Smith, S. F., Hannon, W. H., Ritchie, R. F., van Es, L., Becker, W., Murkowitz, H., Gauldie, F., and Anderson, S. G. (1978). Progress towards international reference standards for human proteins. *J. Biol. Stand.* **6**, 133–158.

Reimer, C. B., Smith, S. J., Wells, T. W., Nakamura, R. M., Keitges, P. W., Ritchie, R. F., Williams, G. W., Hanson, D. J., and Dorsey, D. B. (1982). Collaborative calibration of the U.S. National College of American Pathologists reference preparations for specific serum proteins. *Am. J. Clin. Pathol.* **77**, 12–19.

Rosner, B., Spiegelman, D., and Willett, W. (1990). Collection of logistic regression relative risk estimates and confidence intervals for measurement error: The case of multiple covariates measured with error. *Am. J. Epidemiol.* **132**, 734–745.

Saiki, R. K., Gelfund, D. H., Stoffe, S., Schard, S. J., Hiquchi, R., Horn, G. T., Mullis, K. B., and Erlich, H. A. (1989). Primer-directed enzymatic amplifications of DNA with semi-stable DNA polymerase. *In* "PCR Technology: Principles and Applications for DNA Amplification" (H. Erlich, ed.). Stockton Press, New York.

Santella, R. M., Watson, A., Perera, F. P., Trivers, G. T., Harris, C. C., Young, T. L., Nguyen, D., Lee, B. M., and Poirer, M. C. L. (1988). Interlaboratory comparison of antisera and immunoassays for benzo-[a]pyrene-diol-epoxide-I-modified DNA. *Carcinogenesis* **9**, 1265–1269.

Shrout, P. E., and Fleiss, J. L. (1979). Intraclass correlations: Uses in assessing rator reliability. *Psychol. Bull.* **86**, 420–428.

Snedecor, G. W., and Cochran, W. G. (1978). "Statistical Methods." Iowa State University Press, Ames.

Stites, D. P. (1991). Laboratory evaluation of immune competence. *In* "Basic and Clinical Immunology" (D. P. Stites and A. I. Terr, eds.) pp. 312–318. Appleton and Lange, Norwalk, Connecticut.

Strachan, D., and Rose, G. (1991). Strategies of prevention revisited: Effects of imprecise measurement of risk factors on the evaluation of "high-risk" and "population-based" approaches to prevention of cardiovascular disease. *J. Clin. Epidemiol* **44**, 1187–1196.

Thompson, W. D., and Walter, S. D. (1988). A reappraisal of the kappa coefficient. *J. Clin. Epidemiol.* **41**, 949–958.

Walker, A. M., and Blettner, M. (1985). Comparing imperfect measures of exposure. *Am. J. Epidemiol.* **41**, 949–958.

Willett, W. (1990). "Nutritional Epidemiology." Oxford University Press, New York.

5

Biologic Monitoring and Pharmacokinetic Modeling for the Assessment of Exposure

Pierre O. Droz

Introduction

Many of the lessons of molecular epidemiology regarding exposure and dose already have been learned during consideration of occupational exposures of workers. In this chapter, that experience will be drawn upon to provide some guidelines that might be useful to molecular epidemiologists. The causal link between exposure and disease is shown schematically in Figure 5.1. Exposure to chemicals, if no total protection is provided to the worker, may result in the absorption of a dose of chemical. This global dose then is distributed among different sites in the body, including the target site where toxic action will occur. The target dose eventually will trigger early biochemical changes that could lead to the development of clinically defined diseases. This chain of events rarely is measured itself, since many of its components are not accessible in epidemiologic studies. Therefore, current practice is to use various indirect tools. For example, as shown on Figure 5.1, monitoring of these events is carried out by three relatively distinct surveillance techniques:

1. air monitoring, which is an indicator of the exposure
2. biologic monitoring, which generally is chosen to measure the dose or the target dose indirectly
3. biochemical and medical surveillance, which is aimed at detecting and measuring health effects

Two additional parameters are shown in Figure 5.1. The relationships between exposure and dose, and between dose and early signs or disease, are modulated by specific characteristics of the workers. These modifiers can be, in some identified cases, detected with physiological or biochemical markers.

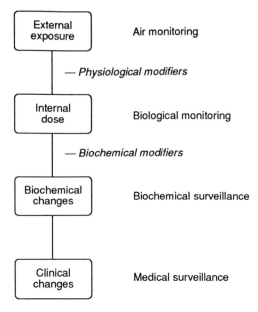

FIGURE 5.1 Schematic representation of the exposure-to-disease continuum with an indication of the various potential monitoring techniques and their objectives.

Three types of biologic measurements are not to be confused, because, although they have different objectives, they often appear quite similar:

1. biologic indicators of the dose
2. biochemical tests, used as a measure of early health effects
3. biochemical measurements, carried out to measure susceptibility of the individual

This chapter considers only biologic indicators of exposure or dose.

Possible Dose Surrogates and Their Characteristics

Exposure Routes

Exposure in industrial settings can occur by several routes. The major exposure route usually considered for chemicals and aerosols is inhalation. In this case, the magnitude of the exposure depends on two parameters: the concentration of chemical or aerosol in the air and the duration of exposure. Other parameters are nevertheless of importance. Physical workload influences the pulmonary ventilation of the worker, thus determining the amount of chemical presented to the body for deposition or absorption. At rest, a worker breaths about 4 m^3 of air for an 8-hr workshift; for a moderate physical workload (50 W), the amount will rise to 10 m^3; with a heavy workload

(100 W), the amount will be about 20 m³ (Droz, 1989). The intensity of physical activity also will affect the deposition of aerosols (Vincent, 1990) and the absorption of gases and vapors into the blood circulation (Åstrand, 1983).

Although gastrointestinal (GI) absorption usually is considered relatively unimportant in occupational exposure, it should not be neglected. A special case in which this route can play an important role is exposure to dust via inhalation. Particulates with large aerodynamic diameters deposit mainly in the ciliated regions of the lungs and are then cleared via the GI tract. GI absorption therefore depends on air concentrations, particulate distribution, pulmonary ventilation, and solubility for penetration into the bloodstream.

Skin exposure often is recognized but rarely is measured. Even an approximate idea of its relative importance with respect to air exposure often is not available in epidemiologic studies. The magnitude of exposure through the skin depends on the time of contact and on the surface of skin exposed. The status of the skin is also very important. Of course, different chemicals will penetrate the skin at different rates or fluxes. Limited experimental data are available for some chemicals. Estimation of these fluxes can be done by a simple model (Fiserova-Bergerova *et al.*, 1990).

$$\text{flux } [\text{mg/cm}^2/\text{hr}] = [C_{\text{sat}} (0.038 + 0.153 \ P) \ e^{-0.016\text{MW}}]/15$$

where C_{sat} is the water solubility in mg/ml, P is the octanol–water partition coefficient, and MW is the molecular weight. This simple model allows an approximate comparison of skin and air exposure for a given chemical.

Potential Biologic Indicators

Several types of biologic indicators of exposure can be distinguished. First, the chemical itself can be measured as a quantitative indication of exposure. Most metals fall into this category; blood or urine is sampled, depending on the metal, and an analysis of the total content of the metal is done (without considering the different chemical forms). Many organic chemicals also can be measured as such in blood, urine, or expired air if they are volatile. Limitations occur when the chemical is biotransformed in the body to a high extent. In this case, concentrations of the parent compounds remain low in body fluids. Therefore, it is often easier to measure the metabolic products. Finally, specific reaction products of the chemicals or their metabolites with macromolecules can be used as indicators of exposure. This approach has been considered especially for genotoxic chemicals.

These biologic tests can be considered specific to a chemical. In the case of exposure to mixtures, it is often of interest to have nonspecific tests, which are indicators of exposure to a group of substances with similar effects or at least similar metabolic pathways. Examples of the different potential biologic indicators are shown in Table 5.1.

TABLE 5.1   Examples of Types of Biological Indicators

Type of indicator	Examples
Chemical specific	
Parent compound	solvents in blood, expired air; metals in urine or blood
Metabolic product	solvent metabolites in urine; arsenic methylated species in urine
Adducts	genotoxic chemicals
Nonspecific	sister chromatid exchanges (SCE); mutagenic activity in urine; thioether urinary excretion; D-glucaric acid

Concepts of Dose

In epidemiologic studies, dose D refers generally to a combination of exposure C and duration T. In its simplest expression,

$$D = \Sigma_i \, C_i \, T_i$$

where i refers to the various homogeneous exposure periods. This approach can be refined to take into account, for example, that time and concentration do not have the same effect on the development of diseases (Checkoway, 1986).

The definition of dose just given is related to exposure only. Another more meaningful approach is the use of a physiological concept for dose definition (Smith, 1991). In this case, dose can take several different forms, for example,

1. amount of chemical absorbed by the body
2. amount of chemical at a target site
3. amount of chemical metabolized or activated

Also, time scale can be introduced in different manners. One can consider the total amount, the average amount, or the peak amount of the chemical. A further refinement would be to use the entire time dependence of the defined amount, by incorporating it into a pharmacodynamic model to obtain an "effect indicator" (Smith, 1991).

Dose is, therefore, not a simple and straightforward concept. Its meaning must be defined in each situation, using any relevant information, such as nature of the toxic effects investigated, exposure situation (history, variability), and monitoring tools (both air and biologic) available.

Importance of Kinetic Behavior of Biologic Indicators

Characteristics of Industrial Environments

Occupational exposure is a process that changes over time. Variations can be described in several categories.

1. The normal work schedule, in most cases is 8 hr potential contact with the contaminant, followed by 15 or 16 hr with no exposure. Further, the period of work can be split into two 4-hr blocks, separated by a 0.5–1-hr break. The workweek is usually 5 days long, followed by a 2-day weekend. This traditional work schedule has changed in recent years to accommodate production needs. There are, therefore, situations with longer work periods followed by longer breaks (Paustenbach, 1985).

2. During a work period of 4, 8, or more hours, exposure is not constant but depends on the production processes, work practices, and controls in operation. The fluctuations in exposure over time very often are approximated statistically by a lognormal distribution, defined by two parameters: the location of the distribution (arithmetic mean, $\bar{x}$; geometric mean, GM) and the variability (variance, s^2; coefficient of variation, CV; geometric standard deviation, GSD) (Rappaport, 1991). It is important to realize that all industrial environments are not lognormal. For example, when exposure is very intermittent, concentrations at or below the limit of detection will be frequent (Hornung and Reed, 1990). Also, when exposure occurs as incidents (leaks, spills), only large peaks of exposure will occur, which is difficult to accommodate with a lognormal distribution (Herrera *et al.*, 1991).

3. Exposure from one day to the next fluctuates for the reasons listed in 2. Therefore, the exposure of a worker probably varies from one day to the next, and can be described by a lognormal distribution also (Rappaport, 1991). Other factors of variations observed in daily exposures are production trends, changes in technology, seasonal trends, and control measure modifications. These parameters must be distinguished from those described earlier. They are not to be considered random factors; they produce systematic changes in exposure, which is no longer stationary. When determining exposure, these factors must be identified and their effect on the relevant dose must be estimated. Pharmacokinetics can provide an important contribution in this aspect.

These sources of variability apply to an individual worker. Of course, when considering a group of workers, another factor comes into play: the between-worker variability. This factor is of interest when studying the kinetics of a contaminant in the body, but in epidemiology, groups of similarly exposed workers are often formed. The variability of exposure in this group often is mixed with the day-to-day variability of individual worker exposure. Analysis of variance has been used on several occasions to separate both sources of variability (Spear *et al.*, 1987).

The magnitude of the variations covers a wide range: GSDs as small as 1.2 were reported for some environments, but extremes as high as 9 can be found (Spear *et al.*, 1987). It is important to realize that variability usually can be reduced by splitting observations into groups with identified differences. Groups can be distinguished by workers, but also by days of the week, seasons, or other characteristics.

Kinetics of Biologic Indicators

Contaminants in body fluids and organs and, more specifically, biologic indicators can be highly affected by fluctuations in exposure. In fact, the influence of these fluctuations depends on the kinetic behavior of the biologic indicator or the contaminant. This behavior often is described by the half-life, which is the time it takes for the contaminant concentration in the body to decrease by a factor of two. That definition usually is restricted to chemicals that appear to be distributed homogeneously in the body; in this case, the concentration can be described by a one-compartment model. Formally,

$$C = C_0 e - kt$$

with

$$t_{1/2} = (\ln 2)/k$$

where C is the concentration at time t, C_0 is the concentration at time 0, k is the first order rate constant, and $t_{1/2}$ is the half-life. The larger the half-life, the longer it will take the biologic indicator or the contaminant to decrease when exposure stops, or to reach a steady-state when exposure is stable.

To illustrate the importance of half-life, Figure 5.2 shows the behavior of four hypothetical biologic indicators during a week of industrial exposure, that is, with a normal work schedule and constant exposure. Half-lives of the biologic indicators vary from 1 to 500 hr. The week shown in Figure 5.2 is at steady state: exposure occurred during the preceding weeks and concentrations do not change from one week to the next. The concentrations in the body of the indicator with the 1-hr half-life react very quickly to exposure changes and follow the exposure level almost permanently. For the chemical with the 5-hr half-life, concentrations rise more slowly, and do not go to 0 before the next exposure. Concentration reaches 0 by the beginning of the

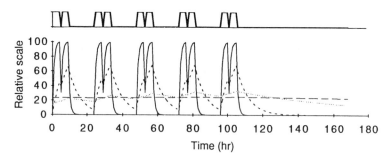

FIGURE 5.2   A week of occupational exposure: 8 hr/day, with 1-hr break, 5 days/wk. Top shows exposure periods. The behavior of four biological indicators or contaminants is shown with half-lives of 1 (–), 5 (– –), 50 (. . .), and 500 (——) hr.

following week. Concentrations in the body of the chemical with the 50-hr half-life change only slightly. Finally, for the 500-hr half-life, exposure fluctuations between work and rest periods do not affect the concentrations of the biologic indicator at all.

In the context of biologic monitoring, the behaviors described in Figure 5.2 are of utmost importance. The results of a biologic indicator will, for short half-lives, largely depend on the timing of sample collection with respect to exposure. For well-established biologic indicators—for example, Biological Exposure Indices (BEI) adopted by the American Conference of Governmental Industrial Hygienists (ACGIH)—recommended levels always are associated to a time of sampling, usually at the end of one shift or before the next shift. In most cases, information on the half-life of the biologic indicator is given also (ACGIH, 1991).

As discussed already, other types of fluctuations in exposure must be considered. Again, half-life is a very useful descriptor to understand how exposure variability will affect biologic indicator and contaminant levels in the body. Figure 5.3 shows several examples of possible fluctuations in exposure: completely random (3A), seasonal (3B), and seasonal with technology change (3C). In this figure, half-lives between 25 and 5000 hr are considered. With half-lives up to 50 hr, biologic indicators follow daily exposure fluctuations rather well and can be considered good indicators of current exposure. With half-lives greater than 500 hr, only large peaks of exposure have an effect on biologic levels; their magnitude is rather small compared to the background level. These indicators can, therefore, be considered dose, or average, exposure indicators.

Figure 5.3B shows that seasonal fluctuations in exposure will have a large effect on biologic indicators and contaminants with biologic half-lives shorter than 50 hr. For the other two examples, results obtained any time during the year give more or less the same information. From this figure, it is evident that the relationships between current exposure (for example, exposure during the summer months) and biologic indicators will be better for those with short half-lives. In this case, if estimation of the long-term average exposure is the objective, it is very important to take into account seasonal fluctuations in exposure, which requires biologic sampling for each identified season.

Figure 5.3C presents a sudden change in exposure conditions due, for example, to an improvement in controls or an improvement in the production process. The corresponding change in exposure is seen immediately in biologic indicators with half-lives of 25 and 50 hr. At the other extreme, the biologic indicator with a 5000-hr half-life is affected only marginally after several years. Again, when assessing exposure, these considerations are important, since some indicators will indicate current exposure preferentially whereas others can be used to investigate past conditions of exposure.

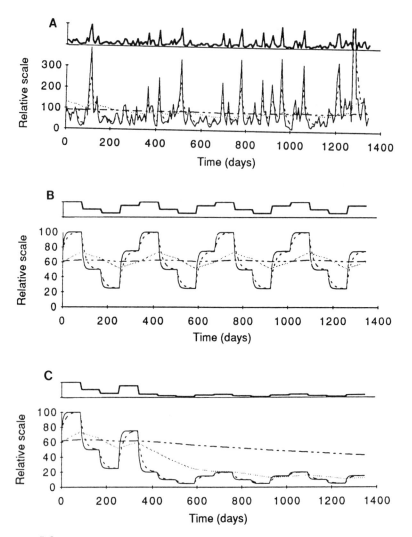

FIGURE 5.3 Influence of exposure variability on the behavior of four biological indicators or contaminants with half-lives of 25 (–), 50 (--), 500 (. . .), and 5000 (——) hr. Top shows the exposure periods and intensities. (A) Random daily exposure fluctuations according to a lognormal distribution of GSD 2.0. (B) Seasonal fluctuations with relative exposures of winter, 1.00; spring, 0.50; summer, 0.25; and autumn, 0.75. (C) Seasonal fluctuations with a sudden decrease in exposure by a factor of four at the beginning of the second winter; seasonal parameters are the same as in B.

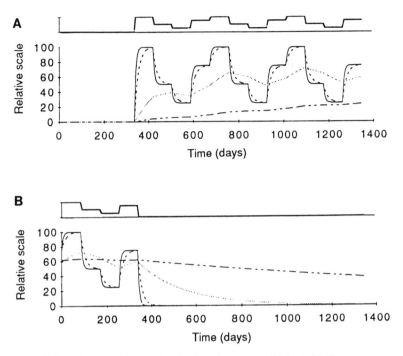

FIGURE 5.4   Influence of start and end of employment on biological indicators or contaminants with half-lives of 25 (–), 50 (--), 500 (. . .), and 5000 (——) hr. Top shows the exposure periods and intensities. Seasonal parameters are the same as in Figure 5.3B. (A) Employment starts after 1 year on the graph. (B) Employment ends after 1 year on the graph.

Figure 5.4 presents two situations often seen in industry: beginning of employment (4A) and termination of employment (4B). When exposure starts for a new employee, biologic levels rise very slowly only for chemicals with long half-lives. For such chemicals, there is a risk of underestimating the exposure. When employment stops, the opposite phenomenon arises. Biologic levels for chemicals with long biologic half-lives can stay elevated for many years.

Examples

Biologic half-life is a critical parameter to describe biologic indicators, but also describes the behavior of contaminants in the body. Table 5.2 gives examples of biologic indicators and their half-lives. Many chemicals do not follow first order kinetic models, because chemicals are distributed in multicompartment systems (see subsequent text). In such cases, the excretion rate will change as a function of time, depending on the relative burden of each compartment. As a consequence, the apparent half-life will be changing, in-

TABLE 5.2   Examples of Biological Half-Lives for Some Biological Indicators

Chemical	Determinant	Media	Timing	Half-life (hr)
aniline	total p-aminophenol	urine	end of shift	4
benzene	phenol	urine	end of shift	5.7
cadmium	cadmium	urine	not critical	20 years
	cadmium	blood	not critical	2400
carbon disulfide	2-thiothiazolidine-4-carboxylic acid	urine	end of shift	5
carbon monoxide	carboxyhemoglobin	blood	end of shift	5
	carbon monoxide	exhaled air	end of shift	5
dimethylformamide	N-methylformamide	urine	end of shift	4
ethyl benzene	mandelic acid	urine	end of shift	4
	ethyl benzene	exhaled air	prior to shift	48
n-hexane	2,5-hexanedione	urine	end of shift	15
	n-hexane	exhaled air	during shift	0.25
lead	lead	blood	not critical	900
	lead	urine	not critical	700
	zinc protoporphyrin	blood	after 1 month	500
methyl chloroform	methyl chloroform	exhaled air	prior to shift	32
	trichloroacetic acid	urine	end of workweek	72
methyl ethyl ketone	methyl ethyl ketone	urine	end of shift	4
organophosphorus	cholinesterase activity	red blood cells	—	700
parathion	p-nitrophenol	urine	end of shift	7
pentachlorophenol	pentachlorophenol	urine	prior to shift	700
perchloroethylene	perchloroethylene	exhaled air	prior to shift	96
	perchloroethylene	blood	prior to shift	96
	trichloroacetic acid	urine	end of workweek	80
phenol	phenol	urine	end of shift	3.5
styrene	mandelic acid	urine	end of shift	4
	styrene	exhaled air	prior to shift	20
toluene	hippuric acid	urine	end of shift	1.5
trichloroethylene	trichloroacetic acid	urine	end of workweek	75
xylenes	methylhippuric acids	urine	end of shift	3.6
alkylating agents	hemoglobin adducts	blood	—	3000[a]
	albumin adducts	blood	—	400[a]

Source: Reprinted with permission from Droz (1989).
[a] Half-lives of hemoglobin and albumin; half-lives of adducts may be lower.

creasing as a function of time after exposure. For simplicity, one can use the apparent half-life at the time the sample is taken. Table 5.2 indicates the sampling time considered.

Possible Pharmacokinetic Models

Objectives of Pharmacokinetic Modeling

As shown in the previous section, chemicals in body tissues and fluids are in constant kinetic change. Even when exposure concentration is stable, biologic levels fluctuate widely because the exposure is intermittent. Fluctuations in exposure make chemical concentrations even more variable. Finally, exposure is only one of the sources of variability: individuals are all biologically different, which will induce large changes in the relationships between biologic indicators and exposure or target-site burden. The consequence is that these fluctuations and variabilities will weaken the relationships between health effects and the dose exposure surrogate.

Pharmacokinetic modeling can help clarify these relationships and the influence of the main factors on them. Models consist of conceptual frameworks that help combine known information about the absorption, deposition, and elimination of chemicals. The models interesting for epidemiology are quantitative. They can be of varying sophistication: one-compartment, multicompartment or physiological. One- and multicompartment models divide the body into homogeneous compartment(s) without specific physiological meanings. On the other hand, physiology-based pharmacokinetic models use biologic concepts to separate the body tissues into compartments.

One-Compartment Pharmacokinetic Models

The simplest type of pharmacokinetic model is a one-compartment model (Figure 5.5A). The uptake of the chemical is proportional to the exposure concentration. Its distribution is only in one compartment; its elimination is via one route and is proportional to the concentration in the compartment. Although very simple, this model can represent many different situations: uptake of gases, solvents, or dusts, if it can be approximated by a simple proportion of the exposure concentration, and elimination by different routes, through urine, feces, expired air, and metabolism. The model also can be applied to metabolites, assuming that the transformation of the parent compound is fast enough.

All predictions in Figures 5.2–5.4 were obtained with a one-compartment pharmacokinetic model. Mathematically, it is a balance equation describing that, during a small time interval (dt), the amount taken up minus the amount excreted is equal to the change in the burden in the compartment (Droz *et al.*, 1991):

$$k_a \, C_{expo} dt - k_e \, C \, dt = V \, dC$$

where k_a is the first order rate constant for the absorption, k_e is the first order rate constant for the elimination, C is the concentration in the compartment of volume V, and C_{expo} is the exposure concentration. This differential equation can be integrated with given initial conditions. For example, the following useful equations can be derived (Roach 1966):

$$\text{during exposure } C = C_{expo} \, k_a \, (1 - e^{(-k_e t)})/k_e$$
$$\text{after exposure } C = C_o \, e^{(-k_e t)}$$

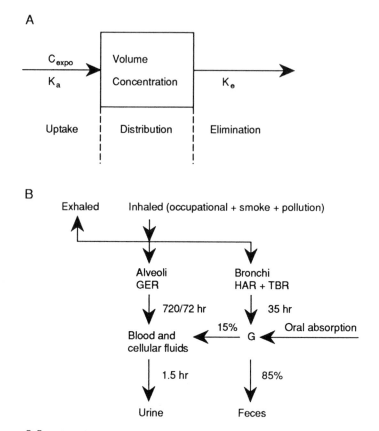

FIGURE 5.5 Schematic description of compartmental models. (A) One-compartment model. (B) Three-compartment model.

k_a and k_e are mathematical concepts. They also can be related to more physiological variables. k_a is dependent on pulmonary or alveolar ventilation and on pulmonary or alveolar retention. For metabolites, k_a also can contain the fraction of the parent chemical transformed into the specific metabolite. k_e is linked to the half-life by the relationship

$$t_{1/2} = (\ln 2)/k_e$$

k_e also can be expressed as a function of the volume of distribution, V, and the clearance, Cl

$$k_e = Cl/V$$

The volume of distribution represents, in principle, the fictitious volume in which the chemical would need to be distributed homogeneously to have the same concentration as in blood (or plasma).

Multicompartment Pharmacokinetic Models

One-compartment models are very useful, mainly for their simplicity, but when more accurate descriptions and predictions must be made, other more complex models should be used. In the category "multicompartment models" are included models of varying complexity containing more than one compartment. In such models, the body is described as a number of compartments. Each is defined as the volume into and out of which the chemical moves at a given rate. A compartment may or may not correspond to a particular tissue and often includes functionally disparate tissues. In other words, the compartment is a conglomeration of tissues with similar input and output rates.

Figure 5.5B represents a tentative multicompartment model for antimony (Droz *et al.,* 1991). The lungs here are described as a two-compartment entity. Absorption also can occur from another compartment (GI). Distribution is represented by only one compartment. Excretion occurs through both urine and feces. Rate constants or percentages are indicated for the transfers between the compartments. In this particular model, compartments already are identified physiologically.

Physiology-Based Pharmacokinetic Models

The primary limitation of the one- and multicompartment models is that their parameters have no direct biologic meaning. Their values are, most of the time, obtained by fitting techniques. Once determined, their predictive value for new situations often is limited.

Physiology-based models are built on physiological concepts. Figure 5.6 presents such a model (Droz and Guillemin, 1983). The body is separated into several compartments representing tissues or groups of tissues with similar perfusion/volume ratios. The entry compartment here represents the lungs, but could also be the skin or a combination of lungs and GI tract for particulates. Internal organs and tissues are shown here in four groups: the muscle group (MG), which contains the muscles and the skin; the fatty group (FG), which includes fat; the vessel rich group (VRG), with well-perfused organs such as the heart, the bladder, and the brain; and, finally, an organ responsible for the metabolism, the liver. Compartments are linked by the blood circulation: arterial on one side, venous on the other side. Usually the assumption is made that transport across the various interfaces (membranes) between alveolar air and blood, and between blood and tissues, is governed mainly by diffusion. Moreover, diffusion has been shown to be very fast for liposoluble compounds; therefore, venous blood generally is assumed to be in equilibrium with the corresponding tissue (Andersen, 1991).

To assemble the model, the following information is needed.

1. *Physiological data* Blood and tissue perfusion ratio and pulmonary ventilation can be found in physiology reference books [International

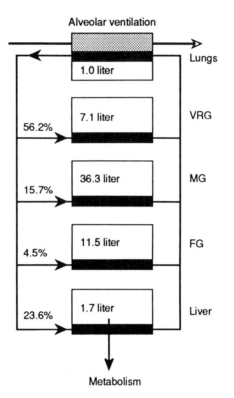

FIGURE 5.6 Example of a physiologically based pharmacokinetic model used for solvents. VRG, vessel rich tissue group; MG, muscle group; FG, fatty tissue group. The blood perfusions (%) and tissue volumes (liter) are indicated also. (Reprinted with permission from Droz and Guillemin, 1983.)

Commission on Radiological Protection (ICRP), 1975]. The representative worker is generally assumed to have a physical workload of 50 W and a reference size of 170 cm and 70 kg. To estimate volumes and perfusion rates for a different situation, equations based on anthropometric measurements are used (Table 5.3) (Droz et al., 1989).

2. *Physicochemical data*   Measures of solubility are blood–gas and tissues–gas partition coefficients, which can be determined by several laboratory techniques or estimated from chemical structure. Partition coefficients for industrial chemicals can be found in the literature (Fiserova-Bergerova and Diaz, 1986; Gargas et al., 1989). Examples of a few partition coefficients are given in Table 5.4.

3. *Biochemical data*   Metabolic clearance can be determined, for example, from the pulmonary uptake at apparent steady state. Dependence of the clearance on the degree of exposure (used in nonlinear models) and, thus,

TABLE 5.3 Physiological Data, Adjustment for Body Build, and Physical Workload[a]

Compartment	Volume (% LBV)	Perfusion rate at rest (ml/min/ml)	Effect of physical workload on perfusion (liter/min)
Lungs			
tissue	0.9	$Q_b^* = \Sigma Q_i^*$	ΣQ_i
air	5.5	$V_{alv}^* = 0.8\,Q_b$	$V_{alv}^* + 22\,\Delta VO_2$
arterial blood	1.9	—	—
Muscles and skin	69	33	$Q_i^* + 7.4\,\Delta VO_2$
Adipose tissues	FBV	22	$Q_i^* + \Delta VO_2$
Brain	2.6	570	unchanged
Kidneys	0.53	4520	$Q_i^*\,(1 - 0.14\,\Delta VO_2)$
Splanchnic	5.8	520	$Q_i^*\,(1 - 0.19\,\Delta VO_2)$
Others	1.0	2830	unchanged
Venous blood	8.3	—	—

Source: Reprinted with permission from Droz *et al.* (1989).

[a] LBV, Lean body volume; FBV, body fat volume; Q_b, cardiac output (liter/min); Q_i, unit perfusion rate of compartment i; V_{alv}, alveolar ventilation (liter/min); *, at rest; ΔVO_2, increase in oxygen consumption above rest (liter/min) (about 1% of workload in Watts).

TABLE 5.4 Examples of Tissue–Gas Partition Coefficients for Some Common Industrial Solvents

Solvent	Blood	Oil	Water	Vessel Rich Group	Muscles and Skin	Adipose tissue
benzene	7	498	2.8	15	10	350
toluene	16	1460	2.5	30	23	1030
m-xylene	34	4321	2.2	80	60	3030
styrene	59	5838	4.9	150	84	4100
methylene chloride	8	157	6.5	17	11	260
chloroform	11	424	3.8	16	11	300
methylchloroform	4	373	0.9	9	6	373
trichloroethylene	9	763	1.5	20	19	600
tetrachloroethylene	14	2072	0.9	45	29	2070

on biologic levels is difficult to study in humans. Qualitative and quantitative differences among the species are the main concern.

4. *Metabolic data* To describe distribution and elimination of metabolites, the physiology-based model is combined with one or more compartments, defined by volume of distribution, and clearance of the metabolites is determined from the elimination curve. Excretion rate and distribution volumes of metabolites are difficult to predict or extrapolate from *in*

vitro or animal experiments. Usually they are estimated in controlled studies on humans.

Applications of Pharmacokinetic Models

Temporal Fluctuations in Exposure and Biologic Indicators

As already discussed for Figures 5.2 and 5.3, biologic indicators are affected by fluctuations in exposure to variable extents, depending mainly on their half-lives. Biologic indicators with long half-lives tend to smooth out the exposure variability and, therefore, be more correlated with average exposure. Such markers also could be useful in some cases to integrate past exposures. An interesting point, when planning to use biologic monitoring, is to have an idea of the exposure period reflected in the sample taken. For example, is the biologic measurement used representing only the current day exposure, or does it show what happened during the last week or month? This issue can be addressed using a simple one-compartment model. Results obtained are summarized in Figure 5.7 (Droz *et al.,* 1991). This plot shows, for a given half-life, the cumulative contribution of each indicated time period (hour, day, week, month, semester) to the biologic indicator. For example, for a biologic determinant with a half-life of 10 hr, the measurement will be influenced only by what happened during the last week, and mainly by the exposure of the last day. The last hour of exposure has a relatively small effect because it is averaged into the entire day's exposure. Results would not be the same for chemicals with shorter half-lives, for which the effect of the last exposure hour is larger. On the other hand, if one considers a biologic deter-

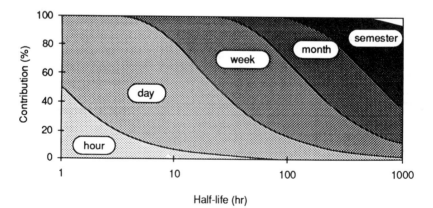

FIGURE 5.7   "Time representativity" of biological indicators as a function of their biological half-lives. Ordinate shows the cumulative contribution of the indicated time periods. (Reprinted with permission from Droz *et al.,* 1991).

minant with a longer half-life (for example, 100 hr), the biologic monitoring results will be influenced by the exposure situation of the last month; the last week would have more importance than the first 3 weeks of the month. In this case, the effect of the last day and hour is relatively small.

Quantitative Description of Individual Variability

The variability observed in biologic monitoring results depends not only on the variability in exposure but also on the variability linked to the individual. To this point, workers were considered average or standard humans. Differences in uptake, deposition, and elimination of chemicals can come from many different sources (Droz et al., 1989):

1. pulmonary ventilation will affect uptake if exposure is via inhalation; pulmonary ventilation is a function of physical workload, but also of body size
2. body build, that is, the size and the proportions of the different tissues (for example, fatty tissues in the case of liposoluble chemicals)
3. liver function will affect the metabolic clearance of metabolized chemicals, involve induction phenomena, but also decrease metabolic capacity due to other competing chemicals
4. kidney function influences chemicals or metabolites because of excretion by the kidneys

Pharmacokinetic modeling can take these sources of variability into account in different ways. In this section they will be treated globally under the term "biologic variability" without identifying the specific factors responsible for this variability.

Individual variability can be described by the coefficient of variation CV_i. In fact, individual results for a biologic parameter generally are distributed lognormally, and their variability is described by a geometric standard deviation that can be transformed into a coefficient of variation. Values for this parameter can be obtained experimentally. For solvents, exposure of groups of subjects under identical conditions allows the study of the variability of some biologic indicators, such as breath or urine concentration. Such data are presented in Table 5.5 for methylchloroform, tetrachloroethylene, and styrene (Droz and Wu, 1991). The CV_i shown vary from 0.20 to 0.85, a rather wide range. Table 5.6 gives an idea of biologic variability based on drugs and several industrial chemicals. Parameters considered are half-life, area under the curve, and peak concentration obtained. Information is summarized by the median CV_i observed, as well as an indication of the range for these CV_is (95% confidence) (Droz and Wu, 1991). Median is situated at about 0.30, and extreme values for the range are from 0.07 to 1.33. This result is quite consistent with the data presented in Table 5.5. For discussion purposes, biologic variability can be described by a median CV_i of 0.30, with

TABLE 5.5   Coefficient of Variation for Some Selected Biological Indicators after Controlled Exposures of Human Subjects[a]

Sample	Methylchloroform (1, TCA; 2, TCE)	Tetrachloroethylene (1, TCA)	Styrene (1, PGA; 2, MA)
Breath ES	0.20	0.23	0.28
Breath PS	0.26	0.22	0.55
Urine ES			
Metabolite 1	0.85	0.36	0.42
Metabolite 2	0.35		0.37

Source: Reprinted with permission from Droz and Wu (1991).
[a] ES, End of shift sample; PS, prior to shift sample; TCA, trichloroacetic acid; TCE, trichloroethanol; MA, mandelic acid; PGA, phenylglyoxylic acid.

TABLE 5.6   Summary Statistics for the Coefficient of Variation of Some Common Pharmacokinetic Parameters

	$T_{1/2}$[a]	AUC[b]	Peak[c]	All
Number of chemicals	44	15	12	75
Median CV	0.26	0.34	0.25	0.28
95% range CV	0.10–0.68	0.11–1.33	0.07–1.15	0.10–0.88

Source: Reprinted with permission from Droz and Wu (1991).
[a] Half-life of drug or chemical.
[b] Area under the curve.
[c] Peak concentration in biological media.

a 95% range of 0.10 to 0.90. To compare these biologic variabilities with other sources (such as environmental variability, which is often expressed in terms of geometric standard deviation), it is helpful to convert the results to GSD (Rappaport, 1991):

$$GSD = e^{(\sqrt{\ln(CV_i^2 + 1)})}$$

With this formula, the median obtained is 1.3, with a range of 1.1–2.2. Note that this variability is rather small compared to exposure variabilities observed in industrial environments.

Comparative Evaluation of Air and Biological Monitoring Data

With the definition and characterization of individual variability given, it is possible to make a comparison with exposure variability itself. This is an important step since, in many cases, the discussion arises in exposure assessment to make use of air monitoring or biologic monitoring techniques. In some situations a definite advantage can be identified for biologic monitoring, for example, possibility of skin exposure or unknown or variable bioavailability, as for some partially soluble metal aerosols. In many cases, however, the selection of the monitoring technique is based on a rough comparison of variabilities.

A more quantitative approach consists of combining the two main sources of variability: exposure and individual variability. This can be done, assuming that the variance is not correlated, by the formula (Droz and Wu, 1991):

$$CV_{total} = \sqrt{(CV_{expo}{}^2 + CV_i{}^2)}$$

CV_{expo}, the contribution of environmental variability, can be calculated from the GSD of the air exposure data and a damping factor, $1/A$, that is a function of the half-life of the biologic indicator, or the corresponding rate constant k (Rappaport, 1991).

$$1/A = (1 - e^{(-24\ k)})\ (1 - e^{(-168k)})/(1 - e^{(-120\ k)})\ \sqrt{(1 - e^{(-48k)})}$$

Using these relationships, air and biologic monitoring approaches can be compared, based on their variabilities. One way to compare is to estimate the number of samples needed for each approach to test the mean biologic or air indicator against a reference level with a given significance and statistical power (Droz and Wu, 1991; Rappaport, 1991). Figure 5.8 shows such a comparison for four industrial environments with variabilities ranging from GSD 1.5 to GSD 3.5 (Droz and Wu, 1991). The dotted line indicates the number of air samples needed to achieve a significance of 0.05 and a power of 0.10. For biologic monitoring, curves are indicated for the same objective, assuming a median of 0.30 and extremes of 0.10–0.90 for individual variabilities. In this case, the number of samples required depends on the half-life of the biologic indicator.

The plots in Figure 5.8 show that the required number of biologic samples is a function of several variables, including the variation in exposure, the biologic variation among individuals, and the half-life of the biologic indicator. Figure 5.8 can be used to compare sample sizes required for air and biologic monitoring. For half-lives of 10 hr or less, there appears to be little or no statistical advantage of biologic monitoring over air monitoring, whatever the exposure GSD is. Thus there is a definite advantage of air monitoring over biologic indicators with relatively high CV_i, in the case of low and moderate environmental variabilities. For biologic indicators with half-lives greater than 10 hr, there is always some advantage of biologic monitoring over air sampling for environmental variabilities of GSD 3.0 and greater, but this advantage is improved only marginally when half-lives of the biologic indicators become large (greater than 1000 hr). This is only true if exposure can be considered stationary. Otherwise, this type of indicator would allow an integration of trends of exposure.

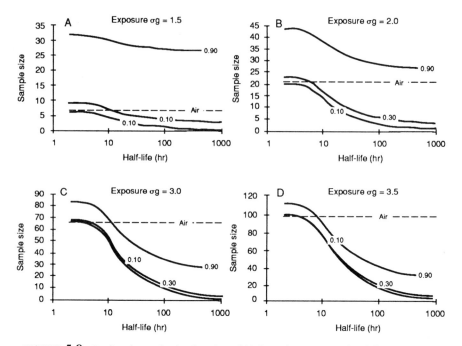

FIGURE 5.8 Predicted sample size for air and biological monitoring for different environmental variabilities (σ_g), individual variabilities (mean $CV = 0.30$, range $0.10-0.90$), and half-lives of the biological indicators.

References

American Conference of Governmental Industrial Hygienists (ed.) (1991). "Threshold Limit Values and Biological Exposure Indices for 1991–1992." Cincinnati, Ohio.

Andersen, M. (1991). Physiological modelling of organic compounds. *Ann. Occup. Hyg.* **35**, 309–321.

Åstrand, I. (1983). Effect of physical exercise on uptake, distribution, and elimination of vapors in man. *In* "Modeling of Inhalation Exposure to Vapors: Uptake, Distribution, and Elimination" (V. Fiserova-Bergerova, ed.), Vol. II, pp. 107–130. CRC Press, Boca Raton, Florida.

Checkoway, H. (1986). Methods of treatment of exposure data in occupational epidemiology. *Med. Lav.* **77**, 48–73.

Droz, P. O. (1989). Biological monitoring. Sources of variability in human response to chemical exposure. *Appl. Ind. Hyg.* **4**, F20–F24.

Droz, P. O., and Guillemin, M. P. (1983). Human styrene exposure. V. Development of a model for biological monitoring. *Arch. Occup. Environ. Health* **53**, 19–36.

Droz, P. O., and Wu, M. M. (1991). Biological monitoring strategies. *In* "Exposure Assessment for Epidemiology and Hazard Control" (S. M. Rappaport and T. J. Smith, eds.), pp. 251–270. Lewis Publishers, Chelsea, Michigan.

Droz, P. O., Wu, M. M., Cumberland, W. G., and Berode, M. (1989). Variability in biological monitoring of solvent exposure. I. Development of a population physiological model. *Br. J. Ind. Med.* **46**, 447–460.

Droz, P. O., Berode, M., and Wu, M. M. (1991). Evaluation of concomitant biological and air monitoring results. *Appl. Occup. Environ. Hyg.* **6**, 465–474.

Fiserova-Bergerova, V., and Diaz, M. L. (1986). Determination and prediction of tissue–gas partition coefficients. *Int. Arch. Occup. Environ. Health* **58**, 75–87.

Fiserova-Bergerova, V., Pierce, J. T., and Droz, P. O. (1990). Dermal absorption potential of industrial chemicals: Criteria for skin notation. *Am. J. Ind. Med.* **17**, 617–635.

Gargas, M. L., Burgess, R. J., Voisard, D. E., Cason, G. H., and Andersen, M. E. (1989). Partition coefficients of low-molecular-weight volatile chemicals in various liquids and tissues. *Toxicol. Appl. Pharmacol.* **98**, 87–99.

Herrera, H., Droz, P. O., and Guillemin, M. P. (1991). Occupational Hygiene Interpretation of Intermittent Dust Exposure. *In:* "Clean Air at Work, New Trends in Assessment and Measurement for the 1990s," (R. H. Brown, M. Curtis, K. I. Saunders and S. Van den driessche, eds.), pp. 255–258. The Royal Society of Chemistry Special Publication No. 108, Cambridge, United Kingdom.

Hornung, R. W., and Reed, L. D. (1990). Estimation of average concentration in the presence of nondetectable values. *Appl. Occup. Environ. Hyg.* **5**, 46–51.

International Commission on Radiological Protection (1975). "Report of the Task Group on Reference Man" (W. S. Snyder, ed.). ICRP Pub. 23. Pergamon Press, Elmsford, New York.

Paustenbach, D. J. (1985). Occupational exposure limits, pharmacokinetics, and unusual work schedules. *In* "Patty's Industrial Hygiene and Toxicology" (L. J. Cralley and L. V. Cralley, eds.), Vol. 3A, pp. 111–277. John Wiley and Sons, New York.

Rappaport, S. M. (1991). Exposure assessment strategies. *In* "Exposure Assessment for Epidemiology and Hazard Control" (S. M. Rappaport and T. J. Smith, eds.), pp. 219–249. Lewis Publishers, Chelsea, Michigan.

Roach, S. A. (1966). A more rational basis for air-sampling programs. *Am. Ind. Hyg. Assoc. J.* **27**, 1–12.

Smith, T. J. (1991). Pharmacokinetic models in the development of exposure indicators in epidemiology. *Ann. Occup. Hyg.* **35**, 543–560.

Spear, R. C., Selvin, S., Schulman, J., and Francis, M. (1987). Benzene exposure in the petroleum refining industry. *Appl. Ind. Hyg.* **2**, 155–163.

Vincent, J. H. (1990). The fate of inhaled aerosols: A review of observed trends and some generalizations. *Ann. Occup. Hyg.* **34**, 623–637.

6

Design Considerations in Molecular Epidemiology

Paul A. Schulte,
Nathaniel Rothman,
and David Schottenfeld

Introduction

Molecular epidemiology does not differ in purpose from epidemiology in general, that is, the study of the distribution and determinants of health-related states and events in populations, and the application of this study to the control of health problems (Last, 1988). Further, molecular epidemiologic studies generally are based on classic epidemiologic designs. What makes molecular epidemiology distinctive, as discussed in Chapter 1, is its ability to look inside the "black box" of the exposure–disease continuum. In the process, it may reduce misclassification of exposure, provide insight into underlying mechanisms, identify gene–gene and gene–environment interactions, and provide information when intervention is potentially more effective.

This chapter considers how biomarkers can be used in the major epidemiologic study designs. We first review the biomarker categories that represent the continuum from exposure to disease. We then explore the initial methodologic work necessary to characterize a newly developed biomarker before it is applied in etiologic and intervention studies. Hulka (1991) has defined these as transitional studies that bridge the gap from the laboratory to population-based studies of disease. We then consider the application of biomarkers to case-control, prospective, and screening and intervention studies. Finally, we present examples of various biological markers in epidemiologic study designs.

Biomarker Categories

Designing studies that incorporate biomarkers requires understanding where along the continuum of exposure to disease a biomarker is located. To determine this location and to develop more effective communication about biomarkers in general, it has been necessary to develop a new vocabulary [National Research Council (NRC), 1987]. The continuum between exposure and disease was initially discussed in Chapter 1 and is presented in Figure 6.1. One way to conceptualize marker categories is to consider exposure (E), internal dose (ID), or biologically effective dose (BED) in one category as markers of exposure and altered structure and function (ASF), clinical disease (CD), and prognostic significance (PS) as markers of effect. Early biologic effects (EBE) are ambiguous markers used prior to study and may be regarded as markers of exposure or markers of effect.

Generally, a study may test if an exposure (E) is related to a marker (M):

$$E \rightarrow M_E \text{ (e.g., paths 1, 7, 8, 9)}$$

if a marker (M) is related to a disease (D):

$$M_D \rightarrow D \text{ (e.g., paths 20, 21)}$$

or if exposures and their markers are related to diseases and their markers:

$$E \text{ or } M_E \rightarrow M_D \text{ or } D \text{ (e.g., 9, 10, 16, 17)}$$

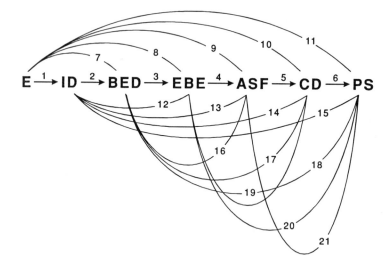

FIGURE 6.1   A continuum of markers between exposure and disease. E, *Exposure;* ID, *internal dose;* BED, *biologically effective dose;* EBE, early biological effects; ASF, *altered structure and function;* CD, *clinical disease;* PS, *prognostic significance.* Numbers refer to studies presented in Table 6.1.

These models are however, admittedly simplistic because they imply a single necessary and sufficient antecedent cause and subsequent effect, and a single mechanistic pathway. In most instances, particularly in chronic diseases, causation is multiple and potentially interactive (Rothman, 1986). Despite its limitations, the continuum of exogenous exposure and resultant disease is a useful way to conceptualize molecular epidemiologic investigations.

Using Biomarkers in Epidemiologic Research

Biomarkers may be used to enhance the assessment of the standard types of risk factors evaluated in epidemiologic investigations, for example, exogenous exposure, demographics, and genetic susceptibility. In addition, biomarkers may be used to evaluate disease status by defining more homogeneous case groups (Hulka and Wilcosky, 1988).

However, a series of new relationships can be evaluated when studying a series of biomarkers that represent the continuum of exposure to disease, that is, the associations between all that precedes a marker in the continuum and all that follows it (Table 6.1). Exploring these associations may provide a better description of disease progression and its determinants, and ultimately may result in improved primary, secondary, and tertiary prevention strategies and application. However, the ultimate significance of biomarkers that represent any point in the continuum up to disease itself can be assessed only by determining the association of these markers with disease.

Molecular Epidemiologic Study Designs

Just as there is a spectrum of biomarker categories, there is also a spectrum of epidemiologic study designs that can make use of the markers (Table 6.2). Matching the most effective and feasible study design with the appropriate biomarker to answer research hypotheses is the fundamental challenge of molecular epidemiology. In this section, we first discuss transitional studies (Hulka, 1991), which lay the foundation for using biomarkers in etiologic studies by evaluating the factors that influence biomarker levels in humans. We then make some general comments on the strengths and limitations of using the major biomarker categories in case-control studies, the most commonly performed etiologic study. We conclude with some comments on methodology of prospective and intervention studies, and on confounding and interaction.

Transitional Studies

Traditionally, external exposure, demographic variables, and family history have constituted the main risk factors in epidemiologic studies. Entire sub-

TABLE 6.1 Potential Studies of Biological Marker Relationships[a] in the Continuum between an Exposure and Disease

Relationship to be studied[b]	Capability
1. E ↔ ID	Validation of exposure marker
2. ID ↔ BED	Assessment of dose
3. BED ↔ EBE	Association of dose and effect
4. EBE ↔ ASF	Determination of pathogenicity of an effect
5. ASF ↔ CD	Validation of preclinical marker
6. CD ↔ PS	Validation of prognostic marker
7. E ↔ BED	Validation of exposure marker
8. E ↔ EBE	Association of exposure with an effect marker
9. E ↔ ASF	Association of exposure with a pathogenic effect
10. E ↔ CD	Association of exposure with disease
11. E ↔ PS	Association of exposure with severity of disease
12. ID ↔ EBE	Association of dose and effect
13. ID ↔ ASF	Association of dose and preclinical effect
14. ID ↔ CD	Association of dose and disease
15. ID ↔ PS	Association of dose and severity of disease
16. BED ↔ ASF	Best indicator of association of dose and preclinical effect
17. BED ↔ CD	Association of dose and disease; indicator of risk
18. BED ↔ PS	Association of dose with severity of disease
19. EBE ↔ CD	Validation of marker of effect regarding disease
20. EBE ↔ PS	Validation of marker of effect of severity of disease
21. ASF ↔ PS	Validation of preclinical marker with severity of disease
22.	Repeat of studies 1–21 with marker of susceptibility as effect modifier

[a] The causal relationship between any two markers is generally from left to right, but the bidirectional arrow indicates that the relationship can be studied prospectively or retrospectively. Numbers correspond to paths in Figure 6.1.
[b] E, Exposure; ID, internal dose; BED, biologically effective dose; EBE, early biological effect; ASF, altered structure/function; CD, clinical disease; PS, prognostic significance.

TABLE 6.2 Typology of Molecular Epidemiologic Studies

	Transitional studies	
Biomarker development studies	Biomarker characterization studies	Disease etiology and intervention studies
Reliability studies	Observational Cross-sectional Longitudinal	Observational Descriptive Analytic
Sample collection and processing studies		Cross-sectional Case-control Longitudinal Hybrid studies
	Intervention	Intervention Risk factor modification Screening Therapeutic

divisions of epidemiology, such as nutritional epidemiology, environmental and occupational epidemiology, and genetic epidemiology, have evolved to develop and validate instruments to evaluate these factors. Similarly, a substantial component of molecular epidemiology involves studies that evaluate biomarkers to determine their optimal use in subsequent population-based etiologic studies. Although transitional studies lay the groundwork for etiologic studies, they do not have the capability to directly assess the predictive value of a biomarker for developing clinically apparent disease.

We propose that transitional studies be further classified as *biomarker development studies,* which assess the reliability (and accuracy, if possible) of the assay to be performed on the specimens and optimize conditions for collecting, processing, and storing biologic specimens prior to an assay; and *biomarker characterization studies,* which evaluate the distribution and the exogenous and endogenous determinants of biomarkers in specified populations (e.g., exposure–marker relationships) to determine how to apply the marker in subsequent etiologic investigations. These studies may be cross-sectional or longitudinal, and may be observational or interventional (e.g., when controlled exposure studies are ethically acceptable, as in dietary metabolic ward studies).

Biomarker Development Studies

Determining the reproducibility of a laboratory assay is one of the most important components of a molecular epidemiology study. Strikingly, it often is overlooked in practice and in the reporting of study results. If an assay is not reproducible, it clearly is not likely to be accurate. If it is reproducible within acceptable parameters, comparison to a standard is appropriate. However, often a standard is not available for a new class of assays. In this case, reproducibility functions as a surrogate for accuracy until alternative methods are developed that represent a more accurate assessment of an underlying process. As noted by Willet, "from the standpoint of most epidemiological applications the consistency of a laboratory is more important than accuracy on an absolute scale" (1987). Several of these issues are described in detail in Chapter 4.

The adverse impact of measurement error cannot be overemphasized, since the resulting misclassification, if nondifferential, generally will bias the associations toward the null hypothesis (Rothman, 1986) and, if differential, can support erroneous conclusions. For example, Schiffman demonstrated that a more reliable assay for measuring the presence of human papilloma virus in cervicovaginal lavage specimens increased the association of detectable virus with cervical dysplasia from an odds ratio of 3.7 (95% confidence interval 2.6–5.3), in a study using an older, less reproducible technique to 20.1 (95% confidence interval 14.4–28.0), using the new technique in a second study of comparable design (M. Schiffman, personal communication).

Assay reproducibility should be addressed before the initiation of a field study. Blind replicates should be sent to a laboratory to assess intralaboratory variation in analysis; if necessary, samples should be sent to a second laboratory for an interlaboratory comparison (Biber *et al.*, 1987; Santella *et al.*, 1988). Statistical methods for evaluating reproducibility studies are discussed in Chapter 4. In addition, blind "spiked" samples containing standards, if available, should be included as a direct assessment of analytic accuracy.

Multiple factors can influence assay results on biologic samples. For example, some of the more obvious variables for assays performed on peripheral blood include the time of day phlebotomy takes place, the use of preservatives and the type of anticoagulant, the time from phlebotomy to processing to storage, and the temperature of storage. Historically, determining the optimum way of collecting and processing biologic samples under field conditions often has been a "learn-as-you-go" process. Not uncommonly, an improved variant on sample processing comes to the attention of the investigator after a study has been completed, reflecting in part the rapid development of new bioassay technologies, but also demonstrating that decisions regarding sample handling often are based on anecdotal information that has not been field tested.

Molecular epidemiology studies are often time consuming and expensive. It behooves the investigator to evaluate and optimize sample collection procedures before the main study takes place. A small, easy-to-institute change in a protocol may have a marked effect on the ease and ultimate success of performing the bioassay. Sample collection procedures and their variants should be evaluated on a sample of individuals who are similar to the intended study participants, in a setting comparable to the field conditions of the main study.

Biomarker Characterization Studies

Biomarker characterization studies evaluate the distribution and determinants of biomarkers in populations. These studies help investigators sift through available markers to select those that are most promising for use in etiologic studies. By providing information on the behavior and determinants of the biomarker, these studies help clarify which etiologic study designs are optimal for biomarker use. Such studies also address how to collect additional information that may be necessary to interpret assay results adequately. Finally, by demonstrating that a xenobiotic compound is absorbed, or causes an early toxic effect, these investigations may provide biologic plausibility for a suspected exposure–disease association. Design considerations for these studies, including selection of study subjects, power calculations, controlling for confounding, data analysis, and interpretation have been discussed previously by several authors, including Hulka (1990a,b, 1991), Hulka and Wilcosky (1988), Wilcosky (1990), Wilcosky and Griffith (1990), Perera (1987), and Schulte (1987, 1989) among others, and are commented on in Chapter 7. Biomarker characterization studies can be grouped into cross-sectional and longitudinal studies.

Cross-sectional studies   A cross-sectional study examines the relationship between a biomarker and other variables of interest as they exist in a defined population at one particular time. The temporal sequence of cause and effect cannot necessarily be determined (adapted from Last, 1988).

Cross-sectional studies can be a useful first step in characterizing the determinants of a biomarker in a specific population. Such studies are particularly valuable for assessing dose–response relationships between contemporaneous external exposures and markers of internal dose, biologically effective dose [e.g., occupational polycylic aromatic hydrocarbon (PAH) exposure and PAH–DNA adduct levels in foundary workers; Perera *et al.*, 1988], and early effect. These studies may be suitable for evaluating variables that modify the effect of the dose–response relationship, for example, other external exposures or host factors (e.g., modification of the association between smoking and 4-aminobiphenyl hemoglobin adducts by acetylation phenotype; Vineis *et al.*, 1990). In addition, cross-sectional studies are useful for evaluating demographic variables that independently modulate biomarker levels, for example, age.

Cross-sectional studies can be used to evaluate the correlation between genotype and phenotypic expression of potential genetic susceptibility markers [e.g., *N*-acetyltransferase gene mutations or activity (Deguchi *et al.*, 1990; Rothman *et al.*, 1992b), or GC alleles for the vitamin D binding protein and glucose tolerance (Iyengar *et al.*, 1989)]. This information can be used to select the appropriate marker, and to use it effectively in subsequent etiologic studies as a confounder or effect modifier.

Cross-sectional studies provide limited information on biomarker kinetics. Although some insight into the kinetics of biomarker persistence can be obtained by analysis, if the time elapsed since the study subjects were last exposed is variable [e.g., time since last consumed charbroiled meat and level of PAH–DNA adducts in white blood cells (Rothman *et al.*, 1992a)], longitudinal studies can assess biomarker accumulation and decline more directly.

Longitudinal studies   In longitudinal studies, subsets of a defined population are identified that are, have been, or in the future may be exposed, not exposed, or exposed in different degrees to a factor or factors hypothesized to influence the biomarker(s) under study (adapted from Last, 1988). In a longitudinal biomarker study, subjects are evaluated two or more times to assess changes in a biomarker level due to internal or external perturbations in the determinants of the biomarker. Longitudinal biomarker studies often are performed when the main exposure reflected in a biomarker varies over the duration of the study period. Studies of worker response to occupational exposures are an example. Depending on the kinetics of the biomarker, workers are evaluated at the beginning and end of a work shift, work week, or vacation period [e.g., decline of PAH–DNA adducts in coke oven workers after a 3-week vacation (Haugen *et al.*, 1986)] or are followed over a period of years during which workplace exposures may decline [e.g., benzene exposure and peripheral blood cell counts (Kippen *et al.*, 1989)].

In some instances, exposure patterns experienced by individuals will vary in a relatively controlled fashion because of an individual's decision to change that exposure [e.g., smoking cessation programs and decline in 4-aminobiphenyl hemoglobin adduct formation (Maclure *et al.*, 1990)] or a consequence of disease that must be treated [e.g., *cis*-platinum–DNA adduct levels in white blood cells in chemotherapy patients (Reed *et al.*, 1990)]. Patterns also may vary due to direct intervention on the part of the investigator when such intervention is ethically acceptable [e.g., effect of controlled exposure to ω3 polyunsaturated fatty acids on blood lipids and platelet function (Sanders and Roshanai, 1983)]. As in cross-sectional studies, markers of genetic susceptibility can be evaluated as effect modifiers of the exposure–marker response relationship. [For example, genetic variation in apolipoprotein genes modifies the response of plasma lipids to dietary change (Xu *et al.*, 1990) (see Example 20).]

Longitudinal studies compare individuals to themselves, thus controlling for genetic differences in effect modification of the exposure–biomarker response. However, repeat measures made on individuals over time are not independent, which violates a primary assumption of many statistical techniques. Consequently, special procedures should be used to analyze data from these studies (Dwyer *et al.*, 1991).

In summary, transitional studies lay the foundation for future etiologic studies. Transitional studies optimize procedures for collecting, processing, and storing biologic samples, and evaluate bioassay reproducibility. These studies assist in selecting appropriate populations for study, provide data for power calculations, and determine what additional information must be collected to interpret biomarker results adequately. Although transitional studies do not have the ability to assess directly the predictive value of a given biomarker and risk of disease, they are a critical first step and may serve to refine an existing hypothesis or generate new study hypotheses for subsequent testing in etiologic studies.

Disease Etiology and Intervention Studies

Disease etiology and intervention studies use biomarkers to study the determinants of disease in specific populations and apply this information to the control of health problems. Subjects either are healthy at entry into the study and are followed forward to disease or are diseased at the time of the study. Schatzkin *et al.* (1990) and Wilcosky and Griffith (1990) have noted that these studies can be used to calculate a biomarker attributable proportion (also known as the etiologic fraction), defined as the proportion of diseased cases that is attributable to an intermediate biomarker. If a marker is causally associated with the disease under study, this measure provides an assessment of what proportion of disease would be eliminated if the biomarker determinants were altered in a way that reduced the marker prevalence in a given population.

Observational Studies

Observational etiologic studies using biomarkers include descriptive, cross-sectional, case-control, and prospective longitudinal (with its hybrid variants) studies. The latter two types of studies have the greatest ability to assess etiologic relationships and will be considered here.

Case control   A case-control study starts with the disease (or other outcome variable) of interest and a suitable control (comparison, reference) group of persons without the disease (or outcome variable). The relationship of an attribute to the disease is examined by comparing the diseased and nondiseased with respect to how frequently the attribute is present or, if the study is quantitative, the levels of the attribute in each of the groups (Last, 1988).

Biomarkers have potential application to several aspects of classic case-control studies in which cases are identified through the presence of clinically apparent disease. Biomarkers may function as dependent variables, further classifying nominal disease categories into more homogeneous subgroups, or as independent risk variables, evaluating exogenous and endogenous exposures, genetic susceptibility, and, in some cases, intermediate states on the pathway from external exposure to disease.

In addition, biomarkers may refine further or, in some instances, provide the sole definition of case and control status in populations undergoing screening (e.g., HIV testing or cholesterol screening). In these instances, "case" status, which has traditionally been based on clinical disease, depends on the presence of a validated marker component that serves as the dependent variable (i.e., the disease surrogate).

Case definition refined   The goal of defining a case usually is to identify individuals who have a disease that is, as far as possible, a homogeneous entity. Biomarkers can be used to subclassify nominal disease categories further into more homogeneous categories. The more refined categories may have stronger associations with exposure or genetic susceptibility factors (Hulka and Wilcosky, 1988; Taylor, 1990a) and may be more or less preventable, screenable, or treatable.

Many examples of this application are seen in cancer research using con-

ventional histologic techniques, such as the strong association between tobacco use and certain lung cancer subtypes (Rosenow and Carr, 1979). Recently developed molecular techniques, however, are providing opportunities
to subclassify tumors based on the presence and nature of mutations in oncogenes and tumor suppressor genes; further, the pattern of mutations in
these genes may have some degree of specificity for exogenous carcinogens.
For example, Suzuki *et al.* (1992) reported that lifetime cigarette use was
strongly associated with the presence of *p53* mutations in non-small cell lung
cancer tumors. These researchers modeled the presence or absence of *p53*
mutations in lung cancer cases as a binomial outcome variable, and evaluated
cigarette use and several demographic and clinical variables for their association with this outcome in a multivariate logistic model. Two studies demonstrated that the pattern of DNA base pair changes in *p53* mutations (the
mutational spectrum) of hepatocellular cancer tumors from patients in geographic areas with heavy aflatoxin contamination was consistent with mutations caused by aflatoxin (Bressac *et al.*, 1991; Hsu *et al.*, 1991). These
studies compared the exposure pattern of cases whose tumors had a given
mutational spectrum in a target gene to the exposure pattern of cases without
the mutational spectrum pattern. Taylor (1990b) suggested that studies of
tumor characteristics and exposure patterns might be called "case–case"
studies, since all study subjects are diseased.

Evaluating exogenous exposures   Biomarkers of chemical, physical, or
biologic exposures have been used extensively as independent variables in
case-control studies for a broad spectrum of disease entities (Rayfield *et al.*,
1988; Ross *et al.*, 1992). Several issues must be considered when using markers of exposure, apart from the obvious issue of the kinetics of biomarker
formation, persistence, and decline (Henderson *et al.*, 1990; see Chapter 3).
The discussion will be limited to the evaluation of incident cases that have
received no prior marker-altering medical therapy.

The interval between exposure to a disease-causing agent and onset of
detectable disease, and whether the exposure is of limited duration or continues up to the time of diagnosis, is one of the most critical issues in evaluating
the usefulness of exposure markers in case-control studies, particularly for
chemical exposures. At one end of the disease latency spectrum are the acute
and subacute diseases that include many infectious diseases and diseases
caused by chemical poisoning. At the other end of the disease latency spectrum are the chronic diseases, such as cancer, chronic infectious diseases,
and most pulmonary and cardiovascular diseases. It is clearly easier to use
markers of exposure for diseases with short latencies since current exposure has occurred over a period of time that is most relevant for disease causation. In addition, markers of exposure with relatively short half-lives
(hours to days) may be usable since the subjects often are currently or recently exposed.

The use of biomarkers of exposure is a more challenging exercise in case-control studies of chronic diseases. In this instance, exposure may have ceased years before the onset of the clinical disease (e.g., an occupational or discrete environmental exposure) or may have continued at some level until case presentation (e.g., diet, smoking). Markers of chemical exposures that have ceased generally do not persist in readily accessible biologic samples such as peripheral blood and urine, but in some instances may accumulate in measurable amounts in other tissue compartments [e.g., (PCBs) polychlorinated biphenyls and dioxins in tissue or fat]. Markers of exposure (e.g., glycophorin A) have been useful in individuals previously exposed to ionizing radiation. Immunologic markers of previous biologic exposures have been useful since certain antibody classes persist for years after exposure has ceased.

Biomarkers of exposures that persist up to the time of case presentation (e.g., biomarkers of diet, smoking, chronic persisting infections) have been used extensively in case-control studies of chronic disease. Even assuming that disease status does not directly or indirectly affect marker status, one must still assess the degree to which current exposure is correlated with historic exposure and integrate this information with knowledge of biomarker persistence.

The other major concern about using markers of exposure in case-control studies is the effect of disease on the marker. Disease status may directly affect the marker through some physiologic perturbation [e.g., total serum cholesterol levels decline immediately after a myocardial infarction (Tibblin and Cramer, 1963)]. Additionally, when exposure occurs up to the time of diagnosis (e.g., diet), case status may influence recent exposure, which would introduce bias in exposure measurement, particularly for markers with relatively short half-lives.

Some of the limitations of using biomarkers of exposure in case-control studies can be addressed by analyzing marker levels by disease stage, to evaluate the effect of disease progression on the marker. Many of the limitations of exposure markers in case-control studies of individuals with clinically apparent disease can be overcome by studying individuals with preclinical disease (e.g., colon adenoma, cervical dysplasia) or elevated levels of intermediate markers (e.g., HIV seropositivity).

Evaluating genetic susceptibility   Genetic contribution to disease has been evaluated primarily by genetic epidemiologists in family and twin studies, study designs that are discussed in Chapter 14. Assessment of the genetic contribution to "sporadic" disease, evaluated in population-based studies, has been limited traditionally to evaluation of family history by questionnaire. Currently available technology allows analysis of some components of genetic susceptibility at the functional, or phenotypic, level and, in some instances, at the DNA, or genotypic, level.

Phenotypic polymorphisms in enzymes that activate procarcinogens have been evaluated in case-control studies of cancer. The association of the slow N-acetyltransferase phenotype with risk of bladder cancer has been assessed in multiple studies in many different countries. (See Vineis and Terracini, 1990, for an overview.) Similarly, the association of the extensive debrisoquine metabolizing phenotype (mediated by P450IID6) with risk of lung cancer has been evaluated in several case-control studies (Caporaso *et al.,* 1990, among others). Polymorphisms in enzymes that conjugate xenobiotics have been described and applied in studies of lung cancer as well [e.g., glutathione S-transferase mu phenotype (Zhong *et al.,* 1991)].

Cardiovascular disease epidemiologists have used lipoproteins, some of which can be considered phenotypic expressions of underlying genetic polymorphisms, to evaluate susceptibility for developing coronary heart disease [e.g., association of Lp(a) with coronary artery disease documented by angiography (Dahlen *et al.,* 1986)]. Similarly, studies of chronic obstructive lung disease have evaluated genetic susceptibility through phenotype assays of alpha 1-antitrypsin activity (Klayton *et al.,* 1975, among others) and evaluation of the blood A antigen (Khoury *et al.,* 1986).

Using phenotype markers to evaluate genetic susceptibility poses some of the same methodologic challenges discussed under markers of exposure, for example, modification of the phenotype by current exposures that do not reflect life-long patterns of exposures (e.g., induction of a P450 enzyme) or by case status itself. Phenotype markers that have a very high concordance with underlying polymorphisms at the genetic level [Lp(a), Nora, 1991; N-acetyltransferase, Deguchi *et al.,* 1990; Rothman *et al.,* 1992b] and are not modified by the effect of current exogenous exposures or by disease status are applied most effectively in case-control studies. However, in these instances, newly developed genotype assays [restriction fragment length polymorphisms (RFLPs), polymerase chain reaction (PCR)] may replace phenotype assays, since they can be performed on a small amount of peripheral white blood cell DNA or on other easily accessible sources of genomic DNA, thus eliminating the need for patients to submit to time-consuming drug-based phenotyping protocols.

Phenotype markers that are affected by recent exposures [e.g., induction of P450IA1 and P450IA2 by PAH (Nebert and Gonzalez, 1987)] are more difficult to apply in case-control studies, particularly when the full spectrum of exposures that alter the phenotype is not known. This problem is particularly applicable in studies of cases with clinically apparent disease for which the pattern of life-long exposures that affected a phenotype is not currently present. The problem may be circumvented partially by applying alterable phenotype markers to studies of subjects with asymptomatic early disease for which the current pattern of relevant exposures may be more representative of life-long exposure.

A growing number of genotype markers have been employed in case-control studies of a broad range of disease categories, for example, HLA-DQ beta chain genotype distributions among insulin-dependent diabetics (Dorman *et al.*, 1990); DNA polymorphisms of the apolipoprotein B gene in patients with premature coronary artery disease (Genest *et al.*, 1990); glutathione *S*-transferase mu locus in patients with lung cancer (Zhong *et al.*, 1991); and NAT2 genotypes in bladder cancer studies (Hayes *et al.*, 1992). These markers are of great utility in case-control studies because disease status and pattern of recent exposures generally do not affect the marker.

One of the greatest potentials for molecular epidemiology studies is evaluating the interaction of exogeneous exposures and genetic susceptibility. Such an interaction has been described between the slow acetylation phenotype and occupational exposure to aromatic amine dyes for risk of bladder cancer (Cartwright *et al.*, 1982). An interaction has been suggested between occupational exposure to asbestos or PAHs and the extensive debrisoquine metabolizing phenotype for risk of lung cancer (Caporaso *et al.*, 1990). Similarly, studies of chronic obstructive lung disease have demonstrated an interaction between the blood A antigen and high levels of smoking (Khoury *et al.*, 1986). In all these cases, individuals with similar exposure histories and a certain phenotype are at a much higher risk of disease than those not having that phenotype.

When an exposure is hypothesized to interact biologically with a given genetic susceptibility, it may increase the study efficiency to oversample cases at the extremes of such an exposure in order to maximize the chances of demonstrating potential biologic interactions.

In the future, genotype assays probably will find extensive use in case-control studies. Researchers should consider obtaining and storing a biologic specimen containing genomic DNA from cases and controls whenever possible. Given the limitations of many current markers of chemical exposures in case-control studies, questionnaires will probably remain the major instrument used to collect information about exposures that modify the expression of underlying genetic polymorphisms or interact with the phenotypic expression of those polymorphisms.

Using biomarkers to define cases detected by screening   Screening programs of healthy individuals can generate cases that can be evaluated in case-control studies. For example, cervical pap smear and sigmoidoscopy screening may detect individuals with early neoplasia and thus define case status for cervical or colon cancer, respectively. An individual who is asymptomatic in a screening study may have less distortion of markers of exposure or markers of susceptibility than would cases that were clinically ascertained. This difference is generated because early disease will be less likely to affect physiology. Further, under some conditions, the association of a potential marker of early biologic effect with case status may be evaluated [e.g., colonic polyamine content and ornithine decarboxylase activity in subjects with and without colonic polyps (McGarrity, *et al.*, 1990)]. Biomarkers of exposure and phenotype assays of genetic susceptibility may be applied more readily in such studies, since misclassification will be less of a problem.

An intermediate marker with a predictive value for developing disease may, in and of itself, be used to define cases from a population undergoing screening [e.g., HIV antibody positivity (Rayfield *et al.*, 1988)]. In such cases, the marker serves to define both cases and controls. Again, markers that represent earlier processes in the exposure–disease continuum can be applied more readily in these instances.

Longitudinal studies (prospective cohort, concurrent cohort)   Longitudinal studies have the potential to use the full range of biomarkers in the continuum from exposure to disease, as long as markers at the target site or appropriate surrogate markers are readily accessible. These are the most powerful molecular epidemiologic designs, yet they are also, in general, the most complex, time consuming, and costly.

The design of longitudinal studies, in which all study subjects provide a biologic specimen on entry into the study, will be determined, in part, by the latency from exposure to disease and the proportion of the study population expected to develop the disease during the span of the study. For any given disease, a study population can be selected that is representative of the general population, thus maximizing the generalizability of the study findings. Alternatively, the study population may be selected to be initially at high risk for developing the disease (e.g., homosexual men followed for development of HIV positivity or high-risk middle-aged men followed for development of coronary artery disease). This selection results in a more efficient study but may produce results that are less generalizable.

A wide range of prospective longitudinal studies has been used to study a spectrum of diseases in which all subjects are sampled and measured at entry. In some instances, repeat sampling and analysis occurs as the cohort is followed over time. Examples of prospective studies include measuring cholesterol levels and other variables in a cohort of healthy subjects that is followed for development of coronary heart disease (Kannel *et al.*, 1971), measuring blood lead levels in a cohort of children monitored *in utero* (via maternal blood lead levels) and followed through the first years of childhood for adverse outcomes such as altered growth patterns and cognitive development (Shukla *et al.*, 1991; Dietrich *et al.*, 1991, among others), and following CD4 lymphocyte levels during the progression of HIV-1 infected subjects to AIDS (Detels *et al.*, 1988; Anderson, *et al.*, 1990; Example 8; among others).

Several methodologic issues arise in prospective studies that sample and measure all members of the cohort over time. Among them are the correlation between repeat measures made on a single subject and laboratory drift over time.

In longitudinal studies, more than one point of evaluation, that is, periodic assessment, is possible. A cohort member initially will have the marker or not, and may go on to develop it or lose it in the follow-up period of the process. The effect marker may appear and disappear or, if the marker is a continuous variable, may increase or decrease at different points of monitoring (Gidlow *et al.*, 1986).

Special procedures must be incorporated into the analysis of repeat measures made on individuals in a cohort. Failure to take into account the correlation of these measures results in inconsistent estimates of precision (Zeger and Liang, 1992). In 1986, Liang and Zeger described techniques for performing longitudinal data analysis using generalized linear models (Liang and Zeger, 1986). This initial work has had substantial use in the analysis of both discrete and continuous outcomes (reviewed in Zeger and Liang, 1992). The analysis of longitudinal data is discussed further in Chapter 7.

When biomarkers are collected and analyzed over extended periods of time, trends in the data may be misleading (Thompson, 1983), especially in prospective studies in which it has not been possible to randomize the order of entry of participants. Often, trends in laboratory data can be the result of long-term drifts in values differentially weighted toward people who have been recruited near the beginning of the study as opposed to those recruited toward the end (Thompson, 1983).

Longitudinal studies are also the most reliable designs for determining the predictive value of markers. Such studies are costly and time consuming, but can be used to resolve questions about predictability. For example, numerous studies of cytogenetic markers, such as chromosomal aberrations, sister chromatid exchanges, and chromosomal micronuclei, have been done yet there still remains a question about whether these markers are predictors of cancer. To answer the question, researchers in Nordic countries are conducting a longitudinal study of individuals for whom cytogenetic tests have been performed. Although the time elapsed has been relatively short (a few years), already a positive trend has been noted for the frequency of chromosomal aberrations and the risk of all cancers (Brogger *et al.*, 1990).

Nested case-control and case-cohort designs (hybrid studies)　Various hybrid designs for epidemiologic studies have been described (Kleinbaum *et al.*, 1982). They may combine elements of two basic designs, extend a design through repetition, or combine elements of a basic observational design with elements of nonobservational designs. Two of the more useful hybrid studies in molecular epidemiology are the nested case-control and case-cohort variants. These designs are efficient and powerful in assessing relative risks because they require exposure, covariates, and specimen collection on only a small subset of a cohort (Wacholder, 1991).

The nested case-control study design, as formalized by Miettinen (1982), involves collecting information on all members of a cohort, storing data and biologic material, following the cohort until the development of the disease, and then analyzing the stored material for the cases and time-matched controls (noncases from the cohort). The case-cohort design is an unmatched variant of the nested case-control design (Wacholder, 1991). In case-cohort studies within a cohort, a random group (the subcohort) that will serve as controls is defined at the outset of the study, whereas in the nested case-control studies, controls are defined after the cases have been identified (Van Noord, 1992).

The nested case-control study design has had broad utility in molecular epidemiologic studies of a wide range of disease categories. It is particularly efficient for evaluating relatively rare disease outcomes, such as leukemia, since only a small portion of the sampled cohort is expected to develop the disease. Some examples include evaluating *Helicobacter pylori* infection and risk of gastric carcinoma (Parsonnet *et al.*, 1991), serum 25-hydroxyvitamin D levels and risk for developing colon cancer (Garland *et al.*, 1989), Epstein–Barr virus exposure and risk for non-Hodgkin's lymphoma (Mueller *et al.*, 1991), urinary aflatoxin biomarkers and risk of hepatocellular carcinoma (Ross *et al.*, 1992), metabolic precursors of essential hypertension (Selby *et al.*, 1990), and cholesterol apolipoproteins and risk of myocardial infarction (Stampfer *et al.*, 1991).

A prototype example of the utility of exposure markers in a nested case-control study is the work of Ross, Groopman, and colleagues (Groopman *et al.*, 1992; Ross *et al.*, 1992). They used assays for urinary aflatoxin B_1, its metabolites, and DNA adducts (which Groopman had previously validated) to assess the relationship between aflatoxin exposure and liver cancer. They matched 22 cases of liver cancer with 5–10 controls per case. Subjects with liver cancer had a 6-fold relative risk of aflatoxin metabolites (RR, 6.2; 95% CI, 1.8–21.5) compared with controls. Additionally, there was a strong interaction between serological markers of chronic hepatitis B infection and aflatoxin exposure in liver cancer risk. This study demonstrates how a marker of exposure can be linked with a disease outcome to confirm an association that was suspected to be causal and to evaluate its interaction with another risk factor.

When the amount and number of the specimens is limited in nested case-control studies, sequential methods are useful and may be more efficient than analyzing all specimens simultaneously. Sequential analysis is a statistical method that allows an experiment to be ended (described by a stopping rule) as soon as an answer of the desired precision is obtained (Armitage, 1977). Sequential methods also are useful when it is important to determine the presence of an effect quickly; the method allows for an earlier determination than would otherwise be possible. Sequential methods in clinical research have a long history (Wald, 1945). Several reasons sequential approaches are attractive for molecular epidemiology in nested case-control studies include the following. (1) Even with few cases, there is a theoretical advantage in presence of an effect quickly; the method allows for an earlier determination than would otherwise be possible. Sequential methods in clinical research have a long history (Wald, 1945). Several reasons sequential approaches are attractive for molecular epidemiology in nested case-control studies include the following. (1) Even with few cases, there is a theoretical advantage in efficiency that can lead to substantial reduction of the length of total follow-up. (2) When sufficient cases are available, the stopping rule for sequential analyses will minimize the number of cases for which biologic samples need to be used up (i.e., entirely analyzed) before a decision about whether an intervention is effective can be made (Van Noord, 1992). The sequential approach involves the application of t tests for each detection of a case and its matched controls using a sequential probability ratio test. More details are described in the study by Van Noord (1992). Despite the potential utility of sequential methods, care should be taken in their application so that biased or inappropriate comparisons of cases and controls are voided.

Intervention Trial Studies

One of the earliest proposed uses of the current generation of biologic markers was to assess the impact of intervention in cohorts at increased risk of cancer and heart disease [Greenwald, 1982; Lipid Research Clinics Program (LRCP), 1984]. For example, in cancer intervention trials, the assumptions were that the marker-indicated cancer was likely to occur and that reduction of the marker was synonymous with control (reduction) of the disease. If such assumptions are true, this type of study is reasonable. The actual correspondence of the marker with the disease (i.e., validity) needs to be evaluated, however, prior to implementing this type of study. Also, the design should insure that statistical significance is reached only if the intervention difference exceeds a specified value of clinical importance (Rumke, 1983). Biologic markers for cancer have been studied widely and provide a profile of the qualities of a useful marker in intervention studies. Ideally, a marker should be useful in risk assessment, aid in the diagnosis of precancerous lesions or early preclinical tumor development, reflect tumor burden, permit prognostic evaluation, serve to monitor the effects of therapy, and permit detection of early recurrence (Nieburgs, 1985). For cancer tumor markers, it seems unlikely that a single marker will be ideal for all possible uses, even with respect to a single type of cancer. Indeed, requirements for different purposes might be mutually exclusive (Bagshawe, 1983). Combinations of different markers, or markers and risk-factor questionnaires, used with various statistical techniques, such as discriminant function or multivariate analyses, might be more useful, therefore, than a single marker (Chisholm *et al.*, 1986).

Another type of intervention study involves the use of biologic markers in the early detection of disease in high-risk groups (Schulte *et al.*, 1986). Various biologic markers can be used, for example, in an occupational cohort at risk of bladder cancer, to differentiate subgroups at risk by using some combination of markers and traditional epidemiologic factors in a discriminant function analysis. Based on this analysis, subgroups can be screened differentially, thus providing prevention for modalities in a cost-effective manner.

In intervention trials, it is important to differentiate direct effects of the interventions (or exposure) from effects relayed through an intermediate variable. The use of the paradigm that a continuum of biologic markers relates an exposure to a disease raises the question of the relative importance of different possible pathways for the effect. Separating the direct effects of an exposure from effects mediated through an intermediate variable (indirect effects) is a problem in epidemiology in general (Robbins and Greenland, 1992) and in molecular epidemiology in particular. Robbins and Greenland (1992) showed that adjustment for the intermediate variable, which is the most common method of estimating direct effects, can be biased. Even in a randomized crossover trial of exposure, direct and indirect effects cannot be separated without assumptions, that is, direct and indirect effects are not separately identifiable when only exposure is randomized. Separation of effects requires a trial randomizing both exposure and intervention. For example, Robbins and Greenland assessed the relationship of hyperlipidemia and cardiovascular disease under different conditions of smoking status. They used the notion of a counterfactual, that is, each subject in the cohort is observed under one set of circumstances and is also considered under so-called counterfactual circumstances—circumstances that, contrary to fact, did not occur. For example, for each subject, they considered whether hyperlipidemia would occur if he smokes, whether hyperlipidemia would occur if he quits smoking, and whether cardiovascular disease would occur under each possible combination of smoking and lipid status. Only when they made assumptions about different probabilities for combinations of effects could they calculate valid estimates of effect. To overcome this problem, they showed that direct (smoking) and indirect (lipid) effects can be separated when there is an intervention that can block the effect of an intermediate cofactor (hyperlipidemia). Thus, hyperlipidemia may be treated with drugs or diet. Randomization of such intervention within exposure levels will allow the calculation of separate effects (Robbins and Greenland, 1992).

Confounding and Effect Modification

Confounding is a classic epidemiologic concept that addresses the mixing of the effect of exposure with that of an extraneous factor (Rothman, 1986). In molecular epidemiologic designs that use variables in the continuum between an exposure and a resultant disease, the determination of confounding may be problematic. Since the continuum represents a serial array of causes (that is, an exposure causes a biologic change that may cause other changes leading to disease), most of the variables in a continuum will be considered risk factors for the downstream markers and, ultimately, for disease. They also will be correlates of predecessor markers leading back to, and correlating with, exposure. Nonetheless, the general wisdom is that any step in the causal chain between exposure and disease is not a confounding factor, since the effect of exposure is mediated through altered molecular, biochemical, or physiological change (Rothman, 1986). In contrast, other causes of the markers, if correlated with exposure, would be confounders.

Assessment of confounding in molecular epidemiologic studies depends on what question is being asked and what subset of the continuum is being evaluated. When confounding is present and uncontrolled, it can lead to an over- or underestimation of an association and should be controlled. In some cases, the wrong variable may be controlled. However, controlling for a nonconfounder does not usually bias an effect estimate but may simply reduce its precision (Checkoway *et al.*, 1989).

Confounding can be controlled in the study design by randomizing subjects or by restricting subject groups to a narrow range of potential confounders (e.g., restricting the study to females with a particular metabolic phenotype and within a narrow age range). It is also possible to match study subjects on potential confounders. These issues have been discussed widely in epidemiology texts (Rothman, 1986; Checkoway *et al.*, 1989; Margetts and Nelson, 1991). For molecular epidemiologic studies, the general principles hold, but some of the details may be confusing. For example, selecting factors to restrict or to match may depend on the extent to which a continuum is defined and an explicit mechanism of causal action is postulated. Rothman (1986) counsels that profound uncertainties about mechanism can justify handling a potential confounding factor as both confounding and not confounding in different analyses. He gives the example of how to treat serum cholesterol levels in evaluating the effect of fatty diet on cancer risk. Cholesterol may be a risk factor for cancer and is associated with a fatty diet, but also possibly mediates the action of a fatty diet on cancer risk. Thus, ". . . one might consider serum cholesterol levels and evaluate the effect of a fatty diet on cancer risk mediated by factors other than serum cholesterol (by controlling confounding by serum cholesterol), in addition to considering it an intermediate cause, and consequently, evaluating the effect of a fatty diet ignoring serum cholesterol in the analyses."

A major strength of molecular epidemiologic research is the elucidation of effect modification. Effect modification historically has been the epidemiologic term to describe statistical interaction, that is, depending on the statistical methods used, two or more risk factors will have joint effects that are either additive or multiplicative. The contribution of molecular epidemiology is not in refining the statistical approach, but in providing variables to acto calculate stratum-specific effect estimates. Molecular and other biologic markers of exposure or susceptibility provide a basis for selecting those strata. The molecular approach depends on finding candidate genes, phenotypes, or genetic "markers" that correlate with disease risk (Taylor, 1990a). The use of susceptibility markers, which may be extremely potent effect modifiers, should make it possible to demonstrate high disease risks for exposed susceptible groups and lower risks for nonsusceptible exposed and unexposed groups (Hulka, 1990). A single gene–environment interaction is likely not to be the only cause of a disease since each single gene may not have a strong effect but, in conjunction with other genes, may shift the risk profile in an unfavorable direction if not counteracted by still other genes or by additional environmental factors (Nora *et al.*, 1991). Multiple susceptibility factors must be accounted for to represent accurately the true dimensions of the interaction between genetic and environmental factors.

Examples of Epidemiologic Study Designs with Various Types of Biomarkers

Examples of each of the components in the continuum are described for four study designs: cross-sectional, cohort, case-control, and intervention trial. Examples are transitional studies, etiologic and prevention studies, and studies that apply previously validated biomarkers for a spectrum of purposes. For each scenario, a brief review and description of the study design is presented, including how the marker was used and whether analytical or interpretive issues were associated with the use of the marker. In some instances, the categorization of a marker is subjective and more than one category might be appropriate. An overview of the studies discussed is shown in Table 6.3.

============================= *Example 1* =============================

Study:	"DNA adducts in humans environmentally exposed to aromatic compounds in an industrial area of Poland" (Hemminki, 1990).
Marker/medium:	DNA adducts/lymphocyte DNA
Assay:	Competitive ELISA with fluorescence detection using antibodies raised against anti-B[a]P-7,8-diol-9,10-epoxide and by ^{32}P-postlabeling of DNA adducts
Design/analysis:	Cross-sectional study using DNA adducts as markers of exposure to aromatic compounds. The levels of adducts were compared among 63 coke oven workers, 19 local urban controls, and 15 rural con-

TABLE 6.3 Examples of Epidemiologic Study Designs with Various Types of Biomarkers[a]

Study design	Dose (ID or BED)[b]	Type of biomarker			
		Early biologic effect	Altered structure/function	Clinical disease	Susceptibility
Cross-sectional	DNA adducts (1)	SCE (2)	DNA hyperploidy (3)	CA 549 (4)	Acetylation phenotype (5)
Cohort	Islet cell antibodies (6)	Prostate specific antigen (7)	CD4, β_2M (8)	Tumor-associated antibody, ACTH, AAT, AFP (9)	HLA, DR3, DR4 haplotypes (10)
Case-control	Peptide sequences HIV1 and HIV2 (11)	Polyamines and ornithine decarboxylase (12)	Nucleotide sequence in HBV (13)	Urokinase-plasminogen activator (14)	CYP2D6 genotype (15)
Intervention trial	4-ABP-adducts (16)	Serum thromboxane B_2 (17)	Glycosylated hemoglobin (18)	Microalbuminuria (19)	Apolipoprotein A1 (20)

[a] Numbers in parentheses associated with each study refer to examples at the end of this chapter.
[b] ID, internal dose; BED, biologically effective dose.

Results:

trols. Data were adjusted by analysis of covariance for age, smoking, and other occupational exposures to aromatic compounds.

Local urban controls exhibited adduct levels and patterns similar to those of the coke oven workers. The rural controls were 2–3 times lower.

Issues pertaining to the use of markers:

Most assays were done in duplicate at two different laboratories and showed a similar pattern that differed by a factor of two ($r=0.66$; $p<0.01$).

Large interindividual variability in the level of DNA adducts: ~10-fold among local and rural controls; ~50-fold among coke oven workers.

═══════════════════ Example 2 ═══════════════════

Study:

"Sister chromatid exchanges in painters recently exposed to solvents" (Kelsey, 1989).

Marker/medium:

Sister chromatid exchange (SCE) peripheral blood lymphocytes

Assay:

Harlequin stain with 5-bromodeoxyuridine (BrdU); blind readers; only mitoses with 40–46 chromosomes

Design/analysis:

Cross-sectional study of a cohort of 117 painters to assess the effect of recent solvent exposure on frequency of SCE. Four groups, smoking and non-smoking exposed and control workers, were compared. SCE frequencies were compared using analysis of variance with linear models. Logarithmic transformation of SCE data was used.

Results:

Significant elevation of SCE frequency was associated only in the workers who were smokers and who currently were exposed to solvents ($r = 0.15$; $p<0.006$). Current exposure was most significant in subjects when estimated by the number of days worked over the month prior to venipuncture.

Issues pertaining to the use of markers:

Cigarette smoking was an important confounder that needed to be controlled.

Kinetics of solvent exposure are important in assessing the genotoxic effect.

═══════════════════ Example 3 ═══════════════════

Study:

"DNA hyperploidy as a marker for biological response to bladder carcinogen exposure" (Hemstreet et al., 1988).

Marker/medium: DNA hyperploidy/exfoliated bladder cells in urine
Assay: Absolute measure of Nuclear Fluorescence Intensity (ANFI)
Design/analysis: Cross-sectional study of 504 workers. DNA hyperploidy was measured with a quantitative fluorescence cytology method using the DNA-binding dye acridine orange. Odds ratios were calculated for the relationship between DNA measurements and various risk factors.
Results: Prevalence of DNA hyperploidy increased in an exposure–response manner from 3.5 to 60% with increasing duration of exposure to 2-naphthylamine. When characterized by exposure and smoking together, just by exposure, just by smoking, or neither, the percentage of DNA positives was 23, 18, 4, and 2, respectively. The odds ratio for the association of exposure and DNA hyperploidy was 7.6 (95% CI, 3.6–16.2).
Issues pertaining to the use of markers: Appearance of the greatest prevalence of DNA hyperploidy in persons with both occupational and smoking exposure, and the lower prevalences in those with no or one risk factor(s) is consistent with what would be predicted on the basis of the relative potency of the risk factors.

═══════════════ *Example 4* ═══════════════

Study: "A new biomarker in monitoring breast cancer." (Beveridge, 1988).
Marker/medium: CA 549/blood
Assay: Radioimmunoassay (RIA)
Design/analysis: Cross-sectional study of 682 randomly selected women with breast cancer, other malignancies, and other benign disorders was conducted to determine the sensitivity and specificity of CA 549. Differences among serum levels in the different groups were assessed by chi-square test, McNemar's test for paired groups, and multiple regression analyses. CA 549 also was compared with Carcinoembryonic Antigen (CEA) as a tumor marker.
Results: CA 549 was found to have a sensitivity of 77% and a specificity of 92% and was statistically significantly better than CEA in indicating a 25% or greater change in women with metastatic disease.

Issues pertaining to   Comparison group of women with no disease was
the use of markers: included.
                          All subjects were selected randomly.
                          Prospective study is still needed to assess the value
                          of CA 549 as a tumor marker for breast cancer.

===================== **Example 5** =====================

Study: "Acetylation phenotype, carcinogen–hemoglobin
                      adducts, and cigarette smoking" (Vineis *et al.*,
                      1990).
Marker/medium: Acetylation phenotype/urine
Assay: Measurement of the urinary excretion of metabo-
                      lites following a test dose of caffeine
Design/analysis: Cross-sectional study of smokers of blonde to-
                      bacco, smokers of black tobacco, and nonsmoking
                      subjects to determine if the level of 4-ami-
                      nobiphenyl (ABP)–hemoglobin adducts in their
                      blood was related to the acetylation phenotype.
                      Groups were compared by analysis of variance and
                      linear regression.
Results: Slow acetylator phenotype is associated with the
                      detection of ABP in adducted hemoglobin.
Issues pertaining to   This study evaluates a variable (acetylation pheno-
the use of markers: type) that modifies a dose (smoking)–response (4-
                         aminobiphenyl-hemoglobin adducts) relationship.

===================== **Example 6** =====================

Study: "Islet cell antibodies as predictive markers for
                      IDDM in children with high background incidence
                      of disease" (Karjalainen, 1990).
Marker/medium: Islet cell antibodies (ICA)-serum
Assay: Standard indirect immunofluorescence test
Design/analysis: Cohort of 1212 Finnish children and adolescents
                      was followed for 8 years for the development of in-
                      sulin-dependent diabetes mellitus (IDDM). Serum
                      samples were collected initially, and 3 and
                      6 years after initiation of the study in all ICA+
                      and in 296 ICA− children. Statistical analysis in-
                      volved univariate statistics and one way analysis of
                      variance for repeated tests in the case of
                      normally distributed variables and Spearman's
                      rank correlation test, Mann–Whitney U test,
                      and Kruskal–Wallis one-way ANOVA for skewed
                      variables.

Results: 50 children (4.1%) were positive for ICA. High persistent levels of ICA [≥ 18 Juvenile Diabetes Foundation units (JDF U)] had a positive predictive value of 50%.

Issues pertaining to the use of markers: Assay had an intrassay coefficient of variation <1.6% and an interassay coefficient of variation of 8.4%.

Long prediabetic period in ICA+ subjects requires a longitudinal study to assess adequately the prevalence of ICAs and evaluate the role of ICAs in predicting IDDM in a nondiabetic population.

Predictive value of low levels (<5 JDF U) of ICA is low, and episodic fluctuation may represent nonspecific immunologic reactions or a consequence of viral infections.

═══════════════════ *Example 7* ═══════════════════

Study: "Measurement of prostate-specific antigen in serum as a screening test for prostate cancer" (Catalona *et al.,* 1991).

Marker/medium: Prostate specific antigen (PSA)/serum
Assay: Immunometric assay
Design/analysis: Two groups were assessed for PSA, a serine protease secreted by prostatic epithelial cells, and followed to determine the incidence of prostate cancer. 1653 ambulatory, healthy male volunteers, 50 years or older, and with no history of prostate disease participated in the study. Serum PSA levels were drawn. The control group consisted of 300 men, 50 years or older, who were evaluated consecutively during the same interval of time with ultrasound-guided biopsy of the prostate for abnormal rectal exam with suspicion of prostate cancer, equivocal rectal exam, signs/symptoms of benign prostatic hypertrophy, elevated levels of serum PSA or acid phosphatase, chronic prostatitis, and miscellaneous other conditions. A log transformation of serum PSA levels was used to normalize the distribution. A multivariate logistic regression analysis was used to determine the extent to which PSA levels predicted prostate cancer, after adjustment for age and findings on rectal exam and ultrasonography.

Results:	Serum PSA had the highest positive predictive value (40%) compared with 33% for rectal examination and 28% for ultrasonography, and the lowest error, 36%.
Issues pertaining to the use of markers:	Histopathologic biopsy used as the reference standard to compare with PSA. However, "blind" and independent comparison of the marker with the standard was lacking.

===================== *Example 8* =====================

Study:	"Use of β_2-microglobulin level and CD4 lymphocyte count to predict development of acquired immunodeficiency virus infection" (Anderson, *et al.* 1990).
Marker/medium:	β_2M/serum; CD4/lymphocytes; HIV antibodies
Assay:	β_2M/RIA; CD4/cell sorter and counter; HIV antibodies/indirect immunofluorescent assay
Design/analysis:	Cohort of 962 single men, recruited by multistage probability sampling from 19 census tracts, was examined, interviewed, and tested every 6 months for 36 months. Progression to AIDS for various levels of β_2M and CD4 was examined using the Kaplein–Meier product limit method. Cox proportional-hazards regression model was used to evaluate the independence of β_2M and CD4 as discrete predictors of progression.
Results:	β_2M was found to be a strong predictor of AIDS in infected persons, either by itself or in combination with the CD4 T lymphocyte count.
Issues pertaining to the use of markers:	β_2M and CD 4 were found to have prognostic implications.
Nomogram summarized the probability of an HIV-infected person developing AIDS in 36 months depending on prevalent levels of serum β_2M and CD4 lymphocytes.

===================== *Example 9* =====================

Study:	"The use of biomarkers in the prediction of survival in patients with pulmonary carcinoma" (Walop, *et. al.*, 1990).
Marker/medium:	Alpha 1-antitrypsin/blood; alpha fetoprotein/blood; CEA/blood; C3/blood; IgA; IgG; cortisol; HCG; insulin; prolactin; ACTH; N-terminal peptide of propriomelanocotin; TAA/blood

Assay: Various techniques: RIA, radial diffusion
                            techniques

Design/analysis: Prognostic value of 16 biomarkers was determined
                            in 119 patients with pulmonary carcinoma. Ques-
                            tionnaire information was collected, and patients
                            were followed during a 3-year period. Survival
                            analysis utilized Kaplan–Meier survival curves and
                            !Cox and forward stepwise regression to determine
                            which biomarkers were predictive of survival.

Results: Elevated levels of TAA and cortisol were associated
                            with longer survival.

Issues pertaining to   Selection of biomarkers by stepwise · regression
the use of markers: needs to be interpreted with caution, especially
                            since the Z scores were found to be dependent on
                            the particular variables included in the model.
                            There is the possibility of collinearity affecting the
                            selection in the forward stepwise procedure.

=================== *Example 10* ===================

Study: "Evolution of the Pittsburgh studies of the epide-
                            miology of insulin-dependent diabetes mellitus"
                            (Kuller, *et al.* 1990).

Marker/medium: HLA DR3 and HLA DR4 haplotypes/lymphocytes

Assay: Total of 367 consecutive newly diagnosed cases

Design/analysis: of insulin-dependent diabetes mellitus (IDDM)
                            were evaluated to determine their HLA DR haplo-
                            types. Both HLA DR3 and HLA DR4 haplotypes
                            were associated with substantially increased risk of
                            IDDM. A prospective study was initiated to deter-
                            mine the incidence of siblings contracting IDDM
                            based on the sharing status in siblings of the HLA
                            DR haplotypes.

                            Life table estimates, using the Cutler–Ederer
                            method, were used to determine the cumulative
                            risk in siblings of IDDM subjects; chi-square analy-
                            sis to determine the significance between expected
                            and observed cases, and relative risk calculations
                            by the standard method to determine risk of con-
                            tracting IDDM when HLA DR haplotypes are
                            present, were employed in the study.

Results: Purpose of the study was to determine if genetic
                            haplotypes of HLA DR were associated with an in-
                            creased incidence of insulin-dependent diabe-
                            tes mellitus (IDDM). Diabetic patients with HLA

DR4 were more likely to be female, to have a more severe onset of disease, and to have had a more recent viral infection than diabetic patients with HLA DR3. The expected vs. observed incidence (%) of IDDM in siblings of diabetic patients stratified by sharing status of haplotypes were 25% vs. 68% (share 2), 50% vs. 26% (share 1), and 25% vs. 6% (share 0).

Haplotypes of HLA DR indicate an overwhelming susceptibility to develop IDDM. Monitoring siblings of diabetic subjects in combination with other markers (antibodies to insulin or β islet cells) may identify patients at an earlier point in the onset of the disease and permit insights into its etiology.

Issues pertaining to the use of markers: HLA DR3 and HLA DR4 haplotype individuals are more likely to develop IDDM than those who do not possess either haplotype. The combination of HLA DR3 and HLA DR4, compared with neither HLA DR3 nor HLA DR4, resulted in an estimated 40- to 60-fold increase in the risk of developing diabetes.

======= *Example 11* =======

Study: "Mixed human immunodeficiency virus (HIV) infection in an individual: Demonstration of both HIV Type 1 and Type 2 proviral sequences by using polymerase chain reaction" (Rayfield *et al.,* 1988).

Marker/medium: Peptide sequences/peripheral blood mononuclear cells

Assay: Western blot analysis of DNA amplified by polymerase chain reaction

Design/analysis: Nested case-control study of sera, from persons seroreactive to HIV-1 and HIV-2 by enzyme immunoassays was selected from a seroprevalence study of 944 persons. Sample of 5 subjects positive for both HIV-1 and HIV-2 was assessed by Western blot and correlated with clinical conditions.

Results: Finding of two subjects with AIDS, infected only with HIV-2, provides support for a causal association.

Issues pertaining to the use of markers: Only a small number of subjects is required to reach the conclusion.

DNA amplification material was derived from

pooled chromosomal material, and was not directed at the chromatin level of a single infected cell. Whether interference occurred at the cellular level was not addressed.

====== *Example 12* ======

Study: "Colonic polyamine content and ornithine decarboxylase activity as markers for adenomas" (McGarrity, *et. al.,* 1990).

Marker/medium: Putrescine, spermidine, spermine/tissue; ornithine decarboxylase (ODC)/tissue

Assay: HPLC (polyamines); radioimmunoassay (ODC)

Design/analysis: Case-control study using patients with and without adenomatous polyps. Subjects were selected from patients scheduled to undergo colonoscopy. Patients with a history of inflammatory bowel disease, polyposis syndrome, prior bowel resection, or cancer were excluded. Comparison between patients with a polyp at colonoscopy, patients with no polyp and no history of polyps, and patients with no polyps at colonoscopy but a history of polyps were made by ANOVA followed by a post hoc student-Neuman–Keuls test. Logistic regression was used to determine the best cut off for ODC and polyamine values between cases and controls.

Results: ODC activity and its direct metabolic product, putrescine, were increased in normal-appearing colonic mucosa of patients with adenomas, compared with the colonic mucosa of patients without adenomas.

Issues pertaining to the use of markers: Study was not conducted blind by laboratory personnel.

No control for age. Patients without adenomatous polyps were approximately 10 years younger than those with polyps.

====== *Example 13* ======

Study: "Application of hepatitis B virus (HBV) DNA sequence polymorphisms to the study of HBV transmission" (Lin *et al.,* 1991).

Marker/medium: Nucleotide sequence in positions 2551–2650 (1: *Eco*RI site)/serum HBV DNA

Assay: Enzyme-linked immunosorbent assay (ELISA); polymerase chain reaction (PCR)

Design/analysis: Case-control-like study of 96 children with hepatitis B e antigen (HB e Ag)-positive and 97 parents tested for HBV serologic markers by ELISA. Hypothesis of random sequences of nucleotides in children and parents was tested.

Results: Intrafamilial transmission of HBV was demonstrated by a nonrandom assortment ($p<0.00005$) of sequences in parents and children.

Issues pertaining to   Some sequences were different by 1% among
the use of markers: family members. The researchers excluded the possibility that such differences resulted from misincorporation of bases in the polymerase chain reaction, the extent of which has been estimated at 0.25% for the products of a 30-cycle amplification. The observed differences could probably be explained by spontaneous mutation after infection or by heterogeneity of the HBVs at the source of infection.

================ *Example 14* ================

Study: "Urokinase-plasminogen activator, a new and independent prognostic marker in breast cancer" (Duffy, *et al.*, 1990).

Marker/medium: Urokinase-plasminogen activator (UK-PA)/breast tumor tissue

Assay: Enzyme-linked immunosorbent assay (ELISA)

Design/analysis: Tumors from 166 patients with breast cancer were assayed for UK-PA. Those with high levels of UK-PA were compared retrospectively with those with low levels for tumor stage, axillary node status, treatment, and survival. Univariate and multivariate survival analyses were used to identify the role of UK-PA as an independent risk factor.

Results: Patients with high levels of UK-PA antigens have significantly higher risk of early disease recurrence (RR=4.1) and shorter overall survival (RR=8.76) than those with lower levels UK-PA. UK-PA was an independent risk factor for both disease-free interval and survival, independent of tumor size, axillary node status, and estradiol receptor.

Issues pertaining to   Cut off point for UK-PA that gave the best prognostic
the use of markers: tic discrimination was 10 ng/mg total protein.

========================= *Example 15* =========================

Study: "Debrisoquine hydroxylase gene polymorphism
                    and susceptibility to Parkinson's disease" (Smith *et
                    al.,* 1991).
Marker/medium: Mutant CYP2D6 alleles/blood, brain tissue
Assay: Polymerase chain reaction (PCR); separation of re-
                    striction fragments by polyacrylamide gel electro-
                    phoresis for G-to-A transition and for base-pair de-
                    letion in exon 5
Design/analysis: 229 patients with clinically diagnosed Parkinson's
                    disease were compared with 720 controls from
                    three widely separate geographic regions. Controls
                    were selected randomly from healthy volunteers
                    or blood samples taken at hospital clinics. Blood
                    samples were processed to provide cell extracts
                    suitable for PCR analysis. Banding patterns of am-
                    plified DNA were determined by restriction frag-
                    ment analysis.
Results: Overall, individuals with a metabolic defect in the
                    cytochrome P450 CYP2D6-debrisoquine hydrox-
                    ylase gene with the poor metabolizer phenotype
                    had a 2.54-fold (95% CI, 1.51–4.28) increased
                    risk of Parkinson's disease.
Issues pertaining to No matching of controls for age or sex since these
the use of markers: are not factors in CYP2D6 genotype.
                    Two analyses were used: one for a G-to-A tran-
                    sition and one for a base-pair deletion.
                    Prior use of pharmacokinetic assays for the de-
                    brisoquine phenotype have yielded equivocal or con-
                    flicting results for detecting an association with
                    Parkinson's disease. Use of DNA-based genetic as-
                    says, such as in this study, could accurately identify
                    90% of the poor metabolizer phenotypes resulting
                    in an unambiguous association.

========================= *Example 16* =========================

Study: "Decline of the hemoglobin adduct 4-aminobi-
                    phenyl during withdrawal from smoking" (Ma-
                    clure *et al.,* 1990).
Marker/medium: 4-aminobiphenyl (ABP)–Hb/erythrocytes; coti-
                    nine/plasma
Assay: Gas chromatography with negative-ion chemical

	ionization mass spectrometry; radioimmunoassay (RIA)
Design/analysis:	4-ABP–Hb and cotinine were used as markers of exposure in a study of 34 smokers enrolled in a withdrawal program. Smokers were assessed at various times, and differences between observed and expected percentage declines in excess adduct formation as a function of time were tested by the one sample *t* test.
Results:	4-ABP–Hb decreased from a mean of 120 ± 6 pg/g after 3 weeks. There was only a weak association between the level of 4-ABP–Hb at the start of the withdrawal program and the daily consumption of cigarettes during the previous months.
Issues pertaining to the use of markers:	Other sources of 4-aminophenyl must be identified, as well as factors that modify the rate of adduct formation with hemoglobin *in vivo*. Weakness of the between-person association of 4-ABP and cigarette smoke intake when compared with the striking associations within individuals suggests that factors such as diet, acetylation phenotype, or enzyme induction may obscure the relation between 4-ABP–Hb levels and 4-ABP intake.

Example 17

Study:	"Biochemical markers of compliance in the physician's health study" (Satterfield, *et al.,* 1990).
Marker/medium:	Thromboxane B_2/serum
Assay:	Double antibody radioimmunoassay
Design/analysis:	Randomized double-blind placebo-controlled trial using a 2×2 factorial design to test the effects of low-dose aspirin on risk of cardiovascular disease. Thromboxane B_2 was measured to assess compliance with aspirin use. The Wilcoxon rank sum test was used to evaluate levels between the aspirin and aspirin placebo groups.
Results:	Thromboxane B_2 levels were significantly different between treatment and placebo groups.
Issues pertaining to the use of markers:	Thromboxane B_2 levels may have a certain amount of bias due to improved compliance in anticipation of blood sampling. Since levels of serum thromboxane B_2 are known to be reduced by aspirin, it represents a biologic ef-

fect that, in this study, is used as a marker of exposure.

========================= *Example 18* =========================

Study: "Effect of long-term monitoring of glycosylated hemoglobin levels in insulin-dependent diabetes mellitus" (Larsen, *et al.*, 1990).

Marker/medium: Glycosylated hemoglobin (hemoglobin A_{1c})/venous blood

Assay: Isoelectric focusing

Design/analysis: Randomized controlled trial in which 240 patients with insulin-dependent diabetes mellitus (IDDM) were assigned to one of two groups and then followed for 1 year during which hemoglobin A_{1c} was measured every 3 months. Hemoglobin values were used to control or modify therapy. Differences between patients were assessed by two-tailed t test. Changes in treatment regimens were compared by chi-square test.

Results: Hemoglobin A_{1c} was a useful marker of altered structure/function as reflected by the fact that the proportion of patients with poor blood sugar control decreased significantly from 46 to 30% in the group whose hemoglobin A_{1c} was known by the physician.

Issues pertaining to the use of markers: Marker was studied using techniques commonly employed in clinical trials; however, in this case, the "therapeutic maneuver" was the knowledge of the HbA_{1c} values.

========================= *Example 19* =========================

Study: "Prevention of diabetic nephropathy with enalapril in normotensive diabetics with microalbuminuria" (Marre, *et al.*, 1988).

Marker/medium: Albumin/urine; arterial pressure; renal function/urine

Assay: Radioimmunoassay of 24-hr urine

Design/analysis: Randomized double-blind placebo-controlled trial of enalapril. Microalbuminuria was used as a marker of diabetic nephropathy. Two-way repeated measures ANOVA was used to assess the effect of enalapril.

Results: Albumin excretion was significantly lower during treatment with enalapril than with placebo.

Issues pertaining to   Microalbuminuria, was used as a marker of dia-
the use of markers: betic nephropathy, was shown to respond to angio-
                        tensin–converting enzyme therapy in diabetics.
                        Therefore, reasonable hypothesis was that it would
                        respond to agents that cause improvement in the
                        permeability of the glomerular basement mem-
                        brane to albumin, after glomerular pressures have
                        improved.

                        Biologic marker exhibited the proper effect un-
                        der controlled conditions and did not exhibit it un-
                        der placebo conditions.

=================================== Example 20 ===================================

Study: "Genetic variation at the apolipoprotein gene loci
                        contribute to response of plasma lipids to dietary
                        change" (Xu et al., 1990).

Marker/medium: Restriction fragment length polymorphisms
                        (RFLPs) of the genes apo B, apo AII, apo E, apo AI-
                        CIII-AZU gene cluster, and the LDC receptor gene/
                        whole blood

Assay: Restriction enzymes; Southern blotting

Design/analysis: Intervention trial with fasting venous blood drawn
                        at the end of a baseline diet and after 6 weeks in-
                        tervention. This was a resampling of 107 individu-
                        als involved in a 3-year trial. A comparison was
                        made between the basal diet and the intervention
                        diet, adjusting by stepwise regression for age and
                        gender.

Results: Major effect in response to dietary changes was
                        seen in the difference in apo AI levels mediated by
                        variation at the apo B gene locus associated with
                        the MspI RFLP. This explains 6.3% of the pheno-
                        typic variance in apo AI change.

Issues pertaining to   Many associations were examined in the study, and
the use of markers: many were not independent of each other. Thus, a
                        method was used to reduce the probability of type
                        I error by limiting significance to the $p < 0.01$ level.
                        Other environmental factors needed to be kept
                        constant, so the impact of genetic variation and di-
                        etary change could be assessed.

                        Biologic marker exhibited the proper effect un-
                        der controlled conditions and did not exhibit it un-
                        der placebo conditions.

References

Anderson, R. E., Lang, W., Shiboshi, S., Royce, R., Jewell, N., Winkelstein, W. (1990). Use of β_2-microglobulin level and CD4 lymphocyte count to predict development of acquired immunodeficiency virus infection. *Arch. Int. Med.* 150, 73–77.

Armitage, P. (1977). "Statistical Methods in Medical Research." Blackwell Scientific Publications, Oxford.

Bagshawe, K. D. (1983). Tumour markers—where do we go from here? *Br. J. Cancer* 48, 167–175.

Bartsch, H., Caporaso, N., Coda, M., Kadlubar, F., Malaveille, C., Skipper, P., Talaska, G., Tannenbaum, S. R., and Vineis, P. (1990). Carcinogen hemoglobin adducts, urinary mutagenicity, and metabolic phenotype in active and passive cigarette smokers. *J. Natl. Cancer Inst.* 82, 1826–1831.

Beveridge, R. A., Chan, D. W., Bruzek, O. (1988). A new biomarker in monitoring breast cancer. *J. Clin. Oncol.* 6, 1815–1821.

Biber, A., Scherer, G., Hoepfner, I., Adlkofer, F., Heller, W.-D., Haddow, J. E., and Knight, G. J. (1987). Determination of nicotine and cotinine in human serum and urine: An interlaboratory study. *Toxicol. Lett.* 35, 45–52.

Bressac, G., Kew, M., Wands, J., and Osturk, M. (1991). Selective G to T mutations of p53 gene in hepatocellular carcinoma from southern Africa. *Nature (London)* 350, 429–431.

Brøgger, A., Hagmar, L., Hansteen, I. L., Heim, S., Hogstedt, B., Knudsen, L., Lambert, B., Linnainmaa, K., Mitelman, F., Nordenson, I., Reuterwall, C., Salomaa, S., Skerfving, S., and Sorsa, M. (1990). An inter-Nordic prospective study on cytogenetic endpoints and cancer risk. *Cancer Genet. Cytogenet.* 45, 85–92.

Bryant, M. S., Skipper, P. L., Tannenbaum, S. R., and Maclure, M. (1987). Hemoglobin adducts of 4-aminobiphenyl in smokers and nonsmokers. *Cancer Res* 47, 602–608.

Caporaso, N. E., Tucker, M. A., Hoover, R. N., Hayes, R. B., Pickle, L. W., Issaq, H. J., Muschik, G. M., Green-Gallo, L., Buivys, D., Aisner, S., Resau, J. H., Trump, B. F., Tollerud, D., Weston, A., and Harris, C. C. (1990). Lung cancer and the debrisoquine metabolic phenotype. *J. Natl. Cancer Inst.* 82, 1264–1272.

Cartwright, R. A., Glashan, R. W., Rogers, H. J., Ahmad, R. A., Barham-Hall, D., Higgins, E., Kahn, M. A. (1982). Role of *N*-acetyltransferase phenotypes in bladder carcinogenesis: A pharmacogenetic epidemiological approach to bladder cancer. *Lancet,* 16, pp. 842–845.

Catalona, W. J., Smith, D. S., Ratliff, T. L., Dodds, K. M., Coplen, D. E., Yuan, J. J., Petros, J. A., and Andriolo, G. L. (1991). Measurement of prostate-specific antigen in serum as a screening test for prostate cancer. *N. Engl. J. Med.* 324, 1656–1561.

Checkoway, H., Pearce, N., and Crawford-Brown, O. J., ed. (1989). "Research Methods in Occupational Epidemiology." Oxford University Press, New York.

Chisholm, E. M., Marshall, R. J., and Brown, D., Cooper, E. H., and Giles, G. R. (1986). The role of questionnaire and four biochemical markers to detect cancer risk in a symptomatic population. *Br. J. Cancer* 53, 53–57.

Dahlen, G. H., Guyton, J. R., Atlar, M., Farmer, J. A., Kautz, J. A., and Gotto, A. M. (1986). Association of levels of lipoprotein Lp(a), plasma lipids, and other lipoproteins with coronary artery disease documented by angiography. *Circulation* 74, 758–765.

Deguchi, T., Mashimo, M., and Suzuki, T. (1990). Correlation between acetylator phenotypes and genotypes of polymorphic arylamine *N*-acetyltransferase in human liver. *J. Biol. Chem.* 265, 12757–12760.

Detels, R., English, P. A., Giorgi, J. V., Visscher, B. R., Fahey, J. L., Taylor, J. M. G., Dudley, J. P., Nishanian, P., Munoz, A., Phair, J. P., Polk, F., and Rinaldo, C. R. (1988). Patterns of CD4+ cell changes after HIV-1 infection indicate the existence of a codeterminant of AIDS. *J. Acq. Immune Def. Syndr.* 1, 390–395.

Dietrich, K. N., Succop, P. A., Berger, O. G., Hammond, P. B., and Bornschein, R. L. (1991). Lead exposure and cognitive development of urban preschool children: The Cincinnati Lead Study cohort at age 4 years. *Neurotoxicol. Teratol.* **13**, 203–211.

Dorman, J. S., LaPorte, R. E., Stone, R. A., and Trucco, M. (1990). Worldwide differences in the incidence of type I diabetes are associated with amino acid variation at position 57 of the HLA-DQ beta chain. *Proc. Natl. Acad. Sci. U.S.A.* **87**, 7370–7374.

Duffy, M. J., Reilly D., O'Sullivan, C., O'Higgins, N., Fennelly, J. J., and Andreasen, P. (1990). Urokinase-plasminogen activator, a new and independent prognostic marker in breast cancer. *Cancer Res.* **50**, 6827–6829.

Dwyer, J. H., Feinleib, M., Lippert, P., and Hoffmeister, H. (1991). "Statistical Models for Longitudinal Studies of Health," Monographs in Epidemiology and Biostatistics 16. Oxford University Press, New York.

Garland, C. F., Comstock, G. W., Garland, F. C., Helsing, K. J., Shaw, E. K., and Gorham, E. D. (1989). Serum 25-hydroxyvitamin D and colon cancer: Eight-year prospective study. *Lancet* **2**, 1176–1178.

Genest, J. J., Ordovas, J. M., McNamara, J. R., Robbins, A. M., Meade, T., Cohn, S. D., Salem, D. N., Wilson, P. W. F., Masharani, U., Frossard, P. M., and Schaefer, E. J. (1990). DNA polymorphism of the apolipoprotein B gene in patients with premature artery disease. *Atherosclerosis* **82**, 7–17.

Gidlow, D. A., Church, J. F., and Clayton, B. E. (1986). Seasonal variations in haematological and biochemical parameters. *Ann. Clin. Biochem.* **2**, 310–316.

Greenwald, P. (1982). New directions in cancer control. *Johns Hopkins Med. J.* **151**, 209–230.

Groopman, J. D., Jiaqi, Z., Donahue, P. R., Pikul, A., Lisheng, Z., Jun-shi, C., and Wogan, G. N. (1992). Molecular dosimetry of urinary aflatoxin-DNA adducts in people living in Guangxi Autonomous Region, People's Republic of China. *Cancer Res.* **52**, 45–52.

Haugen, A., Becher, G., Benestad, C., Vahakangas, K., Trivers, G. E., Newman, M. J., and Harris, C. C. (1986). Determination of polycyclic aromatic hydrocarbons in the urine, benzo(a)pyrene-diol-epoxide–DNA adducts in lymphocyte DNA, and antibodies to the adducts in sera from coke oven workers exposed to measured amounts of polycyclic aromatic hydrocarbons in the work atmosphere. *Cancer Res.* **46**, 4178–4183.

Hayes, R., Bi, W., Rothman, N., Broly, F., Caporaso, N., Feng, P., You, X., Yin, S., Woosley, R. L., and Meyer, U. (1992). A phenotypic and genotypic analysis of *N*-acetylation and bladder cancer in benzidine-exposed workers (*Submitted.*)

Hemstreet, G. P., Schulte, P. A., Ringen, K., Stringer, W., and Altekruse, E. B. (1988). DNA hyperploidy as a marker for biological response to bladder carcinogen exposure. *Int. J. Cancer* **42**, 817–820.

Hemminki, K., Grybowska, E., Chorazy, M., Twardowska-Saucha, K., Sroczynski, J. W., Putman, K. L., Randerath, K., Phillips, D. H., Hewer, A., Santella, R. M., Young, T. L., and Perera, F. P. (1990). DNA adducts in humans environmentally exposed to aromatic compounds in an industrial area of Poland. *Carcinogenesis* **11**, 1229–1231.

Henderson, R. F., Bechtold, W. E., Bond, J. A., and Sun, J. D. (1989). The use of biological markers in toxicology. *Crit. Rev. Tox.* **20**, 65–82.

Hsu, I., Metcalf, R., Sun, T., Welsh, J., Wang, N., and Harris, C. (1991). Mutational hotspot in the p53 gene in human hepatocellular carcinomas. *Nature (London)* **350**, 427–428.

Hulka, B. S. (1990a). Methodologic issues in molecular epidemiology. *In* "Biological Markers in Epidemiology" (B. S. Hulka, T. C. Wilcosky, and J. D. Griffith, eds.), pp. 214–226. Oxford University Press, New York.

Hulka, B. S. (1990b). Overview of biological markers. *In* "Biological Markers in Epidemiology" (B. S. Hulka, T. C. Wilcosky, and J. D. Griffith, eds.), pp. 1–15. Oxford University Press, New York.

Hulka, B. S. (1991). Epidemiological studies using biological markers: Issues for epidemiologists. *Cancer Epidemiol. Biomark. Prev.* **1**, 13–19.

Hulka, B. S., and Wilcosky, T. (1988). Biological markers in epidemiologic research. *Arch. Environ. Health* 43, 83–89.

Igengar, S., Hamman, R. F., Marshall, J. A., Majumder, P. P., and Ferrell, R. E. (1989). On the role of vitamin D binding globulin in glucose homeostasis: Results from the San Luis Valley Diabetes Study. *Genet. Epidemiol.* 6, 691–698.

Kannel, W. B., Castelli, W. P., Gordan, T., and McNamara, P. M. (1971). Serum cholesterol, lipoproteins, and the risk of coronary heart disease. *Ann. Int. Med.* 74, 1–12.

Karjalainen, J. K. (1990). Islet cell antibodies as predictive markers for IDDM in children with high background incidence of disease. *Diabetes* 39, 1144–1150.

Kelsey, K. (1989). Sister chromatid exchange in painters recently exposed to solvents. *Environ. Res.* 50,

Khoury, M. J., Beaty, T. H., Newill, C. A., Bryant, S., and Cohen, B. H. (1986). Genetic-environmental interactions in chronic airways obstruction. *Int. J. Epidemiol.* 15, 65–72.

Kippen, H. M., Cody, R. P., and Goldstein, B. D. (1989). Use of longitudinal analysis of peripheral blood counts to validate historical reconstructions of benzene exposure. *Environ. Health Perspect.* 82, 199–206.

Klayton, R., Fallat, R., and Cohen, A. B. (1975). Determinants of chronic obstructive pulmonary disease in patients with intermediate levels of alpha 1-antitrypsin. *Am. Rev. Resp. Dis.* 112, 71–75.

Kleinbaum, D. G., Kupper, L. L., and Morganstern, H. (1982). "Epidemiologic Research: Principles and Quantitative Methods." Wadsworth, Belmont, California.

Kuller, L. H., Becker, D. J., Cruic, K., Shanks, K. J., Dorman, J. S., Eberhardt, M. S., Drash, A. L., LuPorte, R. E., Lipton, R., Moy, C., O'Leary, L. A., Orchard, T. J., Rewers, M., Songer, T., Tajima, N., Trucco, M., and Wagener, D. (1990). Evolution of the Pittsburgh studies of the epidemiology of insulin-dependent diabetes. *Genet. Epidemiol.* 7, 105–119.

Larsen, M. L., Hørder, M., and Mogensen, E. F. (1990). Effect of long-term monitoring of glycosylated hemoglobin levels in insulin-dependent diabetes mellitus. *N. Engl. J. Med.* 323, 1021–1025.

Last, J. M. (ed.) (1988). "A Dictionary of Epidemiology." 2nd Edition, Oxford University Press, New York.

Liang, K.-Y., and Zeger, S. L. (1986). Longitudinal data analysis using generalized linear models. *Biometrika* 73, 13–22.

Lin, H. S., Lai, C. L., Lauder, I. J., Wu, P. C., Lau, T. K., and Fong, M. W. (1991). Application of hepatitis B virus (HBV) DNA sequence polymorphisms to the study of HBV transmission. *J. Infect. Dis.* 164, 284–288.

Lipid Research Clinics Program (1984). The Lipid Research Clinics Coronary Primary Prevention Trial results. II. The relationship of reduction in incidence of coronary heart disease to cholesterol lowering. *J. Am. Med. Assoc.* 248, 1465–1477.

McGarrity, T. J., Peiffer, L. P., Bartholomew, M. J., and Pegg, A. E. (1990). Colonic polyamine content and ornithine decarboxylase activity as markers for adenomas. *Cancer* 66, 1539–1543.

Maclure, M., Bryant, M. S., Skipper, P. L., and Tannenbaum, S. R. (1990). Decline of the hemoglobin adduct of 4-aminobiphenyl during withdrawal from smoking. *Cancer Res.* 50, 181–184.

Margetts, B. M., and Nelson, M. (eds.) (1991). "Design Concepts in Nutritional Epidemiology." Oxford University Press, New York.

Marre, M., Chatellier, G., Leblanc, H., Guyene, T. T., Menard, J., and Pussa, P. (1988). Prevention of diabetic nephropathy with enalapril in normotensive diabetics with microalbuminuria. *Br. Med. J.* 297, 1092–1095.

Miettinen, O. S. M. (1982). Design options in epidemiologic research: An update. *Scand. J. Work Environ. Health* **Suppl. 1**, 7–14.

Mueller, N., Mohar, A., Evans, A., Harris, N. L., Comstock, G. W., Jellum, E., Magnus, K., Orentreich, N., Polk, B. F., and Vogelman, J. (1991). Epstein-Barr virus antibody patterns preceding the diagnosis of non-Hodgkin's lymphoma. *Int. J. Cancer* **49**, 387–393.

National Research Council (1987). Biological markers in environmental health research. *Environ. Health Perspect* **74**, 1–191.

Nebert, D. W., and Gonzalez, F. J. (1987). p450 genes: Structure, evolution, and regulation. *Ann. Rev. Biochem.* **56**, 945–993.

Nieburgs, H. E. (1985). Introduction. Proceedings of the 2nd International Conference on Human Tumor Markers (Vienna 1984). *Cancer Detect. Prev.* **8**, iii.

Nora, J. J., Berg, K., and Nora, A. H. (1991). Cardiovascular diseases. "Genetics, Epidemiology, and Prevention." Oxford University Press, New York.

Parsonnet, J., Friedman, G. D., Vandersteen, D. P., Chang, Y., Vogelman, J. H., Orentreich, N., and Sibley, R. K. (1991). *Helicobacter pylori* infection and the risk of gastric carcinoma. *N. Engl. J. Med.* **325**, 1127–31.

Perera, F. P. (1987). Molecular cancer epidemiology: A new tool in cancer prevention. *J. Natl. Cancer Inst.* **78**, 887–898.

Perera, F. P., Hemminki, K., Young, T.-L., Brenner, D., Kelly, G., and Santella, R. M. (1988). Detection of polycyclic aromatic hydrocarbon-DNA adducts in white blood cells of foundry workers. *Cancer Res.* **48**, 2288–2291.

Rayfield, M., DeCock, K., Heyward, W., Goldstein, L., Krebs, J., Kwok, S., Lee, S., McCormick, J., Moreau, J. M., Odehoari, K., Schochetman, G., Sninsky, J., and Ou, C. Y. (1988). Mixed human immunodeficiency virus (HIV) infection in an individual: Demonstration of both HIV Type 1 and Type 2 proviral sequences by using polymerase chain reaction. *J. Infect. Dis.* **158**, 1170–1176.

Reed, E., Ostchega, Y., Steinberg, S. M., Yuspa, S. H., Young, R. C., Ozols, R. F., and Poirier, M. C. (1990). Evaluation of platinum–DNA adduct levels relative to known prognostic variables in a cohort of ovarian cancer patients. *Cancer Res.* **50**, 2256–2260.

Robbins, J. M., and Greenland, S. (1992). Identifiability and exchangeability for direct and indirect effects. *Epidemiology* **3**, 143–155.

Rosenow, E. C., III, and Carr, D. T. (1979). Bronchogenic carcinoma. *CA Cancer J. Clin.* **29**, 233–246.

Ross, R. K., Yuan, J.-M., Yu, M. C., Wogan, G. N., Qian, G. S., Tu, J. D., Groopman, J. D., Gao, Y. T., and Henderson, B. H. (1992). Urinary aflatoxin biomarkers and risk of hepatocellular carcinoma. *Lancet* **339**, 943–946.

Rothman, K. (1986). "Modern Epidemiology." Little, Brown, Toronto.

Rothman, N., Poirier, M. C., Correa-Villasenor, A., Ford, D. P., Hansen, J. A., O'Toole, T., and Strickland, P.T. (1992a). Association of PAH–DNA adducts in peripheral white blood cells with dietary exposure to PAHs. *Environ. Health Perspec.* (in press).

Rothman, N., Hayes, R. B., Bi, W., Broly, F., Woosley, R. L., Caporaso, N., Feng, P., You, X., Yin, S., and Meyer, U. (1992b). "Correlation between Genotypes of Polymorphic *N*-Acetyltransferase (NAT2) and Acetylator Phenotypes in a Case-Control Study of Benzidine-Induced Bladder Cancer in China." Paper presented at the American Association for Cancer Research Annual Meeting, San Diego, California, May, 1992.

Rumke, C. L. (1983). Critical reflections on clinical trials. *Stat. Med.* **2**, 175–181.

Sanders, T. A. B., and Roshanai, F. (1983). The influence of different types of ω3 polyunsaturated fatty acids on blood lipids and platelet function in healthy volunteers. *Clin Sci.* **64**, 91–99.

Santella, R. M., Weston, A., Perera, F. P., Trivers, G. T., Harris, C. C., Young, T. L., Nguyen, D., Lee, B. M., and Poirier, M. C. (1988). Interlaboratory comparison of antisera and immunoassays for benzo(*a*)pyrene-diol-epoxide-I-modified DNA. *Carcinogenesis* **9**, 1265–1269.

Satterfield, S., Goeco, P. J., Goldhaber, S. Z., Stampfer, M. J., Swartz, S. L., Stein, E. A., Kuplan,

L., and Hennekens, C. H. (1990). Biochemical markers of compliance in the physician's health study. *Am. J. Prev. Med.* **6**, 290–294.

Schatzkin, A., Freedman, L. S., Schiffman, M. H., and Dawsey, S. M. (1990). Validation of intermediate end points in cancer research. *J. Natl. Cancer Inst.* **82**, 1746–1752.

Schulte, P. A. (1987). Methodologic issues in the use of biologic markers in epidemiologic research. *Am. J. Epidemiol* **126**, 1006–1016.

Schulte, P. A. (1989). A conceptual framework for the validation and use of biologic markers. *Environ. Res.* **48**, 129–144.

Schulte, P. A., Ringen, K., and Hemstreet, G. P. (1986). Optimal management of asymptomatic workers at high risk of bladder cancer. *J. Occup. Med.* **28**, 13–17.

Selby, J. V., Friedman, G. D., and Quesenberry, C. P., Jr. (1990). Precursors of essential hypertension: Pulmonary function, heart rate, uric acid, serum cholesterol, and other serum chemistries. *Am. J. Epidemiol.* **131**, 1017–27.

Shukla, R., Dietrich, K. N., Bornschein, R. L., Berger, O., and Hammond, P. B. (1991). Lead exposure and growth in the early preschool child: A follow-up report from the Cincinnati Lead Study. *Pediatrics* **88**, 886–92.

Smith, C. A. D., Gough, A. C., Leigh, P. N., Summers, B. A., Harding, A. E., Maranganore, D. M., Starman, S. G., Schapira, A. H. V., Williams, A. C., Spurr, N. K., and Wolf, C. R. (1992). Debrisoquine hydroxylase gene polymorphism and susceptibility to Parkinson's disease. *Lancet* **339**, 1375–1377.

Spiegel, M. R. (1976). "Statistics." McGraw-Hill, New York.

Stampfer, M. J., Sacks, F. M., Salvini, S., Willett, W. C., and Hennekens, C. H. (1991). A prospective study of cholesterol apolipoproteins, and the risk of myocardial infarction. *N. Engl. J. Med.* **325**, 373–381.

Suzuki, H., Takahashi, T., Kuroishi, T., Suyama, M., Ariyoshi, Y., and Ueda R. *Cancer Res.* **52**, 734–736.

Taylor, J. A. (1990a). Epidemiologic evidence of genetic susceptibility to cancer. *In* "Detection of Cancer Predisposition: Laboratory Approaches" March of Dimes Birth Defects Foundation, White Plains, New York. (L. Spatz, A. D. Bloom, and N. W. Paul, eds.), pp. 113–127.

Taylor, J. A. (1990b). Oncogenes and their applications in epidemiologic studies. *Am. J. Epidemiol.* **130**, 6–13.

Thompson, S. G. (1983). A method of analysis of laboratory data in an epidemiological study where time trends are present. *Stat. Med.* **2**, 147–153.

Tibblin, G., and Crammer, K. (1963). Serum lipids during the course of an acute myocardial infarction and one year afterwards. *Acta. Med. Scand.* **174**, 451–435.

Van Noord, P. A. H. (1992). "Selenium and Human Cancer Risk: Nail Keratin as a Tool in Metabolic Epidemiology." Thesis Publishers, Amsterdam.

Van Schooten, F. J., Van Leeuwen, F. E., Hillebrand, M. J. X., de Rijke, M. E., Hart, A. A. M., van Veen, H. G., Oosterink, S., and Kriek, E. (1990). Determination of benzo[a]pyrene-diol-epoxide–DNA adducts in white blood cell DNA from coke-oven workers: The impact of smoking. *J. Natl. Cancer Inst.* **82**, 927–933.

Vineis, P., and Terracini, B. (1990). Biochemical epidemiology of bladder cancer. *Epidemiology* **1**, 448–452.

Vineis, P., Caporaso, N., Tannenbaum, S. R., Skipper, P. L., Glogowski, J., Bartsch, H., Coda, M., Talaska, G., and Kadlubar, F. (1990). Acetylation phenotype, carcinogen–hemoglobin adducts, and cigarette smoking. *Cancer Res.* **50**, 3002–3004.

Wacholder, S. (1991). Practical consideration in choosing between the case-cohort and nested case-control designs. *Epidemiology* **2**, 155–158.

Wald, A. (1945). Sequential tests of statistical hypotheses. *Ann. Math. Stat.* **June,** 118–175.

Walop, W., Chretien, M., Colman, N. C., Fraser, R. S., Gilbert, F., Hildvegi, R. S., Hutchinson,

T., Kelley, B., Lis, M., and Spitzer, W. O. (1990). The use of biomarkers in the prediction of survival in patients with pulmonary carcinoma. *Cancer* **65**, 2033–2046.

Weston, A., Caporaso, N. E., Taghizadeh, K., Hoover, R. N., Tannenbaum, S. R., Skipper, P. L., Resau, J. H., Trump, B. F., and Harris, C. C. (1991). Measurement of 4-aminobiphenyl hemoglobin adducts in lung cancer cases and controls. *Cancer Res.* **51**, 5219–5223.

Wilcosky, T. C. (1990). Criteria for selecting and evaluating markers. *In* "Biological Markers in Epidemiology" (B. S. Hulka, T. C. Wilcosky, and J. D. Griffith, eds.), pp. 28–55. Oxford University Press, New York.

Wilcosky, T. C., and Griffith, J. D. (1990). Applications of biological markers. *In* "Biological Markers in Epidemiology" (B. S. Hulka, T. C. Wilcosky, and J. D. Griffith, eds.), pp. 16–27. Oxford University Press, New York.

Willett, W. (1987). Nutritional epidemiology: Issues and challenges. *Int. J. Epidemiol.* **16**, 312–317.

Xu, C. F., Boerwinkle, E., Tikkanen, M. J., Huttunen, J. K., Humphries, S. E., and Talmud, P. J. (1990). Genetic variation at the apolipoprotein gene loci contribute to response to plasma lipids to dietary change. *Genet. Epidemiol.* **7**, 261–275.

Zeger, S. L., and Liang, K.-Y. (1992). An overview of methods for the analysis of longitudinal data. *Stat. Med.* (*in press*).

Zhong, S., Howie, A. F., Ketterer, B., Taylor, J., Hayes, J. D., Beckett, G. J., Wathen, C. G., Wolf, C. R., and Spurr, N. R. (1991). Glutathione *S*-transferase mu locus: Use of genotyping and phenotyping assays to assess association with lung cancer susceptibility. *Carcinogenesis* **12**, 1533–1537.

7

Statistical Methods in Molecular Epidemiology

Vicki Stover Hertzberg
and Estelle Russek-Cohen

Introduction

Several different laboratory methods used in molecular epidemiology have their own statistical requirements, yet the broad perspective of the area requires the use of many of the techniques in the statistical armamentarium. In this chapter, we first discuss issues pertinent to the laboratory. Next we move into more general statistical considerations for the use of soundly collected laboratory data in epidemiologic investigations. Finally, we describe a number of issues common to both the laboratory and the epidemiologic question being raised.

Laboratory Methods

Overview

Many different laboratory techniques are used in molecular epidemiology. The methods we discuss here will not constitute a complete list. However, some statistical issues are common to all these techniques. With any assay, we are concerned with repeatability of the results as well as with accuracy. Since it is rare for all the assays of a study to be completed within a short span of time, the stability of the assay over time as well as with respect to factors such as different laboratories, technicians, or batches of solutions used becomes an important feature to evaluate before and during the course of the study.

In this section, we focus on five types of assays, illustrating the role of statistics in each. In addition, we include a discussion of quality control that provides guidelines common to all assays. We select radioimmunoassay (RIA) as an example of an assay requiring the use of a standard curve. Enzyme-linked immunosorbent assay (ELISA) techniques also may involve the use of a standard curve, but many biologists often use monoclonal antibodies with ELISA to measure the presence or absence of a specific protein or antigen. cDNA probes may be used in a similar fashion. HPLC and GC (high performance liquid chromatography and gas chromatography) data often are recorded as multivariate vectors of observations that reflect the relative concentrations of multiple substances, for example, fatty acid composition. One- and two-dimensional gels rely on the distance a band has migrated on a gel. Cytopathological data can be regarded as an ordinal response variable since patients often are ranked on an ordinal scale (e.g., benign, abnormal, and malignant).

Finally, with the increased use of DNA sequence data, statistics can be used as a tool to identify coding and noncoding regions, to evaluate the potential success of a restriction enzyme in dividing DNA into fragments of appropriate length for sequencing, and to put the pieces back together to sequence an entire gene.

Radioimmunoassay and Assays Involving Standard Curves

From a statistical perspective, RIAs rely on the use of a standard curve. The standard curve is computed by analyzing samples of known levels of the substance to be analyzed along with the samples with unknown levels. A regression method is used to formulate the relationship of the known dose levels (X values) and the response variable (Y), counts per minute in the case of RIAs and optical density in the case of ELISA. For RIAs, the value of Y decreases with increasing dose or log dose (X). In RIAs and in many other assays, the relationship of Y to X is nonlinear. However, once the mathematical relationship of Y with X is described, a standard procedure such as the NLIN (nonlinear regression procedure) in the Statistical Analysis System (SAS) (SAS Institute, 1985) can be used to estimate the parameters of the curve and compute an estimate of the average value of Y for each X within the range of the standard curve. Then, the response Y for each "unknown" sample can be measured and used to "predict" the value of X. This process is called inverse prediction because, in the original assay, the X values are known and "fixed" by the experimenter, whereas in the prediction we observe Y and use that value to predict X (Draper and Smith, 1981; Finney, 1978).

Two strategies often are seen in the literature in the fitting of a standard curve; often investigators fail to report their method in any level of detail. The first strategy consists of fitting a nonlinear regression function such as the four-parameter logistic equation (Rodbard and Frazier, 1975; Ratkowsky and Reedy, 1986).

The four-parameter logistic equation consists of the model

$$Y = a + \frac{b-a}{1 + \exp\ (c - dx)} + \epsilon$$

where a is the minimum value of Y, b is the maximum (i.e., $b - a$ is the range), c and d are the parameters of the model, and x is the logarithm of the dose. This model, although commonly used in RIA and ELISA, is not the only equation available. A curve should be selected that correctly fits the data. An excellent reference that describes a number of nonlinear curves useful in bioassays is Ratkowsky (1983). A second strategy is to linearize the equation. This procedure tends to have several problems when the model assumptions of the linear regression, namely independence and homogeneous errors, are incorrect.

Once the formula for the standard curve has been determined, a lack of fit test can be computed using the data used to construct the standard curve. If the curve fit is a polynomial, one easily can assess lack of fit directly using SAS or some other general linear model procedure (see Draper and Smith, 1981). If the curve is nonlinear, one can use a likelihood ratio procedure to assess lack of fit (e.g., see Volund, 1979). Running standard curves with 5–10 tubes per dose for the purpose of assessing model adequacy would increase the sensitivity of the test. A less common procedure that has been considered is the use of monotone splines (Ramsey, 1988). This approach avoids the need to specify the nonlinear function relating X to Y. However, RIAs and most other assays typically are run with too few points in the standard curve to yield a reliable curve in this manner.

An alternative method for generating a standard curve is to consider a weighted regression (e.g., see Raab, 1981). However, one should examine the data for heterogeneity of variance carefully before selecting weights. No model is required for the standard curve to test for heterogeneity. By repeating measurements of known doses several times, one can generate variance estimates for Y separately for each dose and use Bartlett's test or some other test for homogeneity of variance (Sokal and Rohlf, 1981).

Typically, linearization procedures are used so widely because of the ability to create homegrown software that analyzes both the standard curve and the unknowns. Many biologists are comfortable with linear regression, but are untrained in the use of nonlinear models. However, the arbitrary use of linearization procedures without consideration of model assumptions should be avoided.

Enzyme-Linked Immunosorbent Assay

ELISA is commonly used to assay for proteins and other "antigens" using monoclonal and sometimes polyclonal antibodies. When known standard doses are available, these are analyzed in a manner similar to that for RIAs, generating a standard curve for optical densities instead of counts per minute. These assays often are interpreted to have a discrete outcome, that is, a high absorbance reading is positive and a low reading is negative. Values in between may generate problems from a statistical perspective. Other problems, such as monoclonal antibodies that are too specific also can arise. For example, in the detection of disease caused by closely related viral or bacterial strains, and these assays falsely may declare a sample as negative if the strain in the sample was not used to construct the monoclonal antibody. Only a sufficient amount of testing with an appropriate range of samples can be used to ascertain sensitivity and specificity. These issues are discussed further in Chapter 4. Note that the sensitivity defined here is not necessarily the same as the one that relates the use of the assay to the underlying disease process (see subsequent section). Assays with sensitivity and specificity less than 100% will result in analyses in which the variables are recorded with measurement error. If the sensitivity and specificity are considerably off, then the power of any subsequent analyses to identify the true exposure–disease relationship will be poor. An excellent review of models that incorporate measurement error is given in Fuller (1987), but is beyond the scope of this chapter.

Chromatograph Data

In gas and liquid chromatography, the response recorded is either the area under a peak or the height of the peak. In a number of applications, the locations of peaks and the percentage of area attributable to each peak is of interest. When used in conjunction with a set of standards, the peak location is thought to be associated with the presence of a particular molecule and the height is thought to reflect the proportion of the solution represented by that molecule. From a statistical perspective, these data may be regarded as multivariate continuous. After examining the data for multivariate normality (Johnson and Wichern, 1982), one can do a multivariate analysis of variance with such factors as technician inconsistency. The biggest weakness of chromatograph data is the establishment of a baseline, so relatively rare substances may be more difficult to measure. When known mixtures are processed using chromatograph data, one can generate the multivariate equivalent of a standard curve. In contrast to RIA and ELISA assays, there is a heavier reliance on linear models to describe the relationships of peak heights and concentrations, although one could consider the use of transformations such as log prior to such analyses. Because the number of peaks may be large relative to the number of samples, some chemometrics authors (Sjostrom *et al.*, 1983; Frank and Kowalski, 1985) suggest the use of partial least squares rather than the usual unbiased estimates of least squares for calibrating the standard curve relationships.

When the percentage of area attributable to each peak is of interest, the data in one observation sum to 100%. This type of multivariate data is, therefore, a member of a restricted class known as compositional data. Some approaches to analyzing these data can be found in Aitchison (1986) and Cornell (1981).

Gels

In gels, the presence or absence of one or more bands or spots on a gel is of interest. One-dimensional and two-dimensional gels rely on the distance a protein (or DNA molecule) can migrate through a gel to identify the bands correctly. That the degree of migration can vary from assay to assay is well recognized; some means of establishing repeatability of the results must be used. A common approach in one-dimensional gels is the use of standard samples run with the unknown samples. The standard samples must generate bands representative of a certain size range for the assay to be considered acceptable. Although it is not common, the distances traveled by these standard samples can be recorded and plotted over time to spot systematic changes using the techniques described in a subsequent section. For two-dimensional gels, both the horizontal and vertical distances, X and Y values, of the spots that can be recognized clearly can quantitate quality control (Neel *et al.*, 1984).

Cytopathological Data

Whittemore *et al.* (1982) propose one approach to analyzing data in which the response is a pathology score that typically is recorded as the most abnormal cell type found in the sample. Whittemore develops a model similar to those for ordinal response variables in the context of a generalized linear model (McCullagh and Nelder, 1989). This method contrasts with the data presented earlier, which consists primarily of binary and continuous data. These ordinal models can be used to identify systematic sources of variability by coding dummy variables for the type of factors just discussed.

Commercially Available Kits

Many commercial kits are available for the assay of such substances as insulin, progesterone, and others. The kits vary in quality and ease of use. However, they must be tested for accuracy, repeatability, and stability over time.

Quality Control

Systematic Sources of Variability

Before using an assay in an epidemiologic framework, one must identify and minimize systematic sources of variability in the assay. For example, the use of both multiple laboratories and multiple technicians may yield undesirable excess variation. The use of standard samples run by all laboratories and all technicians and subsequently analyzed using an analysis of variance or some other appropriate procedure will identify potential problems. Because one may regard factors such as laboratories as fixed effects and repeat runs and duplicate readings on the same sample as random, one may find that the resulting model is a mixed model. This procedure is implemented easily in SAS using GLM and the RANDOM statement to compute expected mean squares (see Milliken and Johnson, 1986, or Sokal and Rohlf, 1981, for a detailed explanation of the procedure). SAS (Version 6.03 or later) will use a Satterwaithe's approximation if no exact F test exists. BMDP 3V is an alternative procedure for a similar mixed-model analysis. The advantage these procedures offer is the ability to identify unwanted excess variation. One can correct the laboratory procedure or run extra samples and average to minimize measurement error.

With qualitative data such as cytopathological data, one may assess the agreement between results from two or more technicians using methods of interrater agreement such as Cohen's κ statistic (Fleiss, 1981). This statistic compares the amount of interrater agreement present to the amount that might be expected by chance alone on a scale of -1 to 1. Negative values indicate that there is less agreement than can be expected by chance and a value of zero indicates that there is no more agreement than would have been predicted by chance. In general, values between .4 and .7 are considered to indicate a moderate amount of agreement, whereas values greater than .7 indicate a high level of agreement. Formulas for the standard error of the κ statistic are given by Bishop, Feinberg, and Holland (1975). Care must be taken in developing confidence intervals since the standard error formula depends on the estimate of κ.

For quantitative variables, simple plots of values from one laboratory against values from another for the same samples may uncover systematic problems at the high and low values. A more rigorous approach would be that of Linnet (1990) or Lin (1989), which both model reproducibility. Other factors may also need to be assessed similarly including different batches of antibodies, buffers, and so forth.

Some substances such as creatinine are known to exhibit circadian rhythms, that is, they vary with the time of day in a systematic way. A simple experiment that takes several samples from each of several individuals at fixed time points can identify such problems. However, these samples should be analyzed considering the repeated measures aspects of the design (see subsequent section). Other sources of intraindividual variation pertaining to level of activity, eating habits, and occupation may influence the value observed; for example, a cholesterol reading can be affected by when the last meal was eaten. In addition, place in the menstrual cycle may influence some measurements in premenopausal women. Thus, the importance of variation with time needs to be considered and, if necessary, evaluated in a designed study. More detail on this topic is given in a later section.

Reliability over Time

Some kit and assay components have a shelf life that is shorter than the length of a study or are sensitive to temperature or a variety of other conditions. Thus maintaining a rigorous quality control (QC) program throughout a study is important. One way of detecting a problem in an assay with standard curves is to maintain sample pools that are known to have high, medium, and low titers of the substance of interest (Kay *et al.*, 1983). These samples typically are run (in duplicate) with the standard curve samples; the values of the mean and the range are plotted over time. These data may be typically lognormal in many situations; thus, the log of the predicted value rather than the actual dose may need to be used. If the data consist of binomial proportions, one can use arcsin ($\sqrt{p}$) rather than p since this value yields data with homogeneous variances. Several QC type charts exist (Ryan, 1989; John, 1990). Some charts, such as the Xbar chart, are designed only to spot values outside the range $\bar{X} \pm 3$ SD, whereas R charts are used with the range. The cumulative sum (cusum) chart is designed specifically to detect shifts in the average value over time. A modified use of the cusum chart is the use of a V-mask that identifies where the shift in mean may have occurred (Rowlands *et al.*, 1983). Multivariate charts (Ryan, 1989) based on Mahalanobis distances can be used to look simultaneously at two or more correlated variables and may be appropriate for chromatography data. Often, to assess quality control in chromatography data, a "cocktail" containing several substances of interest is mixed. The location of identifiable peaks is recorded. These distances, reflecting the time since injection, can be regarded as a multivariate vector, provided a cocktail of roughly the same composition is used for each run.

Software to generate many of these charts is readily available in packages such as NCSS, SAS/QC, and Statgraphics. Before using any of the charts based on normal continuous data, one may need to consider the use of transformations. One possibility is to use the first 50 points or so to check for normality using one of the tests described in a later section.

Some common-sense issues exist irrespective of the assay being run. In general, it is important that the technician is blinded to the true classification of the sample as from a diseased or a nondiseased individual or from an exposed or an unexposed individual. In addition, the usual guidelines for conducting a study are important, for example, the need to process cases and controls simultaneously. If it is financially feasible to do so, samples should be run more than once rather than in duplicate because this generates a more repeatable result. In addition, the data should be screened for outliers as the study progresses, not after the study is completed. Assays involving some subjectivity should be standardized as much as possible.

The limits of detection of the assay need to be determined. To arbitrarily set to missing a value that falls below the level of detection is not valid. In fact, arbitrarily setting the value to zero will bias the mean of the variable downward, whereas ignoring the value will bias the mean upward. One possibility is to set the value to half the limit of detection. A more accurate assessment would be to regard the value as left censored and build all subsequent analyses accordingly (Shumway *et al.,* 1989). Analyzing data in which a substantial number of the observations are below the limits of detection is typically messy from a statistical perspective.

DNA Sequence Data

As the ability to sequence DNA rapidly increases, statistics plays a role in several aspects. The number of available sequences has been growing exponentially, spurred on by the human genome project. Also, numerous articles and texts have been written on several aspects on the statistical treatment of DNA data (e.g., Weir, 1988; 1990).

DNA can be thought of as a linear sequence composed of four types of bases (A, C, T, G). Because these sequences can be thousands of bases in length, sequencing an entire gene in a single run or gel is impossible. Restriction enzymes that cut a long sequence whenever a particular subsequence occurs are used to subdivide the sequence (e.g., the enzyme *Eco*RI recognizes GAATTC and CTTAAG and cuts between the G and the A). Since a typical gel reader may read no more than 500 base pairs in a gel, it is reasonable to select a restriction enzyme that is most likely to clip the sequence into segments of the right length. Several algorithms (Weir, 1988; Karlin and Macken, 1991) have been proposed to predict these properties in advance.

DNA can be thought of as a language in which triplets (codons) encode amino acids or a stop. Portions of the DNA follow this structure and should not be regarded as random. Other portions of the DNA do not have this structure; there is some debate about whether these noncoding regions are "nonsense" or serve some function. Therefore a method for identifying coding and noncoding portions of the sequence would be useful. Michel (1986) has proposed a technique based on principal components that permits separation of coding and noncoding regions. Many of the statistical procedures for DNA analysis assume that bases are distributed randomly in the sequence. Some of these assumptions would be better met if the first, second, and third positions of each codon were treated as separate regions and the results somehow combined.

Putting smaller pieces of DNA together to get a longer sequence is another area in which statistics can play a role. However, this problem has received very little attention thus far (Karlin and Macken, 1991).

Determining the Number of Subjects

The ability to quantify exposure and effect on a continuous scale by using one or more biomarkers allows the investigator to detect more subtle shifts in the exposure–disease relationship than was previously possible. Nevertheless, the need still exists to design a study with a number of subjects sufficient to generate adequate statistical power when null hypotheses of no difference among groups or perhaps no association among markers are to be tested.

For the situation in which two groups (i.e., cases and controls) are compared on the basis of presence or absence of some condition (e.g., exposure), the formulation for calculation of sample size for a given combination of significance level (i.e., α), power, and size of expected difference as measured in terms of relative risk is well known (Schesselman, 1982).

Several different scenarios exist in which two or more groups may be compared on the basis of the level(s) of one or more markers; the sample size formulas are specific to each scenario. When the groups are being compared for shifts in central tendency (most commonly, shifts in mean), the sample size is calculated according to a well-known formula given in elementary textbooks (for example, Rosner, 1982) that gives the sample size as a function of α, power, the size of the difference to be detected (i.e., $\mu_{cases} - \mu_{controls}$), and the variance of marker levels within groups (σ^2). Alternatively, the investigator may wish to infer the relative risk of a disease as a function of the incremental change in the level of a given marker. The sample size is given by:

$$m = \frac{k+1}{k} \frac{\{z_\alpha \sigma_x + z_\beta \ [(k\sigma_1^2 + \sigma_0^2)/(k+1)]^{1/2}\}^2}{(\mu_1 - \mu_0)^2}$$

where m is the number of cases, $n = mk$ is the number of controls, α and β are the Type I and Type II error rates, respectively, μ_1 and μ_0 are the mean levels, σ_1^2 and σ_0^2 are the variances in cases and controls, respectively, and

$$\sigma_x^2 = \frac{(\sigma_1^2 + k\sigma_0^2)}{(k+1)} + (\mu_1 - \mu_0)^2 \frac{k}{(k+1)^2}$$

(Lubin *et al.*, 1988). In general, it must be noted that a smaller sample size is needed when a continuous-valued marker is compared among the groups than when a discrete-valued marker (i.e., presence or absence) is being compared, particularly when the marker is measured without too much error (see laboratory section).

Sensitivity, Specificity, and Misclassification

When using markers to evaluate the underlying disease process, one can extend the definition of sensitivity and specificity. In the context of a laboratory assay such as a probe for a specific virus strain, sensitivity would be defined as the probability that the marker is positive when the virus is present. In an epidemiologic context, sensitivity may be defined as the probability that a person with the condition will be classified correctly as having the condition on the basis of the marker, and specificity may be defined as the probability that a person without the condition will be classified correctly as not having the condition. Several authors (e.g., Bross, 1954; Greenland, 1980) have demonstrated that, depending on the existence and direction of the differential in misclassification of disease and exposure status and on the specificity and sensitivity, the estimated effect measure may be biased either away from or toward the null. An excellent review of the statistical literature is given by Kleinbaum, Kupper, and Morgenstern (1982).

With continuous-valued markers, the misclassification errors can be estimated, so classification rules can be derived that can optimize the misclassification rates, even taking into account the possibility of different costs of misclassification from one population into another. This rule assigns a multivariate observation X to population i over population j if

$$X'\Sigma^{-1}(\mu_i - \mu_j) - \frac{1}{2}(\mu_i + \mu_j)'\Sigma^{-1}(\mu_i - \mu_j) > \ln \frac{\pi_i C(i,j)}{\pi_j C(j,i)}$$

where μ_i and μ_j are the means of the ith and jth groups, Σ is the common within-group covariance matrix; π_i and π_j are the *a priori* probabilities that an observation will come from the ith or jth population, and $C(i,j)$ is the cost associated with misclassifying an observation from population j into population i. In case-control studies with equal numbers of subjects, a reasonable choice of $\pi_i(i=0,1)$ is 0.5. Computer packages such as SAS, BMDP, and SPSS all have algorithms that substitute sample means and variances in place of population values. A discussion of this method, called linear discriminant analysis, can be found in Lachenbruch (1975) and, for the case of two groups, in Kleinbaum, Kupper, and Muller (1988). This rule assumes that the observations are drawn from a multivariate normal population. However, serious bias may result if the assumption of normality is not met (Press and Wilson, 1978) or if the number of variables is large relative to the number of observations (Russek *et al.*, 1983). Nonparametric procedures as well as alternative parametric models are available, should this assumption seem unwarranted.

Interactions

In studies of biomarkers, the concept of statistical interaction depends on the outcome being studies, the factors to be related to outcome, and to which class (i.e., outcome or factor) the marker belongs. If the marker is the presence or absence of a particular variation of a *p53* oncogene, an expectation that smokers with this marker will experience increased risk of lung cancer corresponds to an interaction among the independent or explanatory variables in the model where lung disease status is the response variable (i.e., a logistic regression model). Alternatively, if the marker itself is the response variable, for example, serum cholesterol, then an interaction may involve smoking and age. In either case, one would need to employ a general linear model if the response variable is continuous or a generalized linear model if the response is binary (e.g., disease present or absent) or a count variable (e.g., number of tumors in a melanoma study). The general linear model, logistic regression, and Poisson regression are special cases of this generalized linear model. In these models, the underlying assumption is that the original factors (e.g., $p53$ and smoking status) are in the model as well as the product of these terms, that is, the interaction term. Both the general linear model and the generalized linear model compare the fit of two models, one with the interactions present in a model (Model II) and the other the same model without the interactions (Model I). However, the generalized linear model relies on large sample theory whereas more exact results are acquirable with the general linear model.

The general linear model is given by

$$Y = X\beta + \epsilon$$

and assumes that the error terms, ϵ, are independent and normally distributed. It may be necessary to transform the data to meet these assumptions, but the transformation may change the conceptual framework of what an interaction may be (Cox, 1984). The statistic used to test the presence of an interaction is a partial F test, which compares the percentage improvement in the error sums of squares for these two models [i.e., the sum of (observed − predicted)^2], namely,

$$\frac{(SSE_I - SSE_{II})/df_I}{SSE_{II}/df_{II}}$$

The numerator degrees of freedom for this test (df_I) is the number of interaction terms eliminated under the null hypothesis, whereas the denominator degrees of freedom (df_{II}) corresponds to the sample size minus the number of parameters with all the interactions included. Power is influenced greatly by the denominator degrees of freedom. A comprehensive review of these methods is available in Kleinbaum and colleagues (1988) or Draper and Smith (1981).

If a discrete-valued response variable is of interest, either that of disease status or that of marker presence, the procedure for modeling interactions among independent variables is not as easily computed but is nevertheless, in most cases, a solvable problem. In this situation, the use of the generalized linear model or software that handles specialized cases such as logistic regression is necessary. The generalized linear model consists of the following three components:

(1) the distribution function $f(y)$ for a single response variable y where $f(\cdot)$ may depend on the mean μ and other parameters
(2) a linear predictor in the dependent variables and $\eta = X\beta$
(3) a link function $g(\mu) = \eta$ that relates the mean μ to the predictor

The general linear model presented earlier is a special case of the generalized linear model in which the distribution function is $N(\mu, \sigma^2)$ and the link function is the identity function, that is, $g(\mu) = \mu = \eta$. Logistic regression assumes that $f(\cdot)$ follows a binomial distribution and the link function is the logit function.

The test of the hypothesis of no interaction uses difference in -2 times the log likelihood instead of the error SS for each of two models (i.e., one model has the interactions present and the other does not). As in the F procedure, df_{I} corresponds to the number of parameters eliminated under the null hypothesis. However, one uses large sample theory and chi-square distributions to test the significance of the hypotheses (i.e., no need for df_{II}). Further details on the use of generalized linear models are given in McCullagh and Nelder (1989). Software for special cases of these models are available in SAS, SPSS, and BMDP. However GLIM is the most comprehensive package for this class of models.

Multiple Markers

It is increasingly common to employ several markers in the context of a single study. For example, with molecular probes designed to identify strains of human papilloma virus, several probes, one specific to each strain, may be employed. These markers are apt to be correlated, which may pose several problems from a statistical perspective. If the markers are employed as independent or explanatory variables, high intercorrelation among the explanatory variables may cloud the interpretation of individual terms in the model. This problem, multicollinearity, is serious if the explanatory variables are too highly intercorrelated. Diagnostics that identify such problems are available in regression software but are not always available when analyzing discrete response variables. One possible approach in this case is the use of the correlation matrix of the estimated coefficients to perform a principle components analysis (Johnson and Wichern, 1982). If the ratio of the largest eigenvalue to the smallest eigenvalue is sufficiently large, there is a serious multicollinearity problem. In models that have interactions, one way to reduce the multicollinearity is to recode the explanatory variables by centering them (e.g., use $x-\bar{x}$ in place of x and $(x-\bar{x})(z-\bar{z})$ in place of the interaction term xz).

A second issue that arises when the number of markers increases, is that the number of statistical tests performed tends to increase. This will tend to increase the chances of finding a false significant result, that is, an increased experiment-wide type I error rate. One possibility is to decrease the value of α used for each test using the Bonferroni inequality (see Kleinbaum *et al.*, 1988). Others have proposed an alternative Bayesian procedure (Thomas *et al.*, 1985).

Repeated Measures and Serially Correlated Data

The availability of rapid methods for detecting marker labels allows one to use markers as monitoring devices. For example, one may examine serially collected Ca125 or prostate-specific antigen levels.

When examining the same subject at different time points, one cannot assume that the observations are independent as many generalized linear model procedures do. Instead, alternative procedures must be considered. Typically, there are two questions of statistical interest. (1) Are there changes across time in this characteristic? (2) What is the relationship of this characteristic to some other covariate or explanatory variable? Under some very restrictive conditions, namely, normally distributed data and measurements of subjects at the same time points (e.g., 1 months and 3 months after study entry) with no missing data and a specialized covariance structure, a special form of a univariate analysis called the repeated measures ANOVA can be applied to the data (Winer, 1971; Milliken and Johnson, 1986). If some of the restrictions on the covariance matrix are dropped, a profile analysis can be performed to answer the same questions (Johnson and Wichern, 1982).

More often, investigators do not have full control over the circumstances under which measurements are made. For instance, a subject may not appear for a scheduled visit or a test tube containing plasma may be dropped or broken during centrifugation. The resulting unbalanced data set usually will not be appropriate for analysis using the methods described. However, the full multivariate model can be applied using multivariate methods for imputation of missing data in designs that are planned to be balanced in which data are missing at random (Orchard and Woodbury, 1972; Beale and Little, 1975; Dempster *et al.,* 1977). When measurements are made at unique or arbitrary times, when the dimension of the covariance matrix is so large as to be intractable, or when the estimation of (random) individual characteristics is desired, the two-stage random effects model can be used. Laird and Ware (1982) describe a general family of such models that includes growth models and repeated measures models as special cases. The first stage is regression of each individual study subject's values on time and obtaining an estimate of the rate of change (rate of decline or increase) for each study subject. The second stage is to fit a linear model, in which the individual rate of change is related to individual covariates. Laird and Ware also give a unified approach for fitting such models based on a combination of empirical Bayes and maximum likelihood procedures, using the EM algorithm. Liang and Zeger (1986) propose alternative procedures suitable for categorical data observed through time, with the same degree of imbalance posed by Laird and Ware.

With the advances in the technology used to measure markers, the possibility of making many measurements of a given marker on an individual across time is real. The methods described in this chapter are suitable for a relatively small number of observations (≤20) per subject. For a larger number of measurements, a time series approach may be more reasonable (see, for example, Brockwell and Davis, 1987). However, these procedures do not work well in the messy unbalanced situations described earlier.

Transformations

In the sections on laboratory assays and epidemiologic studies, an assumption of normality is associated with the more common techniques of regression and ANOVA. The results may be misleading if the normality assumptions fail to hold. Transforming the data by means of a log or square root often remedies the problem. An illustration is given in Figures 7.1 and 7.2. The first figure depicts a situation in which the mean and the standard deviation are related linearly a situation that arises when the data result from a Poisson distribution. The second figure depicts these data after application of a square root transformation. A common approach, easily implemented in modern statistics packages, is to output the residuals after fitting a linear model such as regression or ANOVA and testing the residuals for normality using one of the standard procedures. One can see whether the skewness is approximately equal to zero, or use a probability plot to see if the residuals follow normality. For smaller n values (≤50), a Shapiro–Wilks test can be used to assess normality, whereas for larger n values, a Kolmogorov–Smirnov type test (i.e., Lillefor's D) can be used (SAS Institute, 1985).

Homogeneity of variance can be assessed using an F statistic corresponding to the ratio of two variances or Bartlett's test for homogeneity of variance (Sokal and Rohlf, 1981). Both tests assume that the variances for each group

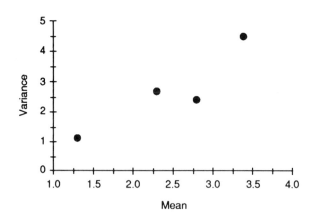

FIGURE 7.1   An illustration of a situation in which mean and variance are related linearly.

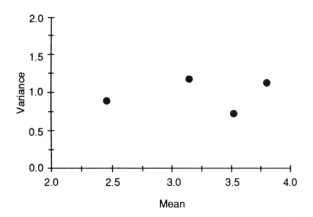

FIGURE 7.2   An illustration of the data of Figure 7.1 after use of a square root transformation.

are independent, which may not be true for some linear models. Often the homogeneity of variance assumption fails because the data are not normally distributed, so for a failure of normality or homogeneity, transformations must be considered.

A general approach to transformations is the Box–Cox family of transformations (Sokal and Rohlf, 1981; Johnson and Wichern, 1982). Here we assume there is a λ such that

$$y = (x^\lambda - 1)/\lambda \qquad \lambda \neq 0$$
$$\ln(x) \qquad\qquad \lambda = 0$$

and y follows a normal distribution. When $\lambda = 0.5$, a square root transform is indicated. When the data are not independent, transformations generally will not remedy the problem.

Conclusions

The explosion of rapid laboratory techniques such as DNA sequencing and ELISA presents the epidemiologist with more sensitive and specific indicators of disease status and exposure status. Nevertheless, the principles of sound epidemiologic design are still applicable, that is, there should be sufficient power of the study to detect the hypothesized associations, the sensitivity and specificity of the assays must be determined, and the data must be examined for the validity of the assumptions underlying the statistical techniques used. In general, the investigator must take care in examining the sources of variability that may generate laboratory artifacts. Moreover, sufficient quality control procedures should be implemented in the laboratory. In summary, we have highlighted many of the key issues in the analysis of molecular data in epidemiologic applications, although our list is by no means complete.

References

Aitchison, J. (1986). "The Statistical Analysis of Compositional Data." Chapman and Hall, London.
Beale, E. M. L., and Little, R. J. A. (1975). Missing values in multivariate analysis. *J. R. Stat. Soc. B.* 37, 129–145.
Bishop, Y., Feinberg, S., and Holland, P. (1975). "Discrete Multivariate Analysis." MIT Press, Boston.
Brockwell, P. J., and Davis, R. A. (1987). "Time Series Theory and Methods." Springer-Verlag, New York.
Bross, I. D. J. (1954). Misclassification in 2×2 tables. *Biometrics* 10, 478–486.
Cornell, J. A. (1981). "Experiments with Mixtures: Designs, Models and the Analysis of Mixture Data." Wiley, New York.
Cox, D. R. (1984). Interactions. *Int. Stat. Rev.* 52, 1–31.
Dempster, A., Laird, N., and Rubin, D. (1977). Maximum likelihood with incomplete data via the E-M Algorithm. *J. R. Stat. Soc. B* 39, 1–38.
Draper, N., and Smith, H. (1981). "Applied Regression Analysis," 2d Ed. Wiley, New York.
Finney, D. J. (1978). "Statistical Methods in Biological Assay," 3d Ed. Griffen, Oxford.
Fleiss, J. (1981). "Statistical Methods for Rates and Proportions," 2d Ed. Wiley, New York.
Frank, I., and Kowalski, B. (1985). Statistical receptor models solved by partial least squares. In "Environmental Applications of Chemometrics," J. J. Breen and P. E. Robinson, eds. pp. 271–279. American Chemical Society, Washington, D.C.
Fuller, W. (1987). "Measurement Error Models." Wiley, New York.
Greenland, S. (1980). The effect of misclassification in the presence of covariates. *Am. J. Epidemiol.* 112, 564–569.
John, P. W. M. (1990). "Statistical Methods in Engineering and Quality Assurance." Wiley, New York.
Johnson, R. A., and Wichern, D. (1982). "Applied Multivariate Statistics." Prentice Hall, Englewood Cliffs, New Jersey.
Karlin, S., and Macken, C. (1991). Some statistical problems in the assessment of inhomogeneities of DNA sequence data. *J. Am. Stat. Assn.* 86, 27–35.
Kay, C., Kemp, K. W., Nix, A. B. J., Rowlands, R. J., Groom, G. V., Wilson, D. W., and Griffiths, K. (1983). The importance of internal quality control in national external quality assessment schemes. Plasma progesterone assays. *Stat. Med.* 2, 123–129.
Kleinbaum, D. G., Kupper, L. L., and Morgenstern, H. (1982). "Epidemiological Research: Principles and Quantitative Methods." Lifetime Learning, Belmont, California.
Kleinbaum, D., Kupper, L. L., and Muller, K. E. (1988). "Applied Regression Analysis and Other Multivariable Methods." PWS-Kent, Boston.
Lachenbruch, A. P. (1975). "Discriminant Analysis." Hafner Press, New York.
Laird, N., and Ware, J. (1982). Random effects models for longitudinal data. *Biometrics* 38, 963–974.
Liang, K.-Y., and Zeger, S. (1986). Longitudinal data analysis using generalized linear models. *Biometrika* 73, 13–22.
Lin, L. I.-K. (1989). A concordance correlation coefficient to evaluate reproducibility. *Biometrics* 45, 255–268.
Linnet, K. (1990). Estimation of the linear relationship between the measurements of two methods with proportional errors. *Stat. Med.* 9, 1463–1474.

Lubin, J. H., Gail, M. H., and Ershow, A. G. (1988). Sample size and power for case-control studies when exposures are continuous. *Stat. Med.* **7**, 363–376.

McCullagh, P., and Nelder, J. (1989). "Generalized Linear Models," 2d Ed. Chapman and Hall, Cambridge.

Michel, C. J. (1986). New statistical approach to discriminate between protein coding and nonprotein coding regions in DNA sequences and its evaluation. *J. Theoret. Biol.* **120**, 223–236.

Milliken, G., and Johnson, D. (1986). "Analysis of Messy Data," Vol. 1, Design of Experiments. Van Nostrand-Reinhold, New York.

Neel, J. V., Rosenblum, B. B., Sing, C. F., Skolnick, M. M., Hanash, S. M., and Steinberg, S. (1984). Adapting two-dimensional gel electrophoresis to the study of human germ-line mutation rates. *In* "Two-Dimensional Gel Electrophoresis of Proteins" (J. E. Celis and R. Bravo, eds.). Academic Press, Orlando, Florida.

Orchard, T., and Woodbury, M. A. (1972). A Missing information principle: Theory and applications. *In* "Proceedings of the Sixth Berkeley Symposium on Mathematical Statistics and Probability" (L. M. LeCam, J. Neyman, and E. L. Scott, eds.), Vol. I, pp. 697–715. University of California Press, Berkeley.

Press, S. J., and Wilson, S. (1978). Choosing between logistic regression and discriminant analysis. *J. Am. Stat. Assn.* **73**, 699–705.

Raab, G. M. (1981). Estimation of a variance function with application to immunoassay. *Appl. Stat.* **30**, 32–40.

Ramsay, J. O. (1988). Monotone regression splines in action (with discussion). *Stat. Sci.* **3**, 425–461.

Ratkowsky, D. (1983). Nonlinear Regression Modeling: A Unified Practical Approach." Marcel Dekker, New York.

Ratkowsky, D. A., and Reedy, T. J. (1986). Choosing near-linear parameters in the four-parameter logistic model for radioligand and related assays. *Biometrics* **42**, 575–582.

Rodbard, D., and Frazier, G. R. (1975). Statistical analysis of radioligand data. *Meth. Enzymol.* **37**, 3–22.

Rosner, B. (1982). "Fundamentals of Biostatistics." Duxbury Press, Boston.

Rowlands, R. J., Griffiths, K., Kemp, K. W., Nix, A. B. J., Richards, G., and Wilson, D. W. (1983). Application of cusum techniques to the routine monitoring of clinical laboratories. *Stat. Med.* **2**, 141–145.

Russek, E., Kronmal, R. A., and Fisher, L. D. (1983). The effect of assuming independence in applying Bayes' theorem to risk estimation and classification in diagnosis. *Computers Biomed. Res.* **16**, 537–552.

Ryan, T. P. (1989). "Statistical Methods for Quality Improvement." Wiley, New York.

SAS Institute (1985). "SAS User's Guide: Basics," Version 5. SAS Institute, Cary, North Carolina.

Schesselman, J. J. (1982). "Case-Control Studies." Oxford University Press, New York.

Shumway, R., Azari, A. S., and Johnson, P. (1989). Estimating mean concentration under transformation for environmental data with detection limits. *Technometrics* **31**, 347–351.

Sjostrom, M., Wold, S., Lindberg, W., Persson, J.-A., and Martens, H. (1983). A multivariate calibration problem in analytical chemistry solved by partial least-squares models in latent variables. *Anal. Chim. Acta* **150**, 61–70.

Sokal, R. R., and Rohlf, F. J. (1981). "Biometry," 2d Ed. Freeman, New York.

Thomas, D. C., Siemiatycki, J., Dewar, R., Robins, J., Goldberg, M., and Armstrong, B. G. (1985). The problem of multiple inference on studies designed to generate hypotheses. *Am. J. Epidemiol.* **122**, 1080–1095.

Volund, A. (1979). Application of the four-parameter logistic model to bioassay: Comparison with slope ratio and parallel line models. *Biometrics* **34**, 357–365.

Weir, B. (1988). Statistical analysis of DNA sequences. *J. Natl. Cancer Inst.* **80**, 395–406.

Weir, B. (1990). "*Genetic Data Analysis.*" Sinauer Associates, Sunderland, Massachusetts.

Weir, B., and Basten, C. J. (1990). Sampling strategies for distances between DNA sequences. *Biometrics* **46,** 551–571.

Whittemore, A., McLarty, J. W., Fortson, N., and Anderson, K. (1982). Regression analysis of cytopathological data. *Biometrics* **38,** 899–906.

Winer, B. (1971). "Statistical Principles in Experimental Design," 2d Ed. McGraw-Hill, New York.

8

Biologic Specimen Banks: A Resource for Molecular Epidemiologic Studies

*Deborah M. Winn
and Elaine W. Gunter*

Introduction

A biologic specimen bank may be defined as "a system which will store one or many types of biologic specimens for later analysis from single or multiple studies under conditions which permit efficient retrieval and optimum stability of the samples" (Winn *et al.*, 1990). Specimen banks serve a fundamental role in molecular epidemiologic studies because they function as the repository of materials from which molecular-based assessments will be made for use in epidemiologic studies. Banks can include a wide range of biologic materials isolated from human subjects. The Directory of On-Going Research on Cancer of the International Agency for Research on Cancer (Coleman and Wahrendorf, 1991) annually lists banks for potential use in studies of cancer. A recent version of the directory suggests the diversity of material that is available, listing cells in culture, feces, red cells, saliva, serum/plasma, tissue, urine, white blood cells, and various other materials (e.g., toenails, DNA, hair, semen, and maternal milk). The assessments made on banked materials can vary from the characteristics of the tissues (e.g., histologic features) to measurements made at the level of the gene (e.g., DNA adducts). Banks sometimes are devised for longitudinal observational studies or community intervention studies that are intended to be longitudinal in design. Repositories of patient material, collected in the course of patient diagnosis or care, can be a rich source of material for markers of prognosis in clinical therapeutic trials or observational studies of patient populations (Whiteside and Herberman, 1990).

Although the bank itself may exist as a single refrigerator in the basement of a laboratory, banks also may exist as sophisticated storage facilities with associated computer databases, security systems, and management structures. The physical integrity of the specimens is the most important function of the bank itself; however, banking of molecular epidemiologic specimens should be considered in a much wider context that begins at the design of the survey, continues through the collection of the specimens, and ends at the reporting of accurate data on molecular or biochemical measurements. In this chapter, we take this wider view of the function of biologic banks. The value of these banks for molecular epidemiologic studies depends on many factors that are described in this chapter.

Utility of a Bank for Molecular Epidemiologic Studies

The value of a biologic specimen bank for epidemiologists is that it provides the opportunity to make biologic measurements on humans. The emerging area of molecular epidemiology requires using measurements of events at the molecular or biochemical level in conjunction with other epidemiologic data. Banks preserve biologic materials obtained at a particular point in time and permit retrospective analysis of newly emerging analytes. A by-product of using banked material is that batching specimens and running laboratory assessments all at once sometimes leads to more efficient laboratory processing of biologic materials and may eliminate drift in analyte levels over time. Drift in laboratory analysis occurs when a systematic deviation in the laboratory protocol occurs due to any number of factors (e.g., technician error, deterioration of equipment or materials) and leads, over time, to erroneous measurements.

Biologic specimen banks capture a specific moment in the life of study participants, in much the same way that an archive of completed health questionnaires or a storeroom of radiographs does. The laboratory assays or the linkage of assay data to other epidemiologic factors may take place later. However, the biologic state of the study participant at the time of acquiring the specimen is preserved. This aspect is especially important in longitudinal studies, in which understanding of temporal sequencing of events is important.

Major Influences on the Utility of Biologic Banks

The two most important factors that influence the value of the biologic materials in the bank are the specimen donor and the study investigator.

Specimen Donor

The characteristics of the donor, including his or her physiological and psychological state genetic make-up, health behaviors, and health status, that converge at the time of specimen donation all may affect measurements made on a biologic specimen. Among the many important physiological parameters are biologic rhythms, which have a profound impact on measurements made on biologic materials. Circadian, menstrual, and other biologic rhythms influence levels of many hormones such as sex hormone-binding globulin (Yie *et al.*, 1990), prolactin (Ashby, 1987), and testosterone (Demetriou, 1987), as well as many other biochemical parameters such as zinc and iron (Morrison *et al.*, 1979), and should be investigated for their potential role in influencing molecular epidemiologic measurements. Interindividual variation occurs for molecular epidemiologic measures as well, although the causes or nature of the variations are generally less well understood (Hulka, 1990; Perera, 1991). For example, the reasons for wide variations among smokers in levels of hemoglobin adducts to tobacco-specific carcinogens are unclear (Carmella *et al.*, 1990) and could include many different physiological parameters. The psychological state of the donor during the blood draw can influence some analytes. For example, stress levels influence serum prolactin (Sassin *et al.*, 1972). The importance of psychological factors should be explored in molecular epidemiologic studies. The role of genetic factors is illustrated by work on apparently dominantly transmitted tumor suppressor genes in the development of selected cancers (Harris *et al.*, 1987). Health behaviors also may modify analyte readings. The nature of the administration of tobacco (cigarettes or snuff) or the route of administration (oral vs inhalation) are factors that may explain why snuff dippers have higher levels of hemoglobin adducts to tobacco-specific nitrosamines than expected (Carmella *et al.*, 1990). As another example, caffeine consumption appears to modify the levels of DNA adducts in the placentas of pregnant smokers (Everson *et al.*, 1988a). Finally, health status influences analyte levels; for example, low folate levels appear to influence the levels of micronucleated red blood cells in persons with minimal spleen function (Everson *et al.*, 1988b; Schreinemachers and Everson, 1991).

Study Investigator

The other profound influence on the utility of stored specimens is the study investigator, who identifies the types of molecular assessments that are likely to be performed on banked materials, determines which biologic materials (e.g., urine, blood) will provide measurements to meet those requirements, acquires the specimens, stores the material, plans for and executes the laboratory analyses, and reports the findings.

Biologic Parameter of Interest

Epidemiologic studies often have used laboratory measurements made on stored materials. Forward-thinking investigators know that, in a rapidly evolving field such as molecular epidemiology, long-term storage of materials will help meet future epidemiologic needs for information. However, identifying which measurements might be made in the future is difficult, since there may be interest in new or different analytes, changes in specimen or resource requirements for the laboratory assay, or improvements in methodologies that offer more accurate or detailed results.

Types of Biologic Specimens

The types of specimens, or matrix, collected and stored for epidemiologic studies can range from blood to tissues to toenails. Trade-offs always exist in the choice of matrix for molecular epidemiologic studies. Some types of specimen are more abundantly available (e.g., urine) than others (e.g., saliva). Some (e.g., serum) can provide markers of many different biologic processes (e.g., immunological, nutritional, carcinogenic). Many analytes can be measured in more than one matrix but, because of human metabolism, the meaning of the level of analyte will vary depending on the matrix. For example, high levels of lead indicate relatively recent exposure when measured in blood, but indicate chronic exposure when assessed in bone and teeth (Shapiro *et al.*, 1978). The meaning of molecular epidemiologic measurements also depends on the matrix. For example, since the half-life of red blood cells is 120 days, hemoglobin adducts to potential carcinogens measure cumulative exposure over that period (Goldring and Lucier, 1990). However, the underlying meaning of differences in various matrices may not be apparent without further detailed investigations. These interrelationships may be complex, as suggested by findings by one research team that hemoglobin and DNA adduct formation in the liver and lung of rats did not exhibit the same pattern in response to dose of carcinogen (Murphy *et al.*, 1990).

Molecular epidemiology has opened up new uses for some matrices and preservation methods. For example, increasing research interest in hemoglobin adducts may make banking of red blood cells more commonplace. The ability to examine DNA from pathology slides (Shibata *et al.*, 1988) extends the usefulness of these specimens beyond microscopic examination.

Many of the newer laboratory assays for molecular-level assessments, such as ^{32}P-postlabeling, are adaptable to many different matrices. For example, using this technique, DNA adducts to potential carcinogens have been examined from the placenta of smokers (Everson *et al.*, 1986), the buccal mucosa of Bombay cigarette smokers (Chacko and Gupta, 1988), and other materials (Harris *et al.*, 1987).

Acquisition Procedure

The invasiveness of the procedure required to obtain the specimen is one particularly important factor in the choice of the matrix used to measure the analyte. Materials (e.g., fat cells) obtained through invasive procedures such as biopsy are usually only available in pathology departments and can be obtained only from individuals on whom such procedures are justified medically. Using minimally invasive procedures is more suited to large-scale epidemiologic studies than to clinical or experimental studies because of the reduced medical risks, the greater study subject compliance, and the lower specimen acquisition costs. The desire for readily obtainable biologic markers is also evident in the growing interest in measurement of hemoglobin adducts. For example, protein adducts in hemoglobin to tobacco-specific nitrosamines are being studied as potential indicators of DNA adduct levels in inaccessible tissues targeted by carcinogens (e.g., lung) (Murphy *et al.*, 1990). The potential for identifying oncogene-related products in an easily obtained fluid, urine, may improve options for studying cancer (Stock *et al.*, 1987; Schwartz, 1990).

Polymerase chain reaction (PCR) can be used to generate analyzable quantities of DNA from small quantities of DNA-containing material from any part of the body (e.g., from a spot of blood from a pricked finger on filter paper) (Gibbs and Caskey, 1989) and has wide potential applicability for epidemiologic studies because of the flexibility in terms of matrix and sample size. Most epidemiologic studies that examine DNA markers will need measurements on large numbers of often healthy subjects, so efficient simple collection protocols are most desirable. Using white blood cells from venipuncture may be the approach of choice in studies that expect to involve DNA probes because specimen collection and analysis methodologies can be highly standardized and involve minimally invasive procedures. For instance, white blood cells from persons participating in the Third National Health and Nutrition Examination Survey now in progress are being stored for later analysis of DNA (National Center for Health Statistics, 1990; Woteki *et al.*, 1990; Wagener *et al.*, 1991).

Storage Characteristics and Duration

Many different types of molecular epidemiologic determinations can be made successfully using properly stored specimens. Goldring and Lucier (1990) state that DNA and protein adducts should be stable at storage temperatures of $-70°$. In one study, hemoglobin adducts to ethylene oxide were measured in smokers and nonsmokers whose red blood cell hemolysates were stored for an unspecified time period, possibly up to 2–3 years, at $-70°C$ (Tornqvist *et al.*, 1986). Hulka (1990) pointed out that pathology and other specimens archived for medical reasons may provide a rich source of specimens for molecular epidemiologic studies. For example, papilloma virus has been detected in slides of formalin- and paraffin-preserved tissue from cervical cancer patients whose specimens were collected decades previously (Shibata *et al.*, 1988; Kiyabu *et al.*, 1989). Banked tumor tissue from women with breast cancer was used in a study to examine the relationship between oncogene amplification and survival (Slamon *et al.*, 1987).

Several reviews outline procedures and policies for biologic specimen banking. Whiteside and Herberman (1990) and the Committee on National Monitoring of Human Tissues (CNMHT, 1991) have provided excellent guidance on establishing well-designed serum and tissue storage banks.

How biologic specimens are transported and stored and the duration of the storage may affect results of laboratory assays. A distressing example of stored biologic materials rendered useless due poor storage practices is described in a report of a tissue archive for toxicologic studies (NMHT, 1991). If receipt and control procedures that document transmission and location of specimens are handled poorly, specimens may be lost or mislabeled. Temperature, freeze/thaw cycles, duration of storage, desiccation, and contamination all affect measurements of toxic chemicals (Aitio and Jarvisalo, 1985). The extent to which nutritional, cardiovascular, chemical, and immunologic parameters are affected by storage varies considerably by type of blood specimen, storage temperature, and other factors (Petrakis, 1985). Degradation can occur with formalin-fixed paraffin-imbedded slides, influencing results of PCR amplification (Shibata *et al.*, 1988).

Laboratory Analysis

The laboratory assay used and the implementation and reporting of the procedure also are important. The investigator makes choices in selecting the assay to meet many considerations, including type and quantity of specimen needed, precision of the assay, procedural aspects of the assay, and quality of the laboratory operation.

Data Reporting

Investigators who analyze data derived from biologic specimens have ethical obligations such as obtaining informed consent and protecting privacy and confidentiality. They also are obliged to notify study participants about findings from biologic materials that may impact on their health and needs for medical care, although in research projects it is often difficult to know the significance of many findings and to determine what should be communicated to the study subject (Schulte, 1987; Ashford, 1991). With stored specimens, these ethical issues are compounded, since some laboratory assessments may be made years or even decades after the collection of the specimen.

Bias in Studies Involving Banked Materials

The same types of bias that can affect any epidemiologic study are operative in molecular epidemiologic studies involving banked materials: measurement errors, selection bias, and confounding. Measurement errors are systematic or random errors in the measurement of the variables under study. Selection bias occurs when inclusion of a study subject in a group is influenced by the factor under study. Confounding factors are those that are related to the exposure and the outcome of interest and may distort estimates of the relationship between the exposure and the outcome. The potential sources of these biases in molecular epidemiologic studies involving banked materials and some possible solutions to these problems are discussed in the following sections.

Using stored specimens for molecular epidemiology studies is a field in infancy. Much of the current work on many types of molecular markers has involved laboratory assessments that were made close to the time of specimen donation; banked materials have not yet been used much in molecular epidemiologic studies. Moreover, current molecular epidemiologic uses depend on the existence, in good condition, of materials collected in the past. Some molecular measurements involve using materials, such as urine or red blood cells, that have not been stored as commonly as serum. Therefore, the discussions of epidemiologic bias that follow this section concentrate generally on nutritional biochemical assessments made with stored material, because of the wealth of epidemiologic studies involving stored material for nutritional assessments. However, the principles regarding stored specimens are the same, regardless of the laboratory technique or specimen type. As biologic banks accrue relevant specimens and the potential of stored material for molecular epidemiologic assessments becomes more apparent, applications involving specimen banks are likely to become more common.

Measurement Errors

Measurement errors can occur at any stage of the biologic banking process, from initial donation to reporting of results (see Table 8.1). The donor may fail to follow instructions about providing the specimen. The specimen may volatilize or become contaminated during handling and transport to the specimen bank. Further problems may occur during storage, including temperature problems, contamination, or desiccation due to use of storage containers inappropriate for the storage temperatures or conditions. Some of the many sources of problems in the collection and storage of human biologic materials for assessment of exposures to toxic chemicals have been described (Aitio and Jarvisalo, 1985) and serve to illustrate the potential problems for one class of substances. Even in a perfectly functioning storage facility, prob-

TABLE *8.1*   Measurement Error in Molecular Epidemiologic Studies
Involving Banked Materials

Major sources of error	Solutions
Donor problems (e.g., failure to follow instructions)	Strict adherence to study procedures
	Document study procedures
Collection equipment failures (e.g., contamination, breakage)	Document and evaluate storage history
	Secure physical storage environment
Collection technician error	Monitor for deterioration of specimens and
Transport and handling errors	effects of storage duration
Storage conditions (e.g., temperature, desiccation, contamination, storage duration)	State-of-the-art laboratory procedures and quality control program
	Good record keeping
Receipt and control errors (e.g., transcription errors)	Statistical adjustment to account for storage effects
From donation site to storage	Comparisons of analyte levels from stored
From storage to laboratory	material with reference values
From laboratory to epidemiologic analysis	

lems may occur as a result of the storage itself and become more pronounced with duration of storage (Petrakis, 1985). Laboratory errors also may lead to misclassification errors. Poor record-keeping systems may result in errors in the identity of a specimen, for example, through poor quality container labels that fall off or through transcription errors.

In any epidemiologic study, nondifferential misclassification results in a weakening of relationships between exposures and outcomes (Rothman, 1986). The analyte being measured in stored material may be the exposure or the outcome in an epidemiologic study, depending on what the biologic marker is intended to measure: susceptibility, internal dose, biologically effective dose, or biologic response (Perera and Weinstein, 1982). In the case of specimens from biologic banks, nondifferential misclassification occurs when errors in the measurement of analytes are equivalent for cases and controls or for exposed and unexposed persons. When misclassification occurs differently between compared groups, estimates are biased in either direction (Rothman, 1986); for example, if deterioration of specimens resulting in a faulty measurement of the analyte occurs differentially in cases than in controls. Nondifferential misclassification could not be excluded entirely as an explanation for the finding in one study that vitamin A levels were lower in cancer cases, whose specimens had been subjected to light exposure and freeze–thaw cycles, than in controls, whose specimens were less disturbed (Kark *et al.*, 1981). Using specimens previously used for another purpose is not always a problem; one report showed that an assay for serum selenium was unaffected by a radioimmunoassay for ferritin done previously on the same samples (Stevens *et al.*, 1986).

Several techniques can be used to determine if measurement errors have occurred. One approach is to examine the levels of analyte observed in the stored materials with reference levels of the analyte. For example, in a study of nutrients measured in stored serum, the investigators determined that the levels in their cases and controls were comparable to those from a large national survey (Willett *et al.*, 1984). These same investigators also used the similarity of findings regarding nutrient determinants from their study to findings from other studies (e.g., carotenoid and tocopherol appropriately were correlated positively with serum lipids) to support their conclusion of an absence of errors of measurement of the nutrients. Mean analyte levels from stored materials can be compared with levels from fresh material to assess the potential for measurement errors. Salonen *et al.* (1985) found evidence of some deterioration of vitamin E in stored samples by comparing them with recently obtained material from a comparable population.

Measurement errors can be minimized by taking proper precautions during the acquisition, storing, and analysis of analytes, as shown in Table 8.1. Study manuals should be written that describe procedures in detail; these procedures must be adhered to strictly. For example, in studies involving blood drawing, the procedures should call for standardized setting and equipment; reasons for nonacquisition of a specimen (e.g., fainting, refusal) should be noted. Posture at the time of blood draw is known to influence plasma protein levels (Koller and Kaplan, 1987); this fact can be used to illustrate the need for good study protocols. It would be prudent to include a required physical positioning of the subject in protocols involving blood drawing and to monitor that phlebotomists adhere to the procedure.

A secure physical storage facility that, ideally, has alarms to detect refrigeration or other equipment failures is important. Periodic examination of stored control specimens, that have been treated the same as study specimens, for desiccation, contamination, and the effects of storage duration may allow the investigator to institute procedures to halt further deterioration of stored specimens and to minimize or eliminate problems in specimens not yet affected.

Good bookkeeping and record keeping is a must. Careful records should be kept of the identity and location of all materials. Documentation on all activities related to the specimens should be maintained; for example, removal of a portion of the specimen from the bank or occurrence of a temperature fluctuation should be recorded.

Assays should be performed consistent with good laboratory practice. The importance of analyzing case and control specimens at the same time has been pointed out by Willett, who notes that spurious associations could result otherwise, due to laboratory drift (Willett, 1987). Blinding the technician to the identity of specimens is important. Blind replicate samples analyzed by the laboratory are an important quality control check (Willett, 1987). Averaging results of duplicate assessments is another technique used by some investigators (Coates *et al.*, 1988). Perera points out the importance of confirmation of an assay against other methodologies, reproducibility, and reduction of interlaboratory differences in results (Perera, 1991).

A global quality assurance program that covers the specimen from collection through reporting of results is critical. Quality control procedures and data should be readily accessible to readers who must evaluate the findings reported in an article. Investigators who have biologic specimen facilities should be encouraged to share information about their procedures, successes, and failures as an aid to other investigators. Procedures used should be well documented in research publications.

Some steps to minimize measurement error can be taken during the data analysis stage. For example, to address measurement error issues in data analysis, investigators in one study applied an adjustment factor, consisting of storage duration time and a rate of increase in the level of the analyte, to the results of a cholesterol assay to adjust statistically for the impact of specimen storage (Kalsbeek *et al.*, 1988).

Selection Bias

Selection bias in the context of epidemiologic studies using material from a biologic specimen bank occurs when groups being compared have an unequal probability of having specimens available for study, leading to differences between groups in the molecular or biochemical parameter being measured. Selection bias may involve the loss or deterioration of all or part of the specimen, so the analyte cannot be measured. The losses may occur at any number of stages, from acquiring the specimen to reporting the results of biochemical or molecular measurements. Specimens may not be obtained at all, may be lost during transmission to the storage facility, during storage, in transmission to the laboratory, or at the laboratory, or may be missing from reported results if data are misplaced or faulty receipt and control procedures allow specimen loss to be overlooked. These potential sources of selection bias are listed in Table 8.2. Many times, the loss may be essentially random, as a result of accident or occasional error (e.g., container breaks, technician failures, donor problems). Of concern, however, is loss that may be associated with the variables under study, that is, when a factor related to the molecular measurement influences whether the specimen will be included in the specimen bank or subjected to laboratory analysis.

Losses associated with the factors under study, resulting in selection bias, can occur at any stage, but seem to occur most frequently by self-selection of donors, through problems surrounding specimen donation, or as a result of the specimen being used for other purposes. Self-selection into a study is one common source of selection bias in epidemiologic studies (Rothman, 1986). Potential donors with particular health characteristics related to the factor under study may be more or less likely to appear or be available to provide

TABLE *8.2*   Selection Bias in Molecular Epidemiologic Studies
Involving Banked Materials

Major sources of error	Solutions
Failure to obtain a specimen related to exposure or disease status	Insure high response rates from donors
Differential availability of specimen for laboratory analysis related to exposure or disease status	Insure adequate documentation of all procedures so potential selection bias can be handled as a confounding problem
Specimen acquisition, storage, and analysis procedures related to exposure or disease status	Keep track of uses of specimens
	Use smallest possible quantities of materials
	Aliquot small portions to containers
	Exert good judgment in handling "withdrawals" from bank
	Appropriate use of nested case-control and case-cohort study designs

the specimen. For example, among the eligible sample persons in the National Household Seroprevalence Survey Feasibility Study, those in geographic areas designated as high HIV risk areas were less likely to provide a blood sample than those in low-risk areas (Horvitz *et al.*, 1990).

Although the study subject may arrive at the donation site prepared to donate the specimen, he or she may refuse to go through with the procedure to give a specimen or may have a physical problem such as fainting or a physical disability that interferes with the proper procedures for acquiring the specimen. When these instances occur differentially among groups to be compared, selection bias is a possibility. In the Scottish Heart Health Study, women with a history of myocardial infarction were more likely to have missing blood specimens, and having a missing blood specimen was correlated with the cardiovascular risk factor body mass. Phlebotomists may have had more difficulty obtaining blood from the peripheral veins of heavier women (Woodward *et al.*, 1991). Sometimes persons otherwise eligible and willing to provide a specimen should not do so because it would impair their health. This restriction can be a problem if the reason for not providing a specimen is related to the topic of interest. For example, pain in the gums of persons with periodontal disease may preclude the acquisition of crevicular fluid from the area just below the gums in a study of the natural history of periodontal disease.

If the procedures used to acquire, transport, store, and analyze specimens are influenced by the exposure or disease status characteristics under study, selection bias is a potential problem. For example, scheduling cases and controls to provide specimens on different days could result in different assay levels if equipment or technician problems develop. This type of problem may be mitigated to some extent if data are collected that fully describe potential differences in study procedures (i.e., if the problem can be handled as an issue of confounding). In this example, comparing findings from different technicians may help uncover and allow for adjustment for bias. However, adjustment for confounding will be inadequate if a critical study procedure is highly correlated with exposure or disease status. The best approach is to avoid processing specimens in groups or batches defined by disease or exposure status at every point from specimen collection to laboratory analysis.

Several additional approaches are available to safeguarding against selection bias in molecular epidemiologic studies; these are listed in Table 8.2. Keeping study subject participation high is one key issue. Major challenges, such as drawing blood for HIV antibody levels from a probability sample of persons in a household setting, can be successful with a coordinated and effective program to insure high response rates. The National Seroprevalence Feasibility Survey employed highly trained phlebotomists, a video tape stressing the importance of the survey with an appearance by the popular Surgeon General, C. Everett Koop, and financial remuneration to achieve a very respectable 82% rate of obtaining blood under difficult circumstances (Horvitz *et al.*, 1990).

Equally important in insuring against selection bias is making sure that biologic specimens from eligible persons are available for laboratory analysis. Availability can become problematic in studies using stored specimens that may be used for many purposes. One research team noted that the 121 men in their study of sex hormones were included because they had frozen serum available, and that 97% of men who would have been eligible otherwise could not be included due to lack of sera, leading to a potential for selection bias (Dai *et al.*, 1988). As another example, in a study of selenium, retinol, and cancer, stored serum and plasma were used (Coates *et al.*, 1988). Of 195 persons with cancer, 38 had no serum or plasma available because the materials were used for two other studies. However, the reasons for lack of materials (i.e., need for a replicate lipid assay and nonparticipation in a swine flu investigation) seem unrelated to the study purposes, and no demographic and cancer-related differences were observed among those with and without specimens available for analysis.

In molecular epidemiologic research, some specimens are destroyed in the process of using them. For example, paraffin-imbedded formalin-fixed specimens can be used only once for PCR (Shibata *et al.*, 1988). This makes susceptibility to selection bias an issue. However, in general, techniques such as PCR, which amplify the quantity of material available, hold considerable promise in avoiding selection bias due to lack of specimen.

Keeping stories of specimens from all subjects available for study is difficult. However, some reasonable steps to take include using the smallest possible quantity of material for analysis, aliquoting specimens into multiple containers so use of the specimen for one assay does not use the entire specimen, and keeping good records on previous uses of specimens so the reasons for nonavailability of specimens are known. Storing aliquots in separate facilities might have avoided the loss due to a freezer failure of half of the study subject specimens in one study (Willett *et al.,* 1984).

Strong leadership control by specimen bank managers over appropriate "withdrawals" of specimens from the bank is also important in reducing the potential for selection bias. The materials in a biologic bank should be used as judiciously as possible. Using assays that minimize the amount of specimen used makes more specimen available for other purposes. For example, careful strategic planning has allowed the National Center for Health statistics to perform 78 separate assays on adult participants in NHANES III and reserve up to 5.0 ml serum for long-term storage to meet future needs (National Center for Health Statistics, 1990).

Two epidemiologic study designs, nested case-control and case-cohort designs, are extremely useful for etiologic studies involving banked materials because the time sequence of health events can be established and the number of samples requiring laboratory analysis is minimized (Prentice *et al.,* 1986a,b). Nested case-control studies, in which cases with a particular health outcome accruing in a cohort followed over time are compared with controls drawn from the cohort also at risk of the health outcome, have been used extensively in studies involving banked specimens. Examples include a series of reports assessing vitamin levels in serum from cancer cases and controls derived from large epidemiologic cohorts established for studying heart disease and other conditions (Menkes *et al.,* 1986; Willett *et al.,* 1984; Salonen *et al.,* 1985). These designs may help conserve precious biologic samples, although they do not specifically protect against selection bias.

Confounding

Confounding occurs when a factor related to the analyte and to the outcome of interest is not dealt with adequately in the design or analysis of the study. In studies of molecular epidemiology, confounding is a potential if a factor related to the molecular or biochemical measurement is also a determinant of the exposure or health outcome of interest. These potential sources of confounding bias and some possible strategies to reducing the potential of confounding bias are listed in Table 8.3.

To address confounding issues, the specimen banker should ensure that adequate data on potential confounders are obtained. Data about the specimen should include donor characteristics, circumstances of the collection of the specimens, and storage events. Identifiers linking specimen data with other epidemiologic data on the same persons (e.g., age, smoking status) also should be maintained so the specimens can be characterized adequately. Having critical confounders available can be problematic when specimens

TABLE 8.3   Confounding Bias in Molecular Epidemiologic Studies
Involving Banked Materials

Sources of confounding	Solutions
Failure to obtain data on potential confounders in study design Failure to use data on potential confounders in analysis	Employ knowledge of potential confounders in designing study (e.g., match on storage duration) Collect data on all relevant aspects of specimen acquisition, transport, storage, and laboratory analysis Analyze data to examine confounding related to storage conditions Adjust for confounders in analysis

from pathology archives are used if medical records for patients contain little information or are no longer available.

In the study design, matching specimens on factors related to storage is one technique used to control for confounding caused by these storage characteristics. For example, in one study, matching controls to cases by the length of time the specimens had been held in storage was an approach used to minimize the impact on selenium measurements of a suspected loss of water vapor from containers used to hold specimens (Coates *et al.*, 1988). Thus, differences in storage duration would not bias interpretations of relative differences in the means. Matching by the date the specimen was obtained provides similar benefits (for examples see Wald *et al.*, 1980; Willett *et al.*, 1984). Wald (1985) points out that useful information sometimes may be obtained even when some deterioration of biologic samples has occurred if compared groups are matched by storage duration.

In data analysis, the influence of confounding often can be minimized. Features related to storage of specimens or other characteristics associated with specimen acquisition can be used as adjustment factors to minimize potential confounding in analysis of data based on stored specimens. In one study, the month of serum collection (presumably an indicator of duration of storage) and the hours before blood-draw since the previous meal were used in a multivariate model with other factors to examine the relationship between age and sex and levels of retinol, beta-carotene, and alpha-tocopherol (Comstock *et al.*, 1987). In another study, vials were broken accidentally during transport (but the material was recovered); values of retinol in the broken vials were lower than values in the unbroken ones. In data analysis, the authors analyzed the relationship between vitamin A and cancer separately by vial breakage status and used statistical standardization to control for confounding due to the sample deterioration caused by breakage (Wald *et al.*, 1980).

Research Needed

Research is needed to increase the utility of specimen banks. One important area requiring attention is the development of an appropriate set of methods for monitoring deterioration of specimens over time that would include testing for desiccation, integrity of the container, and stability of the analytes. Specimens that will not be needed for analysis should be included with study specimens; on-going monitoring of specimen integrity using the extra material then will not compromise the study specimens.

Studies of analyte stability over long storage periods are needed. Although there is some scientific understanding of stability for many biochemical analytes (Petrakis, 1985), data is limited for newer molecular-level assays.

A greater understanding of intra- and interindividual variability and the natural history of newer biomarkers is needed. As it accrues, this new knowledge needs to be incorporated into protocols of epidemiologic studies so key information related to analyte levels (e.g., biorhythms, age) can be collected. Better understanding of the reasons for variability also may guide future data collection and procedures for storage of biologic material. For example, studies of the variability of papilloma virus findings within buccal mucosal smears taken from mouth scrapings in different anatomic locations should provide useful data on the minimum number and sites of smears for reliable determinations of papilloma virus infection.

In addition, greater interchange between laboratory specialists, specimen bankers, and epidemiologists is needed. Several books and articles (Hulka, 1990; Whiteside and Herberman, 1990; Yach, 1990) encourage such interchange and are intended to help bridge these gaps between professions. Proper development and use of biologic specimen storage facilities for molecular epidemiologic studies promises to expand vastly our understanding of disease processes.

Acknowledgments

The authors gratefully acknowledge the advice and manuscript reviews provided by Marsha Reichman, Diane Wagener, and John Horm.

References

Aitio, A., and Järvisalo, J. (1985). Biological monitoring of occupational exposure to toxic chemicals: Collection, processing, and storage of specimens. *Ann. Clin. Lab. Sci.* **15,** 121–139.

Ashby, C. D. (1987). Prolactin. *In* "Methods in Clinical Chemistry" (A. J. Pesce and L. A. Kaplan, eds.), p. 264. Mosby, St. Louis, Missouri.

Ashford, N. A. (1991). Medical screening for carcinogens: Legal and ethical considerations. *In* "Molecular Dosimetry and Human Cancer: Analytical, Epidemiological, and Social Considerations" (J. D. Groopman and P. L. Skipper, eds.), pp. 417–442. CRC Press, Boca Raton, Florida.

Carmella, S. G., Kagan, S. S., Kagan, M., Foiles, P. G., Palladino, G., Quart, A. M., Quart, E., and Hecht, S. S. (1990). Mass spectrometric analysis of tobacco-specific nitrosamine hemoglobin adducts in snuff dippers, smokers, and nonsmokers. *Cancer Res.* **50**, 5438–5445.

Chacko, M., and Gupta, R. C. (1988). Evaluation of DNA damage in the oral mucosa of tobacco users and non-users by ^{32}P-adduct assay. *Carcinogenesis* **9**, 2309–2313.

Coates, R. J., Weiss, N. S., Daling, J. R., Morris, J. S., and Labbe, R. F. (1988). Serum levels of selenium and retinol and the subsequent risk of cancer. *Am. J. Epidemiol.* **128**, 515–523.

Coleman, M., and Wahrendorf, J. (1991). Directory of on-going research in cancer epidemiology. *IARC Sci. Publ.* **110**, 733–744.

Committee on National Monitoring of Human Tissues, Board on Environmental Studies and Toxicology, Commission on Life Sciences (1991). "Monitoring Human Tissues for Toxic Substances." National Academy Press, Washington, D.C.

Comstock, G. W., Menkes, M. S., Schober, S. E., Vuilleumier, J-P., and Helsing, K. J. (1988). Serum levels of retinol, beta-carotene, and alpha-tocopherol in older adults. *Am. J. Epidemiol.* **127**, 114–123.

Dai, W. S., Gutai, J. P., Kuller, L. H., and Cauley, J. A. (1988). Cigarette smoking and serum sex hormones in men. *Am. J. Epidemiol.* **128**, 796–805.

Demetriou, J. A. (1987). Testosterone. In "Methods in Clinical Chemistry" (A. J. Pesce and L. A. Kaplan, eds.), p. 268. Mosby, St. Louis, Missouri.

Everson, R. B., Randerath, E., Santella, R. M., Cefalo, R. C., Avitts, T. A., and Randerath, K. (1986). Detection of smoking-related covalent DNA adducts in human placenta. *Science* **231**, 54–57.

Everson, R. B., Randerath, E., Santella, R. M., Avitts, T. A., Weinstein, I. B., and Randerath, K. (1988a). Quantitative associations between DNA damage in human placenta and maternal smoking and birth weight. *J. Natl. Cancer Inst.* **80**, 567–576.

Everson, R. B., Wehr, C. M., Erexson, G. L., and MacGregor, J. T. (1988b). Association of marginal folate depletion with increased human chromosomal damage *in vivo*: Demonstration by analysis of micronucleated erythrocytes. *J. Natl. Cancer Inst.* **80**, 525–529.

Gibbs, R. A., and Caskey, C. T. (1989). The application of recombinant DNA technology for genetic probing in epidemiology. *Ann. Rev. Public Health* **10**, 27–48.

Goldring, J. M., and Lucier, G. W. (1990). Protein and DNA adducts. In "Biological Markers in Epidemiology" (B. S. Hulka, T. C. Wilcosky and J. D. Griffith, eds.), pp. 173–195. Oxford University Press, New York.

Harris, C. C., Weston, A., Willey, J. C., Trivers, G. E., and Mann, D. L. (1987). Biochemical and molecular epidemiology of human cancer: Indicators of carcinogen exposure, DNA damage, and genetic predisposition. *Environ. Health Perspect.* **75**, 109–119.

Horvitz, D. G., Weeks, M. F., Visscher, W., Folsom, R. E., Hurley, P. L., Wright, R. A., Massey, J. T., Ezzati, T. M., and Horvitz, D. G. (1990). A report of the findings of the National Household Seroprevalence Survey Feasibility Study. (1990) *Proc. Sect. Surv. Res. Meth. Am. Stat. Assoc.* 150–159. American Statistical Association, Alexandria, Virginia.

Hulka, B. S. (1990). Methodologic issues in molecular epidemiology. In "Biological Markers in Epidemiology" (B. S. Hulka, T. C. Wilcosky, and J. D. Griffith, eds.), pp. 214–226. Oxford University Press, New York.

Kalsbeek, W. D., Kral, K. M., Wallace, R. B., and Rifkind, B. M. (1988). Comparing mean levels of total cholesterol from Visit 2 of the Lipid Research Clinics Prevalence Study with the Second National Health and Nutrition Examination Survey. *Am. J. Epidemiol.* **128**, 1038–1053.

Kark, J. D., Smith, A. H., Switzer, B. R., and Hames, C. G. (1981). Serum vitamin A (retinol) and cancer incidence in Evans County, Georgia. *J. Natl. Cancer Inst.* **66**, 7–16.

Kiyabu, M. T., Shibata, D., Arnheim, N., Martin, W. J., and Fitzgibbons, P. L. (1989). Detection of human papillomavirus in formalin-fixed, invasive squamous carcinomas using the polymerase chain reaction. *Am. J. Surg. Path.* **13**, 221–224.

Koller, A., and Kaplan, L. A. (1987). Total serum protein. In "Methods in Clinical Chemistry" (A. J. Pesce and L. A. Kaplan, eds.), p. 1135. Mosby, St. Louis, Missouri.

Menkes, M. S., Comstock, G. W., Vuilleumier, J. P., Helsing, K. J., Rider, A. A., and Brookmeyer, R. (1986). Serum beta-carotene, vitamins A and E, selenium, and the risk of lung cancer. *N. Engl. J. Med.* **315**, 1250–1254.

Morrison, B., Shenkin, A., McLelland, A., Robertson, D. A., Barrowman, M., Graham, S., Wuga, G., and Cunningham, K. J. (1979). Intra-individual variation in commonly analyzed serum constituents. *Clin. Chem.* **25**, 1179–1805.

Murphy, S. E., Palomino, A., Hecht, S. S., and Hoffmann, D. (1990). Dose-response study of DNA and hemoglobin adduct formation by 4-(methylnitrosamino)-1-(3-pyridyl)-1-butanone in F344 rats. *Cancer Res.* **50**, 5446–5452.

National Center for Health Statistics (1990). "Third National Health and Nutrition Examination Manual for the Laboratory Team." Westat, Hyattsville, Maryland.

Perera, F. P. (1991). Validation of molecular epidemiologic methods. In "Molecular Dosimetry and Human Cancer: Analytical, Epidemiological, and Social Considerations" (J. D. Groopman and P. L. Skipper, eds.), pp. 53–76. CRC Press, Boca Raton, Florida.

Perera, F. P., and Weinstein, I. B. (1982). Molecular epidemiology and carcinogen-DNA adduct detection: New approaches to studies of human cancer causation. *J. Chron. Dis.* **35**, 581–600.

Petrakis, N. L. (1985). Biologic banking in cohort studies, with special reference to blood. *Natl. Cancer Inst. Monogr.* **7**, 193–198.

Prentice, R. L., Self, S. G., and Mason, M. W. (1986a). Design options for sampling within a cohort. In "Modern Statistical Methods in Chronic Disease Epidemiology" (S. H. Moolgavkar and R. L. Prentice, eds.), pp. 50–62. John Wiley and Sons, New York.

Prentice, R. L., Moolgavkar, S. H., and Farewell, V. T. (1986b). Biostatistical issues and concepts in epidemiologic research. *J. Chron. Dis.* **39**, 1169–1183.

Rothman, K. J. (1986). "Modern Epidemiology." pp. 76–97. Little, Brown, Boston.

Salonen, J. T., Salonen, R., Lappetelainen, R., Mäenpää, P. H., Alfthan, G., and Puska, P. (1985). Risk of cancer in relation to serum concentrations of selenium and vitamins A and E: Matched case-control analysis of prospective data. *Br. Med. J.* **290**, 417–420.

Sassin, J. F., Franzt, A. G., Weitzman, E. D., and Kapen, S. (1972). Human prolactin: 24-Hour pattern with increased release during sleep. *Science* **177**, 1205–1207.

Schreinemachers, D. M., and Everson, R. B. (1991). Effect of residual splenetic function and folate levels on the frequency of micronucleated red blood cells in splenectomized humans. *Mutat. Res.* **263**, 63–67.

Schulte, P. A. (1987). Methodologic issues in the use of biologic markers in epidemiologic research. *Am. J. Epidemiol.* **126**, 1006–1016.

Schwartz, G. G. (1990). Oncogenes: A primer for epidemiologists. In "Biological Markers in Epidemiology" (B. S. Hulka, T. C. Wilcosky, and J. D. Griffith, eds.), pp. 173–195. Oxford University Press, New York.

Shapiro, I. M., Burke, A., Mitchell, G., and Block, P. (1978). X-ray fluorescence analysis of lead in teeth of urban children *in situ*: Correlation between the tooth lead level and the concentration of blood lead and free erythroporphyrins. *Environ. Res.* **17**, 46–52.

Shibata, D., Martin, W. J., and Arnheim, N. (1988). Analysis of DNA sequences in forty-year-old paraffin embedded thin-tissue sections: A bridge between molecular biology and classical histology. *Cancer Res.* **48**, 4564–4566.

Slamon, D. J., Clark, G. M., Wong, S. G., Levin, W. J., Ullrich, A., and McGurie, W. L. (1987). Human breast cancer: Correlation of relapse and survival with amplification of the HER-2/neu oncogene. *Science* **235**, 177–182.

Stevens, R. G., Morris, J. S., Hann, H. L., Pulsipher, B., and Stahlhut, M. W. (1986). Serum selenium assay following serum ferritin assay. *Am. J. Epidemiol.* **124**, 329–331.

Stock, L. M., Brosman, S. A., Fahey, J. L., and Liu, B. C.-S. (1987). Ras related oncogenic protein as a tumor marker in transitional cell carcinoma of the bladder. *J. Urol.* **137**, 789–792.

Tornqvist, M., Osterman-Golkar, S., Kautiainen, A., Jensen, S., Farmer, P. B., and Ehrenberg, L. (1986). Tissue doses of ethylene oxide in cigarette smokers determined from adduct levels in hemoglobin. *Carcinogenesis* **7**, 1519–1521.

Wagener, D. K., Reidy, J. A., Steinberg, K. K., Chen, A. T. L., and McQuillan, G. M. (1991). "Screening in a Heterogeneous Population: The American Public." Paper presented at the Eighth International Congress of Human Genetics, October 6–11, Washington, D.C.

Wald, N. J. (1985). Use of biological sample banks in epidemiological studies. *Maturitas* **7**, 59–67.

Wald, N., Idle, M., Boreham, H., Bailey, A. (1980). Low serum vitamin A and subsequent risk of cancer. *Lancet* **2**, 813–815.

Whiteside, T. L., and Herberman, R. B. (1990). Serum and tissue banks for biological markers. *Immunol. Ser.* **53**, 55–68.

Willett, W. (1987). Nutritional epidemiology: Issues and challenges. *Int. J. Epidemiol.* **16**, 312–317.

Willett, W. C., Polk, F. B., Underwood, B. A., Stampfer, M. J., Pressel, S., Rosner, B., Taylor, J. O., Schneider, K., and Hames, C. G. (1984). Relation of serum vitamins A and E and carotenoids to the risk of cancer. *N. Engl. J. Med.* **310**, 430–434.

Winn, D. M., Reichman, M. E., and Gunter, E. (1990). Epidemiologic issues in the design and use of biologic specimen banks. *Epidemiol. Rev.* **12**, 56–70.

Woodward, M., Smith, W. C. S., and Tunstall-Pedoe, H. (1991). Bias from missing values: Sex differences in implication of failed venepuncture for the Scottish Heart Health Study. *Int. J. Epidemiol.* **20**, 379–383.

Woteki, C. E., Wagener, D. K., McQuillan, G., Gunter, E., and Chen, A. (1990). "NHANES III Specimen Bank for Genetic and Seroepidemiology Research." Paper presented at the FA-SEB meeting, April, Washington, D.C.

Yach, D. (1990). Biological markers: Broadening or narrowing the scope of epidemiology. *J. Clin. Epidemiol.* **43**, 309–310.

Yie, S. M., Wang, R., Zhu, Y. X., Liu, G. Y., and Zheng, F. X. (1990). Circadian variations of serum sex hormone binding globulin binding capacity in normal adult men and women. *J. Steroid Biochem.* **36**, 111–115.

9

Interpretation and Communication of Molecular Epidemiologic Data

Paul A. Schulte

Practitioners of molecular epidemiology are likely to be of laboratory or epidemiologic disciplines. In neither case are they likely to be trained in effective risk communication procedures; indeed they may not even relish such actions. Nonetheless, as in all epidemiology, molecular epidemiology is a part of public health. Thus, it is a discipline that involves human subjects and is targeted directly to making a difference in the health of populations. Hence, molecular epidemiologists must be able to interpret and communicate their results with some proficiency.

The interpretation and communication of molecular epidemiologic results is the responsibility of researchers. It begins with the recruitment of subjects. Patients and research subjects have a right to give informed consent when participating in molecular epidemiologic research. The Nuremberg Code and the Helsinki Convention affirm and reaffirm the right of an individual to be given full disclosure about potential risks and benefits of experimental biomedical procedures (Duncan *et al.*, 1977). These principles might be extended to cases in which researchers have information about potential health risks within identifiable populations (National Commission, 1978; Schulte and Ringen, 1984; Gordis, 1991; Lerman *et al.*, 1991; Schulte, 1991). In these situations, subjects may be able to claim a "right to know" assay and study results. This "right" is based on the view that the right to self-determination is a fundamental democratic principle. In this context, individuals are considered best able to protect their health, lives, and interests if they are informed about a known risk. When the meaning of molecular epidemiologic information is uncertain, the required actions are less defined and more of a problem exists in communication of the information. Issues pertaining to the interpretation and communication of molecular epidemiologic information are discussed in this chapter.

Interpretation

The use of biologic markers in epidemiologic research presents some unique issues regarding the interpretation and communication of results. These issues can arise for markers of exposure, effect, and susceptibility, but are more prevalent for the latter two, especially in studies in which the health risk associated with a marker is not known. Until the sensitivity, specificity, and predictive value of an effect marker are known, it is probably more practical to consider it an indicator of exposure rather than of effect. Markers of susceptibility also present major challenges to interpretation and communication. Widespread use of these markers may have ethical and legal implications that include discrimination in employment or obtaining insurance, and may be used to trigger litigation or requirements in laws pertaining to reporting of health effects.

Questions of interpretation also arise for markers of exposure. For example, how is the contribution of DNA adducts from an environmental hazard distinguished from those from a home hobby? To what extent does a DNA adduct constitute DNA damage? Most markers of exposure can have an implied if not explicit risk interpretation, but the researcher may be uncertain about the meaning or unwilling to push the interpretation to the level of implication for a subject's health.

Generally, subjects may accept that a study only asks a research question or that the investigator is uncertain of the findings. Still, when biomarkers are used in studies of controversial topics, investigators may have to take extra time to discuss results with subjects or community groups. To avoid or reduce confrontational reactions by subjects and affected groups, it may be necessary to involve them in the planning of investigations rather than simply communicate results to them.

Much of the uncertainty in molecular epidemiologic research derives from the newness of the field and the inherent qualities of biologic markers, such as variability, the fact that they are individual-specific measures, and the expectations of research subjects to want definitive answers. These issues will be discussed in subsequent sections.

Variability of Results

Within groups of subjects, early research using biologic markers has shown great variability in results. All interpretations will be influenced by how well sources of variability can be identified and accounted for. Biologists long have recognized this inherent variability among individuals (Mayr, 1982).

With the exception of identical twins, no two people have the same genetic make-up.

In this century, biologists have accepted the notion of biochemical individuality and rejected the notion of essentialism (the belief that everything is a product of a limited number of fixed unchanging forms) (Mayr, 1982). "Population thinking" is a phrase that has been used to represent this acceptance. As Mayr (1982) describes:

> Population thinkers stress the uniqueness of everything in the organic world. What is important for them is the individual, not the type. They emphasize that every individual in sexually reproducing species is uniquely different from all others, with much individuality even existing in uniparentally reproducing ones. There is no typical individual, and mean values are abstractions. Much of what in the past has been designated in biology as classes are populations consisting of unique individuals.

One might question whether this view precludes statistical analysis of grouped data. Brain (1988) has observed that most statistical tests focus on measures of central tendency and their variability; few tests focus on the extremes. He suggests that we need to develop techniques to determine the extent to which observed variations reflect measurement error or true differences in susceptibility. Moreover, we may need to seek new means to accommodate the interpretation of biologic phenomena at the molecular level.

In practice, these views should not preclude statistical analysis and interpretation of biomarker research. However, they should serve as warnings for researchers to design studies that account for major sources of variability and to evaluate research to determine whether this has been done. If it has not, appropriate caveats or alternative interpretations should be included.

Group Results Compared with Individual Results

Interpreting studies with biologic markers has proved difficult for other reasons also. The traditional paradigm that epidemiologic research pertains to a group leaves individual study subjects at a loss regarding the meaning of results for them. Subjects may be able to learn about significant group risks, but may not be able to obtain any meaningful information about individual risk unless investigators have developed risk functions that will calculate individual risk. Still, institutional review boards often require that study subjects receive their own test results along with some explanation or interpretation. Epidemiologists have come to no agreement about the language for these communications.

One of the major potential advances of molecular epidemiology is the ability to obtain individual-specific information that may be predictive of risk (Albertini *et al.,* 1991; Shields and Harris, 1991; Thilly, 1991). This ability is not new to epidemiologic research (e.g., Truett *et al.,* 1967), but the exquisite sensitivity of individual risk functions based on gene assessments puts researchers and society in difficult positions with respect to interpretation of results, privacy, and confidentiality.

Subjects may misconstrue the purpose of research and reduce the research question to whether or not they are "all right." Clearly, this attitude could pose a problem in most biomarker studies that may assess only a marker's validity in providing information that is useful in an epidemiologic, rather than a clinical, sense (Schulte and Singal, 1989). However, some biomarker studies may identify clinically relevant findings.

Any positive study that uses markers that are considered biologic changes capable of being part of a disease process should trigger considerations of the need for medical surveillance. Although this is a prudent policy that may involve surveillance of some subjects with false-positive test results, it will at least allow for identification of early disease in subjects who are candidates (true positives) for early intervention or therapy. Beyond that, researchers still should make a strong effort to describe the limitations of biologic markers, to counsel subjects, and, in some cases, to provide the subjects' personal physicians with information regarding the state of knowledge about the markers. One suggestion for handling uncertainty in molecular epidemiologic research is to couple such research with conventional screening of high-risk groups (Schulte, 1986). At a minimum, this alternative offers the opportunity to provide study subjects with some information (acquired from the conventional screening) that can be interpreted with a known degree of certainty. In such a setting, people often are willing to provide additional specimens for research purposes.

Of highest importance to study subjects and the general population is whether or not epidemiologic information indicates disease risk. Technically, this describes the question of predictive validity discussed in Chapter 3. What is the predictive value of a marker assay? This question pertains primarily to markers of effect and susceptibility. The predictive value of an assay is often misinterpreted because it is presented as a percentage (Gambino, 1989). As Gambino notes,

> It is intuitive to assume that low percentages (70% or less) are bad and high percentages are "good." A positive predictive value of 20% was cited as proof that a test should not be used even though the positive likelihood ratio for the same test was 50. A likelihood ratio of 50 means that the posttest odds will be 50 times higher than the pretest odds of disease. Now that is a large increase in the odds.

In a molecular epidemiologic study, typically only one or few markers are used. A single marker assay rarely should be interpreted in isolation. On an individual basis, the findings should be confirmed by a repeat test given at some later date. Other confirmatory studies should be sought for group results. When possible, batteries of markers may provide a fuller picture than would be seen with one or a few markers.

A danger exists that molecular epidemiology practitioners or, more likely, those in the public or media who interpret such research will be misled by the exquisite sensitivity of some assays to believe that the assays have, for example, identified conclusively a person who has incurred a gene mutation as the result of environmental or occupational exposure to mutagens. Until these findings are validated and the predictability of the marker assessed, indefinite interpretations can cause a range of problems for the individual and the society.

Validated and invalidated marker results can present a variety of societal problems for study subjects. Individuals correctly or incorrectly classified in the tails of a distribution of marker frequencies may be in danger of prejudice and discrimination in terms of insurance, job security, and obtaining loans, and at risk of negative interpersonal responses [Office of Technology Assessment (OTA), 1990; Ashford, 1991; Weatherall, 1991].

To reduce these untoward reactions, it is incumbent on researchers to take care in interpreting and communicating molecular epidemiologic research and, moreover, to anticipate untoward use of such research and participate in societal discussions of such questions.

Communication

Molecular epidemiologic information may need to be communicated to study subjects in at least three different instances as well as to the general public and to patients in clinical settings.

Study Subjects

The three instances in which study subjects may need molecular epidemiologic information are (1) during recruitment when obtaining informed consent; (2) when communicating biomarker test results; and (3) when communicating molecular epidemiologic study results.

Obtaining Informed Consent

The process of communicating information from molecular epidemiologic studies actually begins with the recruitment of subjects. A review of informed consent of hospital patients who participated in research demonstrated that they had low levels of recall, understanding, and knowledge subsequent to the administration of informed consent (Silva and Sorrell, 1988). The reasons given for this lack of comprehension involved clarity, language, and formats used. The ethical principle underlying recruitment of subjects into research studies is that the participants should be provided with a true understanding of what the study entails, the benefits to be derived, and the risks involved. It has been argued that candidate subjects for molecular epidemiologic studies rarely will be able to give truly informed consent because of the complexity of the research and because the researchers themselves are uncertain about the meaning of potential results (Samuels, 1992).

Maximizing the understanding of potential study subjects could result in increasing selection or volunteer bias. However, in a democratic society, this is the cost of contemporary research. Researchers may need study designs to balance such bias among study groups.

Communication of Test Results

Researchers display a difference of opinion about whether the results of biomarker tests should be communicated to study subjects. Some researchers, institutions, and Institutional Review Boards (IRBs) require such communication whereas others do not. When communication is required, an approach sometimes used provides individual results and the average for the group. An example is shown in Figure 9.1.

Although the letter satisfies the spirit of communicating results in a relatively clear manner, it still is quite technical. The problem for research communicators is the implied tension between the need to communicate complex results in an understandable manner and the potential liability of not providing complete information. Communications experts would recommend a brief bulleted message over the letter with its caveats, footnotes, and technical terms. How does the researcher balance these pressures? Certain features in the letter represent attempts to overcome the dilemma of trying to communicate complex molecular epidemiologic data. These features include striving to use simple language, giving group comparisons, putting results in the context of an occupational health standard, and, most importantly, giving a contact person for follow-up discussion.

During the recruitment phase, the participant in a molecular epidemiologic study should be informed of the extent to which the marker assays will reflect individual health. Most molecular epidemiologic studies will have little information about a subject's health. The purpose of such studies may be to validate a marker or elucidate a mechanism; only with validated markers will studies show an association indicative of increased risk of disease [National Research Council (NCR), 1991]. Still, subjects may forget this fact between the time they sign the informed consent document and the time they receive test results. At that time, they generally are interested in knowing whether they are "all right" or have any problems. Because of the limitations or purpose of the study, such information generally cannot be determined from biomarker assays. Some studies may be performed only to confirm exposure, for example, by detecting DNA adducts.

Assays may have various limitations that should be mentioned when informing subjects of results. Since biomarkers often reflect exposures over multiple routes and from different sources, it is important to describe these possibilities to account for variation in findings. Marker results may be confounded or modified by exposures or conditions not actually part of a study.

FIGURE 9.1 *FIGURE 9.1* Example of a letter notifying subjects of the results of molecular epidemiologic test

Date

Dear _____.

Thank you again for participating in the medical study on ethylene oxide exposure. Your individual results from the laboratory tests on your blood are shown below. These results were delayed because they were done in various countries, and some of the tests took longer than expected. The final report of group results will be sent to you in a few months.

As you know, the study was done to see if measurements of ethylene oxide in red blood cells and changes in the genetic material of white blood cells are related to a person's work exposure to ethylene oxide. The measurements of ethylene oxide in red blood cells are called hemoglobin adducts and the genetic changes are called sister chromatid exchanges, chromosomal micronuclei, and HPRT mutations. We compared a group of workers with exposure to ethylene oxide (sterilizer workers) to a group of workers with no ethylene oxide exposure (other hospital workers). We believe that workers with ethylene oxide exposure will have more of these changes. In the following table, your results are shown along with the average values found in each of these two groups:

		Worker group	
	Your results	**Sterilizer workers**	**Other hospital workers**
Hemoglobin adducts[a]			
(Average number in moles per gram)	0.17	0.13	0.07
Sister chromatid exchanges[b]			
(Average number per cell)	7.22	6.15	4.93
Chromosomal micronuclei[c]			
(Average number per cell)	1	0.74	1
HPRT[d]			
(Number of "mutations")	112×10^{-6e}	124×10^{-6}	724×10^{-6}

[a] Hemoglobin adducts are attachments of environmental molecules, in this case ethylene oxide, to the hemoglobin molecule.
[b] Sister chromatid exchanges occur when pieces of chromosomes are exchanged from one chromosome to another.
[c] Chromosomal micronuclei are pieces of chromosomes found in cells.
[d] HPRT is the abbreviation for a gene on the X chromosome.
[e] "$\times 10^{-6}$" means "times one one-millionth."

Hemoglobin adducts of ethylene oxide apparently are found in everyone, and may be due to cigarette smoking or even produced naturally in humans. In order to see how well hemoglobin adducts are associated with exposure, we compared workers with ethylene oxide exposure to those without such exposure. It is not known if the presence of an increased amount of hemoglobin adducts or of the chromosome and gene changes increases a person's risk of disease. This was not the intent, and it cannot be determined in this study.

Additionally, we have measured the potential exposure to ethylene oxide among employees in your department. At least some of the measured exposures in your department were above the NIOSH recommendation for exposure to EtO. All measured exposures were below the Occupational Safety and Health Administration (OSHA) federal standard for exposure to EtO.

If you wish to discuss these results further, please call me at (phone #) or send inquiries to me at the following address: (address)

Sincerely yours,
Study Investigator's Signature

An example of how confounding influence can be addressed is seen in the following excerpt from test-result notification (Schulte and Singal, 1989):

> The type of semen quality changes observed in this study can be found in everyone and happen for a number of reasons. Some of these reasons have been linked with human health problems and some have not. Factors in everyday life, like smoking and

the use of certain medications, can also change semen quality measurements. There are also day-to-day changes in measures of semen quality.

Some assays cannot be interpreted easily because they are research tools and have no established normal range. This fact was expressed in one results letter (Schulte and Singal, 1989) in the following manner:

> We should comment upon the source of the "normal" or reference range for tests. For some tests the reference range is listed as N.A. (not available). Usually these tests are research tools for which reference ("normal") values have not been well established. Research tests provide useful information about groups of workers, but the results are difficult to interpret in individuals. Other tests, such as urine creatinine, are used mainly to standardize other tests and are not meaningful themselves. We report the results for your information but do not comment on them.

The inability of a research test to predict disease generally should be expressed with no attempt to blur the uncertainty. An example (Schulte and Singal, 1989) is a study involving exposure to 4,4′-methylenebis(2-chloroaniline) (MBOCA), a suspected bladder carcinogen, in which subjects were administered both the routine Papanicolaou cytology test and the experimental quantitative fluorescence image analysis test. The letter read:

> QFIA (Quantitative Fluorescence Image Analysis) measures the amount of genetic material in cells, and may detect cells which have an increased content of genetic material. Increased content of genetic material is one difference between tumor cells and normal cells. QFIA is an experimental test, and we still don't know its value in predicting who is likely to get bladder cancer.

Researchers quite often may be uncertain about the meaning of biomarker assays. Not to communicate the true extent of uncertainty about assay results may be unethical. It is important to communicate test results to provide information to the study subjects and to do it in an informative and understandable way.

Communication of Study Results

Perhaps the most important information to communicate are the overall results of the study after assessing the association between exposures and outcomes. At this stage, the researcher can attempt to explain the results. For molecular epidemiologic results, despite the collection of individual samples, the results still may be group specific and not indicative of risk on the individual level. However, if an individual risk function is calculated, as with lipid and other risk factor information in multivariate models of cardiovascular disease or with assays for a particular gene mutation, it may be possible to describe individual risk (Truett *et al.*, 1967; Shields and Harris, 1991). Nonetheless, disclosing group results can be beneficial to subjects by providing them with general risk information about an environmental or occupational hazard. Group information also provides a framework for subjects to know the extent of a troubling finding, that is, it may provide a perspective.

For test and study results, researchers may bear a responsibility to warn subjects of the need for a preventive action or to recommend ongoing monitoring, medical surveillance, or early treatment. Research that uses biologic markers may have the effect of establishing ad hoc "exposure," "risk," or preclinical disease registries (Samuels, 1982; Schulte *et al.*, 1986; Schulte and Kaye, 1988). Whether or not these registries are maintained will depend on the availability of funds, the political will of the registry members, and the interests of the investigator. Generally, it is not current practice to maintain such registries, although federal law (CERCLA, PL 96-510) has prescribed the formation of "exposure registries."

The communication of study results that indicate risk to a cohort may define and highlight a high-risk group in society (Schulte and Ringen, 1984; Schulte 1986). This label has various meanings to different sectors of society. To the subjects, it has the exaggerated meaning that they are entered into a lottery involving their health. Some individuals may be able to put a balanced perspective on this knowledge; others may need additional counseling or support. All may need certain periodic medical surveillance to determine whether or not they have developed signs of the disease for which they are at risk. The ethical responsibilities of researchers and research institutions (including corporate research groups) are only beginning to be well delineated. However, some useful models can be considered (Fischhoff, 1987; Murray, 1988; Lerman *et al.*, 1991; Sandman, 1991).

General Public

Researchers who publish and otherwise disseminate molecular epidemiologic information to subjects may be called on to interpret the meaning of such findings for the general public. When validated markers are available, molecular epidemiologic information may be able to provide individual risk information (Shields and Harris, 1991). When communicating molecular epidemiologic information, care should be taken to distinguish between validated markers and those not yet validated. The current state of research in this field is, essentially, one of validating markers, not of identifying etiologic factors or conducting risk assessments.

For communication of molecular epidemiologic information to be effective, it must comply with established principles of risk communication. These principles include viewing risk communication in the framework of risk management and demonstrating sensitivity to the needs and reactions of the involved parties. Molecular epidemiologic information often contains elements that will be provocative to the general population. Thus, terms such as "mutation," "gene rearrangement," or "DNA adduct" conjure up meanings and fears that may not be warranted by the actual research. The normal background frequency of spontaneous mutations in the genome and the extensive capacity for repair (see Figure 9.2 and Table 9.1) often are not consid-

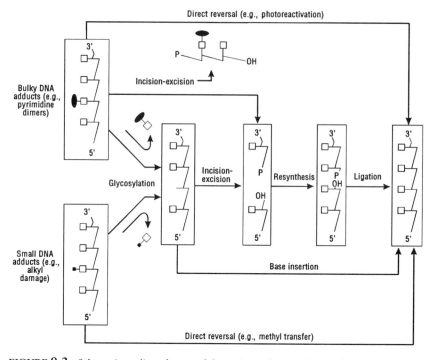

FIGURE 9.2 Schematic outline of some of the major pathways of DNA damage repair. (Reprinted with permission from Vijg *et al.*, 1986, in Baan, 1987. Copyright © 1987 by Springer-Verlag.)

ered when the public hears a message about a molecular epidemiologic study involving measures of mutations or DNA adducts.

The complexity of molecular epidemiologic information perhaps will make the information harder to convey than the usual outcomes of epidemiologic studies. Because molecular biologic research is proceeding so rapidly, it is unlikely that the general population understands the basic principles of cell growth and regulation against which to evaluate a communication.

The results of molecular epidemiologic studies are likely to involve the impact of genetic *and* environmental factors in disease. In discussions of environmental disease, it has been difficult to address genetic risk factors since there have been tendencies to put the emphasis either entirely on the genetics (and blame the victim) or entirely on the environment. For most diseases, a combination of endogenous and exogenous factors is likely, so risk communications must be performed within these parameters. Molecular epidemiologic research does have the potential for clarifying the "nature–nurture" debate, but such clarification still taxes risk communicators.

In "Monitoring the community for exposure and disease: Scientific, legal, and ethical considerations," Ashford *et al.* (1991) identified many of the

TABLE 9.1 Estimated DNA Damage Rates and Repair Rates in Human Cells at 37°C

	Damage			Repair	
Endogenous	No./hr	Exogenous	No./hr	Type	No./hr
Depurination	580	Pyrimidine dimer formation (noon Texas sun)	$.5 \times 10^4$	Single strand breaks	2×10^5
Single strand breaks	2300			Thymine glycol removal	10^5
		Single strand breaks (background radiation)		Pyrimidine dimer removal	
O^6-Methylguanine formation	130		10^{-4}	Normal cells	5×10^4
				XP C cells	5×10^3
Thymine glycol formation	13	Thymine glycol formation (background radiation)	$10^{-5}-10^{-4}$	O^6-Methylguanine removal	$\sim 10^4$

Source: Reprinted with permission from Setlow (1988).

issues pertinent to molecular epidemiology practice and information dissemination. These authors and others (Schulte, 1987; OTA, 1990; NRC, 1991) have addressed this topic and drawn the following conclusions.

- Molecular epidemiologic data should have a relatively large signal-to-noise ratio, that is, there should be as little ambiguity as possible that the "effect" found is above a background rate.
- Communication of molecular epidemiologic information to a community cannot be separated from the social or political use of the data.
- Dissemination of risk information from molecular epidemiologic studies has implications for citizens' and employees' rights to privacy, confidentiality, and nondiscrimination with respect to employment, insurance, and acceptability for loans.

The Centers for Disease Control/Agency for Toxic Substances and Disease Registry [Centers for Disease Control (CDC), 1990] established a subcommittee to review currently available laboratory tests for their suitability to measuring human organ-specific biomarkers. They concluded that:

> . . . when a biomarker is included in a study, it must be evaluated against established batteries of biomarkers. A separate statistically valid evaluation of the new marker must be conducted. The marker assay results produced in this evaluation should be used only for marker description and evaluation, and they should not be presented to the study subjects as individual marker assay results until all relevant data have been compiled and reviewed. Results released before the physiologic significance of the marker is thoroughly assessed could cause unnecessary public alarm and spur demand for the test before the meaning of the results is fully understood. (CDC, 1990)

The subcommittee also concluded that the evaluation process to find new markers should be conducted anonymously, with informed consent of the subjects and coding of specimens to delete identification of all study subjects. Before a test is considered to have completed the investigative phases, the biochemical or physical abnormality associated with the marker should be identified, and the probability that the abnormality will progress to disease as well as the nature of that disease should be known (CDC, 1990).

Various government agencies, universities, and corporate researchers have an array of practices (CMA, 1991). The National Institute for Occupational Safety and Health routinely informs subjects of molecular epidemiologic studies of all test and study results.

The societal response to people with "abnormal" levels of markers can give rise to ethical issues pertaining to discrimination, need for medical follow-up, and removal of workers or residents from areas of imminent danger (Schulte, 1987, 1991; Rothstein, 1989; Ashford et al., 1990). The question is whether people with a certain biologic marker of susceptibility have rights to protection against discrimination as do people with more visible physical handicaps (Rothstein, 1989; OTA, 1990). Increasingly, these types of questions will be asked by individuals who live or work near hazardous waste sites, and who receive biologic monitoring as part of an epidemiologic study or a routine medical surveillance.

The interpretation of biologic monitoring data also can impact litigation concerning alleged health effects and exposure to such environmental concerns as hazardous wastes. Ashford *et al.* (1990) assert that human monitoring data have the potential to bring about a change in the nature of evidence used in such cases.

> Typically, the evidence offered to prove causation in chemical exposure cases is premised on a *statistical* correlation between disease and exposure. Whether the underlying data are from epidemiologic studies, toxicological experiments, or the results of a complicated risk-assessment model, they usually are *population*-based. This places the plaintiff at the mercy of the attributable risk (expressed as the percentage of cases of the disease attributable to the exposure) for the study population. Unless the attributable risk is greater than 50%—that is, unless the incidence rate among those exposed to the chemical is more than double the background rate—the plaintiff cannot prove, *on the basis of the available statistical evidence*, that it is more likely than not that his or her particular case of the disease was caused by the chemical exposure.

Responsibilities of Molecular Epidemiologists

Sandman (1991) has identified communication responsibilities of epidemiologists that apply to molecular epidemiologists as well:

1. Tell the people who are most affected what you have found—and tell them first.
2. Make sure people understand what you are telling them, and what you think the implications are.
3. Develop mechanisms to bolster the credibility of your study and your findings.
4. Acknowledge uncertainty promptly and thoroughly.
5. Show respect for public concerns even when they are not scientific.
6. Involve people in the design, implementation, and interpretation of the study.
7. Decide that communication is part of your job, and learn the rudiments.

Clinical Settings

Genetic and cancer-risk counseling are representative of areas in which molecular epidemiologic information may be communicated presently and in the near future (Weatherall, 1991). In these areas, little is known about the impact of the communication, but there is great potential for help or harm. By communicating molecular epidemiologic information in clinical settings, it is possible to motivate high-risk individuals, for example, to adhere to cancer prevention and surveillance recommendations or to make informed decisions about family planning. However, these communications may have adverse psychological and social consequences (Lerman *et al.*, 1991). Notification of cancer risks that involve genetic susceptibility and environmental exposures—in whatever proportions—are fraught with ethical dilemmas. For example, how do you tell a person that he or she is genetically susceptible to a particular disease? What kind of follow-up, support, and counseling should be anticipated? As Lerman *et al.* (1991) note,

> Nonmaleficence, or obligation of health professionals to do no harm, also is one of the cornerstones of ethical practice. For example, without an action plan for avoiding negative psychosocial sequelae, the mere transmittal of genetic information could cause more harm than good—particularly if no known means are available to reduce risk. This may differ from decisions to communicate to study subjects who have a right to know of test and study results. There too, however, anticipation of untoward effects should be considered.

Banked Specimens

The collection and storage of biologic specimens for future molecular epidemiologic research presents unique communication issues. Initially, there is the question of whether specimens can be used for a purpose or assay different from the one for which they were collected. This question has not yet been widely addressed. Initial discussions in the early 1980s by an ethics subcommittee of the CDC (CDC, 1983) suggested that the use of such specimens for purposes other than those for which they were collected is unethical. Does a broad informed consent statement, such as "Specimens will be stored for additional cancer studies as assays are developed," provide adequate information to subjects from which to make a judgment to allow future assays? Despite the extent of specimen banking worldwide, this question has not been addressed completely in the ethical or scientific literature.

Other questions discussed in this chapter need to be considered with respect to research on banked specimens. Who owns the specimens and, by extension, who owns the human genome (Weatherall, 1991)? To what degree are researchers responsible to communicate results to subjects who provided specimens years earlier? What if, in the intervening years, the marker that was studied originally is found to be predictive of disease? Is there an obligation to inform subjects who gave stored specimens? Many of these questions have not been addressed and will confront researchers in the near future.

Two principles seem to cover this type of research: (1) When researchers obtain health risk information from a molecular epidemiologic study, they have a responsibility to communicate the information to subjects. (2) When collecting biologic specimens for banking, researchers should indicate in the informed consent process the fact that specimens will be banked as well as the range of research that might be conducted.

Access to banked specimens for research should be controlled. When informed consent does not seem to cover subsequent research, one possible solution is to convene a surrogate committee to represent interests of subjects and other parties, to review research proposals, and to determine the extent to which efforts should be made to go back and inform subjects. Subsequent research on banked specimens should adhere to current accepted practices for maintaining confidentiality and privacy of subjects.

References

Albertini, R. J., O'Neill, J. P., Nicklus, J. A., Allegretta, M., Recio, L., and Skopek, T. R. (1991). Molecular and clonal analysis of *in vivo hprt* mutations in human cells. *In* "Human Carcinogen Exposure, Biomonitoring, and Risk Assessment" (R. C. Garner, P. B. Farmer, G. T. Steel, and A. S. Wright, eds.), pp. 103–126. IRL Press, Oxford.

Ashford, N. (1991). Monitoring the community for exposure and disease: Scientific, legal, and ethical considerations. U.S. DHHS; Atlanta PHS/ATSDR.

Ashford, N. A., Spadafor, C. J., Hattis, D. B., and Caldart, C. C. (1990). "Monitoring the Worker for Exposure and Disease. Scientific, Legal, and Ethical Considerations in the Use of Biomarkers." Johns Hopkins University Press, Baltimore.

Baan, R. A. (1987). DNA damage and cytogenetic endpoints. *In* "Cytogenetics" (G. Obe and A. Baslor, eds.), pp. 327–337. Springer-Verlag, Berlin.

Brain, J. D. (1988). Introduction. *In* "Variations in Susceptibility to Inhaled Pollutants" (J. D. Brain, B. D. Beck, A. J. Warren, and R. A. Shaikh, eds.), pp. 1–5. Johns Hopkins University Press, Baltimore.

Centers for Disease Control/Agency for Toxic Substances and Disease Registry (1990). "Subcommittee on Biomarkers of Organ Damage and Dysfunction. Summary Report." August 27. Atlanta, Georgia.

Centers for Disease Control Ethics Subcommittee (1983). "Deliberations: Right to Know and Worker Notification." November 30. Atlanta, Georgia.

Chemical Manufacturers Association (1991). Guidelines for Good Epidemiology Practices for Occupational and Environmental Epidemiological Research. Chemical Manufacturers Association, Washington, D.C.

Duncan, A. S., Dunstan, G. R., and Welbourn, R. B. (eds.) (1977). "The Dictionary of Medical Ethics." Darton, Longman, Dodd, London.

Fischhoff, B. (1987). Treating the public with risk communications: A public health perspective. *Sci. Tech. Human Val.* **12**, 13–19.

Gambino, R. (1989). The misuse of predictive value—Or why you must consider the odds. *Lab. Rep.* **11**, 65–71.

Gordis, L. (1991). Ethical and professional issues in the changing practice of epidemiology. *J. Clin. Epidemiol.* (*Suppl.*) **44**, 95–135.

Lerman, C., Rimer, B. K., and Engstrom, P. F. (1991). Cancer risk notifications. *J. Clin. Oncol.* **9**, 1275–1282.

Mayr, E. (1982). "The Growth of Biological Thought." Harvard University Press, Cambridge.

Murray, T. H. (1988). Efficiency, liberty, and justice in screening for phenotypic variation. *In* "Phenotypic Variation in Populations" (A. D. Woodhead, M. A. Bender, and R. C. Leonard, eds.), pp. 271–280. Plenum Press, New York.

National Commission for the Protection of Human Subjects of Biomedical and Behavioral Research (1978). "The Belmont Report: Ethical Principles and Guidelines for the Protection of Human Subjects of Research." DHEW Publ. No. (OS) 78–0012. U.S. Government Printing Office, Washington, D.C.

National Research Council (1991). "Environmental Epidemiology." National Academy Press, Washington, D.C.

Office of Technology Assessment (1990). "Genetic Screening and Monitoring of Workers." GTA-BH-455. U.S. Government Printing Office, Washington, D.C.

Rothstein, M. A. (1989). "Medical Screening of Workers." Bureau of National Affairs, Washington, D.C.

Samuels, S. W. (1979). Communicating with workers (and everyone else). *Am. Ind. Hyg. Assoc.* 40, 1159–1163.

Samuels, S. W. (1982). The management of populations at high risk in the chemical industry. *Ann. N.Y. Acad. Sci.* 381, 328–343.

Samuels, S. (1993). The ethics of choice and the rights to know and act in the struggle against industrial disease. *Am. J. Ind. Med.* 23, 43–52.

Sandman, P. M. (1991). Emerging communication responsibilities of epidemiologists. *J. Clin. Epidemiol.* 49, 541–550.

Schulte, P. A. (1986). Problems in notification and screening of workers at high risk of disease. Journal of Occupational Medicine 28, 951–957.

Schulte, P. A. (1987). Methodologic issues in the use of biological markers in epidemiologic research. *Am. J. Epidemiol.*, 126, 1006–1016.

Schulte, P. A. (1991). Ethical issues in the communication of results. *J. Clin. Epidemiol.* 44, 551–556.

Schulte, P. A., and Kaye, W. E. (1988). Exposure registries. *Arch. Environ. Health* 43, 155–161.

Schulte, P. A., and Ringen, K. (1984). Notification of workers at high risk: An emerging public health problem. *Am. J. Public Health* 74, 485–491.

Schulte, P. A., and Singal, M. (1989). Interpretation and communication of the results of medical field investigations. *J. Occup. Med.* 31, 589–594.

Setlow, R. B. (1988). Relevance of phenotypic variation in risk assessment: The scientific viewpoint. *In* "Phenotypic Variation in Populations" (A. D. Woodhead, M. A. Bender, and R. C. Leonard, eds.), pp. 3–5. Plenum Press, New York.

Shields, P. G., and Harris, C. C. (1991). Molecular epidemiology and the genetics of environmental cancer. *J. Am. Med. Assoc.* 246, 682–687.

Silva, M. C., and Sorrell, J. M. (1988). Enhancing comprehension of information for informed consent: A review of empirical research. *IRB: Rev. Human Subj. Res.* 10, 10–5.

Thilly, W. G. (1991). Mutational spectrometry: Opportunity and limitations in human risk assessment. *In* "Human Carcinogen Exposure, Biomonitoring, and Risk Assessment" (R. C. Garner, P. B. Farmer, G. T. Steel, and A. S. Wright, eds.), pp. 127–134. IRL Press, Oxford.

Truett, J., Cornfield, J., and Kannel, W. (1967). A multivariate analysis of coronary heart disease in Framingham. *J. Chron. Dis.* 20, 511–524.

Weatherall, D. J. (1991). "The New Genetics and Clinical Practice." Oxford University Press, Oxford.

10

Use of Biomarkers in Risk Assessment

*Dale Hattis
and Ken Silver*

Introduction and General Philosophical Issues

Breaking Open the Black Box—Why?

To many researchers in the health sciences, the proposal to use biologic markers in risk assessment seems to suggest the mobilization of an immature and unreliable set of tools for a purpose that is substantially dubious to begin with. Risk assessment is done, of course, precisely when social policy decisions are in dispute, when the health consequences of alternative policies in question are not subject to direct measurement (at least on a time scale that is helpful for decision making), and when the usual tendency of scientists is to reserve judgment pending further research. Under these circumstances, the standardized procedures used in most risk assessments are intended at least to be relatively consistent and simple [National Research Council (NRC), 1983; Ruckelshaus, 1983]. Use of biomarkers in risk assessment often will present an excellent opportunity to compromise these characteristics.

Nevertheless, it is desirable to use both pharmacokinetic models and biologic markers (intermediate parameters that signal important events along or near the causal pathway to disease) to open the "black box" between exposure and effect for three basic reasons.

- This approach can lead to a more complete scientific understanding, incorporating more relevant information about causal mechanisms, than a simple input–output analysis.
- This approach offers the eventual prospect of better mechanism-based projection of risk beyond the range of possible direct observations, and

better estimation of the magnitude and detailed source of uncertainties in these projections (Hattis, 1991; Hattis and Goble, 1991).

- This approach offers the possibility of greater sensitivity of detection and quantification of adverse effects in some cases, progressing from the organism to the cell as the unit of analysis.

However, realizing the potential of pharmacokinetic modeling and biomarkers requires overcoming some basic philosophical aversion to mathematical theorizing, on the part of experimental biologists and statisticians and epidemiologists. Experimental scientists in Baconian tradition (Kuhn, 1977) are reluctant to build elaborate mathematical models, having been conditioned to view such theoretical efforts as unproductive speculations that divert attention from the necessary job of making measurements of natural phenomena as they really are (Hattis and Smith, 1987). On the other hand, epidemiologists and biostatisticians, in seeking to reduce problems to forms to which their traditional curve-fitting tools are most easily applicable, often are led to produce data fits that satisfy statistical criteria, possibly at the expense of the attainable biologic or mechanistic realism that is required to use the results in risk assessment modeling. [This issue will be illustrated using data from two studies on ethylene oxide and hydroxyethyl hemoglobin adducts (Mayer *et al.*, 1991; Schulte *et al.*, 1992).]

Part of the basic philosophical difficulty in making biomarker data useful for risk assessment relates to the conception of the goals of analysis in basic science as opposed to risk assessment. The primary object of most biomedical research has been seen as the generation of qualitative information on physical characteristics and causal connections—to address whether and how (by what causal pathway) specific phenomena are related. Qualitative facts such as "the shinbone is connected to the kneebone" certainly provide an essential framework for understanding phenomena. However, much more information is required for a full understanding of why, to use an example from sports, marathons run under 2.5 hr are relatively common, marathons run under 2.2 hr are relatively rare, and marathons run under 2 hr are practically unknown, even for professionals who have trained all their lives for the task. To understand marathon performance quantitatively, quantitative facts about the rate and efficiency of the cardiovascular system in delivering oxygen to the muscles and the rate of work output the muscles can develop and sustain must be available. Moreover, a quantitative theory about how specific facts of these kinds are likely to be related to marathon performance is required.

Contemporary biomedical research can be enriched by directing more research attention to questions of "how much," "how fast," and "according to what quantitative functional relationships" among the series of causal intermediate processes involved. The roles of models and biomarkers are to help us understand and to measure the causal intermediate processes at as many steps as possible along the causal chain. With this enriched mechanistic understanding and the improved measurement tools that the biomarkers provide, the mathematical models used for risk assessment and social decision making can be improved correspondingly.

In addition to these discipline-based difficulties, modeling and biomarker projects generally must overcome serious challenges of "validation." To prove that a model or biomarker really is a good indicator of a pathologic process leading to disease or impairment generally requires cross-disciplinary projects that

- transcend different levels of organization of the affected system and fundamentally different kinds of end points (e.g., biochemical, morphological, and functional)
- transcend differences in the methodologies that are available for study of humans compared with other species *in vivo*
- transcend differences in the observations that can be made *in vitro* compared with *in vivo*
- transcend differences in the time scale over which different events occur (for example, a putative biomarker measuring the ongoing loss of neurons of the substantia nigra might only be capable of ultimate validation as a predictor of Parkinsonism with the aid of a long-term prospective epidemiological study)

Organizing research teams, obtaining grants, and publishing results all present unusual hurdles for investigators wishing to pursue these interactive types of efforts. Adaptations in scientific institutional arrangements may be required to foster the kinds of cross-cutting research that can validate relationships between models and biomarkers and human risk.

Model Complexity and Lakatos' Criteria for Theory Change

When shaping a particular application of biomarkers and the accompanying dynamic models to risk assessment, every investigator will be faced with the issue of selecting the appropriate degree of complexity that will be represented in the system. How many different parameters and processes will be included? An increase in theory complexity to accommodate new data can be a "progressive" or a "degenerating" change [in the terminology of Imre Lakatos (1970)]. An increase in complexity is clearly degenerating if the only effect is to explain away a set of otherwise conflicting observations, for example, "These laws of motion apply to all heavenly bodies except for the sixteen objects we have recently seen in sector 17b." Lakatos offers the following criteria for a shift from an old theory T to a new theory T' to be "progressive."

- Excess empirical content: "T' predicts novel facts, that is, facts improbable in the light of, or even forbidden by, T."

- "T' explains the previous success of T, that is, all the unrefuted content of T is included (within the limits of observational error) in the content of T'."
- "Some of the excess content of T' is corroborated."

Overview of Types of Uses of Biomarkers in Risk Assessment

The discussion here surveys a number of specific uses of biomarkers in risk analysis:

- use of biomarkers as dosimeters (e.g., to help measure individual dose in epidemiologic studies, to assist in the projection of risk per biologically equivalent dosage between species, or to test the predictions of pharmacokinetic models that make predictions about internal dose as a function of external exposure)
- use to help define and measure the degree of interindividual variability in susceptibility to the adverse effects of an external exposure
- use of indicators as intermediate parameters to predict effects of concern that are difficult to measure directly
- use to help identify previously unsuspected agents involved in causing specific effects

Use of Biomarkers as Dosimeters

Use of Indicators to Help Measure Individual Dose in Epidemiologic Studies

Measures of Short-Term Dose from Routes that Are Not Easy to Measure Directly—Use of Ethoxyacetic Acid in Urine to Estimate Combined Dermal and Inhalation Absorption of Ethoxyethanol

Ethoxyethanol (EE) and some other glycol ethers are toxic to the male reproductive system, inhibiting specific steps in sperm production (Creasy and Foster, 1984; Creasy *et al.*, 1985). As are other alcohols, EE is metabolized to the corresponding acid (ethoxyacetic acid, EAA) via formation of an aldehyde. This metabolism appears to be essential to the toxicity of glycol ethers (Moss *et al.*, 1985), but it is unclear whether the aldehyde or the acid is primarily responsible for the effects.

As part of a study of potential sperm-reducing effects of EE in human workers, Hattis and Berg (1988) modeled the kinetics of urinary excretion of EAA following EE exposure, initially based on clinical experiments of Groeseneken *et al.* (1986a,b). The modeling indicated that, in humans, EAA is excreted quite slowly from the system via urine; the half-life is on the order of 33–70 hr. These kinetic models were applied to interpret data on the output of EAA per gram of creatinine in workers' urine, based on pre- and postshift sampling on several successive days in a group of 36 workers studied by National Institute for Occupational Safety and Health (NIOSH) researchers McManus (1987) and DeBord and Lowry (1986). Simultaneous industrial hygiene air measurements, assessments of individual respirator use, and qualitative assessments of postshift dermal exposure allowed estimates to be made of (1) the dosage delivered by inhalation and dermal routes for the workers as a group, and (2) the in-practice reduction in dose that was attributable to the use of respirators in those workers who used them (somewhat less than 2-fold in this case).

An unusual methodologic feature of this analysis was that, in many cases, the data allowed two separate estimates to be made of EE absorption on a particular workshift for individual workers—one estimate based on a comparison of preshift urinary EAA with postshift urinary EAA and another estimate based on comparison of preshift urinary EAA with EAA excretion in preshift urine the next day. In each case, the dynamic models[1] were used to predict the EAA excretion expected at various subsequent times if there had been no further exposure to EE during the workshift. The observed excess of urinary EAA over this prediction was used with the same model to infer the indicated moles of EE that had been absorbed during the workshift.

Unfortunately, when the initial results of the two sets of estimates were compared, it was found that estimates of EE absorption from the preshift–postshift comparison constituted only 31–33% of the corresponding estimates made from the preshift–next day preshift comparison. This difficulty was reduced greatly when the estimation procedures were corrected for diurnal changes in urinary creatinine excretion using the observations of Lakatua *et al.* (1982). After correction, the aggregate absorption estimated by the models for the preshift–postshift comparison was 70–118% of the aggregate absorption estimated for the preshift–next day preshift comparison (Hattis and Berg, 1988). This lesson is important for the future use of comparisons of biomarkers that may change as a function of time of day, length of time from meals, or other factors. The distorting effects were particularly important in this case because of the long half-life of EAA in the body and the large resultant buildup of EAA excretion from day to day over the course of a work week.

Measures of Long-Term Past Dose—The Frequency of Glycophorin A Variants in Red Cells as a Measure of Past Exposure to Ionizing Radiation

A serious problem in current retrospective epidemiologic studies (of cohort and case-control designs) is the great difficulty in quantitatively estimating past exposure to putative causal agents. Industrial hygiene records and other measurements are generally spotty, providing, in the best of cases, an

---

[1] Four different model variants were used in this analysis. In general, it is good practice to do parallel analyses with multiple models to see whether reasonable alternative interpretations of the pharmacokinetic data yield appreciably different conclusions.

imprecise set of estimates of individual- and group-average exposures. Dose–response relationships derived with these imprecise estimates of the x variable tend to be biased toward the null hypothesis, that is, toward lower values of the slope of the dose–response relationship and higher values of the intercept of background exposure, if the underlying dose–response relationship is linear. If the underlying dose–response relationship is other than linear, it is difficult to predict the precise distortions in the observed dose–response relationship. However, it generally can be expected that the observed relationship will be different and weaker than the one that would be observed if the independent variable were measured without error. Improved biomarker-based dosimeters therefore, theoretically, have the potential to lead to a profound improvement in our ability to detect and accurately measure human dose–response relationships for a variety of long-term effects.

Unfortunately, hardly any systems are now available that show promise of allowing epidemiologists to assess accurately the magnitude of past exposures to substances with mutagenic or other effects. An exception is an assay system for variant cells based on a common blood group antigen in erythrocytes (the MN blood group system, known as the glycophorin A locus) (Langlois *et al.*, 1987). In individuals who contain genes for both blood group variants (MN heterozygotes, approximately 50% of the population), nearly all the red cells can be stained with fluorescent antibodies to both the M and the N cell-surface markers. However, a few red cells can be found that stain only with one marker or the other—M0 or 0N variants, respectively—suggesting that the activity of the gene producing one of the antigens has been lost in the ancestors of the variant cells. As indicated in Figure 10.1b (which shows group-average rather than individual results), such assay measurements in groups of atomic bomb survivors show a clear linear relationship between variant frequency and estimated radiation dose, although the measurements were made decades after the exposure. On the other hand, the individual data points show enormous scatter, probably because of fluctuations resulting from the fact that very few hematopoietic precursors are active in red cell production at any one time. This result clearly indicates serious limitations to using the test to estimate individual dosage based on measurements taken at a single point in time. There is some hope, however, that a series of repeated measurements, taken over a period of years when different sets of hematopoietic precursors might be active, would result in more stable estimates of individual dose.

Use of Indicators to Assist in Projection of Risk per Equivalent Dosage between Species

The issue of how precisely to project equivalent doses and consequent carcinogenic risks across species is one of the most important outstanding questions in current carcinogenesis risk assessment. Different approaches for

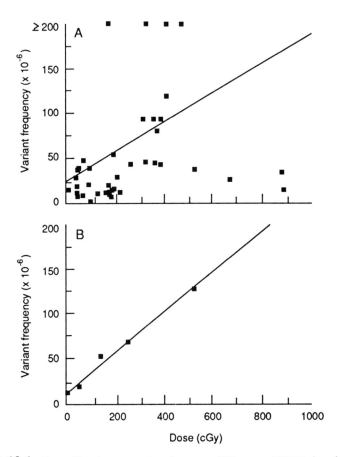

FIGURE 10.1   Plots of hemizygous variant frequency (VF) versus T65DR dose for control and exposed donors. The VF calculated for each donor is the mean value of all N_0 and M_0 measurements made on that donor. (A) Individual data. (B) Group data. (Reprinted from Langlois *et al.*, 1987. Copyright © 1987 by the AAAS.)

scaling equivalent dosage and risk in different agencies are the source of an important portion of the quantitative differences in cancer potencies assessed currently by different federal agencies[2] (Institute of Medicine, 1991; Chapter 6).

[2] Standard practice of the U.S. Environmental Protection Agency is to project risks on the basis of the "surface area" rule, assuming that lifetime daily average doses will be equipotent among species when expressed per (body weight)$^{2/3}$. The U.S. Food and Drug Administration, on the other hand, favors a (body weight)^1 projection rule. Recent interagency discussions aimed at reconciling these different approaches have tended to center on a (body weight)$^{3/4}$ projection rule originally derived from the scaling of general metabolism and clearance rates (Boxenbaum, 1982).

Previous reviews (Hattis, 1988, 1991), have discussed work using physiology-based pharmacokinetic (PBPK) models (and associated measurements) of internal concentrations of putative genetically active agents as aids in high dose-to-low dose and interspecies projection of carcinogenic risks for perchloroethylene, ethylene oxide, and butadiene (Hattis *et al.*, 1986; Hattis, 1987; Hattis and Wasson, 1987). One scientific virtue of PBPK models is that, by using real physical locations as their "compartments" and measureable blood flows and partition coefficients as determinants of transfer, they allow much more detailed testing of the many more points of contact between the real world and the model. Such models are increasingly accepted in public discussions of the risks of specific carcinogens, at least for agents that are relatively well studied.

However, the use of a pharmacokinetic model does not, in itself, resolve the issue of how to define "equivalent dosage" among species for purposes of projecting risk. The models essentially help determine the *adjustment* to a standard risk or dose projection rule that is indicated by the detailed facts available for a particular compound in people as opposed to experimental animals. Thus, if one believes that

- the standard projection rule should be that risks will be equal when lifetime average daily internal dose is expressed per (body weight)^n, and
- the modeling effort yields the result that humans will have x times more internal dose of active carcinogen at low doses at the putative site(s) of action per unit of administered dose [expressed in units of (body weight)^n] than the animals used in the carcinogenesis bioassay,

the modeling should yield an x-fold adjustment of projected human risk relative to what would have been projected directly from the administered dose data.

Use of Indicators to Validate Pharmacokinetic and Pharmacodynamic Models—Ethylene Oxide–Hemoglobin Adduct Formation

The pharmacokinetic models mentioned in the previous section are natural partners for the enterprise of making and interpreting measurements of various biomarkers of internal dose and early effects of carcinogenic compounds in people. Ideally, (1) the models make predictions about the absolute relationships between external dose and specific internal biomarkers (e.g., the relationship between ethylene oxide exposure and hydroxyethyl hemoglobin adduct levels) that are testable with the aid of measurements in humans, (2) the models guide interpretation of the results of the biomarker studies with respect to the likely dynamic relationships between exposure and biomarker generation and loss and to the mathematical form of the overall relationship between external dose and internal biomarker levels, and (3) based on relationships between biomarkers and internal dose, quantitative inferences of human risk can be made from the biomarker data in some cases.

To achieve these results, however, the collection, calibration, and analysis of the biomarker information must be compatible with the pharmacokinetic models. Two studies on hemoglobin adducts and other biomarkers in relation to ethylene oxide (ETO) exposure (Meyer *et al.*, 1991; Schulte *et al.*, 1992) chose to use logarithmic transformations of some of the variables in the construction of statistical models to analyze data on the relationships between hemoglobin adducts, ETO exposure, and a large number of putative confounding factors. For example, in the case of Schulte *et al.* (1991),

$$\ln(\text{Hb adducts}) = B_0 + B_1 \ln(\text{ppm} \cdot \text{hr ETO exposure/4 months}) + B_2(\text{age}) + \ldots B_n(\text{other terms})$$

where B_n multiplies other terms related to gender, race, smoking, tea/coffee consumption, and education.) After expressing exponentially,

$$\text{Hb adducts} = e^{[B_0 + B_2(\text{age}) + \ldots B_n(\text{other terms})]} \times (\text{ppm} \cdot \text{hr ETO})^{B_1}$$

where B_0 is a constant and B_{1-n} are regression coefficients whose magnitude depends on the strength of the assumed additive and multiplicative relationships between each parameter and the hemoglobin adduct level. In the case of Mayer *et al.* (1991), a similar model was used, except that only the hemoglobin adduct data were log transformed (estimated ETO exposure was left as a linear parameter).

For the purposes that Schulte *et al.* (1991) and Mayer *et al.* (1991) had in mind (developing empirical relationships among biomarkers and more traditional measures of exposure), these transformations probably are not unreasonable. However, for purposes of risk assessment modeling, the choice to log transform the dependent variable in these cases alters the form of the hypothesized relationship between ETO exposure and hemoglobin adduct formation away from the linear form indicated by the available pharmacokinetic and mechanism information (Hattis, 1987).

For very good reasons, all pharmacokinetic models must "go linear" in terms of the relationship between internal dose (and resulting adduct levels) and external dose, at the limit of low external dosage (Hattis, 1988, 1990). Moreover, it is almost inconceivable that there could be serious nonlinearities in hemoglobin adduct formation as a function of ETO exposure at the low cumulative ETO levels measured in these studies.

In both cases, the logarithmic transformations used are justified on purely statistical rather than physical or biologic grounds.[3] We would suggest

---

[3] The discussion of the issue in Mayer *et al.* (1991) is "Biomarker data were log transformed to stabilize the variance and to obtain a more symmetric distribution" (Armitage, 1971). Schulte *et al.* (1991) say, "In order to satisfy the assumptions underlying linear regression analysis, the natural log transformation was used for each marker, correcting the extreme skewness in the marker data. Each exposure variable was also log transformed to normalize a skewed distribution and provide a better fit to the regression models."

that, to help make these results usable for risk assessment modeling, other ways should be found to avoid the statistical difficulties alluded to, for example, (1) differential weighting of the data points in an adapted regression procedure (e.g., weighting high dose/high biomarker points less in proportion to the inverse of their estimated variance),[4] and (2) Monte Carlo simulation to assess the biasing effects of quantitative uncertainties in the independent variables (such as ETO exposure) on exposure–response relationships (Goble *et al.*, 1990). With respect to the latter point, we speculate that one reason Schulte *et al.* (1991) seem to have observed an apparent flattening of the relationship between ETO exposure and hemoglobin adducts at high observed exposure levels (making the log transformation fit better) could have been appreciable lognormally distributed errors in the estimation of individual exposure.[5] Recall that ordinary regression procedures assume that independent variables are measured without error. Unfortunately, few analyses at present seek to assess the effects of such errors quantitatively for the apparent form and magnitude of relationships studied.

From our perspective, to produce results that can be related to physical models of biologic processes, biostatisticians should be encouraged to choose their model forms first in terms of physical and biologic plausibility, and only then in terms of the conformance to assumptions of ordinary regression analysis. Until this happens, one must be pessimistic about whether useful information for risk analysis modeling will often result from the hard work invested in gathering such biomarker data in humans.

Use of Indicators to Help Define and Measure Interindividual Variability in Susceptibility

The degree of spread of interindividual variability in susceptibility in a mixed population of exposed people is an important determinant of individual and population aggregate risks, especially for noncancer effects. Here we discuss two different efforts that use biomarkers in different ways to estimate the likely magnitude of human interindividual variability in susceptibility to the effects of specific agents. The first uses maternal hair mercury during pregnancy as a biomarker of dose, and quantal response variables. Interindividual variability is assessed as the inverse of the slope of a simple probit dose–response regression line.

---

[4] This would enable the analyst to avoid departures from the usual regression assumption of homoscedasticity (no relationship between the variance of the dependent variable and the value of the ETO independent variable) without changing the basic linear form of the relationship being modeled.

[5] In the case of Schulte *et al.* (1991), the basic data available for each hospital were 3 to 20 measurements. Based on these measurements, relevant individual exposures were projected as a function of work activities performed over 4 months.

The second analysis described uses a continuous biomarker of early response, the lung function parameter FEV_1. The form of the distribution of deviations of the individual values of this parameter from the values expected as a function of reported cigarette exposure gives clues to the magnitude of interindividual variability in this response.

Analysis of Quantal Response Data Using a Biomarker of Dose

Marsh *et al.* (1987) published detailed information on the incidence of a variety of fetal methyl mercury effects in relation to the maximal levels of mercury found in the hair of the mothers during gestation. (The observations come from an Iraqi mass poisoning incident that resulted from the distribution of methyl mercury-treated Green Revolution seed grain.) Maximum mercury concentrations were assessed by a series of sequential measurements along the hair shafts during fetal development. Log probit dose–response fits[6] to these data (e.g., Figures 10.2, 10.3) indicate very large amounts of interindividual variability in response. A probit slope of about 1 (corresponding to a geometric standard deviation of 10), as suggested by the plot in Figure 10.2, would imply that 95% of the population had thresholds for effect spread out over a 10,000-fold span of dosage, from 100-fold lower to 100-fold higher than the dose that would cause the effect in people of median susceptibility in an exposed population. A probit slope of 2 would suggest less, but still appreciable, variability; thresholds of 95% of the population would spread over a 100-fold range in dosage, from 10-fold lower to 10-fold higher than the threshold for the median person. Such a large amount of interindividual variation as that seen here for fetal effects could suggest appreciable risks at the much lower dosages that are present in the diets of people who consume relatively large amounts of fish with relatively large methyl mercury concentrations.

In addition to statistical uncertainties in the determination of these slopes from limited data, however, there are questions of biologic interpretation. A conclusion that the relationships represented in Figures 10.2 and 10.3 represent true interindividual variability depends on an assumption that the biomarker of exposure used in this case—the maximum hair mercury found at any time during gestation—is the most appropriate direct causal

---

[6] The basic assumptions underlying log probit analysis are that (1) the effect in question occurs when a specific threshold dose is exceeded in each individual, and (2) the thresholds of different individuals in a heterogeneous population are distributed lognormally. A lognormal distribution of individual thresholds would be expected to result if the factors that caused individuals to differ were numerous and tended to have multiplicative effects in determining each individual's threshold (e.g., the influences of differences in individual elimination half-lives would be expected to act in a multiplicative fashion with individual differences in breathing or food consumption rates, which determined individual differences in exposure per unit of concentration in environmental media).

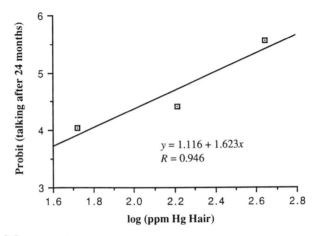

FIGURE 10.2 Log probit dose–response relationship between talking after 24 months and maternal hair mercury content. (Data of Marsh *et al.,* 1987.)

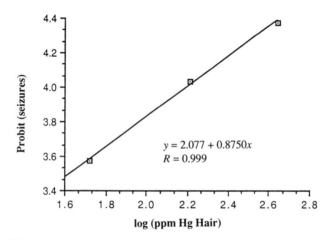

FIGURE 10.3 Log probit dose–response relationship between seizures in children and maternal hair mercury content. (Data of Marsh *et al.,* 1987.)

predictor of response that can be developed. Other possibilities might include the concentration of mercury at a specific sensitive time during gestation, or a weighted sum of concentrations × duration over a specific set of sensitive periods. Accurate assessment of the degree of interindividual variability in susceptibility in humans, and consequent low-dose risks, might depend on quantitative measurement and modeling of the causal processes involved in this case, and reanalysis of the data according to the most likely causally predictive summary measure of delivered dose. A similar, but more extensive, analysis of these data has appeared in a report assessing the safety of seafood contaminants (Institute of Medicine, 1991).

Analysis of the Distribution of Individual Departures from Expected Values for a Continuous Response Variable

The chronic changes in lung function that are caused by cigarette smoking offer an important opportunity to study human variability in susceptibility to a chronic cumulative response of a continuous effect variable, FEV_1. Relatively large studies have been done of FEV_1 in relation to cigarette smoke and other inhaled pollutants that cause chronic obstructive pulmonary disease; the population aggregate response of "FEV_1 residuals"[7] to accumulated pack–years of cigarette smoking has the convenient property of being almost perfectly linear (Dockery *et al.*, 1988).

We have made use of cross-sectional data from the latter study and a smaller earlier study in Tucson (Burrows *et al.*, 1977) to draw inferences regarding human interindividual variability in response to cigarette smoke. Figure 10.4 shows the distribution of age- and height-adjusted FEV_1 residuals from Dockery *et al.* (1988) for 3303 never-smokers (Figure 10.4A) and 4888 smokers in categories of 20 cumulative pack–years (Figure 10.4B). These data suggest a pattern of more and more spreading of the distribution of FEV_1 residuals with increasing dose, as well as the expected trend toward more negative average FEV_1 residuals. A similar pattern can be seen in the percentage of predicted FEV_1 in 1103 smokers and 945 never-smokers by Burrows *et al.* (1977) (not shown). Using these data, our analysis (Silver and Hattis, 1990) asked:

- Is the degree of spreading of FEV_1 residuals with increasing cigarette dose greater than would be expected on the basis of (1) the "baseline" variability of FEV_1 residuals seen in never-smokers and (2) the likely variability in cigarette dose within the categories shown in Figure 10.4B?
- What degree of interindividual variability is most compatible with the data (expressed as a geometric standard deviation for an assumed log-normal distribution of individual rates of decline of FEV_1 per pack–year of cigarette smoking)?
- What are the confidence limits around our estimates of interindividual variability in susceptibility to FEV_1 decline, considering each data set separately and combined?

Briefly, our methodology was to simulate the effects of the different sources of variability mentioned first, and judge the statistical significance of the resulting departures of expected from observed numbers of people within vari-

---

[7] By "FEV_1 residuals" we mean the departures of the values of FEV_1 observed for individual people from the values that would be expected for those people on the basis of their age, height, and gender.

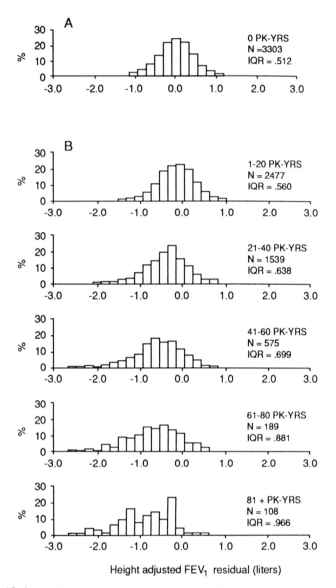

FIGURE 10.4 Distribution of age- and height-adjusted FEV$_1$ residuals in (A) 3303 never-smokers and (B) 4888 smokers. (Reprinted with permission from Dockery *et al.*, 1988.)

ous intervals of FEV$_1$ residual using a simple χ^2 analysis. After finding that there were indeed larger departures than could be accounted for by chance, we estimated the amount of lognormally distributed interindividual variability in susceptibility to the effects of cigarette smoking that would provide the best fit of the FEV$_1$ residual distribution [a geometric standard deviation of 1.9 for the Dockery *et al.* (1988) data set; appreciably more for the Burrows *et al.* (1977) data]. Several biases may have inflated the estimate of interindividual variability artificially in the latter case.

Because of a number of complexities and assumptions in the analysis, and the fact that we were working with data summaries rather than the full raw data sets, our numerical results in this study must be regarded as crude initial estimates. Our analysis is primarily significant because of the question we raise and the potential we show for using a commonly collected type of cross-sectional data to address a central issue in the assessment of risk for noncancer effects.

Use of Indicators to Project Effects of Concern that are Difficult to Measure Directly

Birth Weight as a Predictor of Infant Mortality Risk

Table 10.1 outlines some idealized components of a full quantitative analysis of a noncancer health effect that is mediated by what we call a "functional intermediate" parameter (Hattis, 1991). Such a parameter should be a continuous variable that is thought to be related closely to processes that have a strong causal influence on performance of an important biologic function (although it generally will not be the sole determinant of performance). The parameter should, in turn, be affected by the toxin or exposure under study. Effects on the final health condition of concern should be reasonably likely to be primarily mediated through effects on the "functional intermediate" parameter. For example, a key "functional intermediate" parameter for some reproductive effects of polychlorinated biphenyls (PCBs) may well

TABLE 10.1   **Elements of a Quantitative Risk Analysis for Noncancer Health Effects Mediated by a Functional Intermediate Parameter**

1. Elucidate the quantitative relationships between internal dose/time of toxin exposure and changes in the functional intermediate parameter.
2. Assess the preexisting background distribution of the functional intermediate parameter in the human population.
3. Assess the relationship between the functional intermediate parameter and diminished physiological performance or adverse health efffects.
4. Assess the magnitude of parameter changes likely to result from specific exposures in humans (taking into account interindividual variability in metabolism and other determinants of pharmacokinetics) and consequent changes in the incidence and severity of health effects.
5. Do not attempt, from the biology alone, to determine acceptable levels of parameter change or exposure. (Let the policy makers decide what changes in the incidence and severity of health effects are acceptable in the context of the modes of exposure and in light of the feasibility of reducing or avoiding the exposure.)

Source: Reprinted with permission from Hattis (1991).

TABLE 10.2   Expected Infant Mortality Effects Associated with a 1% (33.66 g)
Reduction in Birth Weights

Weight range (g)	Fraction of births		Mortality risk per 1000 births in category	Fraction of births × mortality risks/1000		
	Original population weight	After 1% reduction		Without birth-weight reduction	After 1% birth-weight reduction	Net change
White infants						
Under 500	0.0006912	0.0007701	1000	0.691	0.770	0.079
500–999	0.002171	0.002335	673.31	1.461	1.572	0.111
1000–1499	0.005249	0.005488	237.85	1.248	1.305	0.057
1500–1999	0.009182	0.009575	76.86	0.706	0.736	0.030
2000–2499	0.029192	0.032804	26.746	0.781	0.877	0.097
2500–2999	0.15164	0.16568	8.3565	1.267	1.384	0.117
3000–3499	0.36237	0.37081	4.2566	1.542	1.578	0.036
3500–3999	0.31749	0.30337	3.0451	0.967	0.924	−0.043
4000–4499	0.10100	0.09027	3.0293	0.306	0.273	−0.032
4500+	0.021021	0.01890	4.941	0.104	0.093	−0.010
Total	1	1		9.074	9.514	0.441
Black infants						
Under 500	0.0026095	0.0028661	1000	2.610	2.866	0.257
500–999	0.006279	0.006666	645.90	4.056	4.306	0.250
1000–1499	0.012709	0.013154	167.98	2.135	2.210	0.075
1500–1999	0.020673	0.02165	57.72	1.193	1.250	0.056
2000–2499	0.067052	0.074351	21.482	1.440	1.597	0.157
2500–2999	0.24894	0.26444	9.832	2.448	2.600	0.152
3000–3499	0.38248	0.37936	6.636	2.538	2.517	−0.021
3500–3999	0.20683	0.19100	5.581	1.154	1.066	−0.088
4000–4499	0.04418	0.04029	5.89	0.260	0.237	−0.023
4500+	0.008253	0.006226	12.33	0.102	0.077	−0.025
Total	1	1		17.936	18.726	0.790

Source: Reprinted with permission from Ballew and Hattis (1989).

be changes in birth weight (Fein *et al.*, 1984; Sunahara *et al.*, 1987; Taylor *et al.*, 1989).

The calculations in Table 10.2 show how modest changes in the population distribution of a key parameter such as birth weight can be reflected in changes in the serious outcome of infant mortality. Birth weights can be seen to be very strongly related to infant mortality. The relationship is continuous: although very low birth weight infants are at dramatically higher risk than infants in the normal weight range, even infants of about 3000 g can be expected to have slightly increased risks because of an agent that causes a marginal change in birth weight. As indicated in the table, because many more infants are born in the 2500–3500 g weight range, the expected population aggregate mortality increase is as large for these categories as the population aggregate mortality increase for infants in the very low birth weight range (500–1500 g).

In principle, use of such intermediate parameters can provide windows on the pathological process because they are more sensitive to the action of potential toxicants earlier in the development of toxicity (compared with attempts to observe actual cases of illness) and more accessible to direct comparative measurement in both animal models and humans. For these purposes, the intermediate parameters chosen should be as close as possible to the actual causal pathway leading to harm. However, even a parameter such as birth weight, which itself may not be directly causally related to infant deaths, may be a sufficiently close indicator of the actual causal processes that it can serve as a useful intermediate predictor variable. Assessment of the uncertainties of such an analysis ideally should include the degree of quantitative uncertainty in how close a predictor the functional intermediate parameter is for the effect(s) of ultimate concern. (For example, if the relationship between the functional intermediate parameter and the effect is mostly due to confounding with the effects of another variable that is actually on the causal pathway to disease, but is not affected directly by the exposure that changes the functional intermediate parameter, then the predicted effects on the health outcomes of concern will tend to overstate the actual effects.)

FEV₁ as a Predictor of Adult Mortality Risk

Olofson *et al.* (1987) conducted an 11-year prospective study of total mortality in 50- and 60-year-old men drawn from the general population in Sweden. Figure 10.5 illustrates the apparent independent contributions of FEV_1 and smoking habits to mortality risk, as well as the graded increase in apparent mortality risk with increasing deficit of observed as opposed to predicted FEV_1. The results were expressed in terms of a logistic regression model; age, smoking habits, height-adjusted FEV_1, and the presence of dyspnea during level walking were explanatory variables. The investigators obtained a coefficient of -0.5 relating mortality risk to height-corrected FEV_1 in liters, implicitly controlled for the influence of the other variables.

In an analysis of coal dust risks (Silver and Hattis, 1991), we used this coefficient in a multiplicative model to develop tentative estimates of the likely mortality effect of the degree of FEV_1 loss associated with permitted exposures for coal miners, presuming that FEV_1 is a good proxy for the causal factors that might mediate between coal dust exposure and effects on overall mortality from nonaccidental causes. Results for average-susceptibility nonsmokers are shown in Table 10.3. Overall, for individuals of average susceptibility, we expect the maximum currently permissible rate of exposure (2 mg/m³) to be associated with somewhat more than a 2% excess probability of death by age 75 if exposure is continuous over a

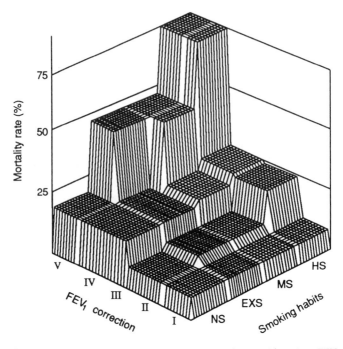

FIGURE 10.5   Mortality risk findings. Mortality rate in subjects with various FEV_1 corrected for height (I–V) and smoking habits in 1973. FEV_1, I: Corrected $FEV_1 >$ predicted normal + 2-SD; FEV_1, II: Corrected FEV_1 between predicted normal + 2-SD and predicted normal; FEV_1, III: Corrected FEV_1 between predicted normal and predicted normal − 2-SD; FEV_1, IV: Corrected FEV_1 between predicted normal − 2-SD and − 3-SD; FEV_1, V: Corrected FEV_1 less than predicted normal − 3-SD. NS, Nonsmokers; ExS, exsmokers; MS, moderate smokers; HS, heavy smokers. (Reprinted with permission from Olofson *et al.*, 1987.)

45-year period. This result translates into a loss of approximately one half year of potential life per exposed worker by age 85. Exposures at lower rates are expected to produce effects that are reduced from these values approximately linearly with cumulative exposure.

Are these results compatible with existing studies of mortality from different causes in coal miners? The most thorough study of coal miner mortality found, in which exposures were estimated quantitatively over an extended period, is the report by Miller *et al.* (1981) for British coal miners. The research was begun in 1953 and included 31,611 men, all those who showed any sign of pneumoconiosis in an initial survey and a 50% stratified sample of the remaining men. The last miner was enrolled in 1958, allowing a 22–26-year period of observation by the end of the follow-up period in 1980. A total of 29,533 men were successfully traced, of which 6,654 were found to have died. Our analysis of the mortality results in this cohort as a function of coal dust exposure (Silver and Hattis, 1991) suggested that mor-

TABLE *10.3*   Predicted Mortality Risks for Nonsmoking
Coal Miners Exposed Continuously[a]

Age	Average Coal Dust Exposure Level[b] (mg/m^3)			
	0[c]	0.5	1	2
Risk of death between age 20 and the age indicated[d]				
55	8.78	8.86	8.94	9.10
65	20.41	20.68	20.94	21.49
75	41.18	41.72	42.26	43.36
Excess risk of death over baseline by age indicated[d]				
55		0.08	0.16	0.33
65		0.26	0.53	1.07
75		0.54	1.08	2.17
Years of potential life lost per 100 exposed people by age indicated				
55		0.7	1.4	2.7
65		2.2	4.5	9.1
75		6.3	12.6	25.5
85		11.7	23.5	47.2

Source: Reprinted with permission from Silver and Hattis (1991).
[a] Prediction based on a model using height-adjusted FEV$_1$ changes resulting from coal dust
  as an intermediate parameter (inferred from epidemiological observations in coal miners)
  and the effects fo FEV$_1$ changes on mortality inferred from the model relationship of Olofson
  et al. (1987).
[b] Up to 45 years exposure by age 65.
[c] Baseline.
[d] Percentage of starting population.

tality effects in this cohort were actually several-fold larger than those predicted from our use of the FEV$_1$ intermediate parameter and the FEV$_1$–
mortality relationship derived by Olofson *et al.* (1987).

The Brave New World of Molecular Pathology—
The Example of Cancer

Advances in our understanding of the molecular pathology of cancer have
the potential to revolutionize the epidemiologic study of carcinogenesis and
carcinogenic exposures in a number of ways. One application that is already
clearly on the horizon is the ability to classify cancers not only by the traditional criteria of site of occurrence and histology, but according to the specific
genes that have been altered and the particular changes in DNA within those
genes that were involved in transforming the ancestral normal cell into a ma-

lignant neoplasm (Hollstein *et al.*, 1991; Jones *et al.*, 1991; Shi *et al.*, 1991). This categorization appears likely to be etiologically significant in some cases, since different mutagenic agents tend to produce different kinds of changes in DNA and show different preferences for the specific location of their "hot spots" in specific genes (although each agent does not always produce a unique spectrum of changes).[8]

Tumor tissues contain many if not all cells clonally derived from a normal ancestor. Thus, with the aid of DNA amplification technology, these tissues can be made to yield large amounts of mutant DNA for precise determination of sequences in relevant genes. In contrast, normal tissues necessarily contain only a minute proportion of cells with specific mutations along specific carcinogenic pathways. This lower "signal" has proved to be an obstacle to attempts to observe the spectrum of DNA changes in unpurified normal tissues directly (Thilly *et al.*, 1989). Nevertheless, if it were possible to rapidly select and count cells from normal tissues with relevant changes in genes along the pathway to carcinogenesis, there would be an enormous potential for identifying the contributions of specific agents to carcinogenesis via counts of the frequencies of relevant precancer mutant cells in target tissues in individuals as a function of age and exposure, rather than via counts of the numbers of individuals with fully developed cancer. Consider the increase in statistical sensitivity that would result from counting 100 or 1000 mutant cells among 10^8 sampled cells from a tissue of an exposed individual,[9] whether or not he or she eventually goes on to develop the tumor in question, rather than "counting" the single tumor that may or may not occur in that person. Scientifically, such studies would have the advantage of being able to test competing theories of the dynamics of the carcinogenic process directly (e.g., the Moolgavkar/Knudson "two-stage" model as opposed to the more traditional multistage model of carcinogenesis). This future "brave new world" of cancer molecular epidemiology, however, awaits some refinements

[8] For example, according to Jones *et al.* (1991), "background" deamination of methylcytosine frequently leads to transitions at CpG dinucleotides; alkylation at the O^6 position of guanine by small molecular weight alkylating agents preferentially gives rise to G to A transitions; alkylation of DNA with bulky adducts derived from benzo(a)pyrene or aflatoxin B_1 tends to lead to G to T transversions, whereas ionizing radiation preferentially causes deletions. Data given by Jones on the relative frequency of these different kinds of genetic changes in mutated *p53* genes from different tumor sites are consistent with a relatively small role for spontaneous deamination mutations in lung cancer, and a relatively large role in cancers of the colon, leukemias, and sarcomas. Bladder cancers appear to show an intermediate pattern, as would be expected from a hypothesis that external chemical agents account for more of the relevant *p53* mutations for this kind of cancer than for lung cancer, but less than for the other tumor types studied. Hollstein *et al.* (1991) provide additional similar data for a few other tumor sites, coming to very similar conclusions.

[9] For accessible tissues such as blood cells and skin, such studies could be done during life for representative people. For tissues such as lung and liver, samples would need to be taken from individuals who died of unrelated causes, or from individuals undergoing biopsies for diagnostic purposes.

in currently available molecular detection techniques to overcome the signal-to-noise ratio problems cited by Thilly *et al.* (1989).

References

Armitage, P. (1971). "Statistical Methods in Medical Research" Blackwell, Oxford.
Ballew, M., and Hattis, D. (1989). "Reproductive Effects of Glycol Ethers in Females—A Quantitative Analysis." CTPID 89-7. M.I.T. Center for Technology, Policy, and Industrial Development, Cambridge, Massachusetts.
Boxenbaum, H. (1982). Interspecies scaling, allometry, physiological time, and the ground plan of pharmacokinetics. *J. Pharmacokinet. Biopharmac.* **10**, 201–227.
Burrows, B., Knudson, R. J., Cline, M. G., and Lebowitz, M. D. (1977). Quantitative relationships between cigarette smoking and ventilatory function. *Am. Rev. Resp. Dis.* **115**, 195–205.
Creasy, D. M., and Foster, P. M. D. (1984). The morphological development of glycol ether-induced testicular atrophy in the rat. *Exp. Mol. Pathol.* **40**, 169–176.
Creasy, D. M., Flynn, J. C., Gray, T. J. B., and Butler, W. H. (1985). A quantitative study of stage-specific spermatocyte damage following administration of ethylene glycol mono-methyl ether in the rat. *Exp. Mol. Pathol.* **43**, 321–336.
DeBord, K. E., and Lowry, L. K. (1986). "Urinary Glycol Ether Acid Metabolite Results: Electric Boat Study." Memo 6-9278378. April 14. National Institute for Occupational Safety and Health, Cincinnati, Ohio.
Dockery, D. W., Speizer, F. E., Ferris, B. G., Ware, J. H., Louis, T. A., and Spiro, A. (1988). Cumulative and reversible effects of lifetime smoking and simple tests of lung function in adults. *Am. Rev. Resp. Dis.* **137**, 286–292.
Fein, G. G., Jacobson, J. L., Jacobson, S. W., Schwartz, P. M., and Dowler, J. K. (1984). Prenatal exposure to polychlorinated biphenyls: Effects on birth size and gestational age. *J. Pediatrics* **105**, 315–320.
Goble, R. L., Hattis, D., and Socolow, R. (1990). "Uncertainties in Population Risk Estimates which Arise from Different Conditions of Exposure to Indoor Radon." Paper presented at the 29th Hanford Symposium on Health and the Environment, Indoor Radon and Lung Cancer: Reality or Myth, October, 1990, Hanford, Washington.
Groeseneken, D., Veulemans, H., and Masschelein, R. (1986a). Respiratory uptake and elimination of ethylene glycol monoethyl ether after experimental human exposure. *Br. J. Ind. Med.* **43**, 544–549.
Groeseneken, D., Veulemans, H., and Masschelein, R. (1986b). Urinary excretion of ethoxy-acetic acid after experimental human exposure to ethylene glycol monoethyl ether. *Br. J. Ind. Med.* **43**, 615–619.
Hattis, D. (1987). "A Pharmacokinetic/Mechanism-Based Analysis of the Carcinogenic Risk of Ethylene Oxide." CTPID 87-1, National Technical Information Service NTIS/PB88-188784. M.I.T. Center for Technology, Policy, and Industrial Development, Cambridge, Massachusetts.
Hattis, D. (1988). The use of biological markers in risk assessment. *Stat. Sci.* **3**, 358–366.
Hattis, D. (1990). Pharmacokinetic principles for dose rate extrapolation of carcinogenic risk from genetically active agents. *Risk Anal.* **10**, 303–316.
Hattis, D. (1991). Use of biological markers and pharmacokinetics in human health risk assessment. *Environ. Health Perspect.* **89**, 230–238.
Hattis, D., and Berg, R. (1988). "Pharmacokinetics of Ethoxyethanol in Humans." CTPID 88-1, National Technical Information Service NTIS/PB88-221528. M.I.T. Center for Technology, Policy, and Industrial Development, Cambridge, Massachusetts.

Hattis, D., and Goble, R. (1991). Expected values for projected cancer risks from putative genetically-acting agents. *Risk Anal.* 11, 359–363 (in press).

Hattis, D., and Smith, J. (1987). What's wrong with quantitative risk assessment. *In* "Quantitative Risk Assessment" (J. M. Humber and R. F. Almeder, eds.), pp. 57–105. Humana Press, Clifton, New Jersey.

Hattis, D., and Wasson, J. (1987). "A Pharmacokinetic/Mechanism-Based Analysis of the Carcinogenic Risk of Butadiene." CTPID 87-3, National Technical Information Service NTIS/PB88-202817. M.I.T. Center for Technology, Policy, and Industrial Development, Cambridge, Massachusetts.

Hattis, D., Tuler, S., Finkelstein, L., and Luo, Z. (1986). "A Pharmacokinetic/Mechanism-Based Analysis of the Carcinogenic Risk of Perchloroethylene." CTPID 86–7, National Technical Information Service NTIS/PB88-163209. M.I.T. Center for Technology, Policy, and Industrial Development, Cambridge, Massachusetts.

Hollstein, M., Sidransky, D., Vogelstein, B., and Harris, C. C. (1991). p53 mutations in human cancers. *Science* 253, 49–53.

Institute of Medicine (1991). "Seafood Safety." National Academy Press, Washington, D.C.

Jones, P. A., Buckley, J. D., Henderson, B. E., Ross, R. K., and Pike, M. C. (1991). From gene to carcinogen: A rapidly evolving field in molecular epidemiology. *Cancer Res.* 51, 3617–3620.

Kuhn, T. S. (1977). Mathematical versus experimental traditions in the development of physical science. *In* "The Essential Tension—Selected Studies in Scientific Tradition and Change," pp. 31–65. University of Chicago Press, Chicago.

Lakatos, I. (1970). Falsification and the methodology of scientific research programs. *In* "Criticism and the Growth of Knowledge" (I. Lakatos and A. Musgrave, eds.), pp. 91–196. Cambridge University Press, London.

Lakatua, D. J., Blomquist, C. H., Haus, E., Sackett-Lundeen, L., Berg, H., and Swoyer, J. (1982). Circadian rhythm in urinary N-acetyl-β-glucosaminidase (NAG) of clinically healthy subjects—timing and phase relation to other urinary circadian rhythms. *Am. J. Clin. Pathol.* 78, 69–77.

Langlois, R. G., Bigbee, W. L., Kyoizumi, S., Nakamura, N., Bean, M. A., Akiyama, M., and Jensen, R. H. (1987). Evidence for increased somatic cell mutations at the glycophorin A locus in atomic bomb survivors. *Science* 236, 445–448.

McManus, K. (1987). "Data Summary for Painters' Study." Unpublished manuscript. National Institute for Occupational Safety and Health, Cincinnati, Ohio.

Marsh, D. O., Clarkson, T. W., Cox, C., Myers, G. J., Amin-Zaki, L., and Al-Tikriti, S. (1987). Fetal methylmercury poisoning. Relationship between concentration in single strands of maternal hair and child effects. *Arch. Neurol.* 44, 1017–1022.

Mayer, J., Warburton, D., Jeffrey, A. M., Pero, R., Walles, S., Andrews, L., Toor, M., Latriano, L., Wazneh, L., Tang, D., Tsai, W. Y., Kuroda, M., and Perera, F. (1991). Biologic markers in ethylene oxide-exposed workers and controls. *Mutat. Res.* 248, 163–176.

Miller, B. G., Jacobsen, M., and Steele, R. C. (1981). "Coalminers' Mortality in Relation to Radiological Category, Lung Function, and Exposure to Airborne Dust." Institute of Occupational Medicine, Edinburgh.

Moss, E. J., Thomas, L. V., Cook, M. W., Walters, D. G., Foster, P. M. D., Creasy, D. M., and Gray, T. J. B. (1985). The role of metabolism in 2-methoxyethanol-induced testicular toxicity. *Toxicol. Appl. Pharmacol.* 79, 480–489.

National Research Council (1983). "Risk Assessment in the Federal Government: Managing the Process." National Academy Press, Washington, D.C.

Olofson, J., Skoogh, B. E., Bake, B., and Svardsudd, K. (1987). Mortality related to smoking habits, respiratory symptoms, and lung function. *Eur. J. Resp. Dis.* 71, 69–76.

Ruckelshaus, W. D. (1983). Science, risk, and public policy. *Science* 221, 1026–1028.

Schulte, P. A., Boeniger, M., Walker, J. T., Schober, S. E., Pereira, M. A., Gulati, D. K., Wojciechowski, J. P., Garza, A., Froelich, R., Strauss, G., Halperin, W. E., Herrick, R., and Griffith, J. (1992). Biologic markers in hospital workers exposed to low levels of ethylene oxide. *Mutat. Res.* **278**, 237–251.

Shi, Y., Zou, M., Schmidt, H., Juhasz, F., Stensky, V., Robb, D., and Farid, N. R. (1991). High rates of *ras* codon 61 mutation in thyroid tumors in an iodide-deficient area. *Cancer Res.* **51**, 2690–2693.

Silver, K., and Hattis, D. (1990). "Human Interindividual Variability in Susceptibility to FEV$_1$ Decline from Smoking. CTPID 90-8. M.I.T. Center for Technology, Policy, and Industrial Development, Cambridge, Massachusetts.

Silver, K., and Hattis, D. (1991). "Methodology for Quantitative Assessment of Risks from Chronic Respiratory Damage: Lung Function Decline and Associated Mortality from Coal Dust." CTPID 90-9. M.I.T. Center for Technology, Policy, and Industrial Development, Cambridge, Massachusetts.

Sunahara, G. I., Nelson, K. G., Wong, T. K., and Lucier, G. W. (1987). Decreased human birth weights after *in utero* exposure to PCBs and PCDFs are associated with decreased placental EGF-stimulated receptor autophosphorylation capacity. *Mol. Pharmacol.* **32**, 572–578.

Taylor, P. R., Stelma, J. M., and Lawrence, C. E. (1989). The relation of polychlorinated biphenyls to birth weight and gestational age in the offspring of occupationally exposed mothers. *Am. J. Epidemiol.* **129**, 395–406.

Thilly, W. G., Liu, V. F., Brown, B. J., Cariello, N. F., Kat, A. G., and Keohavong, P. (1989). Direct measurement of mutational spectra in humans. *Genome* **31**, 590–593.

Tornqvist, M., Osterman-Golkar, S., Kautiainen, A., Jensne, S., Farmer, P. B., and Ehrenberg, L. (1986). Tissue doses of ethylene oxide in cigarette smokers determined from adduct levels in hemoglobin. *Carcinogensis* **7**, 1519–1521.

Part II

PRACTICAL APPLICATIONS

11

Carcinogenesis

Frederica P. Perera
and Regina Santella

Overview of Markers

Molecular epidemiology holds considerable promise for cancer prevention in two specific areas: the qualitative identification of carcinogenic hazards or potential risks and the fine-tuning of quantitative risk assessments. These advances are now possible through the measurement, directly in humans, of the internal dose or biologically effective dose of carcinogens, their resultant biologic effects, and factors that influence individual susceptibility (see Figure 11.1). As discussed in Chapter 1, the first three terms have been used to classify biologic markers in terms of the point at which they occur on the continuum between initial exposure to a carcinogen and overt clinical disease. Markers in the fourth category reflect susceptibility to any or all of the events occurring along the continuum. This chapter provides an overview of markers in the current armamentarium, followed by three case studies designed to illustrate the strengths and the problems and challenges of using biologic markers.

Exposure or Dose

To clarify terminology in molecular cancer epidemiology, internal dose refers to the measurement of the amount of a carcinogen or its metabolites present in cells, tissues, or body fluids. Examples of internal dosimeters include DDT and polychlorinated biphenyls (PCBs) in serum and adipose tissue from environmental contamination, plasma or salivary cotinine from cigarette smoking, urinary aflatoxin indicative of dietary exposure, N-nitroso compounds in urine from dietary sources and cigarette smoke, and mutagenicity of urine, reflective of exposure to various genotoxicants. Such internal do-

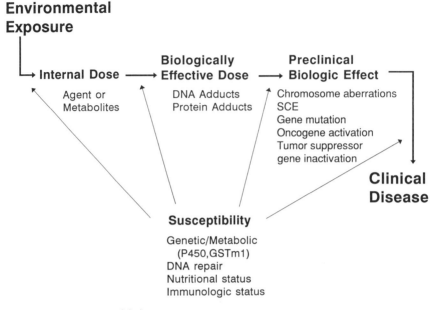

FIGURE 11.1   Factors that influence individual susceptibility.

simeters have the advantage of being comparatively easy to monitor. However, they can demonstrate only that exposure has resulted in uptake and bioactivation of carcinogens and cannot provide quantitative data about interactions with cellular targets. Table 11.1 provides examples of compounds and exposure sources analyzed using each type of internal dosimeter, as well as biologic samples and populations studied.

In contrast, biologically effective dose reflects the amount of carcinogen that has interacted with cellular macromolecules at a target site or with an established surrogate. As such, this class of markers is more mechanistically relevant to carcinogenesis than internal dose but poses more challenging analytical problems. Examples include carcinogen–DNA and carcinogen–protein adducts. (See Table 11.2 for examples of exposures and populations studied to date.) The biologic basis for measuring DNA adducts derives from extensive experimental data supporting their role in the initiation and possibly in the progression of cancer (Miller and Miller, 1981; Weinstein et al., 1984; Harris et al., 1987; Yuspa and Poirier, 1988). Despite this impressive body of information, however, much remains to be uncovered about the quantitative relationship between adduct formation (as a necessary but insufficient event) and cancer risk. Factors that are likely to play a role in cancer risk include the specific type and biologic effectiveness of adducts formed (e.g., location in tissue, cell type, and site of binding on the genome), the rate

TABLE 11.1   **Internal Dose**

Compound analyzed	Exposure source	Biologic sample	Population	Reference
Nitrosamino acids	N-nitroso compounds in diet	Urine	Chinese residing in areas of low and high cancer risk	Lu *et al.* (1986)
N-Nitrosoproline	Cigarette smoke	Urine	Smokers, nonsmokers, unexposed	Garland *et al.* (1986); Hoffman *et al.* (1985)
1-Hydroxypyrene	Coal tar products	Urine	Workers, smokers, coal tar treated patients	Bos and Jongeneelen (1988)
Aflatoxin M_1 Aflatoxin B_1	Diet	Urine	Chinese residing in a high exposure area	Zhu *et al.* (1987); Ross *et al.* 1992
Mutagenicity of urine	Cigarette smoke, various industrial exposures	Urine	Smokers, workers	Vainio *et al.* (1984); for review, Everson *et al.* (1986)
Thymine glycol	Agents that cause oxidative damage to DNA	Urine	Volunteers	Cathgart *et al.* (1984)
MeIQx[a]	Diet	Urine	Fried beef consumers	Murray and Gooderham (1989)

[a] MeIQx, 2-amino-3,8-dimethylimidazo[4.5-f]quinoxaline.

and accuracy of repair prior to cell replication, and the presence of endogenous or exogenous promoting agents or cocarcinogens. Ideally, therefore, this type of detailed characterization is desired for each chemical–DNA adduct being measured. In sharp contrast, most of the available methods provide information on total or multiple adducts and are incapable of pinpointing the "critical" adducts on DNA. Considerable basic research is needed to identify cell- and site-specific adducts that are most biologically effective and to develop adequately sensitive methods for human monitoring. The first steps have, however, been taken in methods development (Olivero *et al.*, 1990; van Schooten *et al.*, 1991).

As evidenced by experimental studies, proteins such as hemoglobin can, in many cases of acute or continuous exposure to carcinogens, serve as a valid surrogate for DNA (Neumann, 1984). However, as a step in validation, it is necessary to establish that relationship for each carcinogen of interest. Interpretation of results also should account for the varying kinetics of DNA and protein adducts, even in steady-state situations.

The larger question, which still remains largely unanswered, is whether

TABLE 11.2 Biologically Effective Dose

Compound analyzed[a]	Exposure source	Biologic sample[b]	Population	Reference
N-3-(2-Hydroxy-ethyl)histidine: N-(2-hydroxy-ethyl) valine	Ethylene oxide	RBC	Workers, smokers, unexposed	Calleman et al. (1978); van Sittert et al. (1985); Farmer et al. (1986); Tornqvist et al. (1886)
Alkylated Hb	Propylene oxide	RBC	Workers	Osterman-Golkar et al. (1984)
4-Aminobiphenyl–Hb	Cigarette smoke	RBC	Smokers, nonsmokers	Bryant et al. (1987, 1988); Maclure et al. (1990)
AFB_1-guanine	Diet	Urine	Chinese and Kenyans residing in a high exposure and high and low risk area, respectively	Groopman et al. (1985); Autrup et al. (1983); Ross et al. (1992)
AFB_1–DNA	Diet	Liver tissue	Taiwanese	Hsieh et al. (1988); Zhang et al. (1992)
3-Methyladenine	Methylating agents	Urine	Unexposed	Shuker and Farmer (1988)
PAH–DNA	PAH in cigarette smoke, in workplace, in air pollution	WBC, lung tissue, placenta	Lung cancer patients, smokers, workers, residents	Santella (1988, 1992); Perera et al. (1992a)

PAH–protein	PAH in workplace, in cigarette smoke	Plasma	Workers, smokers	Sherson et al. (1990); Weston et al. (1989); Lee et al. (1991)
O^6-Methyldeoxyguanosine	Nitrosamines in diet, smoking	Esophageal and stomach mucosa, placenta	Chinese and European cancer patients, smokers	Umbenhauer et al. (1985); Wild et al. (1986); Foiles et al. (1988)
Cisplatinum–DNA, Cisplatinum–protein	Cisplatin chemotherapy	WBC, Hb, plasma proteins	Chemotherapy patients	Reed et al. (1986, 1988); Fichtinger-Schepman et al. (1987, 1990); Mustonen et al. (1988); Parker et al. (1991); Perera et al. (1992b)
8-Methoxypsoralen–DNA	8-Methoxypsoralen chemotherapy	Skin	Psoriasis patients	Santella et al. (1988)
Spectrum of DNA adducts	Betel and tobacco chewing, smoking, industrial exposures, wood smoke	Placenta, lung tissue, oral mucosa, WBC, bone marrow, colonic mucosa	Smokers, workers, volunteers	Santella (1988, 1992)
NNK, NNN–Hb	Cigarette smoke	RBC	Smokers	Carmella et al. (1990)

[a] AFB_1, aflatoxin B_1; PAH, polycyclic aromatic hydrocarbons; NNK, 4-(methylnitrosamino)-1-(3-pyridyl)-1-butanone; NNN, N^1-nitrosonornicotine.
[b] RBC, red blood cells; WBC, white blood cells.

tissue accessible for biomonitoring, for example, peripheral blood cells, is a reasonable surrogate for the target tissue, for example, the lung. Experimental studies with benzo[a]pyrene (BP) suggest that comparable levels of adducts are formed in white blood cells and other tissues (Stowers and Anderson, 1985), but comparisons in humans are only now underway in lung cancer patients and controls. Illustrating the potential problem of organ, tissue, and even cell specificity of DNA adducts is the observation that bronchial cells of smokers showed clear cell-type specific induction of BP–DNA adducts (van Schooten et al., 1991). Therefore, studies are needed to understand the biologic relevance of adducts in surrogate tissue such as peripheral blood cells.

Effect

The third category comprises markers that indicate an irreversible biologic effect resulting from a toxic interaction, either at the target or at an analogous site, that is known or believed to be linked pathogenically to cancer. Many biomarkers fall into this category, from DNA single-strand breaks to mutational spectra. Table 11.3 provides examples of exposures and populations biomonitored for these end points. None of these markers are chemical or exposure specific. Hence, extensive information on other factors (life-style and environmental) that could affect these end points and act as confounding variables in a molecular epidemiologic study is required.

Susceptibility

Individual susceptibility to cancer may result from several factors, including differences in metabolism, DNA repair, altered proto-oncogene or tumor suppressor gene expression, and nutritional status. Since most carcinogens require metabolic activation before binding to DNA, individuals with elevated metabolic capacity may be at elevated risk. One example is the relationship of specific cytochrome P450 enzyme activity to lung cancer risk. CYP2D6 metabolizes debrisoquine, a β-adrenergic receptor blocking agent used to treat hypertension. A urine test can determine metabolic capacity based on excretion of debrisoquine and its 4-hydroxy metabolite. Extensive metabolizers have 10–200 times higher rates of metabolism than poor metabolizers, who constitute 5–10% of the Caucasian population in the United States. Although poor metabolizers are at increased risk of adverse drug reactions, extensive metabolizers are at 4 to 6 fold increased risk of lung cancer (Ayesh et al., 1984; Caporaso et al., 1990). The demonstration of the ability of the P4502D6 protein to metabolize the tobacco-specific nitrosamine 4-methylnitrosoamino)-1-(3-pyridyl)-1-butanone (NNK) suggests a mechanism for the association of extensive metabolizers of debrisoquine with elevated lung cancer risk (Crespi et al., 1991). The development of polymerase

TABLE 11.3 Early Biologic Effect or Response[a]

Compound analyzed	Exposure source	Biologic sample	Population	Reference
Single strand breaks	Styrene	WBC	Workers	Walles et al. (1988)
Unscheduled DNA synthesis	Propylene oxide	WBC	Workers	Pero et al. (1982)
Sister chromatid exchange	Various industrial exposures, radiation, air pollution	WBC	Workers, residents	Carrano and Moore (1982); Wilcosky and Rynard (1990), for review; Perera et al. (1992)
Micronuclei	Organic solvents, heavy metals, cigarette smoke, betel quid	WBC, oral mucosa	Workers	Hogstedt et al. (1983); Stich and Dunn (1988)
Chromosomal aberrations	Various industrial exposures, radiation, air pollution	WBC	Workers, residents	Evans (1982) for review; Perera et al. (1992)
DNA hyperploidy	Aromatic amines	Bladder and lung cells	Workers	Hemstreet (1988)
HPRT mutation	Chemotherapeutic agents, radiation	WBC	Patients, workers	O'Neill et al. (1987); Messing et al. (1986); McGinniss et al. (1990); Ostrosky-Wegman et al. (1990)
GPA mutation	Chemotherapeutic agents, radiation	RBC	Patients, Japanese atom bomb survivors	Langlois et al. (1987); Jensen et al. (1986); Bigbee et al. (1990); Kyoizumi et al. (1989)
Mutation in tumor suppressor genes	AFB$_1$	Tumor tissue	Patients	Hsu et al. (1991); Bressac et al. (1991)
Oncogene activation	PAH, cigarette smoke	Serum	Workers, cancer patients	Brandt-Rauf (1988b); Perera et al. (1988a)

[a] AFB$_1$, aflatoxin B$_1$; GPA, glycophorin A; HPRT, hypoxanthine guanine phosphoribosyl transferase; PAH, polycyclic aromatic hydrocarbons; RBC, red blood cells; WBC, white blood cells.

chain reaction methods for genotyping individuals should facilitate studies of this risk factor greatly (Heim and Meyer, 1990).

A second metabolizer phenotype related to increased cancer risk is the ability to N-acetylate aromatic amines. First identified in isoniazid-treated tuberculous patients, slow acetylators more slowly inactivate the drug than fast acetylators do. This bimodality is based on a homozygous recessive mutation carried by approximately half the population. As discussed here, since N-acetylation is a detoxifying pathway for amines, rapid acetylators are at decreased risk for bladder cancer. (See more detailed discussion below.)

CYP1A1, a P450 enzyme with aryl hydrocarbon hydroxylase (AHH) activity, catalyzes the oxygenation of polycyclic aromatic hydrocarbons (PAHs) such as benzo[a]pyrene. Inducibility of this enzyme has been associated with higher risk of lung cancer in smokers (reviewed in Vahakangas and Pelkonen, 1989). For example, the low inducibility phenotype (determined *in vitro* on lymphocytes) constituted 45% of controls and only 4% of cases, whereas the high inducibility phenotype constituted 9% of controls but 30% of cases (Kellermann *et al.*, 1973). Although not completely characterized, high CYP1A1 inducibility probably represents differences in the *Ah* receptor gene rather than differences in the structural gene (Nebert, 1991). A restriction fragment length polymorphism for CYP1A1 has been described that is reported to correlate with lung cancer risk in Japanese patients (Kawajiri *et al.*, 1990).

Whereas "phase I" enzymes are involved in the activation of chemical carcinogens, "phase II" enzymes conjugate metabolites with glucuronide, glutathione, or sulfate to produce hydrophilic products for excretion. Glutathione S-transferases are a family of proteins classified by their isoelectric points into α (basic), μ (neutral), and π (acidic) forms. Whereas several α forms are present in all human livers, only about 50% express the μ isozyme. Enzyme activity has been measured in lymphocytes using *trans*-stilbene oxide as a substrate (Seidegard *et al.*, 1986). Studies in bronchial carcinoma patients and controls matched for age and smoking history showed that controls had a greater likelihood of having μ enzyme activity (59%) than patients (35%) (Seidegard *et al.*, 1986). A follow-up study on a total of 383 subjects confirmed the previous study; 58% of noncancer smokers expressed the enzyme, compared with 36% of lung cancer patients (Seidegard *et al.*, 1990). A 3-fold greater risk for developing adenocarcinoma in individuals lacking μ activity was reported also (Strange *et al.*, 1991).

Genetic predisposition to cancer induction also may be related to inherited mutations in tumor suppressor genes that regulate cell growth and terminal differentiation (for review, see Marshall, 1991). Inactivation or altered function results in increased risk for development of tumors. These genes are inherited in a recessive form, requiring the loss of both copies for the phenotype to be expressed. Retinoblastoma is a dramatic example of this process. In the familial form of the disease, patients inherit one defective allele of the

Rb gene and the other is lost through later somatic mutation. In the spontaneous form, both copies of the gene are lost through gene deletions or mutations (Marshall, 1991).

A second example of inherited predisposition to cancer related to tumor suppressor genes involves *p53*, a phosphoprotein controlling cell proliferation. In Li–Fraumeni syndrome, patients with inherited mutations in *p53* have up to 1000-fold increased risk for cancers at multiple sites (Li, 1990).

Another factor that may influence susceptibility to certain cancers is nutritional status. The role of specific dietary constituents such as fat and fiber, vitamins, and minerals in various human cancers has proved elusive; epidemiologic studies frequently show weak or inconsistent associations (National Research Council, 1982; Cohen, 1987; Higginson, 1988). The association between dietary fat and breast cancer, although limited, appears strongest, followed by that between fat intake and colon cancer (Cohen, 1987, Kelsey and Berkowitz, 1988; Willett, 1989). There long has been intense interest in the role of vitamins in cancer (Willett and MacMahon, 1984; Marx, 1991). The retinoids are known to induce cell maturation and inhibit cell proliferation. β-Carotene is the most abundant carotenoid with vitamin A activity and is a precursor of retinol. The majority of studies have shown a protective effect of β-carotene, most strongly for lung cancer. Vitamin C is postulated to act via several mechanisms, including an antioxidant effect and blocking the carcinogenic conversion of nitrates and nitroso compounds. Similarly, vitamin E is an intracellular antioxidant. For the antioxidant vitamins (C, E and beta-carotene), experimental and epidemiological data regarding a chemopreventive effect are increasingly persuasive (Block, 1992). The field of chemoprevention is so promising that more than 40 clinical trials are now underway at the National Cancer Institute involving agents such as vitamins, fiber, calcium, and aspirin (Marx, 1991).

Many of the same methodologic issues encountered with the previous three types of markers apply to markers of susceptibility. They engender some unique ethical concerns as well (Ashford, 1986; Schulte, 1989).

Case Studies: Methods, Problems, Challenges, and Study Design Considerations

Polycyclic Aromatic Hydrocarbon–DNA Adducts

Several methods have been developed for quantitation of PAH–DNA adducts. Immunologic methods have used antibodies developed against DNA modified *in vitro* with benzo[*a*]pyrene-diol-epoxide I (BPDE-I). These antisera have been used in a number of studies to investigate adduct formation in human populations (reviewed in Poirier *et al.*, 1990; Santella, 1992). Antisera cross-react with DNAs containing adducts produced by other PAH *trans*-diol epoxides (Santella *et al.*, 1987; Weston *et al.*, 1987). Thus, when used to monitor adducts in human populations with exposure to complex mixtures of PAHs, multiple adducts may be detected. Absolute quantitation of adduct levels is impossible, since the identity of the individual adducts being measured is unknown. Differences in antibody affinity with modification level of the DNA also can interfere with accurate quantitation of adducts (van Schooten *et al.*, 1987; Santella *et al.*, 1988). Thus, thorough antisera characterization, including cross-reactivity with structurally related adducts and varied DNA modification levels, are essential before application to human samples.

PAH–DNA adducts have been found in lung tissue of cancer patients (Perera *et al.*, 1982; van Schooten *et al.*, 1990b) and in placental and white blood cell DNA of smokers and nonsmokers. Although adduct levels were higher in placental DNA of smokers than of nonsmokers, the difference was not significant (Everson *et al.*, 1986). Most studies of adducts in white blood cell DNA have reported nonsignificant or borderline differences in adducts related to smoking exposure (Perera *et al.*, 1987). However, significant smoking-related differences were seen by van Schooten *et al.* (1990b). Clear increases in PAH–DNA adduct levels have been seen in lymphocyte DNA from smokers, compared with those seen in nonsmokers (Santella *et al.*, in press). Dietary sources may be a major source of PAH exposure (Rothman *et al.*, 1990). Dramatic differences have been seen in workers occupationally exposed to high levels of PAHs. Elevated levels of white blood cell adducts were found in foundry workers, roofers, and coke oven workers (Harris *et al.*, 1985; Shamsuddin *et al.*, 1985; Haugen *et al.*, 1986; Perera *et al.*, 1988a; van Schooten *et al.*, 1990a). Exposure to PAH in environmental pollution also has resulted in significant increases in adduct levels (Hemminki *et al.*, 1990b; Perera *et al.*, 1992). PAH–DNA adducts were correlated with chromosomal mutation, linking molecular dose with a genetic effect of air pollution (Perera *et al.*, 1992).

PAH–DNA adducts also have been measured by ^{32}P-postlabeling (Randerath *et al.*, 1981; Gupta *et al.*, 1982). In this assay, DNA is digested to nucleoside 3'-monophosphates, then labeled with [^{32}P]ATP and kinase to produce nucleoside 3',5'-bisphosphates. Thin layer chromatography on polyethyleneimine cellulose is used to fingerprint adducts. To increase sensitivity, several methods are available to decrease the amount of normal nucleotides present, including extraction of hydrophobic adducts with solvents (Gupta, 1985) and digestion of normal nucleotides with nuclease P1 (Reddy and Randerath, 1986). As with the enzyme-linked immunosorbent assay (ELISA) method, limitations of the ^{32}P-postlabeling technique have included lack of standardization across laboratories and inadequate quality control.

The first study using PAH–DNA adduct levels in human samples found an unidentified adduct in placental DNA that was strongly related to maternal smoking during pregnancy (Everson *et al.*, 1986). BPDE-I-DNA adducts also have been identified in placental tissue with the nuclease P1 method (Manchester *et al.*, 1990). Adducts have been found in lung, bronchus, kidney, bladder, esophagus, heart, liver, and larynx from smokers (Cuzick *et al.*, 1990; Phillips *et al.*, 1988a,b; Randerath *et al.*, 1986,1989). A diagonal zone of radioactivity (DZR) was found in smokers' lung DNA, suggesting a large number of adducts; lower but detectable amounts of adducts were seen in nonsmokers. A linear relationship between adduct levels and daily or lifetime cigarette consumption has been seen in lung tissue (Phillips *et al.*, 1988b).

Several ^{32}P-postlabeling studies on DNA adducts levels in total white blood cells of both smokers and nonsmokers revealed no exposure-related differences (Phillips *et al.*, 1986,1990b). Whereas one study found no difference in adducts in lymphocytes isolated by Ficoll–Hypaque from smokers and nonsmokers (Jahnke *et al.*, 1990), another was able to demonstrate smoking-related adducts in monocytes (Holz *et al.*, 1990). Research comparing adducts in lymphocytes and granulocytes found elevated adducts in lymphocytes of smokers (Savela and Hemminki, 1991). Thus, studies on smoking-related adducts have produced conflicting results. Postlabeling studies of occupational exposure to PAH have demonstrated elevated levels of white blood cell DNA adducts in foundry workers, roofers, coke oven workers, and aluminum plant workers (Phillips *et al.*, 1988a; Hemminki *et al.*, 1990a,b; Herbert *et al.*, 1990; Schoket *et al.*, 1991).

Because of their intrinsic fluorescence, BPDE-I-DNA adducts have been measured by release of BP tetrols from DNA by acid treatment followed by synchronous fluorescence spectroscopy. Positive results were seen with white blood cell DNA from coke oven workers (Harris *et al.*, 1985; Haugen *et al.*, 1986). Because of the broad peaks observed in most human samples, probably resulting from multiple adducts, data were not quantitative.

Although the information available on PAH–DNA adducts in humans has increased dramatically over the past few years, some important questions remain unanswered. Most studies have been on single samples from each subject. Thus, the intraindividual variability in adduct levels is unknown. Prior studies, suggesting seasonal variability in adduct levels, must be expanded since the time of sample collection may influence results (Perera *et al.*, 1989). Most studies have used readily available blood samples but the relationship of blood DNA adducts to those in target tissue of humans has yet to be established. The question is complicated further by the differences found in studies looking at total white blood cells from those that used lymphocytes. Both the immuno and ^{32}P-postlabeling assays quantitate multiple PAH adducts, which may result from exposure to several sources including diet. This effect may confound studies trying to investigate the relationship between adducts and a specific source of exposure.

Acquired Gene Mutation

Researchers have developed methods to detect both large- and small-scale mutations in human genes that result from environmental chemicals and radiation. One of these techniques, a clonal assay, selects T cells with mutations at the *HPRT* (hypoxanthine–guanine phosphoribosyl transferase) gene locus (Albertini, 1985). Further steps include Southern blot analysis to identify independent mutations and sequencing of those mutations (Marx, 1989). Similarly, it is possible to probe mutations causing alterations in surface proteins (glycophorins) on red blood cells (Langlois *et al.*, 1987). These assays have detected significant increases in human populations exposed to ionizing radiation, patients on chemotherapy, and cigarette smokers (Marx, 1989). Work is also underway to determine mutational spectra characteristic of specific environmental agents, using DNA separation techniques and polymerase chain reaction (Thilly, 1985).

The search for tumor suppressor genes, genes that limit cell growth and regulate terminal differentiation, has suggested a link between a specific carcinogen exposure and target organ specificity (Bressac *et al.*, 1991; Hsu, 1991). A total of 26 liver tumors was obtained from patients living in regions of China and southern Africa characterized by high exposure to aflatoxin B_1 (AFB_1) and high prevalence of liver cancer. Of the tumors, 11 (43%) exhibited the same G-to-T mutation at codon 249 of the tumor suppressor gene, *p53*. This signal mutation is produced by AFB_1 when administered to experimental animals. Collectively, the human and experimental data strongly implicate AFB_1 as the causal agent in a large proportion of the liver cancers studied. They also indicate the mechanism by which the carcinogen may be exerting its effect. The molecular epidemiologic data are, therefore, consistent with both classical epidemiologic and experimental data (Marx, 1991).

Drawbacks of these studies are the small number of subjects, the lack of individual exposure histories, and direct measurement of AFB_1 or AFB_1–DNA adducts in tissues of the patients. The theoretical possibility exists in any such retrospective study that biologic markers in tumor tissue may reflect the disease rather than the exposure. Another major drawback is that the *p53* mutation in question is found only in tumor cell DNA, not in normal liver or more accessible tissues for screening purposes. Thus, although this elegant research is important in establishing causal links and understanding mechanisms of action, this type of study is limited in terms of providing an early warning of cancer risk (Marx, 1991).

For that reason, the use of surrogate samples rather than target tissues is a better preventive approach. In addition, a more desirable study design to establish the chain of causation is the nested case-control study. In this instance, questionnaire data and blood and urine specimens from a representative sample of the Chinese and African populations at risk would be stored. Once diagnosed with liver cancer, cases would be matched to appropriate controls and their stored samples would be analyzed for AFB_1 or AFB_1–DNA adducts. The predictive value of the specific markers thus could be determined in biologic samples collected prior to clinical disease and, therefore, not subject to the concern that the markers rather than the disease reflect events on the causal pathway. Tumor tissue would be analyzed for both

AFB_1–DNA adducts and $p53$ alterations. A recent example of this approach has revealed relative risks of 6.2 for ABB_1 metabolites and 4.9 for AFB_1-guanine adducts in urine and liver cancer, as well as an interaction between hepatitis B infection and aflatoxin-related biomarkers (Ross *et al.*, 1992).

Acetylation Phenotype

Exposure to carcinogenic arylamines, such as 4-aminobiphenyl and 2-naphthylamine, results from several sources including cigarette smoke, diet, and occupation, and is associated with bladder cancer. Arylamines are activated by N-hydroxylation, primarily in the liver. N-Acetylation is a competing detoxification reaction catalyzed by N-acetyltransferase. This enzyme is noninducible and under autosomal dominant genetic control. Slow acetylators are homozygous for the slow acetylator gene and fast acetylators are either heterozygous or homozygous for the fast gene.

Acetylation phenotype can be measured by administration of several different compounds, including sulfamethazine, an antibiotic, and caffeine, given as a measured dose of coffee (Blum *et al.*, 1991). The molar ratio of acetylated and nonacetylated metabolites in urine or blood is then determined (reviewed in Weber and Hein, 1985). The identification of the primary mutations (M1 and M2) in two alleles of the gene for the isozyme NAT2, which account for 90% of slow acetylator phenotypes in European populations, has allowed the development of a simple DNA amplification assay to genotype individuals (Blum *et al.*, 1991).

Slow acetylators are at high risk of bladder cancer, especially those individuals who are exposed occupationally (Cartwright *et al.*, 1982; Evans *et al.*, 1983; Karkaya *et al.*, 1986; Mommsen and Aagaard, 1986). In a study of smokers of blond and black tobacco, slow acetylators had higher levels of 4-aminobiphenyl hemoglobin adducts for the same type and quantity of cigarettes smoked than fast acetylators (Bartsch *et al.*, 1990; Vineis *et al.*, 1990). The Bartsch study also measured the rate of N-hydroxylation of caffeine; those with fast N-oxidation and slow N-acetylation phenotype had the highest hemoglobin adduct levels. Thus, determination of both phenotypes may provide a better prediction of risk.

Increased risk of fast acetylators for colon cancer has been seen in two studies with odds ratios of 2.5–3.8 (Ilett *et al.*, 1987; Wohlleb *et al.*, 1990). Although it is assumed that N-acetylation is a detoxification step, a potential activation role exists involving the formation of N-acetoxy arylamines, either by O-acetylation of N–OH arylamines or by N,O-acetyltransfer of arylhydroxamic acids (Kirlin *et al.*, 1991). However, a lack of association of acetylator phenotype with colon cancer also has been reported (Ladero *et al.*, 1991). Studies of association with lung, breast, and lymphoid cancers also have been negative (reviewed in Ladero *et al.*, 1991).

Because the interaction of multiple genetic factors can influence DNA and protein adduct levels, future studies should investigate as many of these factors as possible. For example, activation and detoxifying phenotypes could be studied in the same individual and correlated with adducts. The development of methods to determine individual DNA repair capacity is also an important goal of future studies and should provide further insight into individual risk for cancer development.

Strengths/Limitations

As shown in Tables 11.1–11.3, biologic markers have been studied in three types of populations: those with defined environmental (including occupational) exposures to carcinogens (cigarette smokers, workers, persons with certain dietary exposures), clinical populations with exposure to carcinogenic drugs, and cancer patients. Each one of the methods listed has shown elevated levels of the particular marker in at least one of the exposed populations. Two consistent findings have been (1) significant interindividual variation in levels of markers among persons with comparable exposure and (2) an apparently important contribution of background exposures manifest in so-called unexposed controls.

Although of small scale and generally limited design, this body of research has demonstrated that most of the methods are adequately sensitive for human studies and can provide valuable information on potential group and individual risks. However, the results often are limited by technical variability in the assays, small sample sizes, lack of appropriate controls, failure to account for confounding variables, and a scarcity of data on exposure. The three case studies reinforce these initial impressions.

Research Needs

In a generic sense, the most immediate research need is the thorough validation of biomarkers that have demonstrated the greatest relevance to carcinogenesis. Validation is the process by which certain basic requirements or criteria for biologic markers are met through a combination of experimental and human studies. Criteria for validation follow.

1. The fundamental characteristics of the assay must be established. Low dose sensitivity and reproducibility are key concerns. Within-laboratory variability also must be established, so that differences in marker concentration are not attributed erroneously to intra- or interindividual variation. There should be confirmation of assay results by other methods. An important tool in understanding the characteristics of an individual assay is the use of corroborative methods on the same sample. The extent to which the marker is chemical or exposure specific (i.e., selective for a particular chemi-

cal, exposure source, or period of exposure) should be known. This information determines applicability and interpretation of results. Biologic markers can integrate exposure by multiple routes (inhalation, oral, dermal), from multiple sources (ambient and indoor air, workplace air, cigarette smoke, diet, drinking water), and across all patterns of exposure (past, current, intermittent, continuous). This is an advantage, since risks can be assumed to be additive. However, it is also a disadvantage because many environmental chemicals are ubiquitous and it is difficult to distinguish the effect of any particular exposure source. Markers vary greatly with respect to source specificity; for example, 4-aminobiphenyl–hemoglobin (4-ABP–Hb) has far fewer non-cigarette-smoking-related "background" sources than do ethylene oxide (EtO), PAH, and *N*-nitroso compounds. The extent to which the marker will document specific time periods of exposure will depend on the pharmacokinetics of the chemical and the persistence of the marker in the biologic sample assayed (itself a function of the turnover rate of the sample and repair processes).

The same criteria of adequate sensitivity, specificity, and predictive value that apply to the validation of screening methods should be met by biomarkers. Is the marker able to distinguish truly exposed (or unexposed) and truly "at-risk" (or not at risk) individuals with a high degree of accuracy? To answer this question, calculation of false positive and false negative rates and positive and negative predictive values should be based on a series of independent samples from the population (Griffith *et al.*, 1992). Although considerable progress has been made in evaluating the relationship between exposure and various biomarkers, their ability to predict risk of cancer has not been assessed in prospective studies, either in laboratory animals or in humans. This is a glaring research need that can best be met by prospective animal bioassays and nested case-control studies that take advantage of banks of stored samples from human subjects followed longitudinally to determine cancer risk.

Finally, how feasible is the marker? That is, how acceptable is it to the public, how cost effective is it, and how stable is it in stored samples?

2. The dose–response relationship and the extent of interindividual and intraindividual variability in humans must be characterized. It is often assumed that to be valid, a marker of exposure or biologically effective dose must be highly correlated with an individual's estimated exposure. This assumption ignores the fact that the search for biologic markers was triggered in large part by the awareness of large differences in individual processing of xenobiotics. Moreover, if this assumption were true, there would be little point in collecting biologic information. Indeed, all evidence to date suggests that there is a high degree of variability in biomarker levels among persons smoking the same amount, among individuals working in the same plant, and even among chemotherapy patients treated with standardized doses of chemotherapy agents (Perera *et al.*, 1991), probably because of variability

in metabolic processing of xenobiotics or repair, as well as because of errors in estimating individual exposure. Biomarkers have significant potential to document interindividual variability in biologic response to environmental agents—hence in potential risk of cancer. They can therefore enhance the power of epidemiologic studies of cancer causation and increase the precision of quantitative risk assessment. In so doing, biomarkers can guide and inform strategies to prevent cancer. This potential provides strong motivation for the required validation efforts.

Acknowledgment

This research was supported by NIEHS Program Project Grant 1 PO1 ES05294.

References

Albertini, R. J. (1985). Somatic gene mutations *in vivo* as indicated by 6-thioguanine-resistant T-lymphocytes in human blood. *Mutat. Res.* 150, 411–422.

Ashford, N. A. (1986). Policy considerations for human monitoring in the workplace. *J. Occup. Med.* 28, 563–568.

Autrup, H., Bradley, K. A., Shamsuddin, A. K. M., Wakhisi, J., and Wasunna, A. (1983). Detection of putative adduct with fluorescence characteristics identical to 2,3-dihydro-2-(7guanyl)-3-hydroxy aflatoxin B1 in human urine collected in Murang'a District, Kenya. *Carcinogenesis* 4, 1193–1195.

Ayesh, R., Idle, J. R., Ritchie, J. C., Crothers, M. J., and Hetzel, M. R. (1984). Metabolic oxidation phenotypes as markers for susceptibility to lung cancer. *Nature (London)* 312, 169–170.

Bartsch, H., Caporaso, N., Coda, M., Kadlubar, F., Malaveille, C., Skipper, P., Talasaka, G., and Tannenbaum, S. R. (1990). Carcinogen hemoglobin adducts, urinary mutagenicity, and metabolic phenotype in active and passive cigarette smokers. *JNCI J. Natl. Cancer Inst.* 82, 1826–1831.

Bigbee, W. L., Wyrobek, A. J., Langlois, R. G., Jensen, R. H., and Everson, B. J. (1990). The effect of chemotherapy on the *in vivo* frequency of glycophorin A 'null' variant erythrocytes. *Mutat. Res.* 240, 165–175.

Block, G. The data support a role for antioxidants in reducing cancer risk. *Nutrition Rev.* 50, 207–213, 1992.

Blum, M., Demierre, A., Grant, D. M., Hein, M., and Meyer, U. A. (1991). Molecular mechanism of slow acetylation of drugs and carcinogens in humans. *Proc. Natl. Acad. Sci. U.S.A.* 88, 5237–5241.

Bos, R. P., and Jongeneelen, F. J. (1988). Nonselective and selective methods for biological monitoring of exposure to coal tar products. *In* "Methods for Detecting DNA Damaging Agents in Humans: Applications in Cancer Epidemiology and Preventions" (H. Bartsch, K. Hemminki, and I. K. O'Neill, eds.). IARC, Lyon, France.

Brandt-Rauf, P. W. (1988). New markers for monitoring occupational cancer: The example of oncogene proteins. *J. Occup. Med.* 30, 399–404.

Bressac, B., Kew, M., Wands, J., and Ozturk, M. (1991). Selective G to T mutations of p53 gene in hepatocellular carcinoma from southern Africa. *Nature (London)* 350, 429–431.

Bryant, M. S., Skipper, P. L., and Tannenbaum, S. R. (1987). Hemoglobin adducts of 4-aminobiphenyl in smokers and nonsmokers. *Cancer Res.* 47, 602–608.

Bryant, M. S., Vineis, P., Skipper, P. L., and Tannebaum, S. R. (1988). Hemoglobin adducts of aromatic amines: Associations with smoking status and type of tobacco. *Proc. Natl. Acad. Sci. U.S.A.* **85**, 9788–9791.

Calleman, C. J., Cherenberg, L., Jansson, B., Osterman-Golkar, S., Segerback, D., Svensson, K., and Wachtmeister, C. A. (1978). Monitoring and risk assessment by means of alkyl groups in hemoglobin in persons occupationally exposed to ethylene oxide. *J. Environ. Pathol. Toxicol.* **2**, 427–442.

Caporaso, N. E., Tucker, M. A., Hoover, R. N., Hayes, R. B., Pickle, L. W., Issaq, H. J., and Muschik, G. M. (1990). Lung cancer and the debrisoquine metabolic phenotype. *JNCI J. Natl. Cancer Inst.* **82**, 1264–1272.

Carmella, S. G., Kagan, S. S., Kagan, M., Foiles, P. G., Palladino, G., Quart, A. M., Quart, E., and Hecht, S. S. (1990). Mass spectrometric analysis of tobacco-specific nitrosamine hemoglobin adducts in snuff dippers, smokers, and nonsmokers. *Cancer Res.* **50**, 5438–5445.

Carrano, A. V., and Moore, D. H. (1982). The rationale and methodology for quantifying sister chromatid exchange frequency in humans. In "Mutageneticity: New Horizons in Genetic Toxicology" (J. A. Heddle, ed.), pp. 267–304. Academic Press, New York.

Cathcart, R., Schweirs, E., Saul, R. L., and Ames, B. N. (1984). Thymine glycol and thymidine glycol in human and rat urine: A possible assay for oxidative DNA damage. *Proc. Natl. Acad. Sci. U.S.A.* **81**, 5633–5637.

Cartwright, R. A., Glashan, R. W., Rogers, J. J., Barham-Hall, D., Ahmad, R. A., Higgins, E., and Kahan, M. A. (1982). The role of N-acetyltransferase in bladder carcinogenesis: A pharmacogenetic epidemiological approach to bladder cancer. *Lancet* **2**, 842–845.

Cohen, L. (1987). Diet and cancer. *Sci. Am.* **257**, 42–48.

Crespi, C. L., Penman, B. W., Gelboin, H. V., and Gonzalez, F. J. (1991). A tobacco smoke-derived nitrosamine, 4-(methylnitrosamino)-1-(3pyridyl)-1-butanone, is activated by multiple human cytochrome P450s including the polymorphic human cytochrome P4502D6. *Carcinogenesis* **12**(7), 1197–1201.

Cuzick, J., Routledge, M. N., Jenkins, D., and Garner, R. C. (1990). DNA adducts in different tissues of smokers and non-smokers. *Int. J. Cancer* **45**, 673–678.

Evans, D. A. P., Eze, L. C., and Whibley, E. J. (1983). The association of the slow acetylator phenotype with bladder cancer. *J. Med. Genet.* **20**, 330–333.

Evans, J. (1982). Cytogenetic studies on industrial populations exposed to mutagens. In "Indicators of Genotoxic Exposure" (B. A. Bridges, B. E. Butterworth, and I. B. Weinstein, eds.). Banbury Report No. 13. Cold Spring Harbor Press, Cold Spring, New York.

Everson, R. B., Randerath, E., Santella, R. M., Cefalo, R. C., Avitts, T. A., and Randerath, K. (1986). Detection of smoking-related covalent DNA adducts in human placenta. *Science* **231**, 54–57.

Farmer, P. B., Bailey, E., Gorg, S. M., Tornqvist, M., Osterman-Golkar, S., Kautiiaine, A., and Lewis-Enright, D. P. (1986). Monitoring human exposure to ethylene oxide by the determination of haemoglobin adducts using gas chromatography-mass spectrometry. *Carcinogenesis* **7**, 637–640.

Fichtinger-Schepman, A. M. J., van Oosterom, A. T., Lohman, P. H. M., and Berends, F. (1987). Interindividual human variation in *cis*-platinum sensitivity, predictable in an *in vitro* assay? *Mutat. Res.* **190**, 59–62.

Fichtinger-Schepman, A. M. J., van der Velde-Visser, S. D., van Dijk-Knijnenburg, H. C. M., van Oosterom, A. T., Baan, R. A., and Berends, F. (1990). Kinetics of the formation and removal of cisplatin–DNA adducts in blood cells and tumor tissue of cancer patients receiving chemotherapy: Comparison with *in vitro* adduct formation. *Cancer Res.* **50**, 7887–7894.

Foiles, P. G., Miglietta, L. M., Akerkar, S. A., Everson, R. B., and Hecht, S. S. (1988). Detection of O6-methyldeoxyguanosine in human placental DNA. *Cancer Res.* **48**, 4184–4188.

Garland, W. A., Kuenzing, W., Rubio, F., Kornychuk, H., Norkus, E. P., and Conney, A. H.

(1986). Urinary excretion of nitrosodimethylamine and nitrosoproline in humans: Interindividual differences and the effect of administered ascorbic acid and α-tocopherol. *Cancer Res.* **46**, 5392–5400.

Griffith, J., Duncan, R. C., Goldsmith, J. R., and Hulka, B. S. (1989). Biochemical and biological markers: Implications for epidemiologic studies. *Arch. Environ. Health* **44(6)**, 375–381.

Groopman, J. D., Donahue, P. R., Zhu, J., Chen, J., and Wogan, G. N. (1985). Aflatoxin metabolism and nucleic acid adducts in urine by affinity chromatography. *Proc. Natl. Acad. Sci. U.S.A.* **82**, 6492–6496.

Gupta, R. C. (1985). Enhanced sensitivity of ^{32}P-postlabeling analysis of aromatic carcinogen: DNA adducts. *Cancer Res.* **45**, 5656–5662.

Gupta, R. C., Reddy, M. V., and Randerath, K. (1982). [^{32}P]-Postlabeling analysis of nonradioactive aromatic carcinogen-DNA adducts. *Carcinogenesis* **3**, 1081–1092.

Harris, C. C., Vahakangas, K., Newman, J. M., Trivers, G. E., Shamsuddin, A., Sinopoli, N., Mann, D. L., and Wright, W. E. (1985). Detection of benzo[*a*]pyrene diol epoxide–DNA adducts in peripheral blood lymphocytes and antibodies to the adducts in serum from coke oven workers. *Proc. Natl. Acad. Sci. U.S.A.* **82**, 6672–6676.

Harris, C. C., Weston, A., Willey, J., Trivers, G., and Mann, D. (1987). Biochemical and molecular epidemiology of human cancer: Indicators of carcinogen exposure, DNA damage, and genetic predisposition. *Environ. Health Perspect.* **75**, 109–119.

Haugen, A., Becher, G., Benestad, C., Vahakangas, K., Trivers, G. E., Newman, M. J., and Harris, C. C. (1986). Determination of polycyclic aromatic hydrocarbons in the urine, benzo[*a*]pyrene diol epoxide–DNA adducts in lymphocyte DNA, and antibodies to the adducts in sera from coke oven workers exposed to measured amounts of polycyclic aromatic hydrocarbons in work atmosphere. *Cancer Res.* **46**, 4178–4183.

Heim, M., and Meyer, U. A. (1990). Genotyping of poor metabolisers of debrisoquine by allele-specific PCR amplification. *Lancet* **336**, 529–532.

Hemminki, K., Grzybowska, E., Chorazy, M., Twardowska-Saucha, K., Sroczynski, J. W., Putnam, K. L., Randerath, K., Phillips, D. H., and Hewer, A. (1990a). Aromatic DNA adducts in white blood cells of coke workers. *Int. Arch. Occup. Environ. Health* **62**, 467–470.

Hemminki, K., Grzybowska, E., Chorazy, M., Twardowska-Saucha, K., Sroczynski, J. W., Putnam, K. L., Randerath, K., Phillips, D. H., Hewer, A., Santella, R. M., Young, T. L., and Perera, F. P. (1990b). DNA adducts in humans environmentally exposed to aromatic compounds in an industrial area of Poland. *Carcinogenesis* **11**, 1229–1231.

Hemstreet, G. P., Schulte, P. A., Ringen, K., Stringer, W., and Altekruse, E. B. (1988). DNA hyperploidy as a marker for biological response to bladder carcinogen exposure. *Int. J. Cancer* **42(b)**, 817–20.

Herbert, R., Marcus, M. W., Wolffe, M., Perera, F. P., Andrews, L., Godbold, J. H., Stefanidis, M., Rivera, M., Lu, X., Landrigan, P. J., and Santella, R. M. (1990). Detection of adducts of deoxyribonucleic acid in white blood cells of roofers by ^{32}P-postlabeling: Relationship of adduct levels to measures of exposure to polycyclic aromatic hydrocarbons. *Scand. J. Work Environ. Health* **16**, 135–143.

Higginson, J. (1988). Changing concepts in cancer prevention: Limitations and implications for future research in environmental carcinogenesis. *Cancer Res.* **48**, 1381–1389.

Hoffmann, D., and Hecht, S. S. (1985). Nicotine-derived *N*-nitroamines and tobacco-related cancer: Current staus and future direction. *Cancer Res.* **45**, 935–944.

Hogstedt, B., Akesson, B., Axell, K., Gullberg, B., Mitelman, F., Pero, R. W., Skerving, S., and Welinder, H. (1983). Increased frequency of lymphocyte micronuclei in workers producing reinforced polyester resin with low exposure to styrene. *Scand. J. Work Environ. Health* **49**, 271–276.

Holz, O., Krause, T., Scherer, G., Schmidt-Preub, U., and Rudiger, H. W. (1990). ^{32}P-Postlabeling analysis of DNA adducts in monocytes of smokers and passive smokers. *Int. Arch. Occup. Environ. Health* **62**, 299–303.

Hsieh, L. L., Hsu, S. W., Chen, D. S., and Santella, R. M. (1988). Immunological detection of aflatoxin B1-DNA adducts formed *in vivo*. *Cancer Res.* **48**, 6328–6331.

Hsu, I. C., Metcalf, R. A., Sun, T., Welsh, J., Wang, N. J., and Harris, C. C. (1991). p53 Gene mutational hotspot in human hepatocellular carcinomas from Qidong, China. *Nature (London)* **350**, 427–428.

Ilett, K. F., David, B. M., Dethon, P., Castleden, W. M., and Kwa, R. (1987). Acetylation phenotype in colorectal carcinoma. *Cancer Res.* **47**, 1466–1469.

Jahnke, G. D., Thompson, C. L., Walker, M. P., Gallagher, J. E., Lucier, G. W., and DiAugustine, R. P. (1990). Multiple DNA adducts in lymphocytes of smokers and nonsmokers determined by ^{32}P-postlabeling analysis. *Carcinogenesis* **11**, 205–211.

Jensen, R. H., Langlois, R. G., and Bigbee, W. L. (1986). Determination of somatic mutations in human erythrocytes by flow cytometry. *In* "Genetic Toxicology of Environmental Chemicals. Part B. Genetic Effects and Applied Mutagenesis" (C. Ramel, B. Lambert, and J. Magnusson, eds.), pp. 177–184. Liss, New York.

Karkaya, A. E., Cok, L., Sardas, S., Gogus, O., and Sardar, O. S. (1986). *N*-Acetyltransferase phenotype of patients with bladder cancer. *Hum. Toxicol.* **5**, 333–335.

Kawajiri, K., Nakachi, K., Imai, K., Yoshiii, A., Shinoda, N., and Watanabe, J. (1990). Identification of genetically high-risk individuals to lung cancer by DNA polymorphisms of the cytochrome P450 1A1 gene. *FEBS Lett.* **263**, 131–133.

Kellerman, G., Shaw, C. R., and Luyten-Kellermann, M. (1973). Aryl hydrocarbon hydroxylase inducibility and bronchiogenic carcinoma. *N. Engl. J. Med.* **289**, 934–937.

Kelsey, J. L., and Berkowitz, G. S. (1988). Breast cancer epidemiology. *Cancer Res.* **48**, 5615–5623.

Kirlin, W. G., Ogolla, F., Andrews, A. F., Trinidad, A., Ferguson, R. J., Yerokun, T., Mpezo, M., and Hein, D. W. (1991). Acetylator genotype-dependent expression of arylamine *N*-acetyltransferanse in human colon cytosol from non-cancer and colorectal cancer patients. *Cancer Res.* **51**, 549–555.

Kyoizumi, S., Nakamura, N., Hakoda, M., Awa, A. A., Bean, M. A., Jensen, R. H., and Akiyama, M. (1989). Detection of somatic mutations at the glycophorin A locus in erythrocytes of atomic bomb survivors using a single beam flow sorter. *Cancer Res.* **49**, 581–588.

Ladero, J. M., Gonzalez, J. F., Benitex, J., Vargas, E., Fernandez, M. J., Baki, W., and Diaz-Rubio, M. (1991). Acetylator polymorphism in human colorectal carcinoma. *Cancer Res.* **51**, 2098–2100.

Langlois, R. G., Bigbee, W. L., Koizumi, S., Nakamura, N., Bean, M. A., Akiyama, M., and Jensen, R. H. (1987). Evidence for increased somatic cell mutations at the glycophorin A locus in atom bomb survivors. *Science* **236**, 445–448.

Lee, B. M., Baoyun, Y., Herbert, R., Hemminki, K., Perera, F. P., and Santella, R. M. (1991). Immunologic measurement of polycyclic aromatic hydrocarbon-albumin adducts in foundry workers and roofers. *Scand. J. Work Environ. Health* **17**, 190–194.

Li, F. P. (1990). Familial cancer syndromes and clusters. *Curr. Prob. Cancer* **49**, 75–113.

Lu, S.-H., Ohsima, H., Fu, H.-M., Tian, Y., Li, F.-M., Blettner, M., Wahrendorf, J., and Bartsch, H. (1986). Urinary excretion of *N*-nitrosamino acids and nitrate by inhabitants of high- and low-risk areas for esophageal cancer in northern China: Endogenous formation of nitrosoproline and its inhibition by vitamin C. *Cancer Res.* **46**, 1486–1491.

McGinnis, M. J., Falta, M. T., Sullivan, L. M., and Albertini, R. J. (1990). *In vivo hprt* mutant frequencies in T-cells of normal human newborns. *Mutat. Res.* **240**, 117–126.

Maclure, M., Bryant, M. S., Skipper, P. L., and Tannenbaum, S. R. (1990). Decline of the hemoglobin adduct of 4-aminobiphenyl during withdrawal from smoking. *Cancer Res.* **50**, 181–184.

Manchester, D. K., Wilson, V. L., Hsu, I.-C., Choi, J.-S., Parker, N. B., Mann, D. L., Weston, A., and Harris, C. C. (1990). Synchronous fluorescence spectroscopic, immunoaffinity chromatographic, and ^{32}P-postlabeling analysis of human placental DNA known to contain benzo[*a*]pyrene diol epoxide adducts. *Carcinogenesis* **11**, 553–559.

Marshall, C. J. (1991). Tumor suppressor genes. *Cell* **64**, 313–326.

Marx, J. (1989). Detecting mutations in human genes. *Science* **243**, 738–738.

Marx, J. (1991). Zeroing in on individual cancer risk. *Science* **253**, 612–616.

Messing, K., Seifert, A. M., and Bradley, W. E. C. (1986). *In vivo* mutant frequency of technicians professionally exposed to ionizing radiation. *In* "Monitoring of Occupational Genotoxicants" (M. Sorsa and H. Norppa, eds.), pp. 87–97. Liss, New York.

Miller, E. C., and Miller, J. A. (1981). Mechanisms of chemical carcinogenesis. *Cancer Res.* **47**, 1055–1064.

Mommsen, S., and Aagaard, J. (1986). Susceptibility in urinary bladder cancer: Acetyltransferase phenotypes and related risk factors. *Cancer Lett.* **32**, 199–205.

Murray, S., and Gooderham, N. J. (1989). Detection and measurement of MeIQx in human urine after digestion of a cooked meat meal. *Carcinogenesis* **10**, 763–765.

Mustonen, R., Hemminki, K., Alhonene, A., Hietanen, P., and Kiilunen, M. (1988). Determination of *cis*-diamminedichloroplatinum (II) in blood compartments of cancer patients. *In* "Methods for Detecting DNA Damaging Agents in Man: Application in Cancer Epidemiology and Prevention." (H. Bartsch, K. Hemminki, and I. K. O'Neill, eds.), pp. 329–332. International Agency for Research on Cancer, Lyon, France.

National Research Council (1982). "Diet, Nutrition, and Cancer." National Academy Press, Washington, D.C.

Neumann, H. G. (1984). Dosimetry and dose-response relationships. *In* IARC Sci. Publ. (A. Berlin, M. Draper, K. Hemminki, H. Vainio, eds.) **59**, 115–126. Lyon France.

Olivero, O. A., Huitfedt, H., and Poirier, M. C. (1990). Chromosome site-specific immunohistochemical detection of DNA adducts in N-acetoxy-2-acetylaminofluorene-exposed Chinese hamster ovary cells. *Mol. Carc.* **3**, 37–43.

O'Neill, J. P., McGinniss, M. J., Berman, J. K., Sullivan, L. M., Nicklas, J. A., and Albertini, R. J. (1987). Refinement of a T-lymphocyte cloning assay to quantify the *in vivo* thioguanine-resistant mutant frequency in humans. *Mutagenesis* **2**, 87–94.

Osterman-Golkar, S., Bailey, E., Farmer, P. B., Gorf, S. M., and Lamb, J. H. (1984). Monitoring exposure to propylene oxide through the determination of hemoglobin alkylation. *Scand. J. Work Environ. Health* **10**, 99–102.

Ostrosky-Wegman, P., Montero, R., Palao, A., Cortinas de Nava, C., Hurtado, F., and Albertini, R. J. (1990). 6-Thioguanine-resistant T-lymphocyte autoradiographic assay. Determination of variation frequencies in individuals suspected of radiation exposure. *Mutat. Res.* **232**, 49–52.

Parker, R. J., Gill, I., Tarone, R., Vionnet, J. A., Grunberg, S., Muggia, F. M., and Reed, E. (1991). Platinum-DNA damage in leukocyte DNA of patients receiving carboplating and cisplatin chemotherapy, measured by atomic absorption spectrometry. *Carcinogenesis* **12**, 1253–1258.

Perera, F. P., Poirier, M. C. Yuspa, S. H., Nakayama, J., Jaretzki, A., Curnen, M. M., Knowles, D. M., and Weinstein, I. B. (1982). A pilot project in molecular cancer epidemiology: Determination of benzo[*a*]pyrene–DNA adducts in animal and human tissues by immunoassays. *Carcinogenesis* **3**, 1405–1410.

Perera, F. P., Santella, R. M., Brenner, D., Poirier, M. C., Munshi, A. A., Fischman, H. K., and VanRyzin, J. (1987). DNA adducts, protein adducts and SCE in cigarette smokers and nonsmokers. *JNCI J. Natl. Cancer Inst.* **79**, 449–456.

Perera, F. P., Hemminki, K., Young, T. L., Santella, R. M., Brenner, D., and Kelly, G. (1988a). Detection of polycyclic aromatic hydrocarbon–DNA adducts in white blood cells of foundry workers. *Cancer Res.* **48**, 2288–2291.

Perera, F. P., Santella, R. M., Brenner, D., Young, T. L., and Weinstein, I. B. (1988b). Application of biological markers to the study of lung cancer causation and prevention. *In* "Methods for Detecting DNA Damaging Agents in Humans: Applications in Cancer Epidemilogy and Prevention" (H. Bartsch, K. Hemminki, I. K. O'Neill, eds.), *IARC* **89**, 451–459.

Perera, F., Mayer, J., Jaretzki, A., Hearne, S., Brenner, D., Young, T. L., Fishman, H., and Grimes, M. (1989). Comparison of DNA adducts and sister chromatid exchange in lung cancer cases and controls. *Cancer Res.* **49**, 4446–4451.

Perera, F. P., Mayer, J., Santella, R. M., Brenner, D., Jeffrey, A., Latriano, L., Smith, S., Warburton, D., Young, T. L., Tsai, W. Y., Hemminki, K., and Brandt-Rauf, P. (1991). Biological markers in risk assessment for environmental carcinogens. Environ. Health *Perspect.* **90,** 247–254.

Perera, F. P., Hemminki, K., Gryzbowska, E., Motykiewicz, G., Michalska, J., Santella, R. M., Young, T., Dickey, C., Brandt-Rauf, P., DeVivo, I., Blaner, W., Tsai, W., and Chorazy, M. (1992a). Molecular and genetic damage in humans from environmental pollution in Poland. *Nature* **360, 19,** 256–258.

Perera, F. P., Motzer, R., Tang, D., Reed, E., Parker, R., Warburton, D., O'Neill, P., Albertini, R., Bigbee, W., Jensen, R., Santella, R., Tsai, W., Simon-Cereijido, G., Randall, C., and Bosl, G. (1992b). Multiple Biological Markers in Germ Cell Tumor Patients Treated with Platinum-based Chemotherapy. *Cancer Res.* **52,** 3558–3565.

Pero, R. W., Bryngelesson, T., Widergren, B., Hogstedt, B., and Welinder, H. (1982). A reduced capacity for unscheduled DNA synthesis in lymphocytes from individuals exposed to propylene oxide and ethylene oxide. *Mutat. Res.* **104,** 193–200.

Phillips, D. H., Hewer, A., and Grover, P. L. (1986). Aromatic DNA adducts in human bone marrow and peripheral blood leukocytes. *Carcinogenesis 7,* 2071–2075.

Phillips, D. H., Hemminki, K., Alhonen, A., Hewer, A., and Grover, P. L. (1988a). Monitoring occupational exposure to carcinogens: Detection by ^{32}P-postlabeling of aromatic DNA adducts in white blood cells from iron foundry workers. *Mutat. Res.* **204,** 531–541.

Phillips, D. H., Hewer, A., Martin, C. N., Garner, R. C., and King, M. M. (1988b). Correlation of DNA adduct levels in human lung with cigarette smoking. *Nature (London)* **336,** 790–792.

Phillips, D. H., Hewer, A., Malcolm, A. D. B., Ward, P., and Coleman, D. V. (1990a). Smoking and DNA damage in cervical cells. *Lancet* **335,** 417.

Phillips, D. H., Schoket, B., Hewer, A., Bailey, E., Kostic, S., and Vincze, I. (1990b). Influence of cigarette smoking on the levels of DNA adducts in human bronchial epithelium and white blood cells. *Int. J. Cancer* **46,** 569–575.

Poirier, M., Santella, R., Weinstein, I. B., Grunberger, D., and Yuspa, S. H. (1980). Quantitation of benzo(a)pyrene-deoxyguanosine adducts by radioimmunoassay. *Cancer Res.* **40,** 412–416.

Randerath, D., Reddy, M. V., and Gupta, R. C. (1981). [^{32}P]-Labeling test for DNA damage. *Proc. Natl. Acad. Sci. U.S.A.* **78,** 6126–6129.

Randerath, E., Avitts, T. A., Reddy, M. V., Miller, R. H., Everson, R. B., and Randerath, K. (1986). Comparative ^{32}P-postlabeling analysis of cigarette smoke-induced DNA damage in human tissues and mouse skin. *Cancer Res.* **46,** 5869–5877.

Randerath, E., Miller, R. H., Mittal, D., Avitts, T. A., Dunsford, H. A., and Randerath, K. (1989). Covalent DNA damage in tissues of cigarette smokers as determined by ^{32}P-postlabeling assay. *JNCI J. Natl. Cancer Inst.* **81,** 341–347.

Reddy, M. V., and Randerath, K. (1986). Nuclease P1-mediated enhancement of sensitivity of [^{32}P]-postlabeling test for structurally diverse DNA adducts. *Carcinogenesis* **7,** 1543–1551.

Reed, E., Yuspa, S., Zwelling, L. A., Oxols, R. F., and Poirier, M. C. (1986). Quantitation of *cis*-diamminedichloroplatinum (II) (cisplatin)-DNA-intrastrand adducts in testicular and ovarian cancer patients receiving cisplatin chemotherapy. *J. Clin. Invest.* **77,** 545–550.

Reed, E., Ozols, R. F., Tarone, R., Yuspa, S. H., and Porier, M. C. (1988). The measurement of cisplatin–DNA adduct level in testicular cancer patients. *Carcinogenesis* **9,** 1909–1911.

Ross, R. K., Yuan, J. M., Yu, M. C., Wogan, G. N., Qian, G. S., Tu, J. P., Groopman, J. D., Gao, Y. T. and Henderson, B. E. Urinary aflatoxin biomarkers and risk of hepatocellular carcinoma (1992). *Lancet,* **339,** 943–946.

Rothman, N., Poirier, M. C., Baser, M. E., Hansen, J. A., Gentile, C., Bowman, E. D., and Strickland, P. T. (1990). Formation of polycyclic aromatic hydrocarbon–DNA adducts in peripheral white blood cells during consumption of charcoal-broiled beef. *Carcinogenesis* 11, 1241–1243.

Santella, R. M. (1988). Application of new techniques for the detection of carcinogen adducts to human population monitoring. *Mutat. Res.* 205, 271–282.

Santella, R. M., Gasparro, F. P., and Hsieh, L. L. (1987). Quantitation of carcinogen-DNA adducts with monoclonal antibodies. *Prog. Exp. Tumor Res.* 31, 63–75.

Santella, R. M. (1992). DNA adducts in humans as biomarkers of exposure to environmental and occupational carcinogens. *Environ. Carc. Rev.*

Santella, R. M., Grinberg-Funes, R. A., Young, T. L., Dickey, C., Singh, V. N., Wang, L. W., and Perera, F. P. Cigarette smoking related polycyclic aromatic hydrocarbon-DNA adducts in peripheral mononuclear cells. *Carcinogenesis,* in press.

Santella, R. M., Weston, A., Perera, F. P., Trivers, G. T., Harris, C. C., Young, T. L., Nguyen, D., Lee, B. M., and Poirier, M. C. (1988). Interlaboratory comparison of antisera and immunoassays for benzo(*a*)pyrene-diol-epoxide-I-modified DNA. *Carcinogenesis* 9, 1265–1269.

Savela, K., and Hemminki, K. (1991). DNA adducts in lymphocytes and granulocytes of smokers and nonsmokers detected by the ^{32}P-postlabeling assay. *Carcinogenesis* 12(3), 503–508.

Schoket, B., Phillips, D. H., Hewer, A., and Vincze, I. (1991). ^{32}P-Postlabeling detection of aromatic DNA adducts in peripheral blood lymphocytes from aluminum production plant workers. *Mutat. Res.* 260(I), 89–98.

Schulte, P. A. (1989). A conceptual framework for the validation and use of biologic markers. *Environ. Res.* 48, 129–144.

Seidegard, J., Pero, R. W., Miller, D. G., and Beattie, E. J. (1986). A glutathione transferase in human leukocytes as a marker for the susceptibility to lung cancer. *Carcinogenesis* 7, 751–753.

Seidegard, J., Pero, R. W., Markowitz, M. M., Roush, G., Miller, D. G., and Beattie, E. J. (1990). Isoenzyme(s) of glutathione transferase (class Mu) as a marker for the susceptibility to lung cancer: A follow up study. *Carcinogenesis* 11, 33–36.

Shamsuddin, A. K. M., Sinopoli, N. T., Hemminki, K., Boesch, R. B., and Harris, C. C. (1985). Detection of benzo[*a*]pyrene: DNA adducts in human white blood clels. *Cancer Res.* 45, 66–68.

Sherson, D., Sabro, P., Sigsgaard, T., Johansen, F., and Autrup, H. (1990). Biological monitoring of foundry workers exposed to polycyclic aromatic hydrocarbons. *Br. J. Ind. Med.* 47, 448–453.

Shuker, D. E. G., and Farmer, P. B. (1988). Urinary excretion of 3-methyladenine in humans as a marker of nucleic acid methylation. *In* "Methods for Detecting DNA Damaging Agents in Humans: Applications in Cancer Epidemiology and Preventions" (H. Bartsch, K. Hemminki, and I. K. O'Neil, eds.), pp. 92–96. IARC Publications, Lyon, France.

Stich, H. F., and Dunn, B. P. (1988). DNA adducts, micronuclei, and leukoplakias as intermediate endpoints in interventions trials. *In* "Methods for Detecting DNA Damaging Agents in Humans: Applications in Cancer Epidemiology and Preventions" (H. Bartsch, K. Hemminki, and I. K. O'Neil, eds.), pp. 661–663. IARC Publications, Lyon, France.

Stowers, S. J., and Anderson, M. W. (1985). Formation and persistence of benzo(*a*)pyrene metabolite–DNA adducts. *Environ. Health Perspect.* 62, 31–39.

Strange, R. C., Matharoo, B., Faulder, G. C., Jones, P., Cotton, W., Elder, J. B., and Deakin, M. (1991). The human glutathione S-transferases: A case-control study of the incidence of the GST1 0 phenotype in patients with adenocarcinoma. *Carcinogenesis* 12, 25–28.

Thilly, W. G. (1985). Potential use of gradient denaturing gel electrophoresis in obtaining mutational spectra from human cells. *Carcinogenesis* 10, 511–528.

Tornqvist, M., Osterman-Golkar, S., Kautiainen, A., Jensen, A., Farmer, P. B., and Ehrenberg, L. (1986). Tissue doses of ethylene oxide in cigarette smokers determined from adduct levels in hemoglobin. *Carcinogenesis* 7, 1519–1521.

Umbenhauer, D., Wild, C. P., Montesano, R., Saffhill, R., Boyle, J. M., Huh, N., Kirstein, U., Thomale, J., Rajewsky, M. F., and Lu, S. H. (1985). O-Methyldeoxyguanosine in oesophageal DNA among individuals at high risk of oesophageal cancer. *Int. J. Cancer* **36**, 661–665.

Vahakangas, K., and Pelkonen, O. (1989). Host-variations in carcinogenic metabolism and repair. In "Genetic Epidemiology of Cancer" (H. T. Lynch and T. Hirayama, eds.), pp. 35–54. CRC Press, Boca Raton, Florida.

Vainio, H., Sorsa, M., and Falck, K. (1984). Bacterial urinary assay in monitoring exposure to mutagens and carcinogens. (A. Berlin, M. Draper, K. Hemminki, and H. Vainio, eds.), pp. 247–258. IARC Publications, Lyon, France.

van Schooten, F. J., Kriek, E., Steenwinkel, M. S. T., Noteborn, H. P. J. M., Hillebrand, M. J. X., and van Leeuwen, F. E. (1987). The binding efficiency of polyclonal and monoclonal antibodies to DNA modified with benzo[a]pyrene diol epoxide is dependent on the level of modification. Implications for quantitation of benzo[a]pyrene adducts *in vivo*. *Carcinogenesis* **8**, 1263–1269.

van Schooten, F., van Leeuwen, F. E., Hillebrand, M. J. X., deRijke, M. E., Hart, A. A. M., van Veen, H. G., Oosterink, S., and Kriek, E. (1990a). Determination of benzo(a)pyrene diol epoxide–DNA adducts in white blood cell DNA from coke-oven workers: the impact of smoking. *JNCI J. Natl. Cancer Inst.* **82**, 927–933.

van Schooten, F. J., Hillebrand, M. J. X., van Leeuwen, F. E., Lutgerink, J. T., vanZandwijk, N., Jansen, H. M., and Kriek, E. (1990b). Polycyclic aromatic hydrocarbon–DNA adducts in lung tissue from lung cancer patients. *Carcinogenesis* **11**, 1677–1681.

van Schooten, F. J., Hillebrand, M. J. X., Scherer, E., den Engelse, L., and Kriek, E. (1991). Immunocytochemical visualization of DNA adducts in mouse tissues and human white blood cells following treatment with benzo[a]pyrene or its diol epoxide. A quantitative approach. *Carcinogenesis* **12**(3), 427–423.

van Sittert, N. J., de Jong, G., Clare, M. G., Davies, R., Dean, B. J., Wren, L. J., and Wright, A. S. (1985). Cytogenetic, immunological, and hematological effects in workers in an ethylene oxide manufacturing plant. *Br. J. Ind. Med.* **42**, 19–26.

Vineis, P., Caporaso, N., Tannenbaum, S. R., Skipper, P. L., Glogowski, J., Bartsch, H., Coda, M., and Talaska, G. F. (1990). Acetylation phenotype, carcinogen-hemoglobin adducts, and cigarette smoking. *Cancer Res.* **50**, 3002–3004.

Walles, S. A. S., Norppa, H., Osterman-Golkar, S., and Maki-Paakkanen, J. (1988). Single-strand breaks in DNA of peripheral lymphocytes of styrene-exposed workers. In "Methods for Detecting DNA Damaging Agents in Humans: Applications in Cancer Epidemiology and Preventions" (H. Bartsch, K. Hemminki, and I. K. O'Neil, eds.), pp. 223–226. IARC Publications, Lyon, France.

Weber, W. W., and Hein, D. W. (1985). N-Acetylation pharmacogenetics. *Pharmacol. Rev.* **37**, 25–79.

Weinstein, I. B., Gatoni-Celli, S., Kirschmeier, P., Lambert, M., Hsiao, W., Backer, J., and Jeffrey, A. (1984). Multistage carcinogenesis involves multiple genes and multiple mechanisms. In "Cancer Cells I. The Transformed Phenotype," pp. 229–237. Cold Spring Harbor Laboratory Press, Cold Spring Harbor, New York.

Weston, A., Trivers, G., Vahakangas, K., Newman, M., and Rowe, M. (1987). Detection of carcinogen–DNA adducts in human cells and antibodies to these adducts in human sera. *Prog. Exp. Tumor Res.* **31**, 76–85.

Weston, A., Rowe, M. I., Manchester, D. K., Farmer, P. B., Mann, D. L., and Harris, C. C. (1989). Fluorescence and mass spectral evidence for the formation of benzo(a)pyrene anti-diol-epoxide–DNA and –hemoglobin adducts in humans. *Carcinogenesis* **10**, 251–257.

Wilcosky, T. C., and Rynard, S. M. (1990). Sister chromatid exchanges. In "Biological Markers in Epidemiology." (B. Hulka, T. C. Wilcosky, and J. D. Griffith, eds.), pp. 28–55. Oxford University Press, New York.

Wild, C. P., Umbenhauer, D., Chapot, B., and Montesano, R. (1986). Monitoring of individual human exposure to aflatoxins (AF) and *N*-nitrosamines (NNO) by immunoassays. *J. Cell Biol.* **30,** 171–179.

Willet, W. C. (1989). The search for the causes of breast and colon cancer. *Nature (London)* **338,** 389–394.

Willet, W. C., and MacMahon, B. (1984). Diet and cancer: An overview. *N. Engl. J. Med.* **310,** 633–703.

Wohlleb, J. C., Hunter, C. F., Blass, B., Kadlubar, F. F., Chu, D. Z. J., and Lang, N. P. (1990). Aromatic amine acetyltransferase as a marker for colorectal cancer: Environmental and demographic associations. *Int. J. Cancer* **46,** 22–30.

Yuspa, S. H., and Poirier, M. C. (1988). Chemical carcinogenesis: From animal models in one decade. *Adv. Cancer Res.* **50,** 25–70.

Zhang, Y.-J., Chen, C-J., Lee, C-S., Haghighi, B., Yang, G.-Y., Wang, L.-W., Feitelson, M., and Santella, R. (1992). Aflatoxin B$_1$-DNA adducts and hepatitis B virus antigens in hepatocellular carcinoma and non-tumorous liver tissue. *Carcinogenesis* 2247–2252.

Zhu, J., Zhang, L., Hu, X., Xiao, Y., Chen, J., Xu, Y., Chu, J. F., and Chu, F. S. (1987). Correlation of dietary aflatoxin B1 levels with excretion of aflatoxin M1 in human urine. *Cancer Res.* **47,** 1848–1852.

12

Infectious Diseases

Lee H. Harrison
and Diane E. Griffin

Introduction

The recent proliferation and availability of biologic markers has facilitated and enhanced the study of the epidemiology of infectious diseases greatly. In this chapter, the types of biologic markers that are used to study infectious diseases are reviewed, with a focus on research applications. In addition, examples of how these tools have been used in the study of the epidemiology of specific infectious diseases is discussed. Although the availability of these new techniques has led to major progress, the interpretation of test results is not always straightforward. Some of the methodologic issues applicable to test interpretation are addressed also.

Overview of Markers Used in Infectious Diseases

Infectious disease researchers and clinicians have a multitude of biologic markers and molecular epidemiologic tools at their disposal. Because of the vast number of available markers, an exhaustive review is beyond the scope of this chapter. What follows, however, is a brief review of the basic types of markers that are currently available, as well as a discussion of their advantages and limitations. Readers interested in exploring specific areas in more detail are directed to the literature referenced in the text.

Biologic markers used in infectious diseases can be divided into three main categories: those that indicate exposure to a pathogen, those that indicate susceptibility to infection, and those that measure infection or illness caused by the pathogen.

Markers of Exposure to Infectious Agents

Exposure to an infectious agent usually elicits an immune response to the agent. Therefore, markers of exposure generally use evidence of humoral or cellular immunity to a specific pathogen. The most commonly used marker is serum antibody, although antibodies can be measured in other body fluids, such as cerebrospinal fluid (CSF), tears, stool, and saliva. All infectious organisms are composed of multiple antigens of varying immunogenicity. In addition, antibodies differ in biologic properties, such as the ability to neutralize infectivity or fix complement. Therefore, antibody assays differ widely in their sensitivity and specificity for different infectious agents. The performance characteristics of each assay must be evaluated using appropriate clinical specimens. Numerous methods are available by which organism-specific antibodies can be measured; some of the most common are discussed here.

Antibody Tests

In the complement fixation test (a hemolytic assay) (Escobar, 1991), patient serum is mixed with a known antigen in the presence of a fixed amount of complement. If antibody specific for the test antigen is present in the patient's serum, the complement is removed from solution by being fixed to the antigen–antibody complex. Sheep red blood cells (RBCs) coated with rabbit serum containing anti-sheep RBC antibodies are then added. Hemolysis does not occur if the complement has been depleted; thus, the lack of hemolysis indicates the presence of antibody in the patient's serum. Not all types of immunoglobulin are capable of fixing complement. Therefore, although this test is performed easily, knowledge of the presence and timing of complement fixing antibodies for the specific organism in question should be kept in mind when interpreting test results.

Enzyme-linked immunosorbent assay (ELISA), also referred to as enzyme immunoassay (EIA), is an extension of earlier technology that used radioisotopes (Conroy *et al.*, 1991). Although many variations have been developed, the antigen of interest usually is fixed to a plastic well and used to "capture" the antibody in the patient fluid, usually serum. Anti-human immunoglobulin (usually IgG or IgM) labeled with an enzyme such as horseradish peroxidase or alkaline phosphatase is added, followed by the enzyme substrate, which is detected colorimetrically. The sensitivity and specificity of EIA depend, in part, on the type and purity of the capture antigen that is used. Single antigen EIAs tend to be more specific than those using whole organisms (Gnann *et al.*, 1987; Schulz *et al.*, 1989) but may not be as sensitive.

Western blot technology is a major advance in serologic testing because it allows the investigator to identify which of many antigens the antibody

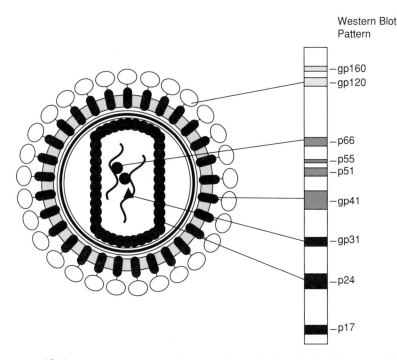

FIGURE 12.1 Conceptual diagram of Western blot technology for the diagnosis of HIV-1 infection. Viral antigens are separated by electrophoresis into a core protein (p17) or core proteins (p55, p24, and p17), envelope proteins (gp160, gp120, and gp41), and polymerase proteins (p66, p51, and p31). The separated viral proteins contained in the gel are blotted onto nitrocellulose paper, which is cut into strips and incubated with the serum sample being tested. The strips are washed and incubated with tagged antihuman antibody. The label is usually horseradish peroxidase, which reacts with chromogenic substrates to produce colored bands. The Centers for Disease Control defines a positive Western blot as the presence of any two of the bands p24, gp41, and gp120/160. The requirement for several positive bands contributes to the high specificity of the test. (Reproduced with permission from Sloand, 1991.)

recognizes (Figure 12.1). In brief, a whole-organism lysate (Centers for Disease Control, 1989b) or specific antigens (Sirisanthana *et al.,* 1988; Harrison *et al.,* 1989b) are separated by electrophoresis on a polyacrylamide gel (Hechemy *et al.,* 1991) and transferred onto a support membrane such as nitrocellulose. EIA is used to detect the presence of antibodies interacting with individual antigen bands on the membrane.

Although more commonly used for antigen detection, agglutination methods use natural or artificial particles, such as *Staphylococcus aureus* or latex beads, respectively, coated with organism-specific antigen that agglutinate when exposed to organism-specific antibodies (Heyward and Curran, 1988; Quinn *et al.,* 1988; Gerber *et al.,* 1990).

Tests of Cellular Immunity

The measurement of cell-mediated immune response to organism-specific antigens by skin testing is another method of identifying exposure to certain infectious pathogens, such as *Mycobacterium tuberculosis* (Snyder, 1982) and pathogenic fungi (Edwards *et al.*, 1969). The test usually is performed by injecting an antigen solution intradermally into the volar aspect of the forearm and measuring the cutaneous induration and erythema 48–72 hr later. A positive result does not differentiate previous exposure from actual infection. *In vitro* assays for the assessment of cell-mediated immune response are also available (Dressler *et al.*, 1991; Ampel *et al.*, 1992; Esolen *et al.*, 1992), but are rarely used in the clinical setting.

Interpretation of Tests

The presence of antibody to a particular organism is presumed to be evidence of infection at some point in time. A single positive antibody titer can signify several possibilities, including past exposure to the organism, exposure to an organism that induces antibodies that cross-react with the organism in question, nonspecific reaction, or current infection resulting in disease. The determination of exposure to organisms among neonates and young infants is difficult because of the presence of maternal IgG in infant serum (Rogers *et al.*, 1990).

Quantification of antibody titer usually is achieved by serially diluting the test serum until the test becomes negative. Interpretation of antibody levels is complicated by the lack of standardization of many assays. This deficit can lead to different readings of the same specimen when performed by different laboratories or even when performed at different times within the same laboratory (Hedberg *et al.*, 1987).

Methods can be used to improve the specificity of serologic tests. The presence of IgM antibody is often an indicator of acute infection, which is exemplified by serologic testing for hepatitis A virus (HAV) among patients with acute hepatitis. The presence of serum IgG against HAV indicates that the patient has, at some time in the past, been exposed to HAV but does not indicate the etiology of the hepatitis (Skinhoj *et al.*, 1977). The presence of IgM antibody, however, is much more specific for recent infection and offers firmer evidence that the current illness is caused by HAV (Roggendorf *et al.*, 1980) (Figure 12.2). In addition, the demonstration of seroconversion, a ≥ 4-fold rise in antibody titer to an organism between sera taken near the onset of the illness and 2–6 weeks later and run simultaneously in the same laboratory, provides firmer evidence that the present illness is caused by the organism in question (Schild and Dowdle, 1975). Certain antibody assays are inherently more specific than others, particularly the Western blot method, in which a positive test often requires the presence of antibodies to several antigens. For certain organisms such as human immunodeficiency virus

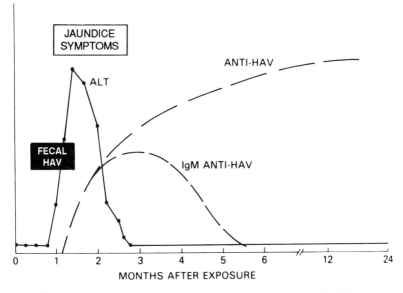

FIGURE 12.2 The serologic course of a typical case of type A hepatitis. The presence of serum IgM for only 3–12 months after the onset of illness makes the measurement of this antibody relatively specific for acute HAV infection. HAV, hepatitis A virus; ANTI-HAV, antibody to hepatitis A virus; ALT, serum alanine aminotransferase. (Reproduced with permission from Hoofnagle et al., 1990. Copyright © 1990 by Churchill Livingstone.)

(HIV), EIA (a sensitive test) is used to screen patients for infection and Western blot (a specific test) is used for conformation of infection (Centers for Disease Control, 1989b).

In addition to these methods used by clinicians to improve the specificity of serologic tests, epidemiologists studying infectious diseases have another method at their disposal: the use of control subjects. For example, interpretation of a single serologic result for influenza can be very difficult because of the ubiquity of influenza virus infection and the persistence of antibodies long after infection has resolved. However, a difference in influenza antibody titers between a group of patients with clinical illness compatible with influenza during an outbreak and a group of carefully selected well controls, even using a single antibody measurement, strongly suggests that the outbreak is due to influenza (Harrison et al., 1991a). Similar methods were used to identify retrospectively an outbreak of Pontiac fever as caused by infection with *Legionella pneumophila* (Glick et al., 1978).

Serology for *Borrelia burgdorferi,* the spirochete that causes Lyme disease (Steere et al., 1983), is a paradigm for the difficulties encountered in interpreting serologic tests. Lyme serology is known to have a low sensitivity in the early stages of the disease (Russel et al., 1984). However, because the principal early cutaneous manifestation, erythema migrans, is so characteristic of Lyme disease, the diagnosis of early Lyme disease usually can be made on clinical grounds. In addition, a small proportion of patients who are treated with antibiotics during the early stages of the illness do not develop antibodies, in spite of the development of chronic Lyme disease (Dattwyler *et al.*, 1988).

However, the principal problem with Lyme serology is the lack of specificity. Up to 10% of individuals in endemic areas may develop antibodies due to asymptomatic infection (Hanrahan *et al.*, 1984; Steere *et al.*, 1986) or antibodies to cross-reacting antigens such as mouth treponemes (Wilske *et al.*, 1984; Hunter *et al.*, 1986; Jones, 1991) or other treponemal and *Borrelia* species (Hunter *et al.*, 1986; Magnarelli *et al.*, 1987). These patients test positive for Lyme disease when presenting with any illness for which the test is ordered. The lack of standardization of laboratories and commercial kits has added to the unreliability of the test (Hedberg *et al.*, 1987; Blank *et al.*, 1990). For example, an evaluation of seven commercial kits (Blank *et al.*, 1990) demonstrated poor agreement among tests, indicating that the test results depend greatly on the commercial kit used. In addition, the estimated specificity of the seven kits ranged from 12 to 60%.

Bayes theorem can be applied to determine the predictive value of a positive Lyme test (Ingelfinger *et al.*, 1983). Assuming that one-third of patients presenting with arthritis to a clinic truly have Lyme arthritis, with a test sensitivity of 95% and a test specificity of 12% or 60%, only 35% and 54%, respectively, of patients with a positive test would have Lyme arthritis. If the true prevalence were 5%, which is probably more typical of clinics that see patients requesting a Lyme test, the predictive values would be 5% and 11%, respectively. If one were to rely on Lyme serology to determine whom to treat in this setting, 100 patients would be treated for every 5–11 who actually have Lyme disease. Until a more specific test becomes available, Lyme disease will remain a clinical diagnosis and serology will continue to be of limited clinical value in most settings.

Biologic Markers of Outcome

First in this section is a brief discussion of issues related to the isolation of organisms in culture and molecular techniques that are used to study organisms that have been isolated. What follows is a description of some of the nonculture methods that are used to identify the presence of an organism in clinical specimens.

Culture Methods

Advantages of culture   Methods for determining outcome in infectious disease focus mainly on determining the presence of organisms in body fluids or tissues. Organism isolation in culture is often the preferred diagnostic method for several reasons. First, the isolation of an organism is often a very specific finding for infection. For example, the isolation of *Neisseria meningitidis* from the CSF of a patient who has clinical meningitis confirms the diagnosis of invasive meningococcal infection. Second, the isolation of a bacterial pathogen permits antimicrobial testing of the isolate, which has tremendous clinical importance for patient care since the result assists physicians in selecting appropriate antibiotic therapy. Third, having the actual organism permits analysis of other potentially important characteristics, such as the production of β-lactamase or the determination of serotype. Fourth, organism isolation in culture usually is needed for studies to evaluate the sensitivity and specificity of diagnostic tests. Culture results are the standard to which the test results are compared.

Interpretation of culture results   The interpretation of culture results, however, is not without difficulties. Although the isolation of an organism from a normally sterile body fluid usually indicates infection, contamination of the culture medium, either from the skin of the patient or health care personnel or from the environment, can give a false-positive result. The isolation of, for example, a coagulase-negative staphylococcus from one of four blood cultures in an ambulatory patient with no intravascular catheters or aspergillus from the otherwise normal CSF of an immunocompetent patient should raise the suspicion of culture contamination. In addition, the interpretation of culture results from nonsterile body fluids is fraught with danger. Although *Haemophilus influenzae, Streptococcus pneumoniae, N. meningitidis, Candida albicans,* and adenovirus can cause serious infection, they also can be isolated from the throats of normal individuals, which can lead to the contamination of expectorated sputum that is sent for culture (Kalinske *et al.,* 1967). In addition, skin flora often can contaminate voided urine specimens (Sobel and Kay, 1990). The isolation of certain bacteria from stool, such as *Escherichia coli,* is not particularly informative because most strains of *E. coli* are not pathogenic. However, the isolation of certain pathogens, such as *M. tuberculosis,* usually indicates true infection, even when isolated from nonsterile fluids such as sputum.

Culture techniques may be relatively insensitive because of special requirements for organism isolation. Although patients with streptococcal endocarditis who have not taken antibiotics almost always have a positive blood culture (Werner *et al.,* 1967), patients with pneumococcal pneumonia often do not (Shann *et al.,* 1984; Wall *et al.,* 1986). Some organisms, such as anaerobic bacteria, require special laboratory conditions for isolation, whereas others, such as certain viruses and parasites, cannot be cultured.

Subtyping of Isolates

Once an organism is isolated in culture, the ability to further distinguish among isolates of the same species is often required, particularly during disease outbreaks. The purpose of subtyping is to determine the modes of transmission and the sources of infection (Nolte *et al.*, 1984; Linnan *et al.*, 1988; Centers for Disease Control, 1989c; Breiman *et al.*, 1990; Lin *et al.*, 1991; Minshu *et al.*, 1991). Before the advent of molecular techniques, subtyping methods included determining antimicrobial susceptibility patterns, serologic typing or grouping, and biochemical typing (Kilian, 1976; Oberhofer and Back, 1979). The principal limitation of these methods is their lack of resolving power. For example, 95% of invasive *H. influenzae* isolates are serotype b, which makes serotype determination of limited epidemiologic utility for this organism.

Advances in molecular biology have led to the use of molecular techniques that are based on the principle that there are phenotypic or genotypic differences from strain to strain but not within a given strain (Tenover, 1991). The ability to subtype organisms using molecular methods has proved to be a powerful addition to traditional epidemiologic methods. The most commonly used methods are discussed next.

Analysis of Chromosomal DNA

Restriction digests of chromosomal DNA   Otherwise known as DNA fingerprinting, this method uses purified chromosomal DNA cleaved by endonucleases; the pieces are separated electrophoretically on an agarose gel (Tenover, 1991). Although it is a relatively crude method, differences among isolates often can be identified, and a wide range of bacteria, such as *M. tuberculosis* and *L. pneumophila*, viruses (Tsilimigras *et al.*, 1989; Stout *et al.*, 1992), and parasites (Hide and Tait, 1991) can be typed using this method. In a modification of this technique (restriction fragment length polymorphism, RFLP), the digested DNA is hybridized with a probe to a variable region of the genome (Daley *et al.*, 1992) (Figure 12.3). An advantage of RFLP testing is that it accentuates differences among strains that may not be recognizable with DNA fingerprinting by analyzing the variable region.

Ribosomal DNA (rDNA) gene restriction patterns (ribotyping)   This method is essentially RFLP testing that uses ^{32}P-labeled *E. coli* 16 + 23 S ribosomal RNA (Grimont and Grimont, 1986) as the probe. Each unique banding pattern represents a rDNA restriction pattern. This method has been used for the typing of numerous bacteria including *H. influenzae* biogroup *aegyptius* (Irino *et al.*, 1988), *Streptococcus agalactiae* (Blumberg *et al.*, 1991), and *S. aureus* (Blumberg *et al.*, 1992). It also has been used for differentiation among fungi (Spitzer *et al.*, 1989) and parasites (Hide and Tait, 1991).

Nucleic acid sequencing   This method is used in epidemiologic studies to determine the genetic relatedness of isolates, usually viruses. A selected portion of the viral genome known to vary among strains is isolated and its nu-

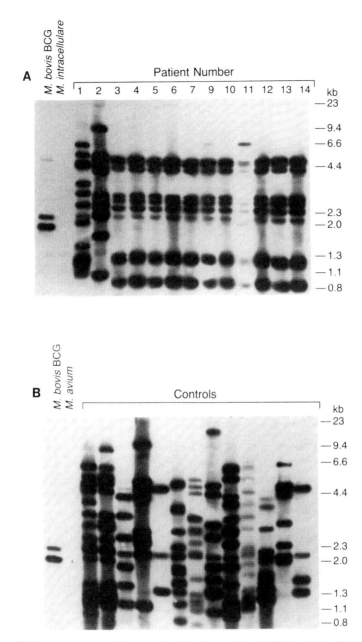

FIGURE 12.3 (A) Restriction fragment length polymorphism (RFLP) analysis of *Mycobacterium bovis* BCG, *M. intracellulare,* and clinical *M. tuberculosis* isolates from residents of an HIV congregate living site. Patients 1 and 2 were receiving antituberculosis therapy when they entered the facility. Tuberculosis developed in the remaining patients while they lived at the facility. Patient 8, who did not have positive cultures, is not shown. (Reproduced with permission from Daley, 1992.) (B) RFLPs of *M. bovis* BCG, *M. avium,* and random clinical isolates of *M. tuberculosis* obtained from patients identified in San Francisco at the time of the outbreak and used as controls. The epidemiologic data and the fact that all 13 of these control isolates had a unique RFLP pattern strongly supports the hypothesis that the outbreak represented a spread of infection from patient 3. (Reproduced with permission from Daley, 1992.)

cleotide sequence is determined (Kew *et al.*, 1990; Trent *et al.*, 1990; Lin *et al.*, 1991). To classify bacteria and define species, however, sequencing the genome is not practical; the genetic relatedness of two isolates is determined by nucleic acid hybridization of the entire genome (Brenner *et al.*, 1982; Brenner, 1991).

Analysis of Plasmid DNA

This method involves determining whether a plasmid is present and, if so, determining its size and restriction endonuclease pattern. Plasmid analysis has been useful in numerous epidemiologic studies of bacterial infections (Taylor *et al.*, 1982; John and Twitty, 1986; Schaberg and Zervos, 1986; Wachsmuth, 1986; Tacket, 1989), although its use is somewhat limited by the fact that many bacteria do not contain plasmids.

Protein Analysis

Outer membrane protein analysis   This procedure involves extracting the bacterial membranes and performing sodium dodecyl sulfate-polyacrylamide gel electrophoresis to determine the electrophoretic pattern of the surface-exposed outer membrane proteins (OMPs). This method has been particularly useful for defining the molecular epidemiology of *H. influenzae* type b (van Alphen and Bijlmer, 1990; Bijlmer *et al.*, 1992) and meningococcal (Frasch *et al.*, 1985) infections.

Multilocus enzyme typing   This method exploits the existence of multiple alleles among cytoplasmic enzymes contained by all strains of a species. Relative electrophoretic mobilities of multiple enzymes are determined in starch gels (Selander *et al.*, 1986) and the genetic relatedness of isolates can be determined (Bibb *et al.*, 1990) (Figure 12.4). This procedure has been used to define the molecular epidemiology of a variety of bacteria, including *H. influenzae* (Musser *et al.*, 1990), *Listeria monocytogenes* (Bibb *et al.*, 1989; Piffaretti *et al.*,1989); *N. meningitidis* (Caugant *et al.*, 1986, 1988; Moore *et al.*, 1989), and fungi (Safrin *et al.*, 1986).

Monoclonal Antibodies

Monoclonal antibodies (Baron and Finegold, 1990) also can be used to subtype infectious agents. This technique has been useful for the subtyping of *L. pneumophila* (Joly *et al.*, 1986).

Phage Typing

This procedure determines the ability of a panel of bacteriophages to lyse a bacterium. This method has been useful for typing isolates of a variety of

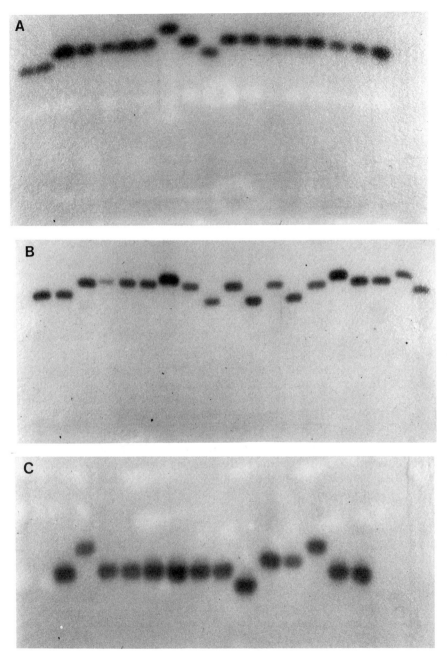

FIGURE 12.4 Multilocus enzyme typing. Gels illustrating electrophoretic variation in three enzymes. (A) Mannitol 1-phosphate dehydrogenase in *Escherichia coli*, 18 isolates. (B) Glucose 6-phosphate dehydrogenase in *Neisseria meningitidis*, 19 isolates. (C) Malate dehydrogenase in *E. coli*, 14 isolates. The classification of isolates into enzyme types relies on the use of multiple enzymes. (Reproduced with permission from Selander, 1986.)

bacteria, including *S. aureus* (Archer and Mayhall, 1983), *Pseudomonas aeruginosa* (Bergan, 1978), *Salmonella* (Holmberg *et al.*, 1984; Ward *et al.*, 1987), and *L. monocytogenes* (Rocourt *et al.*, 1985). The principal limitations are that few laboratories are equipped to perform phage typing and that many isolates cannot be phage typed.

Interpretation of Subtyping Information

Although these methods have revolutionized the study of the epidemiology of infectious disease, several important methodologic issues should be emphasized. First, there must be substantial heterogeneity among isolates of the same species in the molecular characteristic being examined for a method to be useful. Second, it follows that some knowledge of the heterogeneity within a species is necessary for the interpretation of test results (Lin *et al.*, 1988), including geographic and temporal variability (Bibb *et al.*, 1990). An example of the concept of heterogeneity is a recent cluster of HIV-1 infection epidemiologically linked to a dentist (Centers for Disease Control, 1990, 1991a; Ciesielski *et al.*, 1991). Study of the DNA sequence of the V3 region of the HIV-1 envelope gene of the dentist and three patients with isolates revealed a difference of only 3.4%, a small difference for HIV. The three patients had no known risk factors for HIV infection. With only these data, it is difficult to determine the probability that the dentist infected these three patients. However, study of the V3 region of isolates from a patient of the dentist that had a known risk factor for HIV infection, 8 randomly selected HIV strains from the same community, and 21 other United States isolates indicate that they were substantially different from the isolates from the dentist and the three patients. These data suggest that these patients were infected with the dentist's strain of HIV-1 during their dental procedures.

Although many molecular subtyping methods are available, the utility of each is organism specific. For example, analysis of OMPs has been useful for subtyping isolates of *N. meningitidis*, in contrast with the study of *H. influenzae* biogroup *aegyptius*, in which the vast majority of isolates are of a single OMP type (Brenner *et al.*, 1988). However, multilocus enzyme typing has been very useful for this organism. In addition, plasmid analysis has little utility in the study of *Salmonella typhi*, since most isolates have plasmids with the same restriction endonuclease digest pattern, in spite of having susceptibility to different phage types (Maher *et al.*, 1986). Finally, the use of several subtyping methods is sometimes needed to differentiate fully among isolates of the same species (Brenner *et al.*, 1988; Morris *et al.*, 1986).

Nonculture Methods

Antigen detection   Problems with the culture methods just outlined have led to a substantial effort to develop nonculture methods for diagnosing infection. Antigen detection is an immunologic method that relies on the identification of the presence of a particular antigen that is specific for the pathogen in question.

Agglutination of antibody-coated latex beads or staphylococci (Coonrod, 1983a,b) is a relatively simple method for the detection of certain bacterial and fungal antigens in body fluids. Agglutination methods are useful for the diagnosis of bacterial meningitis among patients who have received antibiotics (Ward *et al.*, 1978) and for patients with pneumonia and negative blood cultures (O'Neill *et al.*, 1989).

Organisms also may be identified by visualization with specific fluorescent or radioisotopic antibodies (Rosebrock, 1991). For example, the direct fluorescent antibody test is a specific, albeit insensitive, method for identifying *L. pneumophila* in pulmonary fluid (Cherry *et al.*, 1978; Nguyen *et al.*, 1991) and respiratory syncytial virus in nasopharyngeal swabs (Hughes *et al.*, 1988). The EIA method discussed earlier for detection of antibodies also is used for antigen detection (Schaffner *et al.*, 1991) (Figure 12.5).

Nucleic acid detection DNA probes are used for two principal purposes: to determine the species identity of an organism growing in culture and to detect the presence of a particular organism in a clinical specimen (Tenover, 1991). For example, species-specific DNA probes permit the identification of *M. tuberculosis* growing in culture in several hours (Gonzalez and Hanna, 1987). DNA probes also have been useful for epidemic investigation of enteric pathogens (Wachsmuth *et al.*, 1991) and identification of enterotoxigenic *E. coli* (Moseley *et al.*, 1982). DNA probes have proved particularly useful in the screening of nonsterile specimens, not only for the species of interest but, by using a probe that is associated with virulence, only for virulent strains of that species (Miliotis *et al.*, 1989).

Polymerase chain reaction (PCR) is a method used to identify the presence of an organism without actually isolating it in culture (Saiki *et al.*, 1988; Carman, 1991; Lew, 1991). This method uses DNA amplification to detect minute quantities of nucleic acid (Figure 12.6). More than one million copies of DNA can be made in less than 3 hr, allowing for the detection of one target molecule of DNA per 10 μl of blood (Sloand *et al.*, 1991). The specificity of the method is achieved by the use of organism-specific nucleic acid primers. PCR is useful in the laboratory for organisms that cannot be cultured routinely (Wakefield *et al.*, 1991) or in field settings where culture is not feasible (Harrison *et al.*, 1991b). Current limitations of this method are its relative unavailability for most clinical laboratories and problems with false–positive results because of laboratory cross-contamination by small quantities of nucleic acid carried over from other specimens (Sloand *et al.*, 1991). In addition, PCR is so sensitive that it may be positive in patients without apparent infection (Bobo *et al.*, 1991), which may limit its specificity for the diagnosis of acute infection by organisms that cause asymptomatic carriage.

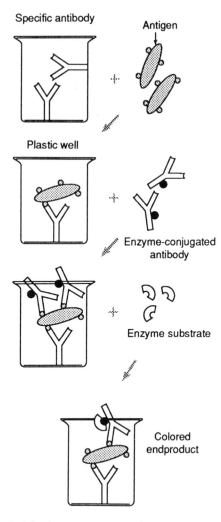

Specific antibody

Antigen

Plastic well

Enzyme-conjugated antibody

Enzyme substrate

Colored endproduct

FIGURE 12.5   The principle of enzyme immunosorbent assay (EIA) for the detection of antigen. A plastic well is coated with specific antibody and the patient fluid is added. If antigen is present in the fluid, it binds to the antibody. A second antibody against the antigen in question, complexed with an enzyme, is added to the system. The enzyme substrate is added, giving a colored end product. (Reproduced with permission from Baron and Finegold, 1990. Copyright © 1990 by C. V. Mosby.)

LPS detection   The *Limulus* lysate test (*Limulus polyphemus,* the horseshoe crab) is a nonculture method for detection of gram-negative bacteria in clinical specimens, usually CSF. The test is based on the ability of lipopolysaccharide (LPS) from the bacterial cell wall to cause gelation of a lysate of limulus amebocytes (Tinghitella and Edberg, 1991). The usefulness of the

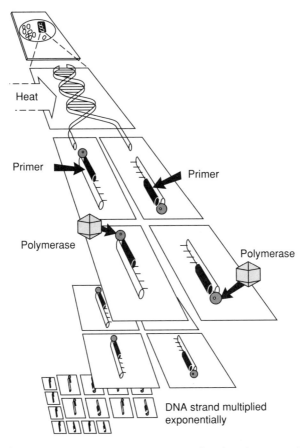

FIGURE *12.6*   The polymerase chain reaction (PCR) is based on the enzymatic amplification of a DNA fragment (target DNA) using a specific oligonucleotide primer, DNA polymerase, and nucleic acids. Repeated cycles of heat denaturation of the target DNA, annealing of the primers to their complementary sequences, and extension of the annealed primers with a DNA polymerase result in the amplification of the segment until detectable levels of organism DNA have been produced. PCR also can be used to amplify RNA sequences. (Reproduced with permission from Sloand, 1991.)

test is limited by the fact that a positive result is not species-specific, although a recent report suggests that specificity may be achieved by using monoclonal antibodies for LPS capture (Mertsola *et al.*, 1991).

Another novel nonculture method for identifying the presence of an infectious agent was described for the hepatitis C virus (Choo *et al.*, 1989; Alter, 1991), an RNA virus previously known as non-A, non-B hepatitis. Investigators extracted the nucleic acid from large volumes of infectious chimpanzee plasma and converted the RNA to complementary DNA by reverse transcription. Using a cloning vector, the DNA was inserted and expressed in *E. coli*. Viral proteins were identified by screening *E. coli* lysates with serum from a patient with non-A, non-B hepatitis. This and subsequent studies have led to the development of serologic tests for hepatitis C that have been effective in reducing the incidence of transfusion-related hepatitis C (Kuo *et al.*, 1989; Esteban *et al.*, 1991). This is a remarkable feat since the virus has never been isolated in culture.

Surrogate markers   Surrogate markers also have been used to identify patients infected with a particular organism. For example, antibodies to hepatitis B core antigen and serum alanine aminotransferase levels (Aach *et al.*, 1981; Stevens *et al.*, 1984) were used to screen blood donors for infection with hepatitis C before a specific serologic test for this virus became available. Although these markers were the only methods available at the time for screening out infected blood donors, this approach was inefficient because of many false positives and negatives (Barrera *et al.*, 1991).

Susceptibility

Two basic types of biologic markers can be used to determine susceptibility to infectious disease: direct immunologic markers and surrogate markers. The absence of specific components of the immune system is known to correlate well with the risk of infection. Isolated immune defects have added to the understanding of host defenses against specific pathogens.

Generalized Susceptibility

Patients with congenital or acquired absence or low levels of serum immunoglobulin are susceptible to a variety of pathogens. Patients with X-linked hypogammaglobulinemia (Bruton, 1952) develop frequent bacterial infections (Lederman and Winkelstein, 1985) and are predisposed to chronic disseminated and central nervous system echovirus infection (Wilfert *et al.*, 1977; Crennan *et al.*, 1986; McKinney *et al.*, 1987). Although these patients are not in general predisposed to parasitic infection, some patients develop severe gastrointestinal infection by *Giardia lamblia* (Ochs *et al.*, 1972). Serum antibody levels following intravenously adminstered immunoglobulin (mostly IgG) correlate with the risk of infection (Roifman *et al.*, 1987), but therapy is not necessarily curative (Crennan *et al.*, 1986). Similar predisposition to bacterial infection is seen among patients with acquired hypogammaglobulinemia (Miller, 1972; Hermans *et al.*, 1976).

Individuals with deficiencies of certain complement components are also susceptible to bacterial infection, particularly invasive meningococcal infection (Ross and Densen, 1984). Of interest is the fact that, although patients with deficiency of terminal components have a higher incidence of meningococcal infection, the case fatality rate is lower than for patients without complement deficiency (Ross and Densen, 1984), suggesting that complement activation plays a role in the pathogenesis of this infection.

Cellular immune defects also leave individuals susceptible to infection. For example, patients with cancer chemotherapy-induced neutropenia are at increased risk of severe bacterial and fungal infection (Singer *et al.*, 1977; De Gregorio *et al.*, 1982). The risk of infection correlates with the absolute granulocyte count, the rapidity of the decline, and the duration of the neutropenia (Bodey *et al.*, 1966; Schimpff, 1990). Monitoring of neutrophil counts among patients receiving chemotherapy is an important component of the care of these patients and is used to determine whether the use of empiric antibiotics is needed in the setting of fever (Klastersky *et al.*, 1988). In addition, CD4 counts correlate with the risk of opportunistic infection among patients infected with HIV.

Specific Susceptibility

Organism-specific antibody levels frequently are measured following immunization to determine response to immunization and whether a protective level has been achieved (Makela *et al.*, 1977). However, determining a protective level is problematic and does not always correlate precisely with the risk of infection in individual patients. In addition, identification of immunologic correlates of protection for certain vaccines has been elusive because the components of the immune response responsible for protection have not been identified. Evaluation of a protective response to pertussis vaccine (Mortimer, 1988) has required the use of the mouse protection test in which immunized mice are inoculated intracerebrally with *Bordetella pertussis* to determine vaccine potency (Kendrick *et al.*, 1947).

Surrogate markers are associated with susceptibility to infection but are generally of limited clinical value. ABO blood group is a marker for risk of and severity of cholera; patients with blood group O are at highest risk (Clemens *et al.*, 1989; Glass *et al.*, 1985). The mechanism for this association is not known. A similar, albeit much weaker, association with blood group O has been reported for susceptibility to enterotoxigenic *E. coli* (Black *et al.*, 1987) but has been difficult to confirm (van Loon *et al.*, 1991). Interestingly, heat labile *E. coli* enterotoxin is biologically and immunologically related to cholera toxin (Field, 1979). HLA-DR2 and HLA-DR4 are associated with the development of chronic Lyme arthritis among patients with any type of Lyme arthritis (Steere *et al.*, 1990; Sebastiani, 1991). It is not clear whether this is due to chronic *B. burgdorferi* infection or an immunogenetic phenomenon (Steere *et al.*, 1990). In addition HLA-B27 is associated with sacroiliitis as well as Reiter syndrome following infection with *Yersinia enterocolitica* (Solem and Lassen 1971; Aho *et al.*, 1973; Leirisalo-Repo, 1988). Certain HLA antigens are associated with less severe malaria in west Africa

(Hill *et al.*, 1991) and diminished responsiveness to hepatitis B vaccine (Marescot *et al.*, 1989).

Case Studies

Exposure: Epidemiology of HTLV-I Infection

Background

Serologic tests can be used to define the epidemiology and clinical spectrum of infectious diseases such as human T-lymphocyte virus type I (HTLV-I) infection. In 1977, a group of 16 Japanese patients with a fulminant form of T-cell leukemia were reported. An infectious etiology was suspected because of geographic clustering: 13 of the cases had been born on the island of Kyushu (Uchiyama *et al.*, 1977). Subsequently, a newly discovered retrovirus, HTLV-I, was isolated from patients with adult T-cell leukemia/lymphoma (ATLL) in the United States (Poiesz *et al.*, 1980) and Japan (Yoshida *et al.*, 1982). The fact that retroviruses were known to cause malignancy in animals (Gallo and Todaro, 1976) and that malignant cells had integrated monoclonally (i.e., integrated in a single site) whole or partial proviral DNA (Clark *et al.*, 1988) suggested that HTLV-I was causally related to this malignancy. Unlike HIV, which causes depletion of CD4 cells, infection with HTLV-I leads to proliferation of both CD4 and CD8 T cells (Gallo, 1991).

The discovery of the first human retrovirus was soon followed by the development of a serologic test for the detection of anti-HTLV-I antibodies. The development of this test allowed seroepidemiologic studies to be performed, with the goals of defining the clinical manifestations associated with HTLV-I infection and the risk factors for infection. Although serology for HTLV-I has been crucial to the definition of these issues, problems have been encountered with the test. Current HTLV-I serology cannot distinguish between infection with HTLV-I and HTLV-II because of the high degree of amino-acid homology between the two viruses, although serologic methods to distinguish the two are being developed (Chen *et al.*, 1990; Lipka *et al.*, 1990). Initial serologic studies were performed with a relatively nonspecific whole-virus EIA (Biggar *et al.*, 1985), which led to an overestimation of infection incidence and indicated the need for confirmatory Western blot serology (Constantine and Fox, 1989). Although viral isolation and PCR are available to detect the presence of HTLV-I, serologic studies are logistically easier. In spite of these limitations, HTLV-I serology has been used to define the epidemiology of HTLV-I infection.

Initial studies focused on identifying a link between ATLL by testing for the presence of HTLV-I antibodies in patients with ATLL and controls. Kalyanaraman *et al.* found that two patients with ATLL had HTLV-I serum antibodies as measured by radioimmune precipitation (Kalyanaraman *et al.*, 1981). Control sera were tested from 21 patients with other T-cell malignancies, 50 normal donors, and 11 close family members of patients with T-cell malignancies. Only one control serum, from the wife of one of the two seropositive ATLL patients, was positive. This and other studies (Posner *et al.*, 1981; Robert-Guroff *et al.*, 1982) led to the conclusion that HTLV-I was etiologically linked to ATLL.

Subsequent seroepidemiologic studies indicated that HTLV-I-related ATLL is endemic in Japan, in the Caribbean (Blattner *et al.*, 1982; Catovsky *et al.*, 1982), among blacks in the southeastern United States (Blayney *et al.*, 1983; Levine *et al.*, 1988; Ratner and Poiesz, 1988), and in various other geographic regions (Catovsky *et al.*, 1982).

Seroprevalence in Endemic Areas

HTLV-I seroprevalence has been estimated in several studies and found to be 2.1% among 95 blacks from Georgia (Blayney *et al.*, 1983), 3.3% among 336 blacks from St. Vincent Island, British West Indies (Blattner *et al.*, 1982), and 28% among 202 residents of three communities of Kyushu (Hinuma *et al.*, 1982). In all studies, the seroprevalence increased with age.

Clinical Manifestations

Although the majority of patients infected with HTLV-I are asymptomatic, seroepidemiologic studies have identified two principal clinical syndromes associated with HTLV-I infection. ATLL is characterized by frequent cutaneous involvement, lymphadenopathy and hepatosplenomegaly, hypercalcemia, and a fulminant terminal leukemic phase. Serologic studies have been used to determine clinical features that differentiate HTLV-I-related lymphoma from other non-Hodgkin's lymphomas. Tests for serum HTLV-I antibodies and clinical evaluations were done for 26 Jamaican patients with non-Hodgkin's lymphoma. Of these patients, 15 were antibody positive; the remaining 11 were negative. Seropositive patients were more likely to have a leukemic phase (80% of seropositives versus 36% of seronegatives), cutaneous involvement (40% versus 0%), and abnormal serum hepatic enzyme levels (67% versus 18%) than seronegative patients; these differences were statistically significant. Hypercalcemia was also more frequent among seropositive patients than controls, but this difference was not statistically significant. It has been estimated that individuals infected with HTLV-I before the age of 20 have a 4% lifetime risk of developing ATLL (Murphy *et al.*, 1989a).

The other principal clinical manifestation of HTLV-I infection is tropical spastic paraparesis (TSP), also referred to as HTLV-I-associated myelopathy (HAM). The pathogenesis is unknown but has been postulated to be caused by an HTLV-I-mediated autoimmune response or direct effects of neural infection by HTLV-I (Kramer and Blattner, 1989). Clinical features include chronic spastic paraparesis, weakness of the lower extremities, and sensory symptoms (Scientific Group, 1989). This illness was first reported from the Caribbean in the mid-1950s (Cruickshank, 1956; Rodgers, 1965) but was not linked to HTLV-I until a study from Martinique reported the serendipitous finding of two HTLV-I seropositive TSP/HAM patients (Gessain, 1985). A subsequent study indicated that 15 of 22 (68%) TSP/HAM patients were seropositive compared with only 13 of 303 (4%) controls. In Japan, all 6 patients with TSP/HAM had positive HTLV-I serology, in contrast to only 15% of controls (Osame *et al.*, 1986a,1987). The high rate of seropositivity among controls reflects the high HTLV-I seroprevalence rate in the community. Other studies in Japan suggest that HAM/TSP occurs in 1 of 1000– 2000 (0.1–0.2%) HTLV-I seropositive individuals, in contrast to 1–5% reported from the Caribbean and South America (Manns and Blattner, 1991).

Although few infectious complications have been associated with HTLV-I infection in the absence of leukemia, infectious dermatitis has been associated with HTLV-I seropositivity among Jamaican children (LaGrenade *et al.*, 1990).

Risk Factors for Infection

Seroepidemiologic studies also have been used to define the ways in which HTLV-I is transmitted. Several studies have demonstrated that HTLV-I is transmitted by sexual intercourse (Murphy *et al.*, 1989b) and blood transfusion (Okochi *et al.*, 1984; Osame *et al.*, 1986b; Sato and Okochi, 1986; Minamoto *et al.*, 1988). In addition, a seroepidemiologic study conducted in Japan (Kajiyama *et al.*, 1986) indicated that none of 82 children with a seropositive father but seronegative mother were seropositive in contrast to 86 of 352 (24%) children with seropositive mothers, suggesting mother-to-child transmission. Available evidence suggests that a primary mode of transmission is through breast milk (Hino *et al.*, 1985; Nagamine *et al.*, 1991; Take *et al.*, 1992).

HTLV-II   Similar studies have been conducted to define the epidemiology of infection with HTLV-II, a related human retrovirus. Although first isolated from two patients with a T-cell variant of hairy cell leukemia (Kalyanaraman *et al.*, 1982; Rosenblatt *et al.*, 1986), the association between this malignancy and HTLV-II was not confirmed in subsequent serologic studies (Rosenblatt *et al.*, 1987; Hjelle *et al.*, 1991). The epidemiology and clinical spectrum of HTLV-II infection remains to be defined. An association between HTLV-I and human disease is biologically plausible for two reasons. First, HTLV-II shares ~60% of its genome with HTLV-I (Myers *et al.*, 1988), and the two viruses are structurally similar. Second, HTLV-II infection also leads to proliferation of CD4 and CD8 T cells (Gallo, 1991), which is thought to be responsible for some of the clinical manifestations of HTLV-I. A study (Hjelle *et al.*, 1991) demonstrating a seroprevalence of 1.0–1.6% among American Indians suggests that this population may be useful for defining the natural history of infection with this organism, although prospective studies of human retroviruses are complicated by a typically long incubation period. Intravenous drug abusers in the United States (Lee *et al.*, 1989) and Guaymi Indians in Panama (Centers for Disease Control, 1992) also have been found to have a relatively high prevalence of HTLV-II infection.

Outcome: Brazilian Purpuric Fever—Epidemiologic Investigation of a New Disease

Molecular techniques have been crucial in elucidating the epidemiology of Brazilian purpuric fever (BPF), a recently recognized pediatric infectious disease. BPF can serve as a paradigm for the utility of classic epidemiologic tools for interpretation of molecular data and how epidemiologic studies can be enhanced greatly by molecular techniques.

BPF first was recognized following an outbreak in late 1984 in the small town of Promissão, São Paulo, Brazil, when 10 children presented to the local emergency room with fever, purpura, and hypotensive shock; all died within several hours (BPF Study Group, 1987a,b). Several case-control studies demonstrated an association with purulent conjunctivitis, which generally had resolved 3–15 days prior to admission (Harrison *et al.*, 1989a). An investigation revealed that a conjunctivitis outbreak occurred during the BPF outbreak. Although the etiology of BPF could not be determined, the conjunctivitis outbreak was determined to be due to *Haemophilus aegyptius,* an organism known previously only for its propensity to cause benign conjunctivitis. In addition, *H. aegyptius* was isolated from a nonaseptically obtained skin scraping of a fatal BPF case.

Although there was no proof that *H. aegyptius* was causally related to BPF, it was clear that this organism was epidemiologically linked to the illness. Initial laboratory efforts sought to determine whether the *H. aegyptius* skin isolate had any unique phenotypic or genotypic features. It had a 24.5-megadalton plasmid, although some stock strains of *H. aegyptius* from several countries also had a plasmid of the same molecular mass (BPF Study Group, 1987a).

A case-control study was conducted to determine whether this plasmid was epidemiologically associated with BPF. Of 9 isolates associated with BPF, all had a 24.5-megadalton plasmid, in contrast to only 7 of 19 *H. aegyptius* isolates from towns experiencing *H. aegyptius* conjunctivitis but not BPF ($p = 0.002$). The suspicion that BPF was due to *H. aegyptius* bacteremia was confirmed in 1986 during a BPF outbreak in Serrana, São Paulo State, where *H. aegyptius* was isolated from the blood of 5 of 6 children who met a clinical case definition for BPF (BPF Study Group, 1987b).

The sudden appearance of outbreaks of bacteremia due to an organism previously known only to cause conjunctivitis suggested either unusual host factors predisposing to bacteremia, a previously unrecognized virulence factor among *H. aegyptius* isoaltes, or both. Investigation of BPF cases and healthy children failed to identify predisposing host factors. Subsequent laboratory investigations using molecular methods focused on determining whether *H. aegyptius* isolates associated with BPF were different from those not associated with BPF (Brenner *et al.*, 1988).

A classic case-control study was used to determine whether there were characteristics that were unique to the *H. aegyptius* isolates from BPF cases (Brenner *et al.*, 1988). Case isolates (15) were from normally sterile sites of BPF cases, with the exception of the one cultured from a skin scraping. These isolates represented substantial temporal and geographic dispersion, since they were isolated over a 2-year period in areas separated by as much as 250 miles. Control isolates (32) were from patients with conjunctivitis in towns with no known BPF cases.

All isolates from BPF cases were found to be of a single enzyme type (type 2), a single OMP type (type 1), and one of two ribotypes (type 3 or 4). In addition, all BPF isolates had identical DNA fingerprints and the 24.5-megadalton plasmid with the same endonuclease pattern (type 3031) (Figure 12.7).

OMP analysis was found to be unhelpful in distinguishing case from control isolates, since almost all control isolates were type 1. However, plasmid type 3031, enzyme type 2, and ribotype 3 or 4 were found to be highly associated with BPF. Only 1 control isolate had these characteristics, making the association highly significant ($p = 2 \times 10^{-11}$). These and subsequent studies have indicated that these and other characteristics appear to be unique to what is now known as the BPF clone (Brenner *et al.*, 1988).

Unfortunately, these studies do not determine which characteristics are causally related to virulence rather than being simple markers for the clone (Carlone *et al.*, 1989; Harrison *et al.*, 1989c). Certain traits, such as multilocus enzyme type, are clearly markers. Others, such as the presence of plasmid 3031, have been postulated to be causally related to virulence. The recent isolation of a plasmidless strain that otherwise has the characteristics of the clone from a child with BPF suggests that the plasmid is not associated with virulence (B. Perkins, Centers for Disease Control, 1991, personal communication). Additional studies determined that plasmid 3031 is not integrated into genomic DNA in this isolate (M. L. C. Tondella, Adolfo Lutz Institute, 1991, personal communication). Several other unique characteristics are being investigated for virulence, including a 38-kDa extracellular protein (Carlone *et al.*, 1989) and a 25-kDa pilin protein (Weyant *et al.*, 1990).

Susceptibility: CD4 Counts as a Clinical Predictor of AIDS Progression

Biologic markers also can be used to determine susceptibility to infectious disease. An example is the measurement of the CD4 (complementarity deter-

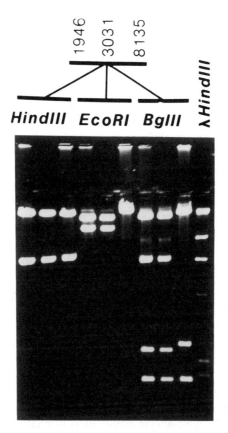

FIGURE 12.7   Restriction endonuclease digests of the plasmids of 3 isolates of *Haemophilus aegyptius:* isolate 1946 from a skin scraping of a child with Brazilian purpuric fever (BPF), Promissão, 1984; isolate 3031 from the blood of a child with BPF, Serrana, 1986; and isolate 8135 from a patient with conjunctivitis, Texas, 1950. *Hind*III failed to distinguish between the three plasmids. Digestion with *Eco*RI and *Bgl*II, however, indicated that the two BPF isolates were the same, and different from the Texas conjunctivitis isolate. This and other subtyping methods indicated that all *H. aegyptius* isolates from Brazilian children with BPF were of a single clone. (Photograph courtesy of Dr. Leonard W. Mayer, Centers for Disease Control.)

mining protein 4) T cell count to identify HIV-infected patients who are susceptible to progression to the acquired immunodeficiency syndrome (AIDS). CD4 cells recognize antigens that have been processed by macrophages, which results in the release of cytokines that contribute to the immune response. The glycoprotein portion of the HIV outer envelope, gp120, binds to the CD4 molecule on T helper cells and infects the cell, causing dysfunction in this aspect of the immune response. HIV also affects other arms of the immune system, including the humoral immune response. Although the precise cytotoxic mechanism is unknown, the depletion of CD4 T cells appears to be largely responsible for the markedly increased susceptibility to certain infectious diseases and malignancies associated with AIDS, an association that has been recognized since the beginning of the AIDS epidemic (Gottlieb *et al.*, 1981; Masur *et al.*, 1981).

Investigators have studied the predictive validity of CD4 lymphocyte counts as a marker of AIDS progression. Validity studies are crucial for determining the utility of biologic markers (Schulte, 1987,1989; Stein *et al.*, 1992). Detels *et al.* (1987) noted that a steady decrease in CD4 lymphocytes preceded the development of AIDS in a small cohort of homosexual men, and that 50% of men who had a CD4 lymphocyte count of <325 developed AIDS during several years of follow-up. *Mycobacterium avium* infection is rare in patients with >100 CD4 cells per mm^3 (Horsburgh, 1991). In a cohort of 1835 HIV seropositive homosexual men, a CD4 count of ≤300 per mm^3 was an independent predictor for progression to AIDS (Polk *et al.*, 1987). Goedert and colleagues showed that, during 3 years of follow-up, 10 of 23 (43%) seropositive patients with CD4 lymphocyte counts <300 per mm^3, 6 of 22 (27%) patients with CD4 counts of 300–399, 2 of 19 (11%) patients with CD4 counts of 400–549, and 1 of 22 (5%) patients with counts >549 developed AIDS, demonstrating that there is a dose–response effect of CD4 counts (Goedert *et al.*, 1987). Among a cohort of 100 HIV seropositive patients, almost all episodes of pulmonary infection with *Pneumocystis carinii*, *Cryptococcus neoformans*, cytomegalovirus, and *M. avium-intracellulare* occurred among patients with CD4 counts <200 cells per mm^3 (Figure 12.8).

Several investigators have explored the prognostic value of a combination of markers. Fahey followed a cohort of HIV seropositive patients and found that the CD4 lymphocyte count in combination with either serum neopterin or β_2-microglobulin levels was the best predictor of the progression to AIDS during a 4-year period (Fahey *et al.*, 1990). Although the exact role of β_2-microglobulin is not known, the α chain is involved in the transport of the β chain of the major histocompatibility complex class I antigen to the cell surface; increased serum levels may reflect increased lymphocyte turnover. The authors suggest that this combination is the most predictive because it includes both a measure of immunodeficiency (CD4 lymphocyte count) and immune activation (neopterin or β_2-microglobulin levels). In another study, CD4 lymphocyte counts and β_2-microglobulin levels were helpful in identifying patients at high risk for AIDS (Anderson *et al.*, 1990). Of 29 HIV seropositive patients with β_2-microglobulin levels above 322 nmol/liter and CD4 lymphocyte counts below 500/µl, 19 (66%) developed AIDS in 3 years. p24 antigenemia, an indication of active HIV replication, is also predictive, albeit insensitive, for progression to AIDS (Moss *et al.*, 1988; MacDonnell *et al.*, 1990). Certain HLA haplotypes have been associated with the rate of disease progression (Louie *et al.*, 1991).

The use of CD4 lymphocyte counts for predicting the development of *P. carinii* pneumonia (PCP) among children has been less helpful and supports

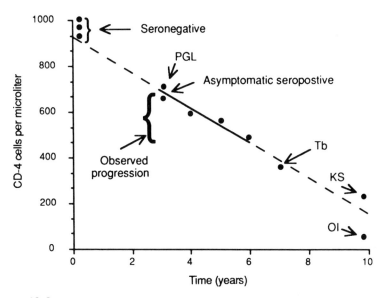

FIGURE 12.8 Conceptualization of the effects of HIV infection on CD4 lymphocytes over time, and susceptibility to disease progression, infection, and malignancy. The solid line represents data observed in a cohort of HIV-infected men (Moss, 1988), and the dashed line represents continuation of this slope both backward and forward in time. The dots indicated by the arrows are the number of CD4 lymphocytes found in seronegative persons and in persons with different HIV-related conditions. PGL, progressive generalized lymphadenopathy; Tb, tuberculosis; KS, Kaposi's sarcoma; OI, opportunistic infection. (Reprinted with permission from Hopewell, 1989.)

the notion that CD4 count alone does not tell the full story of AIDS progression. For example, Leibovitz *et al.* (1990) found that 8 of 22 (36%) infants presenting with PCP had CD4 lymphocyte counts above $450/\mu l$. These and other data have led to age-specific CD4 counts for the determination of when to begin PCP chemoprophylaxis; the recommended cutoffs for children 0–11 months, 12–23 months, 2–5 years, and >6 years are 1500, 750, 500, and 200, respectively (Centers for Disease Control, 1991b). In addition, although CD4 lymphocyte counts are very helpful in assessing the risk of disease progression, they are not always helpful in individual patients (Polis and Masur, 1990). Interestingly, if one compares the mortalities of patients receiving either zidovudine or placebo, it is lower in zidovudine recipients than in placebo recipients with the same CD4 count (Lagakos, 1991), underscoring that factors other than CD4 count are responsible for clinical progression. The use of a CD4 lymphocyte functional assay is being evaluated as a marker for disease progression (Lucey *et al.,* 1991).

Recognition of the value of the CD4 lymphocyte count as a prognostic indicator has had substantial impact on how intervention trials have been designed, implemented, and analyzed (Fischl *et al.*, 1987,1990; Volberding *et al.*, 1988). It also has had a direct impact on patient care since CD4 levels are used to monitor HIV-infected patients and to determine when to begin antiretroviral drug therapy and chemoprophylaxis for certain infectious diseases (Centers for Disease Control, 1989a,1991b). This strategy appears to have increased the life-span of patients infected with HIV (Moore *et al.*, 1990; Graham *et al.*, 1991).

Research Needs

The great strides that have been made in the last decade highlight the need for additional research. Laboratory methods for the measurement of biologic markers should be simplified to increase availability in clinical laboratories in both developed and developing countries. The development of better markers for AIDS progression among HIV-infected individuals would help refine the approach to antiretroviral therapy and chemoprophylaxis against opportunistic infections. Future research will determine whether CD4 function is a better marker than the CD4 count. In addition, better immunologic markers of protection following vaccination theoretically could reduce the number of clinical efficacy studies needed in different populations. Methods for distinguishing virulence markers from virulence factors will facilitate epidemiologic studies and increase the understanding of the pathogenesis of infectious diseases.

More specific nonculture methods are needed for the diagnosis of infectious disease, including serologic and antigen detection methods. Currently, many of these tests perform poorly when applied in populations with a low prevalence of infection. Although PCR is a novel method, additional work is needed to make the method feasible for clinical laboratories and to reduce the incidence of false-positive tests due to nucleic acid contamination of negative specimens.

References

Aach, R. D., Szmuness, W., Mosley, J. W., Holinger, F. B., Kahn, R. A., Stevens, C. E., *et al* (1981). Serum alanine aminotransferase of donors in relation to the risk of non-A, non-B hepatitis in recipients: The transfusion-transmitted viruses study. *N. Engl. J. Med.* **304**, 989–994.
Aho, K., Ahvonsen, P., Lassus, A., *et al.* (1973). HL-A antigen and reactive arthritis. *Lancet* **2**, 157.
Alter, H. J. (1991). Descartes before the horse: I clone, therefore I am: The hepatitis C virus in current perspective. Ann. Intern. Med. **115**, 644–649.
Ampel, N. M., Bejarano, G. C., Salas, S. D., and Galgiani, J. N. (1992). *In vitro* assessment of cellular immunity in human coccidioidomycosis: Relationship between dermal hypersen-

sitivity, lymphocyte transformation, and lymphokine production by peripheral blood mononuclear cells from healthy adults. *J. Infect. Dis.* **165**, 710–715.

Anderson, R. E., Lang, W., Shiboski, S., Royce, R., Jewell, N., and Winkelstein, W. (1990). Use of β₂-macroglobulin level and CD4 lymphocyte count to predict development of acquired immunodeficiency syndrome in persons with human immunodeficiency virus infection. *Arch. Intern. Med.* **150**, 73–77.

Archer, G. L., and Mayhall, C. G. (1983). Comparison of epidemiological markers used in the investigation of an outbreak of methicillin-resistant *Staphylococcus aureus* infections. *J. Clin. Microbiol.* **18**, 395–399.

Balows, A., Hausler, W. J., Herrmann, K. L., Isenberg, H. D., and Shadomy, H. J. (eds.) (1991). "Manual of Clinical Microbiology," 5th Ed. American Society for Microbiology, Washington, D.C.

Baron, E. J., and Finegold, S. M. (1990). Nontraditional methods for identification and detection of pathogens or their products. *In* "Bailey & Scott's Diagnostic Microbiology" (S. M. Finegold and E. J. Baron, eds.), 8th Ed., pp. 127–141. Mosby, St. Louis, Missouri.

Barrera, J. M., Bruguera, M., Ercilla, G., Sánchez-Tapias, J. M., Gil, M. P., Gil, C., Costa, J., Gelabert, A., Rodés, J., and Castillo, R. (1991). Incidence of non-A, non-B hepatitis after screening blood donors for antibodies to hepatitis C virus and surrogate markers. *Ann. Intern. Med.* **115**, 596–600.

Bergan, T. (1978). Phage typing of *Pseudomonas aeruginosa*. *In* "Methods in Microbiology" (T. Bergan and J. Norris, ed.), Vol. 10, pp. 169–199. Academic Press, London.

Bibb, W. F., Schwartz, B., Gellen, B. G., Plikaytis, B. D., and Weaver, R. E. (1989). Analysis of *Listeria monocytogenes* by multilocus enzyme electrophoresis and application of the method to epidemiologic investigations. *Int. J. Food Microbiol.* **8**, 233–239.

Bibb, W. F., Gellen, B. G., Weaver, R., Schwartz, B., Plikaytis, B. D., Reeves, M. W., Pinner, P. W., and Broome, C. V. (1990). Analysis of clinical and food-borne isolates of *Listeria monocytogenes* in the United States by multilocus enzyme electrophoresis and application of the method to epidemiologic investigations. *Appl. Environ. Microbiol.* **56**, 2133–2141.

Biggar, R. J., Gigase, P. L., Melbye, M., Kestens, L., Sarin, P. S., Bodner, A. J., Demedts, P., Stevens, W. J., Paluku, L., Delacollette, C., and Blattner, W. A. (1985). ELISA HTLV retrovirus antibody reactivity associated with malaria and immune complexes in healthy Africans. *Lancet* **2**, 520–523.

Bijlmer, H. A., van Alphen, L., den Broek, L. G., Greenwood, B. M., Valkenburg, H. A., and Dankert, J. (1992). Molecular epidemiology of *Haemophilus influenzae* type b in The Gambia. *J. Clin. Microbiol.* **30**, 386–390.

Black, R. E., Levine, M. M., Clements, M. L., Hughs, T., and O'Donnell, S. (1987). Association between O blood group and occurrence and severity of diarrhoea due to *Escherichia coli*. *Trans. R. Soc. Trop. Med. Hyg.* **81**, 120–123.

Blank, E. C., Quan, T. J., Mayer, L. W., Craven, R. B., Bailey, R., Dennis, D. T., Campbell, G. L., and Gubler, D. J. (1990). Lyme disease serology test kit evaluations. *In* "Proceedings of the First National Conference on Lyme Disease Testing, Dearborn, Michigan," pp. 79–89.

Blattner, W. A., Kalyanaraman, V. S., Robert-Guroff, M., Lister, T. A., Galton, D. A., Sarin, P. S., Crawford, M. H., Catovsky, D., Greaves, M., and Gallo, R. C. (1982). The human type-C retrovirus, HTLV, in blacks from the Caribbean region, and relationship to adult T-cell leukemia/lymphoma. *Int. J. Cancer* **30**, 257–263.

Blayney, D. W., Blattner, W. A., Robert-Guroff, M., Jaffe, E. S., Fisher, R. I., Bunn, P. A., Patton, M. G., Rarick, H. R., and Gallo, R. C. (1983). The human T-cell leukemia-lymphoma virus in the southeastern United States. *JAMA J. Am. Med. Assoc.* **250**, 1048–1052.

Blumberg, H. M., Stephens, D. S., Licitra, C., Pigott, N., Swaminathan, B., and Wachsmuth, I. K. (1991). Molecular epidemiology of group B streptococcal infections: Use of restriction endonuclease analysis of chromosomal DNA and DNA restriction fragment length polymorphisms of rRNA genes. *Abstr. 31 Intersci. Conf. Antimicrob. Agents Chemother.* **1057, 275.**

Blumberg, H. M., Rimland, D., Kiehlbauch, J. A., Terry, P. M., and Wachsmuth, I. K. (1992). Epidemiologic typing of *Staphylococcus aureus* by DNA restriction fragment length polymorphisms of rRNA genes: Elucidation of the clonal nature of a group of bacteriophage-nontypeable, ciprofloxacin-resistant, methicillin-susceptible *S. aureus* isolates. *J. Clin. Microbiol.* **30,** 362–369.

Bobo, L., Munoz, B., Viscidi, R., Quinn, T., Mkocha, H., and West, S. (1991). Diagnosis of *Chlamydia trachomatis* eye infection in Tanzania by polymerase chain reaction/enzyme immunoassay. *Lancet* **338,** 847–850.

Bodey, G. P., Buckley, M., Sathe, Y. S., and Freireich, E. J. (1966). Quantitive relationships between circulating leukocytes and infection in patients with acute leukemia. *Ann. Intern. Med.* **64,** 328–340.

Brazilian Purpuric Fever Study Group (1987a). Brazilian purpuric fever. Epidemic purpura fulminans associated with antecedent purulent conjunctivitis. *Lancet* **ii,** 757–761.

Brazilian Purpuric Fever Study Group (1987b). *Haemophilus aegyptius* bacteremia in Brazilian purpuric fever. Lancet **ii,** 761–763.

Breiman, R. F., Fields, B. S., Sandern, G. N., Volmer, L., Meier, A., and Spika, J. S. (1990). Association of shower use with Legionnaires' disease. Possible role of amoebae. *J. Am. Med. Assoc.* **263,** 2924–2926.

Brenner, D. J. (1991). Taxonomy, classification and nomenclature of bacteria. *In* "Manual of Clinical Microbiology" (A. Balows, W. J. Hausler, K. L. Herrmann, H. D. Isenberg, and H. J. Shadomy, eds.), 5th Ed., pp. 209–215. American Society for Microbiology, Washington, D.C.

Brenner, D. J., McWhorter, A. C., Leete Knutson, J. K., and Steigerwalt, A. G. (1982). *Escherichia vulneris:* A new species of Enterobacteriaceae associated with human wounds. *J. Clin. Microbiol.* **15,** 1133–1140.

Brenner, D. J., Mayer, L. W., Carlone, G. M., Harrison, L. H., Bibb, W. F., Brandileone, M. C. C., Sottnek, F. O., Irino, K., Reeves, M. W., Swenson, J. M. Birkness, K. A., Weyant, R. S., Berkley, S. F., Woods, T. C., Steigerwalt, A. G., Grimont, P. A. D., McKinney, R. M., Fleming, D. W., Gheesling, L. L., Cooksey, R. C., Arko, R. J., and Broome, C. V. (1988). Biochemical, genetic, and epidemiologic characterization of *Haemophilus influenzae* biogroup aegyptius (*Haemophilus aegyptius*) strains associated with Brazilian purpuric fever. *J. Clin. Microbiol.* **26,** 1524–1534.

Bruton, O. C. (1952). Agammaglobulinemia. *Pediatrics* **9,** 722–730.

Carlone, G. M., Gorelkin, L., Gheesling, L. L., Erwin, A. L. Hoiseth, S. K., Mulks, M. H., O'Connor, S. P., Weyant, R. S., Myrick, J., Rubin, L., Munford, R. S., White, E. H., Arko, R. J. Swaminathan, B., Graves, L. M., Mayer, L. W., Robinson, M. K., and Caudill, S. P. (1989). Potential virulence-associated factors in Brazilian purpuric fever. *J. Clin. Microbiol.* **27,** 609–614.

Carman, W. F. (1991). The polymerase chain reaction. *Q. J. Med.* **287,** 195–203.

Catovsky, D., Greaves, M. F., Rose, M., Galton, D. A., Goolden, A. W., McCluskey, D. R., White, J. M., Lampert, I., Bourikas, G., Ireland, R., Brownell, A. I., Bridges, J. M., Blattner, W. A., and Gallo, R. C. (1982). Adult T-cell lymphoma-leukaemia in blacks from the West Indies. *Lancet* **1,** 639–643.

Caugant, D. A., Froholm, L. O., Bovre, K., Holten, E., Frasch, C. E., Mocea, L. F., Zollinger, W. D., and Selander, R. K. (1986). Intercontinental spread of a genetically distinctive complex of clones of *Neisseria meningitidis* causing epidemic disease. *Proc. Natl. Acad. Sci. U.S.A.* **83,** 4927–4931.

Caugant, D. A., Kristiansen, B. E., Fromholm, L. O., Bovre, K., and Selander, R. K. (1988). Clonal diversity of *Neisseria meningitidis* from a population of asymptomatic carriers. *Infect. Immun.* **56,** 2060–2068.

Centers for Disease Control (1989a). Guidelines for prophylaxis against *Pneumocystis carinii* pneumona for persons infected with human immunodeficiency virus. *Morbid. Mortal. Wkly. Rept.* **38**, (S-5), 1–9.

Centers for Disease Control (1989b). Interpretation and use of the Western blot assay for sero-diagnosis of human immunodeficiency virus type 1 infections. *Morbid. Mortal. Wkly. Rep.* **38**, (S-7), 1–7.

Centers for Disease Control (1989c). Listeriosis associated with consumption of turkey franks. *Morbid. Mortal. Wkly. Rep.* **38**, 267–268.

Centers for Disease Control (1990). Possible transmission of human immunodeficiency virus to a patient during an invasive dental procedure. *Morbid. Mortal. Wkly. Rep.* **39**, 389–493.

Centers for Disease Control (1991a). Update—Transmission of HIV infection during an invasive dental procedure—Florida. *Morbid. Mortal. Wkly. Rep.* **40**, 21–33.

Centers for Disease Control (1991b). Guidelines for prophylaxis against *Pneumocystis carinii* pneumonia for children infected with human immunodeficiency virus. *Morbid. Mortal. Wkly. Rep.* **40**, (RR-2), 1–13.

Centers for Disease Control (1992). Human T-lymphotropic virus type II among Guaymi Indians—Panama. *Morbid. Mortal. Wkly. Rep.* **41**, 209–211.

Chen, Y. A., Lee, T. H., Wiktor, S. Z., Shaw G. M., Murphy, E. L., Blattner, W. A., and Essex, M. (1990). Type-specific antigens for serological discrimination of HTLV-I and HTLV-II infection. *Lancet* **336**, 1153–1155.

Cherry, W. B., Pittman, B., Harris, P. P., Herbert, G. A., Thomason, B. M., Thacker, S. L., and Weaver, R. E. (1978). Detection of Legionnaire's disease bacteria by direct immunofluorescent staining. *J. Clin. Microbiol.* **8**, 329–338.

Choo, Q. L., Kuo, G., Weiner, A. J., Overly, L. R., Bradley, D. W., and Houghton, M. (1989). Isolation of a cDNA clone derived from a blood-borne non-A, non-B viral hepatitis genome. *Science* **244**, 359–362.

Ciesielski, C. A., Ou, C. Y., Economou, N., Furman, L., Myers, G., and Witte, J. J. (1991). Transmission of HIV to patients in a dental practice. *Abstr. 31 Intersci. Conf. Antimicrob. Agents Chemother.* **980**, 262.

Clark, J. W., Curgo, C., Franchini, G., Gibbs, W. N., Lofter, W., Neuland, Mann, D., Saxinger, C., Gallo, R. C., and Blattner, W. A. (1988). Molecular epidemiology of HTLV-1-associated non-Hodgkin's lymphomas in Jamaica. *Cancer* **61**, 1477–1482.

Clemens, J. D., Sack, D. A., Harris, J. R., Chakraborty, J., Khan, M. R., Hupa, S., Ahmed, F., Gomes, J., Rao, M. R., and Svennerholm, A. M. (1989). ABO blood groups and cholera: New observations on specificity of risk and modification of vaccine efficacy. *J. Infect. Dis.* **159**, 770–773.

Conroy, J. M., Stevens, R. W., and Hechemy, K. E. (1991). Enzyme immunoassay. *In* "Manual of Clinical Microbiology" (A. Balows, W. J. Hausler, K. L. Herrmann, H. D. Isenberg, and H. J. Shadomy, eds.), 5th ed., pp. 87–92. American Society for Microbiology, Washington, D.C.

Constantine, N. T., and Fox, E. (1989). Need to confirm HTLV-1 screening assays (letter). *Lancet* **1**, 108–109.

Coonrod, J. D. (1983a). Urine as an antigen reservoir for diagnosis of infectious disease. *Am. J. Med.* 7585–7592.

Coonrod, J. D. (1983b). Diagnosis of pneumonia by agglutination techniques. *In* "The Direct Detection of Microorganisms in Clinical Samples" (J. D. Coonrod, L. J. Kunz, and M. J. Ferraro, eds.), pp. 153–157. Academic Press, Orlando, Florida.

Crennan, J. M., Van Scoy, R. E., McKenna, C. H., and Smith, T. F. 1986. Echovirus polymyositis in patients with hypogammaglobulinemia: Failure of high-dose intravenous gammaglobulin therapy and review of the literature. *Am. J. Med.* **81**, 35–42.

Cruickshank, E. K. (1956). A neuropathic syndrome of uncertain origin. Review of 100 cases. *West Indian Med. J.* **5**, 147–158.

Daley, C. L., Small, P. M., Schecter, G. F., Schoolnik, G. K., McAdam, R. A., Jacobs, W. R., and Hopewell, P. C. (1992). An outbreak of tuberculosis with accelerated progression among persons infected with the human immunodeficiency virus: An analysis using restriction-fragment-length polymorphisms. *N. Engl. J. Med.* **326,** 231–235.

Dattwyler, R. J., Volkman, D. J., Luft, B. J., Halperin, J. J., Thomas, J., and Golightly, M. G. (1988). Seronegative Lyme disease. Dissociation of specific T- and B-lymphocyte responses to *Borrelia burgdorferi. N. Engl. J. Med.* **319,** 1441–1446.

De Gregorio, M., Lee, W., Linker, C., Jacobs, R. A., and Ries, C. A. (1982). Fungal infections in patients with acute leukemia. *Am. J. Med.* **73,** 543–548.

Detels, R., Visscher, B. R., Fahey, J. L., Sever, J. L., Gravell, M., Madden, D. L., Schwartz, K., Dudley, J. P., English, P. A., Powers, H., Clark, V. A., and Gottlieb, M. S. (1987). Predictors of clinical AIDS in young homosexual men in a high-risk population. *Int. J. Epidemiol.* **16,** 271–276.

Dressler, F., Yoshinari, N. H., and Steere, A. C. (1991). The T-cell proliferative assay in the diagnosis of Lyme disease. *Ann. Intern. Med.* **115,** 533–539.

Edwards, L. B., Acquaviva, F. A., Livesay, V. T., Cross, F. W., and Palmer, C. E. (1969). An atlas of sensitivity to tuberculin, PPD-B and histoplasmin in the United States. *Am. Rev. Resp. Dis.* **99(Suppl),** 1–132.

Escobar, M. R. (1991). Hemolytic assays: Complement fixation and antistreptolysin O. *In* "Manual of Clinical Microbiology" (A. Balows, W. J. Hausler, K. L. Herrmann, H. D. Isenberg, and H. J. Shadomy, eds.), 5th Ed., pp. 73–86. American Society for Microbiology, Washington, D.C.

Esolen, L. M., Fasano, M. B., Flynn, J., Burton, A., and Lederman, H. M. (1992). *Pneumocystis carinii* osteomyelitis in a patient with common variable immunodeficiency. *N. Engl. J. Med.* **326,** 999–1001.

Esteban, J. I., Gonzales, A., Hernandez, J. M., Viladomin, L., Sanchez, C., Lopez-Talavbra, J. C., Lucea, D., Martin-Vela, C., Vidal, X., and Esteban, R. (1990). Evaluation of antibodies to hepatitis C virus in a study of transfusion-associated hepatitis. *N. Engl. J. Med.* **323,** 1107–1112.

Fahey, J. L., Taylor, J. M. G., Detels, R., Hofmann, B., Melmed, R., Nishanian, P., and Giorgi, J. V. (1990). The prognostic value of cellular and serologic markers in infection with human immunodeficiency virus type 1. *N. Engl. J. Med.* **322,** 166–72.

Field, M. (1979). Mechanisms of action of cholera and *Escherichia coli* enterotoxins. *Am. J. Clin. Nutr.* **32,** 189–196.

Fischl, M. A., Richman, D. D., Grieco, M. H., Gottlieb, M. S., Volberding, P. A., Laskin, O. L., Leedom, J. M., Groopman, J. E., Mildvan, D., Schooley, R. T., Jackson, G. G., Durack, D. T., and King, D. (1987). The efficacy of azidothymidine (AZT) in the treatment of patients with AIDS and AIDS-related complex. A double-blind, placebo-controlled trial. *N. Engl. J. Med.* **317,** 185–191.

Fischl, M. A., Richman, D. D., Hansen, N., Collier, A. C., Carey, J. T., Para, M. F., Hardy, D., Dolin, R., Powderly, W. G., Allan, J. D., Wong, B., Merigan, T. C., McAuliffe, W. G., Hyslop, N. E., Rhame, F. S., Balfour, H. H., Spector, S. A., Volberding, P., Pettinelli, C., and Anderson, J. (1990). The safety and efficacy of zidovudine (AZT) in the treatment of subjects with mildly symptomatic human immunodeficiency virus type 1 (HIV) infection. A double-blind, placebo-controlled trial. *Ann. Intern. Med.* **112,** 727–737.

Frasch, C. E., Zollinger, W. D., and Poolman, J. T. (1985). Proposed schema for identification of serotypes of *Neisseria meningitidis. In* "The Pathogenic *Neisseria*" (G. K. Schoolnik, ed.), pp. 519–524. American Society for Microbiology, Washington, D.C.

Gallo, R. C. (1991). Human retroviruses: A decade of discovery and link with human disease. *J. Infect. Dis.* **164,** 235–243.

Gallo, R. C., and Todaro, G. J. (1976). Oncogenic RNA viruses. *Semin. Oncol.* **3,** 81–95.

Gerber, M. A., Caparas, L. S., and Randolph, M. F. (1990). Evaluation of new latex agglutination test for detection of streptolysin O antibodies. *J. Clin. Microbiol.* **28,** 413–415.

Gessain, A., Cernant, J. C., Marus, L., Barin, F., Gout, O., Calender, A., and de Thé, G. (1985). Antibodies to human T-lymphotropic virus type-I in patients with tropical spastic paraparesis. *Lancet* **2**, 407–409.

Glass, R. I., Holmgren, J., Haley, C. E., Khan, M. R., Svennerholm, A. M., Stoll, B. J., Hossain, K. M. B., Black, R. E., Yunus, M., and Barua, D. (1985). Predisposition for cholera of individuals with O blood group. Possible evolutionary significance. *Am. J. Epidemiol.* **121**, 791–796.

Glick, T. H., Gregg, M. B., Berman, B., Mallison, G., Thodes, W. W., and Kassanoff, I. (1978). Pontiac fever. An epidemic of unknown etiology in a health department: I. Clinical and epidemiologic aspects. *Am. J. Epidemiol.* **107**, 149–60.

Gnann, J. W., McCormick, J. B., Mitchell, S., Nelson, J. A., and Oldstone, M. B. A. (1987). Synthetic peptide immunoassay distinguished HIV type 1 and HIV type 2 infections. *Science* **237**, 1346–1349.

Goedert, J., Biggar, R. J., Melbye, M., Mann, D. L., Wilson, S., Gail, M. H., Grossman, R. J., DiBioia, R. A., Sanches, W. C., Weiss, S. H., and Blattner, W. A. (1987). Effect of T4 count and cofactors on the incidence of AIDS in homosexual men infected with human immunodeficiency virus. *J. Am. Med. Assoc.* **257**, 331–334.

Gonzalez, R., and Hanna, B. A. (1987). Evaluation of Gen-Probe DNA hybridization systems for the identification of *Mycobacterium tuberculosis* and *Mycobacterium avium* complex from patients with acquired immunodeficiency syndrome. *J. Clin. Microbiol.* **25**, 1551–1552.

Gottlieb, M. S., Schroff, R., Schanker, H. M., Weisman, J. D., Fan, P. T., Wolf, R. A., and Saxon, A. (1981). *Pneumocystis carinii* pneumonia and mucosal candidiasis in previously healthy homosexual men: Evidence of a new acquired cellular immunodeficiency. *N. Engl. J. Med.* **305**, 1425–1431.

Graham, N. M. H., Zeger, S. L., Park, L. P., Phair, J. P., Detels, R., Vermund, S. H., Ho, M., and Saah, A. J. (1991). Effect of zidovudine and *Pneumocystis carinii* pneumonia prophylaxis on progression of HIV-1 infection in AIDS. *Lancet* **ii**, 265–269.

Grimont, F., and Grimont, P. A. D. (1986). Ribosomal ribonucleic acid gene restriction patterns as potential taxonomic tools. *Ann. Inst. Pasteur/Microbiol.* **137B**, 165–175.

Hanrahan, J. P., Benach, J. L., Coleman, Bosler, E. M., Morse, D. L., Cameron, D. J., Edelman, R., and Kaslow, R. A. (1984). Incidence and cumulative frequency of endemic Lyme disease in a community. *J. Infect. Dis.* **150**, 489–496.

Harrison, L. H., da Silva, G. A., Pittman, M., Fleming, D. W., Vranjac, A., and Broome, C. V. (1989a). Minireview: The epidemiology and clinical spectrum of Brazilian purpuric fever. *J. Clin. Microbiol.* **27**, 599–604.

Harrison, L. H., Ezzell, J. W., Abshire, T. G., Kidd, S., and Kaufmann, A. F. (1989b). Evaluation of serologic tests for diagnosis of anthrax following an outbreak of cutaneous anthrax in Paraguay. *J. Infect. Dis.* **160**, 706–710.

Harrison, L. H., and Broome, C. V. (1989c). Summary of a symposium: Brazilian purpuric fever—Progress but unanswered questions. *Pediatr. Infect. Dis.* **8**, 248–249.

Harrison, L. H., Armstrong, C. W., Jenkins, S. R., Harmon, M. W., Ajello, G. W., Miller, G. B., and Broome, C. V. (1991a). A cluster of meningococcal disease on a school bus following epidemic influenza. *Arch. Intern. Med.* **15**, 1005–1009.

Harrison, L. H., da Silva, A. P. J., Gayle, H. D., Albino, P., Rayfield, M. A., George, R., Lee-Thomas, S., Del Castillo, F., and Heyward, W. L. (1991b). Risk factors for HIV-2 infection in Guinea-Bissau. *J. AIDS* **4**, 1155–1160.

Hechemy, K. E., Stevens, R. W., and Conroy, J. M. (1991). Immunoelectroblot techniques. *In* "Manual of Clinical Microbiology," (A. Balows, W. J. Hausler, K. L. Herrmann, H. D. Isenberg, and H. J. Shadomy, eds.), 5th Ed., pp. 93–98. American Society for Microbiology, Washington, D.C.

Hedberg, C. W., Osterholm, M. T., Macdonald, K. L., and White, K. E. (1987). An interlaboratory study of antibody to *Borrelia burgdorferi*. *J. Infect. Dis.* **155**, 1325–1327.

Hermans, P. E., Diaz-Buxo, J. A., and Stobo, J. D. (1976). Idiopathic late onset immunoglobulin deficiency. Clinical observations in 50 patients. *Am. J. Med.* **61**, 221–232.

Heyward, W. L., and Curran, J. W. (1988). Rapid screening tests for HIV infection (editorial). *JAMA J. Am. Med. Assoc.* **260**, 542.

Hide, G., and Tait, A. (1991). The molecular epidemiology of parasites. *Experientia* **47**, 128–142.

Hill, A. V. S., Allsop, C. E. M., Kwiatkowski, D., Bennett, S., Brewster, D., McMichael, A. J., Greenwood, B. M., Anstey, N. M., Twumasi, P., Rowe, P. A. (1991). Common West African HLA antigens are associated with protection from severe malaria. *Nature (London)* **352**, 595–600.

Hino, S., Yamaguchi, K., Katamine, S. *et al.* (1985). Mother-to-child transmission of human T cell leukemia virus type I. *Japan J Cancer Res (Gann)* **76**, 474–480.

Hinuma, Y., Komoda, H., Chosa, T., Kondo, T., Kohakura, M., Takenara, Kikuchi, M., Ichimaru, M., Yunoki, K., Sato, I., Matsuo, R., Takiuchi, Y., Uchino, H., and Hanaoka, M. (1982). Antibodies to adult T-cell leukemia-virus-associated antigen (ATLA) in sera from patients with ATL and controls in Japan: A nationwide seroepidemiologic study. *Int. J. Cancer* **29**, 631–635.

Hjelle, B., Mills, R., Swenson, S., Mertz, G., Key, C., and Allen, S. (1991). Incidence of hairy cell leukemia, mycosis fungoides, and chronic lymphocytic leukemia in first known HTLV-II-endemic population. *J. Infect. Dis.* **163**, 435–440.

Holmberg, S. D., Wachsmuth, I. K., Hichmann-Brenner, F. W., and Cohen, M. L. (1984). Comparison of plasmid profile analysis, phage typing, and antimicrobial susceptibility testing in characterizing *Salmonella typhimurium* isolates from outbreaks. *J. Clin. Microbiol.* **19**, 100–104.

Hoofnagle, J. H. (1990). Acute viral hepatitis. In "Principles and Practice of Infectious Disease (G. L. Mandell, R. G. Douglas and J. E. Bennett, eds.), pp 1001–1017. Churchill Livingstone, New York.

Hopewell, P. C. (1989). Human immunodeficiency virus—associated lung infection: an overview. *Sem. Resp. Inf.* **4**, 73–74.

Horsburgh, C. R. (1991). *Mycobacterium avium* complex infection in the acquired immunodeficiency syndrome. *N. Engl. J. Med.* **324**, 1332–1337.

Hughs, J. H., Mann, D. R., and Hamparian, V. V. (1988). Detection of respiratory syncytial virus in clinical specimens by viral culture, direct and indirect immunofluorescence and enzyme immunoassay. *J. Clin. Microbiol.* **26**, 588–591.

Hunter, E. F., Russell, H., Farshy, C. E., Sampson, J. S., and Larsen, S. A. (1986). Evaluation of sera from patients with Lyme disease in the fluorescent antibody-absorption test for syphilis. *Sex. Transm. Dis.* **13**, 232–236.

Ingelfinger, J. A., Mosteller, F., Thibodeau, L. A., and Ware, J. H. (1983). Diagnostic testing: Introduction to probability. *In* "Biostatistics in Clinical Medicine," pp. 1–24. Macmillan, New York.

Irino, K., Grimont, F., Casin, I., and Grimont, P. A. (1988). rRNA gene restriction patterns of *Haemophilus influenzae* biogroup *aegyptius* strains associated with Brazilian purpuric fever. *J. Clin. Microbiol.* **26**, 1535–1538.

John, J. F., and Twitty, J. A. (1986). Plasmids as epidemiologic markers in nosocomial gramnegative bacilli: Experience at a university and review of the literature. *Rev. Infect. Dis.* **5**, 693–704.

Joly, J. R., McKinney, R. M., Tobin, J. O., Bibb, William F., Watkins, I., and Ramsay, D. (1986). Development of a standardized subtyping scheme for *Legionella pneumophila* serogroup 1 using monoclonal antibodies. *J. Clin. Microbiol.* **23**, 768–771.

Jones, J. M. (1991). Serodiagnosis of Lyme disease. *Ann. Intern. Med.* **114**, 1064.

Kajiyama, W., Kashiwagi, S., Ikematsu, H., Hayashi, J., Nomura, H., and Okochi, K. (1986). Intrafamilial transmission of adult T cell leukemia virus. *J. Infect. Dis.* **154**, 851–857.

Kalinske, R. W., Parker, R. H., Brandt, D., and Hoeprich, P. (1967). Diagnostic usefulness and safety of transtracheal aspiration. *N. Engl. J. Med.* **276**, 604–608.

Kalyanaraman, V. S., Sarngadharan, M. G., Bunn, P. A., Minna, J. D., and Gallo, R. C. (1981-). Antibodies in human sera reactive against an internal structural protein of human T-cell lymphoma virus. *Nature (London)* **294,** 271–273.

Kalyanaraman, V. S., Sarngadharan, M. G., Robert-Guroff, M., Miyoshi, I., Golde, D., and Gallo, R. C. (1982). A new subtype of human T-cell leukemia virus (HTLV-II) associated with a T-cell variant of hairy cell leukemia. *Science* **218,** 571–573.

Kendrick, P. L., Eldering, G., Dixon, M. K., and Misner, J. (1947). Mouse protection tests in the study of pertussis vaccine: A comparative series using intracerebral route for challenge. *Am. J. Publ. Health* **37,** 803–810.

Kew, O. M. Nottay, B. K., Rico-Hesse, R., and Pallansch, M. (1990). Molecular epidemiology of wild poliovirus transmission. *In* "Virus Variability, Epidemiology, and Control" (E. Kurstak, ed.), pp. 199–221. Plenum Medical Books, New York.

Kilian, M. (1976). A taxonomic study of the genus *Haemophilus,* with the proposal of a new species. *J. Gen. Microbiol.* **93,** 9–62.

Klastersky, J., Zinner, S. H., Calandra, T., Gaya, H., Glauser, M. P., Meunier, F., Rossi, M., Schimpff, S. C., Tattersall, M., Viscoli, C., and EORTC (European Organ. Research & Treatment of Ca). (1988). Empiric antimicrobial therapy for febrile granulocytopenic cancer patients: Lessons from four EORTC trials. *Eur. J. Cancer Clin. Oncol.* **24,** S35–S45.

Kramer, A., and Blattner, W. A. (1989). The HTLV-1 model and chronic demyelinating neurologic diseases. *In* "Concepts in Viral Pathogenesis" (M. B. A. Oldstone and A. L. Notkins, eds.), pp. 204–214. Springer-Verlag, New York.

Kuo, G., Choo, Q. L., Alter, H. J., Gitnick, G. L., Redeker, A. G., Purcell, R. H., Miyamura, T., Dienstag, J. L., Alter, M. J., Stevens, C. E., Tegtmeier, G. E., Bonino, F., Columbo, M., Lee, W.-S., Kuo, C., Berger, K., Schuster, J. R., Overby, L. R., Bradley, D. W., and Houghton, M. (1989). An assay for circulating antibodies to a major etiologic virus of human non-A, non-B viral hepatitis genome. *Science* **244,** 362–364.

Lagakos, S. W. (1991). Of what value are surrogate markers? Paper presented at 31st Conference on Antimicrobial Agents and Chemotherapy. Chicago, September 29.

LaGrenade, L., Hanchard, B., Fletcher, V., Cranston, B., and Blattner, W. (1990). Infective dermatitis of Jamaican children: A marker for HTLV-I infection. *Lancet* **336,** 1345–1347.

Lederman, H. M., and Winkelstein, J. A. (1985). X-linked agammaglobulinemia: An analysis of 96 patients. *Medicine* **64,** 145–156.

Lee, H., Swanson, P., Shorty, V. S., Zack, J. A., Rosenblatt, J. D., and Chen, I. S. Y. (1989). High rate of HTLV-II infection in seropositive IV drug abusers in New Orleans. *Science* **244,** 471–475.

Leibovitz, E., Rigaud, M., Pollack, H., Lawrence, R., Chandwani, S., Krasinski, K., and Borkowsky, W. (1990). *Pneumocystis carinii* pneumonia in infants infected with the human immunodeficiency virus with more than 450 CD4 T lymphocytes per cubic millimeter. *N. Engl. J. Med.* **323,** 531–533.

Leirisalo-Repo, M. (1988). Ten-year follow-up study of patients with *Yersinia* arthritis. *Arthritis Rheum.* **31,** 533–537.

Levine, P. H., Jaffe, E. S., Manns, A., Murphy, E. L., Clark, J., and Blattner, W. A. (1988). Human T-cell lymphotropic virus type I and adult T-cell leukemia, lymphoma outside Japan and the Caribbean basin. *Yale J. Biol. Med.* **61,** 215–222.

Lew, A. M. (1991). The polymerase chain reaction and related techniques. *Curr. Opin. Immunol.* **3,** 242–246.

Lin, F. C., Morris, J. G., Trump, D., Tilghman, D., Wood, P. K., Jackman, N., Israel, E., and Libonati, J. P. (1988). Investigation of an outbreak of *Salmonella enteritidis* gastroenteritis associated with consumption of eggs in a restaurant chain in Maryland. *Am. J. Epidemiol.* **128,** 839–844.

Lin, H. J., Lai, C., Lauder, I. J., Wu, P., Lau, T. K., and Fong, M. (1991). Application of hepatitis B virus (HBV) DNA sequence polymorphisms to the study of HBV transmission. *J. Infect. Dis.* **164,** 284–288.

Linnan, M. J., Mascola, L., Lou, X. D., Goulet, V., May, S., Salminen, C., Hird, D. W., Yonekura, L., Hayes, P., Weaver, R., Audurier, A., Plikaytis, B. D., Fannin, S. L., Kleks, A., and Broome, C. V. (1988). Epidemic listeriosis assocaited with Mexican style cheese. *N. Engl. J. Med.* **319**, 823–828.

Lipka, J. J., Buji, K., Reyes, G. R., Moeckli, R., Wiktor, S. Z., Blattner, W. A., Murphy, E. L., Shaw, G. M., Hanson, C. V., Sninsky, J. J., Foung, S. K. H. (1990). Determination of a unique and immunodominant epitope of human T-cell lymphotropic virus type I. *J. Infect. Dis.* **162**, 353–357.

Louie, L. G., Newman, B., and King, M. (1991). Influence of host genotype on progression to AIDS among HIV-infected men. *J. AIDS* **8**, 814–818.

Lucey, D. R., Melcher, G. P., Hendrix, C. W., Zajac, R. A., Goetz, D. W., Butzin, C. A., Clerici, M., Warner, R. D., Abbadessa, S., Hall, K., Jaso, R., Woolford, B., Miller, S., Stocks, N. I., Salinas, C. M., Wolfe, W. H., Shearer, G. M., and Boswell, R. N. (1991). Human immunodeficiency virus infection in the U.S. Air Force: Seroconversions, clinical staging, and assessment of a T helper cell functional assay to predict change in CD4+ T cell counts. *J. Infect. Dis.* **164**, 631–637.

MacDonnell, K. B., Chmiel, J. S., Poggensee, L., Wu, S., and Phair, J. P. (1990). Predicting progression to AIDS: Combined usefulness of CD4 lymphocyte counts and p24 antigenemia. *Am. J. Med.* **89**, 706–712.

McKinney, R. E., Katz, S. L., and Wilfert, C. M. (1987). Chronic enteroviral meningoencephalitis in agammaglobulinemic patients. *Rev. Infect. Dis.* **9**, 334–356.

Magnarelli, L. A., Anderson, J. F., and Johnson, R. C. (1987). Corss-reactivity in serological tests for Lyme disease and other spirochetal infections. *J. Infect. Dis.* **156**, 183–188.

Maher, K. O., Morris, J. G., Gotuzzo, E., Ferreccio, C., Ward, L. R., Benavente, L., Black, R. E., Rowe, B., and Levine, M. M. (1986). Molecular techniques in the study of *Salmonella typhi* in epidemiologic studies in endemic areas: Comparison with VI Phage typing. *Am. J. Trop. Med. Hyg.* **35**, 831–835.

Makela, P. H., Peltola, H., Kayhty, H., Jousimies, H., Pettay, O., Ruoslahti, E., Sivonen, A., and Renkonen, O. V. (1977). Polysaccharide vaccines of group A *Neisseria meningitidis* and *Haemophilus influenzae* type b: A field trial in Finland. *J. Infect. Dis.* **136(Suppl)**, S43–50.

Manns, A., and Blattner, W. A. (1991). The epidemiology of the human T-cell lymphotropic virus type I and type II: Etiologic role in human disease. *Transfusion* **31**, 67–75.

Marescot, M. R., Budkowska, A., Pillot, J., and Debre, P. (1989). HLA linked immune response to S and pre-S2 gene products in hepatitis B vaccination. *Tissue Antigens* **33**, 495–500.

Masur, H., Michelis, M. A., Green, J. B., Onorato, I., Vande Stouwe, R. A., Holzman, R. S., Wormser, G., Brettman, L., Lange, M., Murray, H., and Cunningham-Rundles, S. (1981). An outbreak of community-acquired *Pneumocytis carinii* pneumonia: Initial manifestation of cellular immune dysfunction. *N. Engl. J. Med.* **305**, 1431–1438.

Mertsola, J., Cope, L. D., Munford, R. S., McCracken, G. H., and Hansen, E. J. (1991). Detection of experimental *Haemophilus influenzae* type b bacteremia and endotoxemia by means of an immunolimulus assay. *J. Infect. Dis.* **164**, 353–358.

Miliotis, M. D., Galen, J. E., Kaper, J. B., and Morris, J. G. (1989). Development and testing of a synthetic oligonucleotide probe for the detection of pathogenic *Yersinia* strains. *J. Clin. Microbiol.* **27**, 1667–1670.

Miller, D. G. (1972). Patterns of immunologic deficiency in lymphomas and leukemias. *Ann. Intern. Med.* **57**, 703–706.

Minamoto, G. Y., Gold, J. W. M, Scheinberg, D. A., Hardy, W. E., Chein, N., Zuckerman, E., Reich, L., Dietz, K., Gee, T., Hoffer, J., Mayer, K., Gabrilove, J., Clarkson, B., and Armstrong, D. (1988). Infection with human T-cell leukemia virus type I in patients with leukemia. *N. Engl. J. Med.* **318**, 219–222.

Minshu, B., Griffin, P. M., Tauxe, R. V., Cameron, D. N., Hutcheson, R. H., and Schaffner, W. (1991). *Salmonella enteritidis* gastroenteritis transmitted by intact chicken eggs. *Ann. Intern. Med.* **115**, 190–194.

Moore, P. S., Reeves, M. W., Schwartz, B., Gellin, B. G., and Broome, C. V. (1989). Intercontinental spread of an epidemic group A *Neisseria meningitidis* strain. *Lancet* **2**, 260–262.
Moore, R. D., Hidalgo, J., Sugland, B. W., and Chaisson, R. E. (1990). Zidovudine and the
natural history of the acquired immunodeficiency syndrome. *N. Engl. J. Med.* **324**,
1412–1416.
Morris, J. G., Lin, F. C., Morrison, C. B., Gross, R. J., Rima, K., Maher, K. O., Rowe, B., Israel,
E., and Libonati, J. P. (1986). Molecular epidemiology of neonatal meningitis due to *Citro-
bacter diversus*: A study of isolates from hospitals in Maryland. *J. Infect. Dis.* **154**,
409–414.
Mortimer, E. A. (1988). Pertussis vaccine. *In* "Vaccines" (S. A. Plotkin and E. A. Mortimer,
eds.), pp. 74–97. Saunders, Philadelphia.
Moseley, S. L., Echeverria, P., Seriwatana, J., Tirapat, C., Chaicumpa, W., Sakuldaipeara, T.,
and Falkow, S. (1982). Identification of enterotoxigenic *Escherichia coli* by colony hybridization using three enterotoxin gene probes. *J. Infect. Dis.* **145**, 863–869.
Moss, A. R., Bacchetti, P., Osmond, D., *et al.* (1988). Seropositivity for HIV and development
of AIDS or AIDS-related condition: Three-year follow-up of the San Francisco General
Hospital Cohort. *Br. Med. J.* **296**, 745–750.
Murphy, E. L., Hanchard, B., Figueroa, J. P., Gibbs, W. N., Lofters, W. S., Campbell, M., Goedert, J. J., and Blattner, W. A. (1989a). Modelling the risk of adult T-cell leukemia/lym-
phoma in persons infected with human T-lymphotropic virus type I. *Int. J. Cancer* **43**,
250–253.
Murphy, E. L., Figueroa, P., Gibbs, W. N., Brathwaite, A., Holding-Cobham, M., Waters, D.,
Cranston, B., Hanchard, B., and Blattner, W. A. (1989b). Sexual transmission of human T-
lymphotropic virus type I (HTLV-I). *Ann. Intern. Med.* **111**, 555–560.
Musser, J. M., Kroll, J. S., Granoff, D. M., Moxon, E. R., Brodeur, B. R., Campos, J., Dabernat,
H., Frederiksen, W., Hamel, J., Hammond, G., Høiby, A., Jonsdottir, K. E., Kabeer, M.,
Kallings, I., Khan, W. N., Kilian, M., Knowles, K., Koornhof, H. J., Law, B., Li, K. I.,
Montgomery, J., Pattison, P. E., Piffaretti, J., Takala, A. I., Thong, M. L., Wall, R. A., Ward,
J. I., and Selander, R. K. (1990). Global genetic structure and molecular epidemiology of
encapsulated *Haemophilus influenzae*. *Rev. Infect. Dis.* **12**, 75–111.
Myers, G., Josephs, S. F., Rabson, A. B., Smith. T. F., and Wong-Staal, F. (1988). "Human
retroviruses and AIDS." Los Alamos National Laboratory, Los Alamos, New Mexico.
Nagamine, M., Nakashima, Y., Uemura, S., Takei, H., Toda, T., Maehama, T., Nakachi, H.,
and Nakayama, M. (1991). DNA amplification of human T lymphotropic virus type I
(HTLV-I) proviral DNA in breast milk of HTLV-I carriers (letter). J. Infect Dis **164**,
1024–1025.
Nguyen, M. H., Stout, J. E., and Yu, V. L. (1991). Legionellosis. *Infect. Dis. N. Am.* **5**,
561–584.
Nolte, F. S., Conlin, C., Roisin, A., and Redmond, S. R. (1984). Plasmids as epidemiological
markers in nosocomial legionnaires' disease. *J. Infect. Dis.* **149**, 251–256.
Oberhofer, T. R., and Back, A. E. (1979). Biotypes of *Haemophilus* encountered in clinical labo-
ratories. *J. Clin. Microbiol.* **10**, 168–174.
Ochs, H. D., Ament, M. E., and Davis, I. D. (1972). Giardiasis with malabsorption in X-linked
agammaglobulinemia. *N. Engl. J. Med.* **287**, 341–342.
Okochi, K., Sato, H., and Hinuma, Y. (1984). A retrospective study on transmission of adult
T cell leukemia virus by blood transfusion: Seroconversion in recipients. *Vox Sang.* **46**,
245–253.
O'Neill, K. P., Lloyd-Evans, N., Campbell, H., Forgie, I. M., Sabally, S., and Greenwood, B. M.
(1989). Latex agglutination test for diagnosing pneumococcal pneumonia in children in
developing countries. *Br. Med. J.* **298**, 1061–1064.
Osame, M., Usuku, K., Izumo, S., Ijichi, N., Amitani, H., Igata, A., Matsumoto, M., and Tara,
M. (1986a). HTLV-1 associated myelopathy, a new clinical entity. *Lancet* **1**, 1031–1032.

Osame, M., Izumo, S., Igata, A., Matsumoto, M., Matsumoto, T., Sonada, S., Tara, M., and Shibata, Y. (1986b). Blood transfusion and HTLV-1 associated myelopathy. *Lancet* 2, 104–105.

Osame, M., Matsumoto, M., Usuku, K., Izumo, S., Ijichi, N., Amitani, H., Igata, A., and Tara, M. (1987). Chronic progressive myelopathy associated with elevated antibodies to human T-lymphotropic virus type I and adult T-cell leukemialike cells. *Ann. Neurol.* 21, 117–122.

Piffaretti, J. C., Kressebuch, H., Aeschbacher, M., Bille, J., Bannerman, E., Musser, J. M., Selander, R. K., and Rocourt, J. (1989). Genetic characterization of clones of the bacterium *Listeria monocytogenes* causing epidemic disease. *Proc. Natl. Acad. Sci. U.S.A.* 86, 3818–3822.

Poiesz, B. J., Ruscetti, F. W., Gazdar, A. F., Bunn, P. A., Minna, J. D., and Gallo, R. C. (1980). Detection and isolation of type-C retrovirus particles from fresh and cultured lymphocytes of patients with cutaneous T-cell lymphoma. *Proc. Natl. Acad. Sci. U.S.A.* 77, 7415–7419.

Polis, M. A., and Masur, H. (1990). Predicting the progression to AIDS. *Am. J. Med.* 89, 701–705.

Polk, B. F., Fox, R., Brookmeyer, R., Kanchanaraksa, S., Kaslow, R., Visscher, B., Rinaldo, C., and Phair, J. (1987). Predictors of the acquired immunodeficiency syndrome developing in a cohort of seropositive homosexual men. *N. Engl. J. Med.* 316, 61–66.

Posner, L. E., Robert-Guroff, M., Kalyanaraman, V. S., Poiesz, B. J., Ruscetti, F. W., Fossieck, B., Bunn, P. A., Minna, D., and Gallo, R. C. (1981). National antibodies to the human T cell lymphoma virus in patients with cutaneous T cell lymphomas. *J. Exp. Med.* 154, 333–346.

Quinn, T. C., Riggin, C. H., Kline, R. L., Francis, H., Mulanga, K., Sension, M., and Fauci, A. (1988). Rapid latex agglutination assay using recombinant envelope polypeptide for the detection of antibody to the human immunodeficiency virus. *J. Am. Med. Assoc.* 260, 510–513.

Ratner, L., and Poiesz, B. J. (1988). Leukemias associated with human T-cell lymphotropic virus type I in a non-endemic region. *Medicine* 67, 401–422.

Robert-Guroff, M., Nakao, K., Notake, K., Ito, Y., Sliska, A., and Gallo, R. C. (1982). Natural antibodies to human retrovirus HTLV in a cluster of Japanese patients with adult T cell leukemia. **Science** 215, 975–978.

Rocourt, J., Audurier, A., Courtieu, A. L., Durst, J., Ortel, S., Schrettenbrunner, A., and Taylor, A. G. (1985). A multi-center study on the phage typing of *Listeria monocytogenes. Zentralbl. Bakteriol. Parasitenkd. Infektionskr. Hyg.* 259, 489–497.

Rodgers, P. E. B. (1965). The clinical features and aetiology of the neuropathic syndrome in Jamaica. *West Indian Med. J.* 14, 36–47.

Rogers, M. F., Ou, C. Y., Kilbourne, B., and Schochetman, G. (1990). Advances and problems in the diagnosis of HIV infection in infants. *In* "Pediatric AIDS—The Challenge of HIV Infection in Infants, Children, and Adolescents" (P. A. Pizzo and C. M. Wilfert, eds.), pp. 159–174. Williams & Wilkins, Baltimore.

Roggendorf, M., Frosner, G. G., Deinhardt, F., and Scheid, R. (1980). Comparison of solid phase test systems for demonstrating antibodies against hepatitis A virus of the IgM class. *J. Med. Virol.* 5, 47–62.

Roifman, C. M., Levison, H., and Gelfand, E. W. (1987). High-dose versus low-dose intravenous immunoglobulin in hypogammaglobulinemia and chronic lung disease. *Lancet* i, 1075–1077.

Rosebrock, J. A. (1991). Labeled-antibody techniques: Fluroescent, radioisotopic, immunochemical. *In* "Manual of Clinical Microbiology" (A. Balows, W. J. Hausler, K. L. Herrmann, H. D. Isenberg, and H. J. Shadomy, eds.), 5th Ed., pp. 79–86. American Society for Microbiology, Washington, D.C.

Rosenblatt, J. D., Golde, D. W., Wachsman, W., Giorgi, J. V., Jacobs, A., Gasson, J. C., and Chen, I. S. Y. (1986). A second isolate of HTLV-II associated with atypical hairy-cell leukemia. *N. Engl. J. Med.* 315, 372–377.

Rosenblatt, J. D., Gasson, J. C., Glaspy, J., Bhuta, S., Aboud, M., Chen, I. S. Y., Golde, D. (1987). Relationship between human T-cell leukemia virus-II and atypical hairy cell leukemia: A serologic study of hairy cell leukemia patients. *Leukemia* 1, 397–401.

Ross, S. C., and Densen, P. (1984). Complement deficiency states and infection: Epidemiology, pathogenesis, and consequences of neisserial and other infections in an immune deficiency. *Medicine (Baltimore)* 63, 243–273.

Russel, H., Sampson, J. S., Schnid, G. P., Wilkinson, H. W., and Plikaytis, B. (1984). Enzyme-linked immunosorbent assay and immunofluorescent assay for Lyme disease. *J. Infect. Dis.* 149, 465–470.

Safrin, R. E., Lancaster, L. A., Davis, C. E., and Braude, A. I. (1986). Differentiation of *Cryptococcus neoformans* serotypes by isoenzyme electrophoresis. *Am. J. Clin. Pathol.* 86, 204–208.

Saiki, R. K., Gelfand, D. H., Stoffel, S., Scharf, S. J., Higuchi, R., Horn, G. T., Mullis, K. B., and Erlich, H. A. (1988). Primer-directed enzymatic amplification of DNA with a thermostable DNA polymerase. *Science* 239, 487–491.

Sato, H., and Okochi, K. (1986). Transmission of human T-cell leukemia virus (HTLV-1) by blood transfusion: Demonstration of proviral DNA in recipient's blood lymphocytes. *Int. J. Cancer* 37, 395–400.

Schaberg, D. R., and Zervos, M. (1986). Plasmid analysis in the study of the epidemiology of nosocomial gram-positive cocci. *Rev. Infect. Dis.* 8, 705–712.

Schaffner, A., Michel-Harder, C., and Yeginsoy, S. (1991). Detection of capsular polysaccharide in serum for the diagnosis of pneumococcal pneumonia: Clinical and experimental evaluation. *J. Infect. Dis.* 163, 1094–1102.

Schild, G. B., and Dowdle, W. R. (1975). Influenza virus characterization and diagnostic serology. *In* "The Influenza Viruses and Influenza" (E. D. Kilbourne, ed.), pp. 315. Academic Press, New York.

Schimpff, S. S. (1990). Infections in the compromised host—An overview. *In* "Principles and Practices of Infectious Diseases" (G. L. Mandell, R. G. Douglas, and J. E. Bennett, eds.), pp. 2258–2265. Churchill Livingstone, New York.

Schulte, P. A. (1987). Methodologic issues in the use of biologic markers in epidemiologic research. *Am. J. Epidemiol.* 126, 1006–1016.

Schulte, P. A. (1989). A conceptual framework for the validation and use of biologic markers. *Environ. Res.* 48, 129–144.

Schulz, T. F., Oberhuber, W., Hofbauer, J. M., Hengster, P., Larcher, C., Gürtler, L. C., Tedder, R., Wachter, H., and Dierich, M. P. (1989). Recombinant peptides derived from the *env*-gene of HIV-2 in the serodiagnosis of HIV-2 infections. *J. AIDS* 3, 165–172.

Scientific Group (1989). Report from the scientific group on HTLV-1 infection and its associated diseases. *Wkly. Epidemiol. Rec.* 49, 382–383.

Sebastiani, G. D. (1991). Chronic Lyme arthritis and HLA alleles. *N. Engl. J. Med.* 324, 129.

Selander, R. K., Caugant, D. A., Ochman, H., Musser, J. M., Gilmour, M. N., and Whittam, T. S. (1986). Methods of multilocus enzyme electrophoresis for bacterial population genetics and systematics. *Appl. Environ. Microbiol.* 51, 873–884.

Shann, F., Germer, S., Hazlett, D., Gratten, M., Linnemann, V., and Payne, R. (1984). Aetiology of pneumonia in children in Goroka Hospital, Papua New Guinea. *Lancet* 2, 537–541.

Singer, C., Kaplan, M. H., and Armstrong, D. (1977). Bacteremia and fungemia complicating neoplastic disease. A study of 364 cases. *Am. J. Med.* 62, 731–742.

Sirisanthana, T., Nelson, K. E., Ezzell, J. W., and Abshire, T. G. (1988). Serological studies of patients with cutaneous and oral-oropharyngeal anthrax from northern Thailand. *Am. J. Trop. Med. Hyg.* 39, 575–581.

Skinhoj, P., Mikkelsen, F., and Hollinger, F. B. (1977). Hepatitis A in Greenland: Importance of specific antibody testing in epidemiologic surveillance. *Am. J. Epidemiol.* 105, 140–147.

Sloand, E. M., Pitt, E., Chiarello, R. J., and Nemo, G. J. (1991). HIV testing—State of the Art. *J. Am. Med. Assoc.* 266, 2861–2866.

Snyder, D. E. (1982). The tuberculin skin test. *Am. Rev. Respir. Dis.* **125**, 108–118.

Sobel, J. D., and Kay, D. (1990). Urinary tract infections. *In* "Principles and Practice of Infectious Diseases" (G. L. Mandell, R. G. Douglas, and J. E. Bennett, eds.), pp. 582–611. Churchill Livingstone, New York.

Solem, J. H., and Lassen, J. (1971). Reiter's disease following *Yersinia entercolitica* infection. *Scand. J. Infect. Dis.* **3**, 83–85.

Spitzer, E. D., Lasker, B. A., Travis, S. J., Kobayashi, G. S., and Medoff, G. (1989). Use of mitochondrial and ribosomal DNA polymorphisms to classify clinical and soil isolates of *Histoplasma capsulatum. Infect. Immun.* **57**, 1409–1412.

Steere, A. C., Grodzicki, R. L., Kornblatt, A. N., Craft, J. E., Barbour, A. G., Burgdorfer, W., Schmid, G. P., Johnson, E., and Malawista, S. E. (1983). The spirochetal etiology of Lyme disease. *N. Engl. J. Med.* **308**, 733–740.

Steere, A. C., Taylor, E., Wilson, M. L., Levine, J. F., and Spielman, A. (1986). Longitudinal assessment of the clinical and epidemiological features of Lyme disease in a defined population. *J. Infect. Dis.* **154**, 295–300.

Steere, A. C., Dwyer, E., and Winchester, R. (1990). Association of chronic Lyme arthritis with HLA-DR4 and HLA-DR2 alleles. *N. Engl. J. Med.* **323**, 219–23.

Stein, D. S., Korvick, J. A., and Vermund, S. H. (1992). CD4[+] lymphocyte cell enumeration for prediction of clinical course of human immunodeficiency virus disease: A review. *J. Infect. Dis.* **165**, 352–363.

Stevens, C. E., Aach, R. D., Hollinger, F. B., Mosley, J. W., Szmuness, W., Kahn, R., Werch, J., and Edwards, V. (1984). Hepatitis B virus antibody in blood donors and the occurrence of non-A, non-B hepatitis in transfusion recipients. An analysis of the transfusion-transmitted viruses study. *Ann. Intern. Med.* **101**, 733–738.

Stout, J. E., Yu, V. L., Muraca, P., Poly, J., Troup, N., and Tompkins, L. S. (1992). Potable water as a cause of sporadic cases of community-acquired Legionnaires' disease. *N. Engl. J. Med.* **326**, 151–155.

Tacket, C. O. (1989). Molecular epidemiology of salmonella. *Epidemiol. Rev.* **11**, 99–108.

Take, H., Umemoto, M., Jitousho, T., Hatae, M., Kuraya, K., and Mochitomi, M. Studies on intrafamilial transmission of HTLV-I: analysis of 186 families. (1992). Proceedings of the Fifth International Conference on Human Retrovirology: HTLV. Kumamoto, Japan, May 11–13, Abstract P-11.

Taylor, D. N., Wachsmuth, I. K., Shangjuan, Y., Schmidt, E. V., Barrett, T. J., Schrader, J. S., Scherach, C. S., McGee, H. B., Feldman, R. A., and Brenner, D. J. (1982). Salmonellosis associated with marijuana. A multistate outbreak traced by plasmid fingerprinting. *N. Engl. J. Med.* **306**, 1249–1953.

Tenover, F. C. (1991). Molecular methods for the clinical microbiology laboratory. *In* "Manual of Clinical Microbiology" (A. Balows, W. J. Hausler, K. L. Herrmann, H. D. Isenberg, and H. J. Shadomy, eds.), 5th Ed., pp. 119–127. American Society for Microbiology, Washington, D.C.

Tinghitella, T. J., and Edberg, S. C. (1991). Agglutination tests and *Limulus* assay for the diagnosis of infectious diseases. *In* "Manual of Clinical Microbiology" (A. Balows, W. J. Hausler, K. L. Herrmann, H. D. Isenberg, and H. J. Shadomy, eds.), 5th Ed., pp. 61–72. American Society for Microbiology, Washington, D.C.

Trent, D. W., Manske, C. L., Fox, G. E., Chu, M. C., Kliks, S. C., and Monath, T. P. (1990). The molecular epidemiology of dengue viruses. Genetic variation and microevolution. *In* "Virus Variability, Epidemiology, and Control" (E. Kurstak, ed.), pp. 293–315. Plenum Medical Books, New York.

Tsilimigras, C. W. A., Rossouw, E., and Schoub, B. D. (1989). Outbreak of poliomyelitis in South Africa investigated by oligonucleotide mapping. *J. Med. Virol.* **28**, 52–56.

Uchiyama, T., Yodoi, J., Sagawa, K., Takatsuki, K., and Uchino, H. (1977). Adult T-cell leukemia: Clinical and hematologic features of 16 cases. *Blood* **50**, 481–492.

van Alphen, L., and Bijlmer, H. A. (1990). Molecular epidemiology of *Haemophilus influenzae* type b. *Pediatrics (Suppl)*, **85**, 636–642.

van Loon, F. P. L., Clemens, J. D., Saci, D. A., Rao, M. R., Ahmed, F., Chowdhury, S., Harris, J. R., Ali, M., Chakraborty, J., Khan, M. R., Neogy, P. K., Svennerholm, A. M., and Holmgren, J. (1991). ABO blood groups and the risk of diarrhea due to enterotoxigenic *Escherichia coli. J. Infect. Dis.* **163**, 1243–1246.

Volberding, P. A., Lagakos, S. W., Koch, M. A., Pettinelli, C., Myers, M. W., Booth, D. K., Balfour, H. H., Reichman, R. C., Bartlett, J. A., Hirsch, M. S., Murphy, R. L., Hardy, W. D., Soeiro, R., Fischl, M. A., Bartlett, J. G., Merigan, T. C., Hyslop, N. E., Richman, D. D., Valentine, F. T., and Corey, L. (1988). Zidovudine in asymptomatic human immunodeficiency virus infection. A controlled trial in persons with fewer than 500 CD4-positive cells per cubic millimeter. *N. Engl. J. Med.* **322**, 941–949.

Wachsmuth, K. (1986). Molecular epidemiology of bacterial infections: Examples of methodology and of investigations of outbreaks. *Rev. Infect. Dis.* **5**, 682–692.

Wachsmuth, I. K., Kiehlbaudh, J. A., Bopp, C. A., Cameron, D. N., Strockbine, N. A., Wells, J. G., and Blake, P. A. (1991). The use of plasmid profiles and nucleic acid probes in epidemiologic investigations of foodborne, diarrheal disease. *Int. J. Food Microbiol.* **12**, 77–90.

Wakefield, A. E., Guiver, L., Miller, R. F., and Hopkin, J. M. (1991). DNA amplification on induced sputum samples for *Pneumocystis carinii* pneumonia. *Lancet* **337**, 1378–1379.

Wall, R. A., Corrah, P. T., Mabey, D. C. W., and Greenwood, B. M. (1986). The etiology of lobar pneumonia in the Gambia. *Bull. WHO* **64**, 553–558.

Ward, J. I., Siber, G. R., Scheifele, D. W., and Smith, D. H. (1978). Rapid diagnosis of *Haemophilus influenzae* type b infection by latex particle agglutination and counterimmunoelectrophoresis. *J. Pediatr.* **93**, 37–42.

Ward, L. R., de Sa, J. D. H., and Rowe, B. (1987). A phage-typing scheme for *Salmonella enteritidis. Epidemiol. Infect.* **99**, 291–294.

Werner, A. S., Cobbs, C. G., Kaye, D., and Hook, E. W. (1967). Studies on the bacteremia of bacterial endocarditis. *JAMA J. Am. Med. Assoc.* **202**, 199–203.

Weyant, R. S., Bibb, W. F., Stephens, D. S., Holloway, B. P., Moo-Penn, W., Birkness, K., Helsel, L., and Mayer, L. (1990). Purification and characterization of a pilin specific for Brazilian purpuric fever-associated *Haemophilus influenzae* biogroup *aegyptius* (*Haemophilus aegyptius*) strains. *J. Clin. Microbiol.* **28**, 756–763.

Wilfert, C. M., Buckley, R. H., Mohanakumar, T., Griffith, J., Katz, S., Whisnant, J., Eggleston, P., Moore, M., Treadwell, E., Oxman, M., and Rosen, F. (1977). Persistent and fatal central nervous system echo virus infections in patients with agammaglobulinemia. *N. Engl. J. Med.* **296**, 1485–1489.

Wilske, B., Schierz, G., Preac-Mursic, V., Weber, K., Pfister, H. W., and Einhaupt, K. (1984). Serologic diagnosis of erythema migrans disease and related disorders. *Infection* **12**, 331–337.

Yoshida, M., Miyoshi, I., and Hinuma, Y. (1982). Isolation and characterization of retrovirus from cell lines of human adult T-cell leukemia and its implication in the disease. *Proc. Natl. Acad. Sci. U.S.A.* **79**, 2031–2035.

13

Cardiovascular Disease

Paul A. Schulte,
Nathaniel Rothman,
and Melissa A. Austin

Introduction

Cardiovascular epidemiology has made use of biomarkers over several decades of research and can, therefore, be viewed as a resource for methods in molecular epidemiology that can be applied to other diseases. This chapter selects examples from the vast literature in this area to illustrate some of the challenges of etiologic research involving biomarkers and, at the same time, provides examples of some of the exciting advances being made in cardiovascular epidemiology.

Cardiovascular diseases span a broad range of pathologic processes, including coronary heart disease (CHD), hypertension, cardiomyopathies, valvular abnormalities, and arrhythmias. This chapter focuses predominantly on the application of lipid and lipoprotein biomarkers to CHD (Table 13.1).

Historically, cardiovascular studies began with the measurement of very broad categories of lipids, such as total cholesterol (Figure 13.1), and progressed to more specific lipid-related biomarkers, such as very low density lipoprotein (VLDL), low density lipoprotein (LDL), high density lipoprotein (HDL), low density lipoprotein subclass phenotypes A and B, and plasma levels of apo AI and B (Keys and Parlin, 1966; Kannel *et al.*, 1971b; Brunzell *et al.*, 1984; Austin *et al.*, 1988,1990b). Genetic markers such as restriction fragment length polymorphisms (RFLPs) for apo B, apo E, and the LDL receptor genes have been used to evaluate directly the inherited contribution to CHD (Humphries, 1987; Friedl *et al.*, 1990b; Dallongeville *et al.*, 1991) (Table 13.2). Thus, a range of lipid markers and their environmental and genetic determinants have been used in epidemiologic studies to evaluate risk of developing coronary heart disease.

TABLE 13.1 **Biologic Markers of Lipid Metabolism Related to Cardiovascular Disease Risk**

Marker type	Example(s)
Plasma lipid levels	Total cholesterol, triglyceride, LDL-C, HDL-C
Plasma apolipoprotein levels	ApoB, ApoAI
Enzyme activity	Lipoprotein lipase
Lipoprotein mass, lipid and protein components	Lp(a)
Subclasses of LDL, subpopulations of HDL	Subclass A and B phenotypes
DNA markers for candidate genes	ApoB, AI–CIII–AIV complex
RFLP VNTR[a] markers	LDL receptor
Genetic protein markers	ApoE isoforms; Apo(a) phenotypes

[a]Variable number of tandem repeats.

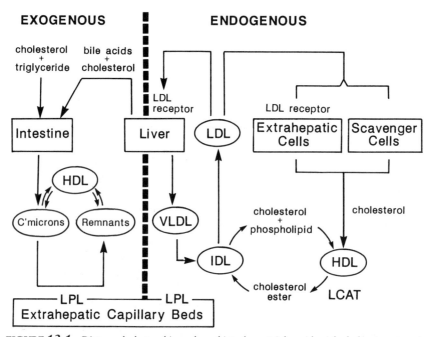

FIGURE 13.1 Dietary cholesterol is packaged into large triglyceride-rich chylomicron particles in the intestinal enterocytes and secreted into the blood stream via the mesenteric lymphatic chain. Lipoprotein lipase in adipose tissue and striated muscle hydrolyses the triglyceride core of the chylomicron, shedding redundant surface phospholipid and protein into HDL. The resulting remnant is assimilated by a hepatic receptor. Hepatic lipids are secreted as triglyceride-rich VLDL that enter a delipidation cascade that reduces their triglyceride content and results in the production of cholesterol-enriched remnants or intermediate density lipoprotein (IDL). The majority of the latter is further remodeled to LDL. This, in turn, is catabolized primarily by receptors present on liver and peripheral tissues. Cholesterol balance at these peripheral sites is maintained by a system of reverse cholesterol transport involving HDL. (Reprinted with permission from Shepherd *et al.*, 1991. Copyright © 1991 by John Wiley & Sons.)

TABLE 13.2   Some Candidate Genes Involved in Lipid Metabolism

Class	Gene
Apolipoproteins	ApoAI
	ApoAII
	ApoB
	ApoCI
	ApoCII
	ApoCIII
	ApoE
Enzymes	HMG CoA reductase
	Lecithin cholesterol acyl transferase
	Fatty acyl-CoA cholesterol acyl transferase
	Endothelial lipoprotein lipase
	Hepatic triglyceride lipase
	Fatty acid synthetase
	Phosphatidic acid phosphohydrolase
	Cholesterol ester hydrolase
	Cholesterol 7-α hydrolase
Transfer proteins	Lipid transfer proteins
Receptors	Low density lipoprotein
	High density lipoprotein

Source: Reprinted with permission from Scott, 1989, in Weatherall, 1991.

Plasma Lipids

Markers of Biologically Effective Dose

Characterization of the role of plasma lipids in CHD can be traced to 1847 when Vogel reported that atherosclerotic plaques contained relatively large amounts of cholesterol (Vogel, 1847). In 1873, Fagge reported a post-mortem examination of a patient with xanthomas (yellow colored nodules or plaques made of lipid that affect the eyelids and other sites) and atheromatous degeneration of the arteries and cardiac disease. Thus, one of the first clues to the importance of plasma lipids in the pathogenesis of CHD literally stared physicians in the face! Familial aggregation of individuals with tendon xanthomas, hypercholesterolemia, and CHD was demonstrated first by Thannhauser (1938) and Muller (1938) (Brunzell *et al.*, 1984) and subsequently has been studied by numerous investigators (Goldstein *et al.*, 1973). The Framingham study was one of the first studies to generalize the finding of hypercholesterolemia as a risk factor in families to the concept that it was also a risk factor for the general population (Kannel *et al.*, 1971b). This observation has been demonstrated repeatedly in a large number of subsequent programs, including the Multiple Risk Factor Intervention screening, the largest to date, which demonstrated a strong dose-dependent association of baseline serum cholesterol levels with subsequent risk of CHD mortality in 356,222 men (Stamler *et al.*, 1986).

The relationship between cholesterol and CHD has been evaluated in case-control, prospective, nested case-control, intervention, twin, and family studies. Of these various study designs, case-control studies are somewhat limited in their ability to study CHD since a substantial number of individuals with myocardial infarctions experience sudden death outside the hospital. Further, the demonstration that cholesterol levels decline in the period immediately following a myocardial infarction has resulted in the need to study cases several months after infarction, further increasing the selection bias of case-control studies that can evaluate only survivors (Tibblin and Cramer, 1963). With the advent of widespread use of coronary angiography, however, it has been possible to perform case-control studies in individuals with and without coronary artery disease (Dahlen *et al.*, 1986). This option has provided an opportunity to evaluate a range of biomarkers associated with the severity of atherogenesis. Overall, prospective observational studies have been perhaps the most important study designs for assessing lipid risk factors for CHD because of their ability to show the predictive value of lipid biomarkers. In addition, angiographic evaluation has strengthened further any knowledge about causal associations between cholesterol level and risk of CHD derived from observational studies (Blankenhorn *et al.*, 1987; Brown *et al.*, 1990).

Thus, using a variety of epidemiologic study designs, total serum cholesterol has been demonstrated to be the first validated biomarker associated with increased risk of CHD. Serum cholesterol conforms to the definition of a marker of biologically effective dose, for it represents a measure of potentially atherogenic lipids that can be delivered to the target organ (atheromas in coronary arteries). It integrates exogenous influences, such as dietary intake of cholesterol and its precursors, and other factors such as exercise, smoking, and medication (e.g., estrogen treatment in women) (Austin *et al.*, 1987; Rose, 1990; Castelli *et al.*, 1990; Nora *et al.*, 1991), and endogenous effects, such as hepatic cholesterol synthesis and genetic influences. Although total plasma cholesterol measures the cholesterol carried by several types of lipoprotein particles that are associated with increased risk (LDL-C) and decreased risk (HDL-C), this overall measurement still remains a relevant risk factor for most individuals in the general population.

Low density lipoprotein cholesterol (LDL-C) is an important risk factor for the development of CHD. Prospective epidemiologic studies (Kannel *et al.*, 1971a; Pekkanen *et al.*, 1990) and intervention trials [Lipid Research Clinics Program (LRCP), 1984; Ross, 1986; Blankenhorn *et al.*, 1987; Manninen *et al.*, 1988] have demonstrated that plasma levels of LDLs, the primary carriers of cholesterol in the bloodstream, are directly related to disease risk.

Several studies have demonstrated decreased risk of CHD with high levels of HDL cholesterol (Stampfer *et al.*, 1985; Miller, 1987; Gordon *et al.*, els of HDL cholesterol (Stampfer *et al.*, 1985; Miller, 1987; Gordon *et al.*, 1989). The ratio of total cholesterol to HDL has been demonstrated to be one of the strongest risk factors for developing CHD, particularly in individuals with total cholesterol levels under 200 (Stampfer *et al.*, 1985). The evidence for a causal association between elevated HDL levels and decreased risk for CHD has been strengthened by intervention studies such as the Helsinki Heart Study, which demonstrated that raising HDL cholesterol levels through drug treatment of dyslipidemic men resulted in a decline in CHD (Manninen *et al.*, 1988).

Markers of Susceptibility

Extensive evidence is available that CHD is mediated by many genetic loci (MacCleur and Kammerer, 1991) (Figure 13.2). Nora and colleagues (1991) have reviewed the environmental and genetic determinants of several of the most important lipids and apoproteins. Nora *et al.* (1991) have concluded that although monogenic hyperlipidemias account for at least 20% of myocardial infarctions prior to age 60, other "genetic dyslipidemias" must account for a portion of excess risk for early onset CHD in the general population, based on data from family studies. Some of the more important potential susceptibility markers follow.

Apoliprotein E Polymorphism

Apoliprotein (apo) E polymorphism was among the first reported genetic polymorphisms that explained part of the normal variation in cholesterol concentrations in humans (Sing and Davignon, 1985). Because apo E is present on VLDL particles and is a ligand for the LDL receptor, it is important in the catabolism of cholesterol. The three common alleles or isoforms, E2, E3, and E4, reflect specific amino acid substitutions in the apo E protein (Davignon and Gregg, 1988). Relative to the most common E3 allele, E2 has been associated consistently with lower levels of plasma cholesterol whereas E4 is associated with increases in LDL cholesterol in a variety of ethnic groups (Sing and Davignon, 1985). Allele variation at the apo E gene was compared in 182 subjects with endogenous hypertriglyceridemia, 98 subjects with familial hypercholesterolemia due to a 10-kb deletion in their LDL receptor gene, and 424 normolipidemic controls from the same environmental background (Dallongeville *et al.*, 1991). The study showed that lipid and lipoprotein concentration variability within the dyslipoproteinemic groups is related to apo E phenotype. The apo E polymorphism has been reported to explain 5–8% of the variance in serum cholesterol levels in populations from several countries (Humphries *et al.*, 1991). Although reports of direct associations between these alleles and atherosclerotic diseases are less consistent (Davignon and Gregg, 1988), a report based on young males who died from accidental causes demonstrated that apo E genotypes were associated with

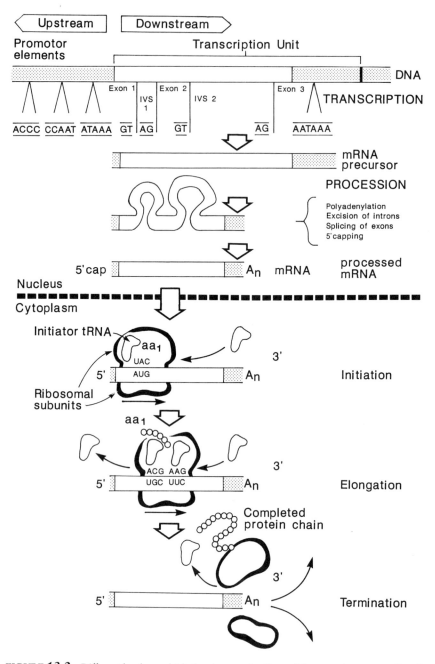

FIGURE 13.2   Different levels at which the phenotypic effects of the mutations are mediated. (Reprinted with permission from Weatherall, *et al.*, 1991.)

ciation persisted after adjustment for total cholesterol, suggesting that apo E alleles may have a direct effect on atherosclerosis.

Lipoprotein(a)

Lipoprotein(a) [Lp(a)] is a unique lipoprotein consisting of two components: a particle similar to LDL, including apo B, and the apo(a) protein linked by a disulfide bond to the apo B molecule (Utermann, 1989). The apo(a) protein is the distinguishing characteristic of Lp(a) and results in a lipoprotein particle that is larger and more dense than LDL. The gene for the apo(a) protein has been mapped to the tip of the long arm of chromosome 6, closely linked to the gene for plasminogen (Frank *et al.*, 1988). These two genes show a high degree of homology, including the protease domain and several sequences that code protein kringle domains (McLean *et al.*, 1987).

Lp(a) has been the focus of intensive research interest since its original discovery by Berg (1963). It is now well established that increased plasma levels of Lp(a) constitute an independent risk factor for CHD that also may be involved in thrombogenesis (Utermann *et al.*, 1987; Scanu, 1991). In 1987, Utermann reported size isoforms of apo(a) that were inherited in families and were associated with Lp(a) concentrations in plasma (Utermann *et al.*, 1987). Mean levels of plasma Lp(a) consistently have been associated inversely with the apo(a) size polymorphism in subsequent studies as well (Gaubatz *et al.*, 1990). Boerwinkle *et al.* (1989) have reported that 41.6% of the variance in Lp(a) levels can be attributed to apo(a) phenotypes in Caucasians, compared with as much as 70% of the variation in Asians (Sandholzer *et al.*, 1991). Lp(a) levels also are known to vary dramatically among ethnic groups (Sandholzer *et al.*, 1991). Lackner *et al.* (1991) have identified 19 alleles at the apo(a) locus using pulsed-field gel electrophoresis. These alleles reflect the number of kringle IV repeats, each of which is approximately 5.5 kb in length; nearly all individuals are heterozygous using this system. Thus, the apo(a) locus is even more polymorphic than can be recognized by apo(a) size isoforms. Further studies leading to an understanding of the genetics of Lp(a) undoubtedly will provide important insights into the role of Lp(a) in atherosclerosis risk among families.

Apolipoprotein B

Apo B is the primary protein component of LDL particles. Elevated plasma apo B levels are another important risk factor for CHD. Although a prospective study of apolipoproteins found only a borderline association between apo B levels and myocardial infarction (Stampfer *et al.*, 1991), case-control studies have shown significant associations (Sniderman *et al.*, 1980; Brunzell *et al.*, 1984). Regression of coronary lesions was associated with reductions in apo B levels as a result of intensive lipid-lowering intervention (Brown *et al.*, 1990). A disorder denoted hyperapobetalipoproteinemia was identified in a study that found a subset of subjects with CHD who had normal LDL cholesterol levels but increased apo B levels (Sniderman *et al.*,

1980). A number of large-scale family studies have examined genetic models for the inheritance of apo B levels. Three of these, using complex segregation analysis, found evidence for major gene effects (Amos *et al.,* 1987; Hasstedt *et al.,* 1987; Pairitz *et al.,* 1988), whereas another found significant polygenic effects (Beaty *et al.,* 1986). An analysis of familial combined hyperlipidemia, the most common of the familial hyperlipoproteinemias, demonstrated that apo B levels were distributed bimodally among family members with LDL subclass phenotype B (Austin *et al.,* 1992), providing additional evidence for genetic influences on apo B in this form of familial hyperlipidemia. Although two specific kinds of relatively rare mutations in the apo B gene on chromosome 2 (familial defective apo B100 and truncated apo B isoforms) are known to alter apo B plasma levels (Assman *et al.,* 1991), associations have been more difficult to establish in larger population-based studies (Galton, 1990).

Apolipoprotein AI

Similar to HDL cholesterol, apo AI, the major protein component of HDL particles, is associated inversely with risk of CHD (Brunzell *et al.,* 1984; Stampfer *et al.,* 1991). Based on a large population-based sample of families, evidence for the influence of a single major gene was found in a subset of the pedigrees (Moll *et al.,* 1989). Lower levels of apo AI also have been reported in children whose fathers had histories of myocardial infarction than in children with no such parental history (Freedman *et al.,* 1986). At least one study of RFLPs in the region of the AI-CIII-AIV gene complex on chromosome 11 has reported an association between apo AI plasma levels and genetic variation at these RFLP sites (Kessling *et al.,* 1988). Thus, decreased apo AI levels are likely to be another genetically influenced factor related to CHD.

LDL-Receptor Gene

The mechanism underlying familial hypercholesterolemia (FH) is the best understood genetic defect of LDL metabolism known to cause atherosclerosis (Brown and Goldstein, 1986). In FH, defects in the LDL receptor gene on chromosome 19 lead to an accumulation of cholesterol in plasma and to premature CHD. Clinically, the presence of tendon xanthomas is a hallmark of this disease. The homozygous form is very rare and very severe: cholesterol levels are often 10 times above normal, and CHD occurs in the teens or twenties (Brown and Goldstein, 1986). However, FH is a relatively rare disorder; the prevalence of heterozygotes and homozygotes for this disease is estimated to be 1 in 500 and 1 in a million, respectively (Goldstein *et al.,* 1973). It is now known that there are few true homozygotes, since a variety of different mutations in the LDL receptor gene exist. With the exception of the French Canadian population (Moorjani *et al.,* 1989), most apparent homozygotes are actually compound heterozygotes (Brown and Goldstein, 1986). The heterozygous form of FH, although less severe, re-

sults in cholesterol elevations 2–3 times greater than normal and premature coronary disease. The metabolism and feedback system for receptor-
mediated cholesterol homeostasis is now well understood and provides some
of the most compelling evidence that elevated LDL cholesterol levels cause
atherosclerosis.

Using four RFLPs of the human LDL receptor (LDL-R) gene, Humphries
et al. (1991) studied 289 normolipidemic individuals in Italy and found that
the *Pvu*II RFLP explained 9.6% of the variance in LDL cholesterol levels.
This RFLP was associated with lower levels of total and LDL cholesterol. The
data suggest it may be associated with increased survival. The 3' half of the
LDL-R gene, showing localization of restriction sites and polymorphic sites,
is shown in Figure 13.3A. Autoradiographs of Southern blot hybridization
showing variable bands of the LDL-R gene observed in individuals with different genotypes for four polymorphisms are shown in Figure 13.3B.

Markers of Thrombosis

In addition to markers related to atherogenesis, another contributor to CHD
is a tendency toward arterial thrombosis. Studies have suggested that variability in factor VII coagulant activity and plasma fibrinogen (Meade *et al.*,
1986), platelet count and platelet aggregation (Thaulow *et al.*, 1991), fast
acting plasminogen activator inhibitor (an indicator of reduced fibrinolytic
capacity; Hamsten *et al.*, 1987), and aspirin intake [Steering Committee of
the Physician's Health Study Research Group (1989)] may affect risk of
CHD. This area of CHD research promises to continue to be active.

Assessing Genetic and Environmental Interactions for Coronary Heart Disease Risk

Multivariate analyses have been used extensively to combine biomarker and
nonbiomarker data to assess the effects of environmental exposures (such as
the role of smoking and diet), blood pressure, and genetic risk factors for risk
of CHD and their potential interactions.

MacCleur and Kammerer (1991) summarized the progression of thinking about genetic and environmental interactions:

> In 1954, Lerner (1954) proposed that genes influence not only the mean levels of
> quantitative traits but also the variation around the mean. He suggested that differ-
> ences in intragenotypic variances may be due to an inability by individuals of a spe-
> cific genotype to "buffer" their phenotypes against other genetic and environmental
> factors. Murphy (1979) extended these concepts, with particular reference to human
> populations. . . . Berg (1988, 1990; also see Berg *et al.*, 1989) has investigated the
> effects of "variability genes" on CHD risk factors. For such genes, the mean value of
> two genotypes may be similar, but the intragenotypic variances differ. Increased phe-

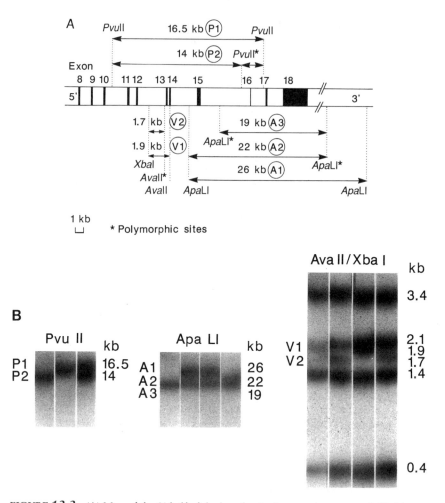

FIGURE 13.3 (A) Map of the 3' half of the low density lipoprotein receptor (LDL-R) gene showing localization of *Pvu*II, *Apa*LI, and *Ava*II restriction sites and polymorphic sites. Also shown are alleles P1, P2, V1, V2, A1, A2, and A3. (B) Autoradiograms of Southern blot hybridization showing variable bands of the LDL-R gene, observed in individuals with different genotypes for four polymorphisms. (Reprinted with permission from Humphries *et al.*, 1991).

notypic variance within a genotype may be due either to genotype × environment or genotype × genotype interaction or to linkage disequilibrium.

Markers of Effect

Few validated biological markers of effect are available for CHD, with the exception of short-term markers (cardiolytic enzymes) associated with acute myocardial infarction. Many of the lipid and lipoprotein markers indicate risk of coronary artery disease, but do not necessarily indicate the presence or degree of atherosclerotic disease. Although often the diagnosis of coronary artery disease can be made from the history and by noninvasive techniques, angiography provides a definitive indication of the disease (Pasternak and Reis, 1991). Angiographic findings represent a biologic marker different than the serum or plasma molecular genetic markers considered elsewhere in this chapter. Nonetheless, angiographic imaging is a biomarker of disease. It is valid because it is a direct representation of arterial stenosis. It allows for an examination of the coronary tree and details of the coronary anatomy, assessment of individual variations in arterial distribution, anatomic or functional pathology (atherosclerosis, thrombosis, congenital anomaly, or focal coronary spasm), and the presence of inter- and intracoronary collateral connections (Baim and Grossman, 1991). Angiographic assessments of disease severity have been used particularly effectively in intervention trials (Blankenhorn *et al.*, 1987; Brown *et al.*, 1990).

Angiography historically involved nonselective injections of contrast medium into the aortic root, with simultaneous specification of both the left and right coronary arteries, and recording of the angiographic images on a conventional sheet of film. Although this approach is still in use, it is being replaced by selective coronary injections using specially designed catheters advanced from the brachial or femoral approach. The outcomes of angiography are images that can be used to quantitate coronary stenosis. The degree of stenosis usually is quantitated by visual evaluation of the percentage of diametric reduction relative to the caliber of the adjacent normal segments (Baim and Grossman, 1991). This assessment usually gives a fairly accurate representation in very mild or very severe stenoses, although overestimation of stenosis severity is common. There is substantial interobserver variability (frequently $\pm$ 20% in the visual quantitation of moderate stenoses between 40 and 82%). This range of stenosis is particularly important because a 50% diameter stenosis (75% cross-sectional area) is barely "hemodynamically significant" at peak coronary flows, but a 70% diameter stenosis (90% cross-sectional area) is restrictive at these same peak flows (Wilson *et al.*, 1987; Baim and Grossman, 1991). Angiography studies can suffer from incomplete, uninterpretable, or misinterpreted findings due to the inexperience of the operator who uses faulty techniques or cannot recognize misleading images. In addition, observations are not always independent, since multiple lesions and arteries may be evaluated within one individual.

Angiographic results are being used more frequently to define "cases" in case-control studies (Maciejko *et al.*, 1983; Dahlen *et al.*, 1986). In addition to traditional cinegraphic angiography are a number of imaging techniques, such as magnetic resonance angiography, Doppler sonography, B mode ultrasonography, and single photon emission computed tomography (Underwood *et al.*, 1990; van der Wall *et al.*, 1991; Polak *et al.*, 1992) that may become useful in epidemiologic research. As Rose (1990) notes:

The future lies with a new style of cardiovascular epidemiology, much more closely linked with clinical and laboratory disciplines. . . . It will seek to integrate epidemiology with the study of intermediary outcomes; and in its measurement of disease, it will avail itself of advanced techniques such as noninvasive imaging, which can so powerfully supplement clinical outcomes as a measure of the occurrence and progress of disease.

Issues in the Analysis of Biomarkers

The biologic markers discussed thus far are part of a web of causes and effects. Clarification of their roles and relationships may be enhanced by multivariate analyses. There are potential difficulties in the simultaneous analysis of multiple markers; markers may exhibit different degrees of laboratory precision in their measurement, different degrees of true intraindividual variations, and some degree of correlation. These problems are not new to epidemiologists, but are illustrated well in cardiovascular epidemiology.

Intraindividual Variation in Biomarker Measurements

Most large-scale prospective studies biologically monitor study participants at only one point in time. Substantial intraindividual variation in biomarker levels may result in marked misclassification of the true mean level of the marker. If the misclassification bias is nondifferential (i.e., does not differ for diseased and control groups), this bias may attenuate measures of association between a given marker and disease (Kelsey *et al.*, 1986).

Investigators have attempted to understand and control some of the determinants of the daily fluctuation of serum cholesterol levels by standardizing phlebotomy conditions (Cooper *et al.*, 1992). Given the multifactorial determinants of serum cholesterol, however, some degree of intraindividual variation is inevitable and may be accounted for in data analysis if it is known. For example, using data from the Western Electric Study, which determined the intraclass correlation for repeat cholesterol measurements at a 2-year interval (Shekelle *et al.*, 1981), Willett (1990) demonstrated how that variability could be accounted for in risk estimates.

Not all lipid measurements show the same degree of intraindividual variability. For example, Figure 13.4 shows that the intraindividual variation in triglyceride values is considerably higher than the variation in plasma cholesterol values (Jacobs and Barrett-Connor, 1982). This difference can be determined by sampling the same individual at two different time points and calculating the retest reliability (equivalent to the standard deviation of the paired values) (Austin, 1991) (Figures 13.4 and 13.5). Based on Lipid Research Clinics data with an average of 2.5 months between measurements, the retest reliability of triglyceride was consistently lower than the retest reliability for cholesterol over a range of values of 100 to more than

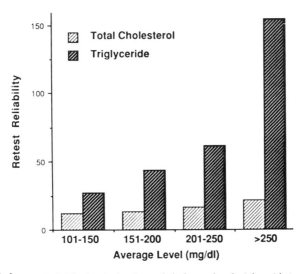

FIGURE 13.4   Intraindividual variation in total cholesterol and triglyceride (mg/dl), based on two measurements from the same individuals. Retest reliability (*y* axis) is equivalent to the SD of paired individual values. Larger values reflect greater intraindividual variation. (Reprinted with permission from Austin, 1991.)

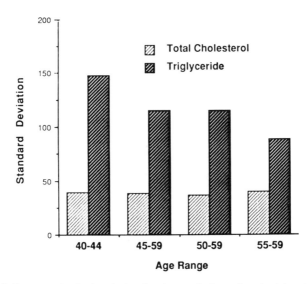

FIGURE 13.5   Interindividual variation in plasma cholesterol and triglyceride, based on sample of middle-aged men. (Reprinted with permission from Austin, 1991.)

250 mg/dl. This difference is likely to result in greater attenuation of the association between triglyceride levels and risk of CHD than of cholesterol levels and CHD. Laboratory measurement error also may attenuate the association between a given biomarker and risk of CHD. Markers measured with less precision will reduce further the ability to detect associations with CHD, especially in comparison with other less variable lipid markers.

Correlations among Lipid Biomarkers

Many studies report significant correlations between triglyceride and other lipid and lipoprotein levels. Such a correlation is illustrated by the relationship between plasma triglyceride and HDL-C levels. Four studies reporting such correlations are summarized in Table 13.3 (Rhoads *et al.*, 1976; Gordon *et al.*, 1977; Albrink *et al.*, 1980; Davis *et al.*, 1980). Consistent, strong, inverse correlations are seen between triglyceride and HDL-C, that is, increased triglyceride levels are associated with decreased HDL-C levels. As seen in the table, the correlations range from −0.2 to −0.7 (Austin, 1991). These statistical correlations no doubt reflect metabolic interrelations of lipoprotein particles involving exchanges in core lipids. In fact, it has been noted that "the close coupling of HDL-C to plasma triglyceride transport makes it difficult indeed to evaluate the distinct roles of hypertriglyceridemia and HDL-C" (Havel, 1988).

A subgroup analysis of the Helsinki Heart Study results has shown that "joint" lipid risk factors can be important for CHD risk as well (Manninen *et al.*, 1992). The results revealed a high-risk subgroup of subjects with LDL-C-to-HDL-C ratio greater than 5.0 and triglyceride of more than 2.3 mmol/liter, that constituted approximately 12% of the study sample. In the placebo group, the relative risk in this subgroup over the 5 years of the study was 3.8 (95% confidence interval, 2.2–6.6), compared with the subgroup of subjects who had a ratio less than or equal to 5.0 and triglyceride less than or equal to 2.3 mmol/liter. Relative risks were close to 1 in the

TABLE 13.3   Correlation of Triglyceride and High Density Lipoprotein Cholesterol in Several Epidemiological Studies

Study/year	Subjects	Correlation coefficient
Honolulu Heart Study, 1976	Men	−0.36
Framingham study, 1977	Men	−0.35
	Women	−0.43
Modesto, California, 1980	Men and women	−0.32
Lipid Research Clinics Prevalence Study, 1980	Men	−0.23 to −0.55
	Women	−0.21 to −0.65

Source: Reprinted with permission from Austin, 1991.

remaining two placebo groups, that is, those with only a high ratio or only high triglyceride. Among the treated subjects, the corresponding high-risk subgroup also received by far the most benefit from gemfibrozil treatment, with a remarkable 71% decrease in incidence of CHD. Much less benefit was seen in other groups. The authors conclude that "serum triglyceride concentration has prognostic value in combination with LDL-C and HDL-C levels" (Austin, 1992).

Determining the relative contribution of lipid subcategories is critical, since specific interventions may raise some and lower other lipid categories. This task can be challenging when biomarkers being evaluated have different degrees of true intraindividual variation, have different degrees of precision in their measurement, and are partially correlated. A modeling exercise of the effects of triglyceride and HDL-C for risk of heart disease suggested that a true association between triglyceride levels and risk of CHD might be masked in the presence of HDL-C due to less intraindividual variation in HDL-C and its partial correlation with triglycerides (Davis and Kim, 1990).

Future Contributions

The key contribution of molecular epidemiologic approaches to the future of cardiovascular research will be in the area of (1) identifying the sources of variation for various blood lipids and lipoproteins, (2) targeting people for specific interventions, and (3) assessing the impact of interventions.

Identifying Sources of Biomarker Variation

Molecular epidemiologic research will assist in disentangling the genetic and environmental factors that account for variation in biomarkers shown to be predictive of increased CHD risk (Berg, 1989). The research will include metabolic ward studies (e.g., Weinberg et al., 1990), international comparison studies (e.g., International Collaborative Study Group, 1986), family studies (e.g., Bodurtha et al., 1991), and population-based studies. These studies will assist in developing prevention strategies that will combine risk factor modification and appropriate screening.

An important consideration in all such studies is the standardization of laboratory measurements. A report from the International Federation of Clinical Chemistry Committee on Apolipoproteins addresses many of the issues involved (Marcovina and Albers, 1991).

Targeting People for Specific Interventions

Cardiovascular research has pioneered the concept of using biologic markers in calculating the probability that an individual with a given combination of risk factors will develop disease (Truett et al., 1967). Truett and colleagues

(1967) applied multivariate statistical procedures to determine the association of cholesterol and other risk factors for coronary heart disease in the Framingham study, which followed 2187 men and 2669 women aged 30–62, free of CHD at entry, for 12 years. Using the logistic function, the researchers calculated the probability of an individual developing coronary heart disease given the baseline serum cholesterol level, age, systolic blood pressure, relative weight, hemoglobin, cigarettes smoked per day, and ECG (normal vs. abnormal). This study represents one of the first attempts to determine the probability of an individual developing disease based on a premorbid biomarker measurement. Using the logistic function, Truett *et al.,* (1967) showed that each person in the sample from the Framingham study had an estimated probability of the event (i.e., CHD) that could be expressed as:

$$y(x) = [1 + e^{-(a + bx)}] - 1$$

Thus, if x is the specific serum cholesterol at baseline, y is the probability that the person will develop CHD during a specified duration of follow-up, given that he or she has the high risk serum level. This model may be expanded to involve a number of risk factors, some characterized by markers and others that are not. The multivariate model is

$$y(x) = [1 + e^{-(a + \Sigma b_i x_i)}] - 1$$

With an efficient set of variables, it is possible to use these "risk functions" to identify "high risk" individuals (that is, individuals whose conditional probabilities are higher than average). The model will inevitably be enhanced as markers of susceptibility [such as the apo E polymorphisms and Lp(a)] are understood further and applied.

Assessing the Impact of Intervention

Observational studies first demonstrated associations between several lipid biomarkers and risk of CHD. These studies, as well as a host of observational and controlled studies of the determinants of these lipid measurements, have given rise to several generations of intervention trials aimed at determining if the lowering of total cholesterol and other relevant lipid levels with diet or medication will decrease individual risk for developing disease.

The Lipid Research Clinics Program conducted an intervention trial in 12 centers in North America. Subjects were assigned randomly to groups receiving a cholesterol-lowering drug or a placebo. A dose–response relationship was observed between the amount of drug and reduction of cholesterol and LDL-C cholesterol. Metabolic and molecular markers applied to intervention trials in the future may allow for a shorter time for such trials, since some markers may reflect intermediate conditions. If these markers have been validated with respect to the prediction of disease, they can be used as dependent variables. Another approach described by Brown *et al.* (1990)

used the percentage change in apolipoprotein B and HDL, among other independent variables, in an assessment of the effect of therapies for hyperlipidemia. The effect of intensive lipid lowering therapy on coronary arteriosclerosis among men at high risk for cardiovascular events was assessed by quantitative arteriography (Brown *et al.*, 1990). Multivariate analysis indicated that a reduction in the level of apolipoprotein B and systolic blood pressure and an increase in HDL cholesterol correlated independently with regression of coronary lesions (see Figure 13.6).

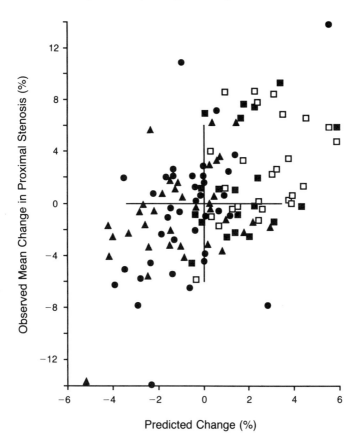

FIGURE 13.6   Results of multivariate statistical analysis. For each patient, the observed mean change in the severity of proximal stenosis ($\Delta\%S_{prox}$) is plotted against a value estimated with the following predictive expression: $\%S_{prox} = 0.07\% \, \Delta apoB - 0.032\% \, \Delta HDL_c + 0.14\% \, \Delta BP_{sys} - 0.7 \, \Delta ST + 1.5$. The expression uses the maximal ST-segment depression (ΔST) during the baseline treadmill exercise test and the percentage change in the apolipoprotein B level ($\%\Delta apoB$), in the HDL cholesterol level ($\%\Delta HDL_c$), and in systolic blood pressure ($\%\Delta BP_{sys}$) during treatment to provide the most accurate estimate of observed change ($r = 0.51$, $P<0.0001$). Open squares indicate that the patient received only placebos, solid squares represent placebo and colestipol, solid triangles represent niacin and colestipol, and solid circles represent lovastatin and colestipol. (Reprinted by permission of the *New England Journal of Medicine* from Brown *et al.*, 1990.)

Researchers are developing computer models to assess the impact of cholesterol and other risk factor modifications on reducing CHD morbidity and mortality (Grover *et al* 1992). Grover noted that the wide variation surrounding estimates of the impact of lowering cholesterol levels for men and women in different age groups demonstrates that future research will need to define better which groups of individuals will gain the most from cholesterol modification (Grover *et al.,* 1992).

References

Albrink, M. J., Krauss, R. M., Lindgren, F. T., Von Der Groeben, V. D., and Wood, P. D. (1980). Intercorrelations among high density lipoprotein, obesity, and triglycerides in a normal population. *Lipids* **15,** 668–678.

Amos, C. I., Elston, R. C., Srinivasan, S. R., Wilson, A. F., Cresana, J. L., Ward, L. S., and Berenson, G. S. (1987). Linkage and segregation analyses of apolipoproteins AI and B, and lipoprotein cholesterol levels in a large pedigree with excess coronary heart disease. *Genet. Epidemiol.* **4,** 115–128.

Assmann, G., von Eckardstein, A., Funke, H., Rust, S., and Sandkamp, M. (1991). Analysis of genetically determined structural polymorphisms in apolipoproteins E, B and (a). *Curr. Opin. Lipidol.* **2,** 367–375.

Austin, M. A. (1991). Plasma triglyceride and coronary heart disease. *Arteriosclerosis Thrombosis* **11,** 2–14.

Austin, M. A. (1992). Joint lipid risk factors and coronary heart disease. *Circulation* **85,** 315–367.

Austin, M. A., King, M. C., Bawol, R. D., Halley, S. B., and Friedman, G. D. (1987). Risk factors for coronary heart disease in adult female twins. *Am. J. Epidemiol.* **125,** 308–318.

Austin, M. A., Breslow, J. L., Hennekens, C. H., Buring, J. E., Willett, W. C., and Krauss, R. M. (1988). Low-density lipoprotein subclass patterns and risk of myocardial infarction. *J. Am. Med. Assoc.* **260,** 1917–1921.

Austin, M. A., King, M. C., Vranizan, K. M., and Krauss, R. M. (1990a). Atherogenic lipoprotein phenotype. A proposed genetic marker for coronary heart disease risk. *Circulation* **87,** 495–506.

Austin, M. A., Brunzell, J. D., Fitch, W. L., and Krauss, R. M. (1990b). Inheritance of low density lipoprotein subclass patterns in familial combined hyperlipidemia. *Arteriosclerosis* **10,** 520–530.

Austin, M. A., Horowitz, H., Wijsman, E., Krauss, R., and Brunzell, J. D. (1992). Biomodality of plasma apolipoprotein B levels in familial combined hyperlipidemia. *Atherosclerosis.* **92,** 61–77.

Baim, D. S., and Grossman, W. (1991). Coronary angiography. *In* "Cardiac Catheterization, Angiography, and Intervention" (W. Grossman and A. S. Baim, eds.), 4th Ed., pp. 185–214. Lea & Febiger, Philadelphia.

Beaty, T. H., Kwiterovich, P. O., Jr., Khoury, M. J., White, S., Banchorik, P. S., Smith, H. H., Teng, B., and Sniderman, A. (1986). Genetic analysis of plasma sitosterol, apoprotein B and lipoproteins in a large Amish pedigree with sitosterolemia. *Am. J. Hum. Genet.* **38,** 492–504.

Berg, K. (1963). A new serum system in man, the LP system. *Acta Pathol. Microbiol. Scand.* **59,** 369–382.

Berg, K. (1988). Variability gene effect on cholesterol at the Kidd blood group locus. *Clin. Genet.* **33,** 102–107.

Berg, K. (1989). Predictive genetic testing to control coronary heart disease and hyperlipidemia. *Arteriosclerosis* 9 (**Suppl. 1**), 140–158.

Berg, K. (1990). Level genes and variability genes in the etiology of hyperlipidemia and atherosclerosis. *In* "From Phenotype to Gene in Common Disorders" (K. Berg, N. Rettersol, and S. Refsum, eds.). Munksgaard, Copenhagen.

Berg, K., Kondo, I., Drayna, D., and Lawn, R. (1989). "Variability gene" effect of cholesterol ester transfer protein (CETP) genes. *Clin Genet.* 35, 437–445.

Blankenhorn, D. H., Nissim, S. A., Johnson, R. L., Sanmarco, M. E., Azen, S. P., Cashin-Hemphill, L. (1987). Beneficial effects of combined colistipol-niacin therapy on coronary atherosclerosis and coronary venous bypass graft. *J. Am. Med. Assoc.* 257, 3233–3240.

Bodurtha, J. N., Chen, C. W., Mosteller, M., Nance, W. E., Schieken, R. M., and Segrest, J. (1991). Genetic and environmental contributions to cholesterol and its subfractions in 11-year old twins. *Arteriosclerosis and Thrombosis* 11, 844–850.

Boerwinkle, E., Menzel, H. J., Kraft, H. G., and Utermann, G. (1989). Genetics of the quantitative Lp(a) lipoprotein trait III. Contribution of Lp(a) glycoprotein phenotypes to normal lipid variation. *Am. J. Hum. Genet.* 82, 73–78.

Brown, G., Albers, J. J., Fisher, L. D., Schaeffer, S. M., Lin, J. T., Kaplan, C., Zhao, X. Q., Bisson, B. D., Fitzpatrick, V., and Dodge, H. T. (1990). Regression of coronary artery disease as a result of intensive lipid lowering therapy in men with high levels of apolipoprotein B. *N. Engl. J. Med.* 323, 1289–1298.

Brown, M. S., and Goldstein, J. L. (1986). A receptor-mediated pathway for cholesterol homeostasis. *Science* 232, 34–47.

Brunzell, J. D., Sniderman, A. D., Albers, J. J., and Kwitecovich, P. O., Jr. (1984). Apoproteins B and A-1 and coronary disease in humans. *Arteriosclerosis* 4, 79–83.

Castelli, W. P., Wilson, P. W. F., Levy, D., Anderson, K., (1990). Serum lipids and risk of coronary artery disease. *Atherosclerosis Rev.* 21:7–19.

Cooper, G. R., Myers, G. L., Smith, J., and Schlant, R. C. (1992). Blood lipid measurements. Variations and practical utility. *J. Am. Med. Assoc.* 267, 1652–1660.

Dahlen, G. H., Guyton, J. R., Attar, M., Farmer, J. A., Kautz, J. A., and Gotto, A. M. (1986). Association of levels of lipoprotein Lp(a), plasma lipids, and other lipoproteins with coronary artery disease documented by angiography. *Circulation* 74, 758–765.

Dallongeville, J., Roy, M., Leboeuf, N., Xhignesse, M., Davignon, J., and Lussier-Cacan, S. (1991). Apolipoprotein E polymorphism association with lipoprotein profile in endogenous hypertriglyceridemia and familial hypercholesterolemia. *Thrombosis* 11, 272–278.

Davignon, J., and Gregg, R. E. (1988). Apolipoprotein E polymorphism and atherosclerosis. *Arteriosclerosis* 8, 1–21.

Davis, C. E., and Kim, H. (1990). Is triglyceride an independent risk factor for CHD? *Circulation* 81, 14.

Davis, C. E., Gordon, D., LaRosa, J., Wood, P. D. S., and Halperin, M. (1980). Correlations of plasma high density lipoprotein cholesterol levels with other plasma lipid and lipoprotein concentrations. *Circulation* 62 (**Suppl IV**), 24–30.

Fagge, C. H. (1873). X. Diseases, etc., of the skin. 1; General xanthelasma or vitiligoidea. *Trans. Pathol. Soc.* 24, 242–250.

Frank, S. L., Klisak, I., Sparkes, R. S., Mohandas, T., Tomlinson, J. E., McLean, J. W., Lawn, R. M., and Lusis, A. J. (1988). The apolipoprotein(a) gene resides on human chromosome 6q26-27, in close proximity to the homologous gene for plasminogen. *Hum. Genet.* 79, 352–356.

Freedman, D. S., Sathanur, R. S., Shear, C. L., Franklin, F. A., Webber, L. S., and Berenson, G. S. (1986). The relation of apolipoproteins A-I and B in children to parental myocardial infarction. *N. Engl. J. Med.* 315, 721–726.

Friedl, W. F., Ludwig, E. H., Paulweber, B., Sandhofer, F., and McCarthy, B. (1990a). Hypervariability in a minisatellite 3′ of the apolipoprotein B gene in patients with coronary heart disease compared with normal controls. *J. Lipid Res.* **31**, 659–665.

Friedl, W., Ludwig, E. H., Balestra, M. E., Arnold, K. S., Paulweber, B., Sandhofer, F., McCarthy, B. J., and Innerarity, T. L. (1990b). Apolipoprotein B gene mutations in Austrian subjects with heart disease and their kindred. *Arteriosclerosis Thrombosis* **11**, 371–378.

Galton, D. J. (1990). DNA variation of the apolipoprotein loci and risk for coronary atherosclerosis. *In* "From Phenotype to Gene in Common Disorders" (K. Berg, N. Retterstol, and S. Refsum, eds.), pp. 60–76. Munksgaard, Copenhagen.

Gaubatz, J. W., Ghanem, K. I., Guevara, J., Jr., Nava, M. L., Patsch, W., and Morrisett, J. D. (1990). Polymorphic forms of human apolipoprotein[a]: Inheritance and relationship of their molecular weights to plasma levels of lipoprotein[a]. *J. Lipid Res.* **31**, 603–613.

Goldstein, J. L., and Brown, M. S. (1983). Familial hypercholesterolemia. *In* "The Metabolic Basis of Inherited Disease" (J. B. Stanbury, D. S. Wynquarden, J. L. Goldstein, and M. S. Brown, eds.), pp. 672–712. McGraw-Hill, New York.

Goldstein, J. L., Schrott, H. G., Hazzard, W. R., Bierman, E. L., and Motulsky, A. G. (1973). Hyperlipidemia in coronary heart disease. II. Genetic analysis of lipid levels in 1766 families and delineation of a new inherited disorder, combined hyperlipidemia. *J. Clin. Invest.* **52**, 1544–1568.

Gordon, D. J., Probstfield, J. L., Garrison, R. J., Neaton, J. D., Castelli, W. P., Knoke, J. D., Jacobs, D. R., Bangdiwala, S., and Tyroler, H. A. (1989). High-density lipoprotein cholesterol and cardiovascular disease: Four prospective American studies. *Circulation* **79**, 8–15.

Gordon, T., Castelli, W. P., Hjortland, M. C., Kannel, W. B., and Dawber, T. R. (1977). High density lipoproteins as a protective factor against coronary heart disease. *Am. J. Med.* **62**, 707–714.

Grover, S. A., Abrahamowicz, M., Joseph, L., Brewer, C., Coupal, L., and Suissa, S. (1992). The benefits of treating hyperlipidemia to prevent coronary heart disease. Estimating changes in life expectancy and morbidity. *J. Am. Med. Assoc.* **267**, 816–822.

Grundy, S. (1990). Cholesterol and coronary heart disease. *J. Am. Med. Assoc.* **264**, 3053–3054.

Hamsten, A., Walldius, G., Szamosi, A., Blomback, M., Faire, U., Dahlen, G., Landou, C., and Bjorn, W. (1987). Plasminogen activator inhibitor in plasma: Risk factor for recurrent myocardial infarction. *Lancet* **87**, 3–8.

Hasstedt, S. J., Wu, L., and Williams, R. R. (1987). Major locus inheritance of apolipoprotein B in Utah pedigrees. *Genet. Epidemiol.* **4**, 67–86.

Havel, R. J. (1988). Lowering cholesterol 1988: Rationale, mechanisms, and means. *J. Clin. Invest.* **81**, 1653–1660.

Hixon, J. E. (1991). Pathobiological determinants of atherosclerosis in youth research group: Apolipoprotein E polymorphisms affect atherosclerosis in young males. *Arteriosclerosis Thrombosis* **11**, 1237–1244.

Humphries, S. E. (1987). The use of gene probes to investigate the aetiology of arterial diseases—Hyperlipidemia as an example. *In* "Immunology and Molecular Biology of Cardiovascular Diseases" (J. F. Spry, ed.), pp. 39–57. MTP Press, Lancaster, England.

Humphries, S., Coviello, D. A., Masturzo, P., Balestreri, R., Orecchini, G., and Bertolini, S., (1991). Variation in the low density lipoprotein receptor gene is associated with differences in plasma low density lipoprotein cholesterol levels in young and old normal individuals from Italy. *Arteriosclerosis Thromb.* **11**, 509–516.

International Collaborative Study Group (1986). Metabolic epidemiology of plasma cholesterol. Mechanisms of variation of plasma cholesterol within populations and between populations. *Lancet* November 1, 991–995.

Jacobs, D. R., Jr., and Barrett-Connor, E. (1982). Retest reliability of plasma cholesterol and triglyceride: The Lipid Research Clinics Prevalence Study. *Am. J. Epidemiol.* **116**, 878–885.

Kannel, W. B., Garcia, M. J., McNamara, P. M., and Pearson, G. (1971a). Serum lipid precursors of coronary heart disease. *Human Pathol.* **12**, 129–151.

Kannel, W. B., Castelli, W. P., Gordon, T., and McNamara, P. M. (1971b). Serum cholesterol, lipoproteins, and the risk of coronary heart disease. *Ann. Int. Med.* **74**, 1–12.

Kelsey, J., Thompson, W. D., and Evans, A. S. (1986). "Methods in Observational Epidemiology." Oxford University Press, New York.

Kessling, A. M., Rajput-Williams, J., Bainton, D., Scott, J., Miller, N. E. Baker, I., and Humphries, S. E. (1988). DNA polymorphisms of the apolipoprotein AII and AI-CIII-AIV genes: A study of men selected for differences in high-density lipoprotein cholesterol concentration. *Am. J. Hum. Genet.* **42**, 458–467.

Keys, A., and Parlin, R. W. (1966). Serum-cholesterol response to changes in dietary lipids. *Am. J. Clin. Nutr.* **19**, 175–181.

Lackner, C., Boerwinkle, E., Leffert, C. C., Rahmig, T., and Hobbs, H. H. (1991). Molecular basis of apolipoprotein (a) isoform size heterogeneity as revealed by pulsed-field gel electrophoresis. *J. Clin Invest.* **87**, 2153–2161.

Lerner, I. M. (1954). "Genetic Homeostasis." Oliver and Boyd, Edinburgh.

Lipid Research Clinics (1980). "Population Studies Data Book," Vol. I, The Prevalence Study. NIH publication No. 80-1527, Department of Health and Human Services, Public Health Service, Bethesda, Maryland.

Lipid Research Clinics Program (1984a). The lipid research clinics coronary primary prevention trial results. II. The relationship of reduction and incidence of coronary heart disease to cholesterol lowering. *J. Am. Med. Assoc.* **251**, 365–374.

Lipid Research Clinics Program (1984b). The lipid research clinics coronary primary prevention trial results. I. The reduction in incidence of coronary heart disease. *J. Am. Med. Assoc.* **251**, 351–364.

MacCleur, J. W., and Kammerer, C. M. (1991). Invited editorial: Dissecting the genetic contribution to coronary heart disease. *Am. J. Hum. Genet.* **49**, 1139–1144.

Maciejko, J. J., Holmes, D. R., Kottke, B. A., Zinsmeister, A. R., Dingh, O. M., and Mao, S. J. T. (1983). Apolipoprotein A-1 as a marker of angiographically assessed coronary artery disease. *N. Engl. J. Med.* **309**, 385–389.

McLean, J. W., Tomlinson, J. E., Kuang, W.-J., Eaton, D. L., Chen, E. Y., Fless, G. M., Scanu, A. M., and Lawn, R. M. (1987). cDNA sequence of human apolipoprotein(a) is homologous to plasminogen. *Nature (London)* **330**, 132–37.

Manninen, V., Elo, M. O., Frick, M. H., Haapa, K., Heinonen, O. P., Heinsalmi, P., Helo, P., Huttunen, J. K., Kaitaniemi, P., Koskinen, P., Maenpaa, H., Malkonen, M., Manttari, M., Norola, S., Pasternak, A. P., Karainen, J., Romo, M., Sjoblom, T., and Nikkila, E. A. (1988). Lipid alterations of decline in the incidence of coronary heart disease in the Helsinki Heart Study. *J. Am. Med. Assoc.* **260**, 641–651.

Manninen, V., Tenkaunen, L., Koskinen, P., Huttunen, J. K., Manttari, M., Heinonen, O. P., and Frick, M. H. (1992). Joint effects of serum triglyceride and LDL cholesterol and HDL cholesterol concentrations on coronary heart disease risk in the Helsinki Heart Study. *Circulation* **85**, 37–45.

Marcovina, S. M., and Albers, J. J. (1991). International Federation of Clinical Chemistry study on the standardization of apolipoproteins A-I and B. *Curr. Opin. Lipidol.* **2**, 355–361.

Meade, T. W., Brozovic, M., Chakrabarti, R. R., Haines, A. P., Imeson, J. D., Mellows, S., Miller, G. J., North, W. R. S., Stirling, Y., and Thompson, S. G. (1986). Haemostatic function and ischaemic heart disease: Principal results of the Northwick Park heart study. *Lancet* **86**, 533–537.

Miller, N. E. (1987). Associations of high-density lipoprotein subclasses and apolipoproteins with ischemic heart disease and coronary atherosclerosis. *Am. Heart J.* **113**, 589–97.

Moll, P. P., Michels, V. V., Weidman, W. H., and Kottke, B. A. (1989). Genetic determination of plasma apolipoprotein AI in a population-based sample. *Am. J. Hum. Genet.* **44**, 124–139.

Moorjani, S., Roy, M., Gange, C., Davignon, J., Brun, D., Toussaint, M., Lambert, M., Campeau, L., Blaicham, S., and Lupien, P. (1989). Homozygous familial hypercholesterolemia among French Canadians in Quebec Province. *Arteriosclerosis* **9**, 211–216.

Muller, C. (1938). Xanthomata, hypercholesterolemia, angina pectoris. *Acta Med Scand* Suppl 89, 75–79.

Murphy, E. A. (1979). Quantitative genetics: A critique. *Soc. Biol.* 26, 126–141.

National Cholesterol Education Program (1988). Report of the expert panel on detection, evaluation, and treatment of high blood cholesterol in adults. *Arch. Intern. Med.* 148, 36–69.

Nora, J. J., Berg, K., and Nora, A. H. (1991). "Cardiovascular Diseases. Genetics, Epidemiology, and Prevention." Oxford University Press, New York.

Pairitz, G., Davignon, J., Maillous, H., and Sing, C. F. (1988). Sources of interindividual variation in the quantitative levels of apolipoprotein B in pedigrees ascertained through a lipid clinic. *Am. J. Hum. Genet.* 43, 311–321.

Pasternak, R. C., and Reis. (1991). *In* "Cardiac Catheterization, Angiography and Intervention" (W. Grossman and A. S. Baim, eds.), 4th ed., pp. 382–402.

Pekkanen, J., Linn, S., Heiss, G., Suchindran, C. M., Leon, A., Rifkind, B. M., and Tyroler, H. A. (1990). Ten-year mortality from cardiovascular disease in relation to cholesterol level among men with and without pre-existing cardiovascular disease. *N. Engl. J. Med.* 322, 1700–1707.

Polak, J. F., Bajakian, R. L., O'Leary, D. H., Anderson, M. R., Donaldson, M. C., and Jolesz, F. A. (1992). Detection of internal carotid artery stenosis: Comparison of MR angiography, color Doppler sonography and arteriography. *Radiology* 182, 35–40.

Rhoads, G. G., Gulbrandsen, C. L., and Kagan, A. (1976). Serum lipoproteins and coronary heart disease in a population study of Hawaiian Japanese men. *N. Engl. J. Med.* 294, 293–298.

Rose, G. (1990). Preventive cardiology: what lies ahead? *Prev. Med.* 19, 97–104.

Ross, R. (1986). The pathogenesis of atherosclerosis—An update. *N. Engl. J. Med.* 314, 488–500.

Sandholzer, C., Hallman, D. M., Saha, N., Sigurdsson, G., Lackner, C., Csaszar, A. Boerwinkle, E., and Utermann, G. (1991). Effects of the apolipoprotein(a) size polymorphism on the lipoprotein(a) concentration in 7 ethnic groups. *Hum. Genet.* 86, 607–614.

Scanu, A. M. (1991). Update on lipoprotein(a). *Currn. Opin. Lipidol.* 2, 259–265.

Scott, J. (1989). The molecular and cell biology of apolipoprotein B. *Mol. Biol. Med.* 6, 65–80.

Shekelle, R. B., Shryock, A. M., Paul, O., Lepper, M., Stamler, J., Liu, S., and Raynor, W. J. (1981). Diet, serum cholesterol and death from coronary heart disease. The Western Electric Study. *N. Engl. J. Med.* 304, 65–70.

Shepherd, J., Gaffney, D., Packard, C. J., Affairs of the heart: cholesterol and coronary heart disease risk (1991). *Disease Markers* 9, 63–71.

Sing, C. F., and Davignon, J. (1985). Role of apolipoprotein E polymorphism in determining normal plasma lipid and lipoprotein variation. *Am. J. Hum. Genet.* 37, 268–285.

Sniderman, A., Shapiro, S., Marpole, D., Skinner, B., Teng, B., and Kwiterovich, P. O., Jr. (1980). Association of coronary atherosclerosis with hyperapobetalipoproteinemia (increased protein but normal cholesterol levels in human plasma low density (B) lipoproteins). *Proc. Natl. Acad. Sci. U.S.A.* 77, 604–608.

Stamler, J. S., Wentworth, D., and Neaton, J. D. (1986). Is the relationship between serum cholesterol and risk of premature death from coronary heart disease continuous or graded? Findings in 356,222 primary screenees of the Multiple Risk Factor Intervention Trial (MRFIT). *J. Am. Med. Assoc.* 256, 2823–2828.

Stampfer, M. J., Buring, J., Willett, W., Rosner, B., Eberlein, K., and Hennekens, C. H. (1985). The 2×2 factorial design: Its application to a randomized trial of aspirin and carotene in U.S. physicians. *Stat. Med.* 4, 111–116.

Stampfer, M. J., Sacks, F. M., Salvini, S., Willett, W. C., and Hennekens, C. H. (1991). A prospective study of cholesterol, apolipoproteins, and the risk of myocardial infarction. *N. Engl. J. Med.* 325, 373–381.

Steering Committee of the Physician's Health Study Research Group (1989). Final report on the aspirin component of the ongoing physicians' health study. *N. Engl. J. Med.* 321, 129–35.

Thannhauser, D. J. and Magendantz, H. (1938). The different clinical groups of xanthomatous diseases: A clinical physiological study of 22 cases. *Ann Intern Med* **11**, 1662–1746.

Thaulow, E., Erikssen, J., Sandvik, L., Stormorken, H., and Cohn, P. F. (1991). Blood platelet count and function are related to total and cardiovascular death in apparently healthy men. *Circulation* **84**, 613–617.

Tibblin, G., and Cramer, K. (1963). Serum lipids during the course of an acute myocardial infarction and one year afterwards. *Acta Med. Scand.* **174**, 451–455.

Truett, J., Cornfield, J., and Kannel, W. (1967). A multivariate analysis of the risk of coronary heart disease in Framingham. *J. Chron. Dis.* **20**, 511–554.

Underwood, S. R., Firmin, D. N., Rees, R. S., and Longmore, D. B. (1990). Magnetic resonance velocity mapping. *Clin. Phys. Physiol. Meas.* (*Suppl. A*) **II**, 37–43.

Utermann, G. (1989). The mysteries of lipoprotein (a). *Science* **246**, 904–910.

Utermann, G., Menzel, H. J., Kraft, H. G., Duba, H. C., Kemmier, H. G., and Seitz, C. (1987). Lp(a) glycoprotein phenotypes inheritance and relation to Lp(a)-lipoprotein concentrations in plasma. *J. Clin. Invest.* **80**, 458–465.

van der Wall, E. E., de Roos H., van Voorthkuisen, A. E., and Bruschke, A. V. (1991). Magnetic resonance imaging: A new approach for evaluating coronary artery disease. *Am. Heart J.* **121**, 1203–1220.

Vogel, J. (1847). The Pathological Anatomy of the Human Body. Cited by T. Crawford (1955). Philadelphia, Lean and Blanchard.

Weatherall, D. J. (ed.) (1991). The New Genetics in Clinical Practice. 3rd Ed. Oxford, Oxford University Press.

Weinberg, R. B., Dantzker, C., and Patton, C. S. (1990). Sensitivity of serum apolipoprotein A-IV levels to changes in dietary fat content. *Gastroenterology* **98**, 17–24.

Willett, W. (1990). "Nutritional Epidemiology." Oxford University Press, New York.

Wilson, R. F., Marcus, M. L., and White, C. W. (1987). Prediction of physiologic significance of coronary arterial lesions by quantitative lesion geometry in patients with limited coronary artery disease. *Circulation* **75**, 723.

14

Genetic Disease

Muin J. Khoury
and Janice S. Dorman

Introduction

Continuing advances in recombinant DNA technology are being applied to the classification, diagnosis, and treatment of genetic diseases (Donis-Keller and Botstein, 1988; Antonorakis, 1989; Gibbs and Caskey, 1989; Reiss and Cooper, 1990). In particular, the human genome project, a systematic effort to map and sequence the human genome (Antonorakis, 1990; Brenner, 1990; Gardiner, 1990; Watson, 1990), is fueling hopes that the molecular basis for most genetic diseases finally will be found. Perhaps as importantly, this project promises an in-depth examination of the genetic basis for susceptibility to common diseases such as coronary heart disease, cancer, and birth defects (Childs and Motulsky, 1988; Easton, 1989; Risch *et al.*, 1989; Khoury *et al.*, 1990; Williamson and Kessling, 1990). The study of genetic factors and their interaction with environmental factors in the occurrence of disease in populations and families—a central theme of the rapidly developing discipline of genetic epidemiology (Morton, 1982; King *et al.*, 1984; Khoury *et al.*, 1993)—clearly interfaces with the field of molecular epidemiology, in which biologic markers of exposure, susceptibility, and outcomes are used.

In this chapter, we present an overview of the principles and applications of biomarkers in the epidemiologic study of genetic diseases. We focus on the methodologic issues involved in conducting epidemiologic studies designed to evaluate (1) the distribution and determinants of classical genetic diseases in populations and (2) the role of specific genetic traits in the occurrence of common diseases in populations. We use examples of biomarkers of exposure, susceptibility, and effect for insulin-dependent diabetes mellitus (IDDM), a common disease with genetic and environmental determinants.

Classification of Genetic Disease

Nature of Genetic Involvement in Disease

Changes in the genetic material play a central role in normal physiologic variation and in susceptibility to disease. In this context, mutations can be defined broadly to include any alteration in genetic material, including simple base substitutions (point mutations), duplications or deletions of genetic material, and derangements of the entire chromosomal complement (e.g., aneuploidies). Although mutations occur in somatic and germ cells, only germ-cell mutations are transmitted across generations and can be associated with inherited pathologic variation.

Classical genetic diseases include Mendelian disorders, which are caused by mutations that occur within a single gene locus and cause a deficiency or alteration of the gene product (generally an enzyme or a structural protein) (McKusick, 1988), and chromosomal disorders, which are caused by mutations that involve observable chromosomal segments and cause numerical or structural chromosomal imbalance. With advances in high resolution chromosome banding techniques, the borders between Mendelian and chromosomal disorders are becoming more and more blurred with the realization that many Mendelian disorders involve submicroscopic deletions of chromosome segments, leading to the loss of several genes, the so-called contiguous gene syndromes (Schmickel, 1986; Emanuel, 1988). Examples include abnormalities in 15q11 in Prader–Willi syndrome (Mattei *et al.*, 1985), deletions of 17p13 in Miller–Dieker syndrome (Schwartz *et al.*, 1988), and deletions in 22q11 in DiGeorge syndrome (Greenberg *et al.*, 1988).

In addition to the classical single-gene and chromosomal disorders, genetic contribution to disease also can involve several gene loci. In particular, polygenic inheritance refers to the special situation in which many independently segregating loci contribute additively to the clinical phenotype (e.g., dermatoglyphic traits). On the other hand, the term multifactorial disorders is best reserved for conditions influenced by both environmental and genetic factors, frequently without specification of the number of loci involved (Khoury *et al.*, 1993, in press).

Genetic Biomarkers: Biomarkers of Disease or Susceptibility?

Chronic diseases now present a major challenge in molecular epidemiology to identify single genes that can, singly or in combination with other genes and environmental factors, influence disease occurrence. For example, many genetic polymorphic traits (such as blood group antigens, HLA antigens, and serum enzyme systems) are under single-locus control and have been associated with many diseases. An example is the strong association between the

TABLE 14.1   Genetic Disease Viewed in the Context of Necessary and
Sufficient Risk Factors

| Factor | | Disease risk by genotype | | | Attributable | |
Necessary	Sufficient	With	Without	Relative risk[a]	fraction[b]	Comment
Yes	Yes	1	0	Infinite	1	Homogeneous genetic disease with complete penetrance
Yes	No	$P(D|G_1)$[c]	0	Infinite	1	Genetic disease with incomplete penetrance
No	Yes	1	$P(D|G_0)$[d]	$1/P(D|G_0)$	AF	Etiologic heterogeneity
Yes	Yes	$P(D|G_1)$	$P(D|G_0)$	$P(D|G_1)/P(D|G_0)$	AF	Etiologic heterogeneity and incomplete penetrance

[a] Relative risk, $P(D|G_1)/P(D|G_0)$.
[b] AF (attributable fraction) is calculated using standard epidemiologic formulas (Rothman, 1986).
[c] $P(D|G_1)$, Disease risk in persons with the susceptible genotype (penetrance).
[d] $P(D|G_0)$, Disease risk in persons without the susceptible genotype.

HLA-B27 allele and ankylosing spondylitis, for which the relative risk is about 100; the percentage of cases "attributable" to the allele is about 90% (Ahearn and Hochberg, 1988; Khoury *et al.*, 1990). Here, the HLA-B27 allele becomes a biomarker of genetic susceptibility rather than a biomarker of the disease, since most individuals with the HLA-B27 allele will never develop ankylosing spondylitis.

The unavoidable overlap between a disease biomarker and a susceptibility biomarker can be understood quantitatively using the concepts of necessary and sufficient risk factors (Khoury *et al.*, 1990; Table 14.1). For well-defined single-gene disorders with complete penetrance and no etiologic heterogeneity, the presence of the susceptible genotype is necessary and sufficient for the development of disease (e.g., Tay-Sachs disease in the homozygous genotype). Here, disease risk in persons with the susceptible genotype is 1, disease risk in persons without the genotype is 0, the relative risk is infinity, and the population attributable fraction is 1. For such disorders, the biomarker is a disease marker. On the other hand, for most common diseases, the susceptible genotype (at one or many loci) is neither necessary (etiologic heterogeneity) nor sufficient (penetrance of the genotype depends on other factors) for disease development. For these conditions, the biomarkers could be considered markers for genetic susceptibility.

Methodologic Issues in the Use of Biomarkers in Epidemiologic Studies of Genetic Diseases

In this section, we consider some methodologic issues in using biomarkers in epidemiologic studies of classical genetic diseases (Mendelian and chromosomal disorders) and complex disorders with a genetic component.

Classical Genetic Diseases

Levels of Identification of Genetic Disease

Biomarkers for Mendelian diseases include measurements of physiologic end points, gene products, or mutation(s) at the DNA level. For example, for phenylketonuria (PKU), an autosomal recessive disorder, biomarkers for the disease include high phenylalanine levels in the blood (physiologic level), phenylalanine hydroxylase deficiency in the liver (gene-product level), and detection of a G-to-A donor splice mutation at intron 12 of the PKU gene on chromosome 12, band q24.1 (one of several DNA mutations) (McKusick, 1988). For chromosomal disorders, biomarkers are based on karyotypic analysis of peripheral blood cells, usually involving banding techniques that may detect small chromosome deletions or rearrangements. The different methods of ascertainment of genetic diseases can affect the accuracy and completeness of ascertainment when conducting epidemiologic studies. Here, the sensitivity and specificity of biomarkers should be considered. Sensitivity refers to the proportion of persons with the genetic condition who manifest the biomarker; specificity refers to the proportion of persons without the genetic condition who do not manifest the biomarker. In PKU, hyperphenylalaninemia occurs rapidly within the first few days of life since affected infants ingest phenylalanine in milk. Therefore, if high blood phenylalanine levels are detected in a newborn's bloodspot (if taken after the first 2 days), those results are fairly sensitive but are not specific, since other causes of hyperphenylalaninemia exist (Scriver *et al.*, 1989). In contrast, the detection of a gene mutation at the molecular level is specific but often not completely sensitive if other mutations are involved at the same locus.

Another important concept is the predictive value of the biomarker. The positive predictive value (PPV) refers to the proportion of individuals with the marker who actually have the disease, and the negative predictive value (NPV) refers to the proportion of individuals without the marker who do not have the disease. PPV and NPV are functions of sensitivity, specificity, and prevalence of the disease. For PKU, the PPV of high phenylalanine level on bloodspots of newborns is low, despite high sensitivity, because the disease is rare (1 in 10,000 newborns).

Frequency of Genetic Disease in Populations

To characterize the frequency of genetic diseases in different populations, population genetic concepts must be merged with epidemiologic methods of counting (Khoury *et al.,* 1993, in press). This merging of methods is helpful in characterizing the population frequency of alleles as well as the frequency of the disease. For example, knowing that the prevalence of PKU at birth is 1 in 10,000, one can estimate the frequency of the PKU allele (1%) by using assumptions of the Hardy–Weinberg equilibrium (Vogel and Motulsky, 1986). In addition, population genetic concepts can be used to evaluate frequencies of new genetic mutations in a population. In this context, the term "mutation rate" refers to "the probability with which a particular mutational event takes place in a fertilized germ cell per generation" (Vogel and Motulsky, 1986). Direct and indirect methods are available to measure mutation rates in humans (Vogel and Rathenberg, 1975). For example, if the prevalence of trisomy 21 is 1 in 1000 births, the estimated mutation rate (nondisjunction of chromosome 21 during meiosis) is about 5 in 10,000 gametes (since the mutation can occur in either parental germ cell). From molecular and cytogenetic data (Antonorakis *et al.,* 1990; Sherman *et al.,* 1990), the parental origin seems to be maternal in 95% of the cases. Thus, one can estimate that the mutation rate for chromosome 21 meiotic nondisjunction is about 9.5 in 10,000 fertilized ova and about 0.5 in 10,000 fertilizing sperms.

Determinants of Genetic Disease in Populations

The frequency of genetic disease in populations is influenced by two categories of determinants: (1) factors that affect the occurrence of germ-cell mutations at the level of parental generation and (2) factors that influence the transmission of alleles across generations.

The design, conduct, and analysis of studies to examine the occurrence of germ-cell mutations fall within the domain of classical epidemiology (mutation epidemiology) (Hook and Porter, 1981). Examples of studies in this area include the possible association between paternal occupational exposures and trisomy 21 in offspring (Olshan *et al.,* 1989) and the well-known association between advanced maternal age and trisomy 21 (Hassold and Jacobs, 1984). Other examples are the studies of adverse pregnancy outcomes in children of atomic bomb survivors in Japan. In particular, recent studies have focused on measuring plasma protein electrophoretic variants to estimate the effect of radiation exposure and its doubling dose. Although no differences have been shown between mutation rates in about 13,000 children of "proximally-exposed" parents (0.5 per 100,000 gametes) and those in about 10,000 children of "distally-exposed" parents (0.6 per 100,000 gametes) (Neel *et al.,* 1988), these studies emphasize sample size considerations and the issue of statistical power when dealing with rare end points. As discussed elsewhere (Khoury *et al.*, 1993, in press), to achieve statistical significance for detecting a relative risk of 2 (doubling dose) with a background mutation rate of 1 per 100,000, more than 2 million exposed and unexposed persons must be examined.

In addition to factors affecting the occurrence of germ-cell mutations, the frequency of genetic diseases in populations also is affected by factors that alter the transmission and persistence of alleles across generations. These determinants include selection forces, chance fluctuations (founder effect/ genetic drift), and deviations from random mating (e.g., due to inbreeding, racial admixture, assortative mating). These factors traditionally are studied within the realm of classical population genetics (Cavalli-Sforza and Bodmer, 1971; Crow, 1986), but can have important implications in epidemiologic studies of genetic diseases (Khoury *et al.*, 1993). For example, when comparing the frequency of autosomal recessive conditions (such as cystic fibrosis and sickle cell anemia) in populations, it is important to go beyond the study of parental risk factors and exposures because, in most instances, persons with recessive disorders have inherited, from each parent, an abnormal allele that could have been hidden in the heterozygous state for several generations. Therefore, to elucidate determinants of these disorders in populations, epidemiologic studies must examine factors affecting the dynamics of the gene pool over generations as well as the interaction of selection forces, drift, and deviations from random mating (e.g., effects of migration, inbreeding, and radial admixture).

A well-known example of selection in genetic diseases is the high frequency of the sickle cell allele in African countries with endemic falciparum malaria. This high frequency has occurred even though individuals with sickle cell anemia (homozygotes for the sickle cell allele) have markedly decreased survival and reproductive fitness. Such an ecologic association has been explained by a compelling body of evidence documenting the survival advantage of individuals with the sickle cell trait (heterozygotes for the sickle cell allele) over individuals with normal hemoglobin alleles in the face of malaria infection (Vogel and Motulsky, 1986). This clear cut, environmentally induced, selective advantage is usually difficult to document for many genetic mutations, because dramatic changes in the environment have occurred over the last few generations.

Common Diseases with Genetic Components

Strategies for Finding Disease Susceptibility Genes

Since most common diseases have complex etiologies and pathogenesis, it is important to address the role of specific genes in the occurrence of these diseases. The search for disease susceptibility genes is conducted using two main methods: (1) the association approach, in which evidence is sought for statistical associations between one or more alleles (or a genotype at one or more loci) and a disease using standard epidemiologic methods, and (2) the linkage approach, in which evidence for co-segregation between a locus and a putative disease susceptibility locus is sought using family studies (Khoury *et al.*, 1990).

Using the association approach, investigators have begun to evaluate the role of specific candidate genes in the pathogenesis of diseases. Such genes play a role in normal physiologic variation and may be affected in pathologic processes. Many such genetic traits are polymorphic (i.e., more than one allele exists and the frequency of the most common allele is less than 0.99) and under single-locus control (e.g., HLA or blood groups). One example of genetic trait disease associations is the alpha-1 antitrypsin polymorphism (*PiZ* locus) and chronic obstructive pulmonary disease (COPD) (Tockman *et al.*, 1985). The presence of the *PiZ* allele in the homozygous state leads to a deficiency in alpha-1 antitrypsin, a protein with apparent protective function against protease-mediated destruction of lung tissue during inflammatory processes (Kimbel, 1985). Individuals can be classified genotypically by identifying genetic mutations at each locus, using either a gene-product approach (evaluation of proteins, enzyme levels, etc.) or a molecular approach (evaluation of DNA sequence differences). As shown in Table 14.2, if two alleles are present, N for normal and S for susceptible, three genotypes are possible, NN, NS, and SS. The risk of disease can be evaluated according to genotype using case-control or cohort methodologies (Table 14.2). If the allele confers dominant disease susceptibility, individuals with NS and SS genotypes are at higher disease risk than individuals with the NN genotype. On the other hand, if the allele confers recessive disease susceptibility, the risk of disease is elevated only among individuals with the SS genotype. Analytic issues and limitations of the association approach are discussed in subsequent text.

Using the linkage approach, researchers can look for marker loci that are genetically linked to the putative susceptibility gene loci (i.e., situated very closely on the same chromosome and generally segregate together dur-

TABLE *14.2*   Epidemiologic Methods to Evaluate the Association between Specific Alleles and Diseases

Genotype at one susceptibility locus[a]	Cohort study		Case-control study		
	Disease risk	Relative risk	No. cases	No. controls	Odds ratio
NN	p0	1	a0	b0	1
NS	p1	p1/p0	a1	b1	a1b0/a0b1
SS	p2	p2/p1	a2	b2	a2b0/a0b2

[a] N, Normal allele; S, susceptible allele.

ing meiosis). Any deviation from perfect co-segregation occurs because of recombination during meiosis, which can be estimated using linkage analysis in family studies. The increasing number of "anonymous" DNA markers scattered on different chromosomes should permit investigators to search for linked disease susceptibility loci. In contrast to the association approach, in linkage analysis specific alleles at a locus are associated with the disease under study only within a particular family. Therefore, when different families from a population are pooled, no association between a particular allele and the disease is expected (Khoury et al., 1990).

Although linkage analysis usually is conducted in extended pedigrees, methods are available to incorporate such an approach into epidemiologic studies. The sib-pair methodology provides an approach to evaluate linkage between genetic marker loci and disease susceptibility loci in sibships (Penrose, 1953; Haseman and Elston, 1972; Day and Simmons, 1976). The basic approach with this method is the comparison of the distribution of alleles identical by descent (ibd) at a marker locus between pairs of relatives with the expected Mendelian distribution without linkage. For example, for affected sib-pairs, the expected Mendelian distribution of alleles ibd in the absence of linkage is 25% for 0 alleles, 50% for 1 allele, and 25% for 2 alleles. This methodology has been developed for quantitative and dichotomous traits (Haseman and Elston, 1972) and can be extended to any pair of relatives (Weeks and Lange, 1988).

In the context of an epidemiologic study design, cohort and case-control approaches can be used (Table 14.3) (Khoury et al., 1992). In a cohort approach, first probands with a disease are identified from a well-defined population and their unaffected siblings are followed for the development of disease. Disease risk in siblings then is stratified according to the number of alleles ibd they share with the proband. In the absence of linkage between the marker locus and the disease susceptibility locus (null hypothesis), disease recurrence risks are expected to be identical in the three groups. Examples of this approach are seen in studies of sibling recurrence risks of IDDM conducted in Pittsburgh (Cavender et al., 1984; Lipton et al., 1992). In these studies, a higher disease risk was found among siblings sharing 1 and 2 haplotypes ibd in the HLA region with the proband than among siblings sharing 0 haplotypes ibd, suggesting genetic linkage between the HLA complex and the IDDM susceptibility locus. Alternatively, a nested case-control study can be used (Table 14.3), in which cases are siblings of the proband that eventually become affected and controls are chosen randomly from the population of siblings. Then the frequency of the number of alleles ibd is compared between cases and controls. Odds ratios obtained here closely approximate values of relative risk obtained from a cohort study. Advantages and limitations of the sib-pair linkage approach have been discussed (Day and Simmons, 1976; Suarez, 1978; Thomson, 1986; Risch, 1988; Khoury et al., 1990, 1992).

TABLE 14.3   Epidemiologic Methods to Evaluate the Linkage between a Specific Gene Locus and a Disease in the Context of a Sibling Recurrence Risk Study

No. haploptypes identical by descent with the proband	Cohort study		Case-control study		
	Sibling recurrence risk	Relative risk	No. affected siblings	No. unaffected siblings	Odds ratio
0	p0	1	a0	b0	1
1	p1	p1/p0	a1	b1	a1b0/a0b1
2	p2	p2/p0	a2	b2	a2b0/a0b2

Pitfalls and Limitations

When conducting association and linkage studies to evaluate the role of genetic factors in disease, the usual epidemiologic considerations apply, including issues related to case definition and representativeness, selection of appropriate control groups, and methods for minimizing biases related to ascertainment, recall, and confounding (Khoury *et al.*, 1993, in press). For linkage studies, a desirable approach in a family study is complete ascertainment of incident cases in well-defined populations, for example, through the use of registries (Dorman *et al.*, 1988). Although the study of single large pedigrees with multiply affected family members provides more information for linkage analysis, such families may not be representative of the universe of cases and could limit generalizability of the findings (Khoury *et al.*, 1990).

Let us remember that demonstrating a statistically significant association between an allele and a disease may not reflect a causal role of the allele. Alternative explanations include confounding, linkage disequilibrium, and type I errors. In addition, in these studies, the presence of gene–environment interaction must be evaluated.

Confounding   Confounding can occur if both the disease and the allele (or other risk factor) of interest are associated with one or more variables that are causally associated with the disease. For example, if one wanted to study whether the risk of trisomy 21 is increased among offspring of older fathers, advanced maternal age is an obvious potential confounder, since it is associated both with the risk of trisomy 21 and with advanced paternal age (Erickson and Bjerkedal, 1981). Maternal age has to be adjusted for, by design, analysis, or both. More generally, in disease–genetic trait association studies, confounding can occur if cases and controls are drawn from different racial or ethnic groups or other population subgroups (Khoury *et al.*, 1992). One example of confounding is provided by the study by Knowler *et al.*

(1988) on the association between a genetic marker, *Gm* (3;5,13,14), and the prevalence of non-insulin-dependent diabetes mellitus in the Pima indians. In this population, the absence of this *Gm* haplotype was associated with a higher prevalence of diabetes (29% compared with 8%). However, this *Gm* haplotype was actually an index of white racial admixture. The higher the level of white admixture (as measured by an index of Indian heritage in their study), the higher the prevalence of the genetic marker and the lower the prevalence of diabetes. The association between *Gm* haplotype and diabetes all but disappeared when the analysis was stratified by, and adjusted for, the level of white admixture (Knowler *et al.*, 1988).

Linkage disequilibrium   Linkage disequilibrium refers to the situation in which an association between a disease and an allele is caused indirectly by the association between the disease and another allele at a closely linked locus, especially in the presence of founder effect or genetic drift or recent occurrence of the mutation (Gibbs and Caskey, 1989). In general, with increasing number of generations, meiotic recombination—even between tightly linked loci—will tend to dilute the association between alleles from neighboring loci (Morton, 1984). Currently, most associations between diseases and DNA markers identified through restriction enzyme analysis should be viewed in the context of linkage disequilibrium rather than direct cause–effect relationship (Cooper and Clayton, 1988), because restriction enzyme analysis can, thus far, only detect gene mutations occurring at restriction enzyme recognition sites. An example of these associations is the case-control study of Ardinger *et al.* (1989), in which an association was found between cleft lip and palate and DNA polymorphisms at the transforming growth factor alpha gene (TGFA). These findings should be complemented with family studies that demonstrate genetic linkage between these marker loci and the disease.

Type I errors   Type I errors have been discussed in the context of general epidemiologic studies (Rothman, 1986). In particular, when multiple genetic markers are examined in relation to diseases, some statistically significant associations at the conventional 0.05 level undoubtedly will occur by chance. Type I errors are especially relevant to DNA markers; as more genes are cloned and sequenced and as more restriction enzymes are used in conjunction with multiple gene probes, more DNA variation is bound to be uncovered in small regions of the human genome. A major challenge for genetic epidemiology in the coming decades is to differentiate the spurious from the biologically meaningful associations between diseases and DNA markers. As in other areas of epidemiology, establishing a cause–effect relationship depends, in part, on the reproducibility of the findings as well as on biologically plausible models to explain them.

TABLE 14.4   Simple Gene–Environment Interaction Model in the Context
of Epidemiologic Studies

Genotype, environment[a]	Cohort study		Case-control study		
	Disease risk	Relative risk	No cases	No controls	Odds ratio
− , −	I	1	a0	b0	1
− , +	IRe	Re	a1	b1	a1b0/a0b1
+ , −	IRg	Rg	a2	b2	a2b0/a0b2
+ , +	IRge	Rge	a3	b3	a3b0/a0b3

Source: Adapted from Khoury *et al.* (1988).
[a] − , Absent; +, present.

Interaction   Finally, when designing epidemiologic studies that examine
the association between diseases and specific alleles, it is important to include
measurements of the interaction between alleles and environmental factors
in the prediction of disease. An epidemiologic framework can be used to look
for evidence of gene–environment interaction (Khoury *et al.*, 1988; Ottman,
1990). As shown in Table 14.4, the relationship between a genetic trait and
a disease can be stratified by the presence or absence of an environmental
factor in cohort and case-control studies. These studies can be extended to
more than one locus and one environmental factor, and can account for exposure dosage. Epidemiologic methods are available to assess interaction in
the context of additive and multiplicative models (Rothman, 1986). Failure
to search for interaction may lead to dilution of the magnitude of the association between alleles and diseases (Khoury *et al.*, 1988,1990). For example,
Cartwright *et al.* (1982) reported an association between the slow acetylator
phenotype and bladder cancer. The relative risk was small (about 1.6), however. When the association was examined among workers possibly exposed
to aryl nitrite compounds, the magnitude of the relative risk was much higher
(about 17), which suggested a possible biologic interaction between carcinogenic compounds and the acetylation pathway.

Examples of Biomarkers for a Common Disease
with Genetic Components

Insulin-dependent diabetes mellitus (IDDM) is one of the leading chronic diseases of childhood, occurring in both developing and industrialized countries
(LaPorte and Cruickshank, 1985). IDDM is associated with many severe
long-term complications, such as blindness and renal disease, that frequently
lead to disability and premature mortality (Deckert *et al.*, 1978; Dorman
et al., 1984).

With the development of standardized population-based incidence registries, it has become apparent that IDDM risk varies considerably across racial groups and countries [Diabetes Epidemiology Research International Study Group (DERI), 1988]. The age-adjusted rates are high in Scandinavian countries (such as Finland), but very low in the Oriental populations (such as Korean and Japanese). Disease risk varies also among whites (DERI, 1990) and among racial subgroups within a population (DERI, 1988). The causes of these striking geographic and ethnic patterns of incidence are being investigated now using population-based epidemiologic approaches with an emphasis on finding possible familial and genetic determinants of the disease.

Susceptibility Biomarkers: HLA Genes

HLA association studies (Wolf *et al.*, 1983; Bertram and Baur, 1984), and linkage studies in families (Green and Woodrow, 1977; Barbosa *et al.*, 1980) have confirmed that IDDM susceptibility genes are located in the HLA region of chromosome 6. These genes encode class I, II, and III molecules as well as several other factors that are important for normal immune response. Specific class II molecules (HLA-DR, DQ, and DP locus antigens) are associated with IDDM and other autoimmune diseases; they function immunologically by presenting antigens to helper T lymphocytes (Trucco and Dorman, 1989). The HLA molecule–antigen complex is important in T-cell recognition and in determining and the T-cell repertoire of an individual (Marrack and Kappler, 1987).

Results of HLA association studies have shown consistently that a higher prevalence of the serologically defined HLA-DR3 and DR4 antigens exists among IDDM subjects than among healthy nondiabetic control subjects (Wolf *et al.*, 1983; Bertram and Baur, 1984). We now know that the presence of DNA sequences encoding an amino acid other than aspartic acid in the position 57 of the DQ beta chain (non-Asp57) is highly associated with susceptibility to IDDM, whereas an aspartic acid in this position (Asp57) appears to confer resistance to the disease (Todd *et al.*, 1987; Morel *et al.*, 1988). With aspartic acid in position 57 of the DQ beta chain, the peptide binding ability of the molecule is affected because of the formation of a salt bridge with the alpha chain. This finding offers a biologic explanation for the importance of specific amino acid sequences in determining susceptibility or resistance to certain autoimmune diseases such as IDDM. Although the presence of a non-Asp57 is unlikely to be the only allele associated with IDDM susceptibility (Erlich *et al.*, 1990; Kwok *et al.*, 1990), it is currently the genetic marker that is related most strongly to the disease.

To begin to test the hypothesis that the worldwide variation in IDDM incidence is related to population differences in the frequency of non-Asp57, standardized, population-based, case-control studies have been initiated around the world. Preliminary data from studies of IDDM cases and nondiabetic controls in five areas that included populations at low, moderate, and high risk for IDDM have shown that HLA-DQ beta genotype distribution significantly differed between subjects with IDDM and nondiabetic controls (Dorman *et al.*, 1990a). Non-Asp57 alleles were associated significantly with IDDM in all areas; population-specific odds ratios for non-Asp57 homozygotes relative to Asp57 homozygotes ranged from 14 to 111. Moreover, the frequency of the non-Asp57 allele in the general population was related directly to disease incidence. This finding suggests that variation in the distribution of non-Asp57 alleles may explain much of the geographic variation in IDDM incidence.

Exposure Biomarkers: Markers of Viral Infections

Although the geographic patterns of IDDM incidence may be related to population differences in the frequency of particular HLA-DQ beta alleles, other etiologic factors clearly play a role in the development of the disease. Many non-Asp57 homozygous individuals do not develop IDDM (Dorman *et al.,* 1990a); the diabetes concordance rates among monozygotic twins is typically low, ranging from 30 to 50% (Barnett *et al.*, 1981). Apparent epidemics of IDDM also have been reported (Rewers *et al.,* 1988). Possible explanations for these findings include the presence of other IDDM susceptibility genes, variation in exposure to environmental risk factors such as viruses (Schattner and Rager-Zisman, 1990), nutrition (Scott, 1990), stress (Leclere and Weryha, 1989), and immunologic differences.

One strategy for investigating the role of other etiologic factors within and across populations is to control first for the strong influence of host susceptibility. This can be done by focusing not only on affected persons, but also on high-risk individuals such as family members (Dorman *et al.*, 1988). Typically, first-degree relatives of IDDM probands are recruited for genetic studies. They represent an excellent populations for analytic investigations of other etiologic determinants of the disease.

Studies have provided evidence that persistent viral infections may act as possible triggers of autoimmune diseases. The presence of human cytomegalovirus (CMV) gene segments in the genomic DNA extracted from peripheral blood lymphocytes is considered a useful marker of chronic CMV infection; CMV genome positivity has been associated significantly with IDDM among newly diagnosed patients (Pak *et al.*, 1988). Persistent CMV infection may lead to the expression of viral or host antigens on the beta cells of the pancreas, causing an autoimmune response (Haywood, 1986). Alternatively, molecular mimicry may result in the production of antibodies that recognize both viral and host antigens.

The contribution of chronic CMV infection to the occurrence of IDDM in families was investigated. Although CMV genome positivity was not found to be an independent risk factor for IDDM among first-degree relatives, a striking interaction between CMV and non-Asp57 was observed (Dorman *et al.*, 1990b). The cumulative risk for non-Asp57 homozygous and CMV genome positive individuals was more than 30% through age 30 years, approximating the recurrence risk in identical twins. Had information regarding host susceptibility not been available for these family members, the significant contribution of CMV genome positivity to familial IDDM would not have been detected. This evaluation illustrates the importance of population-based family studies to estimate the risks and interactions associated with the genetic and environmental determinants of disease.

Effect Biomarkers: Islet Cell Antibodies

Organ-specific autoantibodies, such as islet cell antibodies (ICA), long have been recognized as one of the hallmarks of autoimmune disease. However, it is unclear whether they play a direct role in the disease process or serve as a marker of tissue damage initiated by other etiologic agents (Bottazzo, 1986). Despite immunologic heterogeneity among ICAs and variation in laboratory methodology, most studies have reported a very high prevalence of ICAs (up to 100%) among patients with newly diagnosed IDDM (Lipton and LaPorte, 1989). After disease onset, however, ICAs rarely persist. The high prevalence of ICAs among IDDM patients at diagnosis contrasts with rates of 2–5% among their first-degree relatives (Orchard and Rosenbloom, 1985; Srikanta *et al.*, 1985), and 0.5% or less among control subjects (Notso *et al.*, 1983; Betterle *et al.*, 1984).

First-degree relatives who are ICA positive have markedly higher rates of IDDM than those who are ICA negative; relative risks range from 50 to 500 (Riley *et al.*, 1987; Lipton *et al.*, 1989; Bruining *et al.*, 1988; Tarn *et al.*, 1988). Islet cell antibodies frequently can be detected many years before the onset of clinical symptoms (Tarn *et al.*, 1987), indicating that ICAs are among the most important predictors of the disease. However, most normoglycemic individuals who are ICA positive will never develop diabetes (Bruining *et al.*, 1988; Bonifacio *et al.*, 1990). Presently, it is not possible to distinguish ICA-positive individuals who will eventually develop IDDM from those who will remain disease free.

Population-based, case-control, and family studies such as those described here can contribute greatly to the understanding of the etiology of IDDM. By evaluating the presence or absence of high-risk IDDM susceptibility genes, organ-specific autoantibodies, and exposure to environmental risk factors, in families and in case-control studies, one can assess the relative and the absolute risks associated with these potential determinants of the disease. More important, the interaction between genetic and environmental factors can be evaluated. The incorporation of molecular biology into traditional epidemiologic research, therefore, represents an important interface for future studies of the etiology of chronic diseases.

Finally, it is appropriate to mention that the human genome project represents a major stimulus for the evolution of molecular epidemiology. This massive effort will result in the mapping and sequencing of all 100,000 genes in the human genome (Congress of the United States, 1988; Stephens *et al.,* 1990; Watson and Cook-Dugan, 1991). The project will provide scientists with a tremendous amount of new information regarding genetic susceptibility to acute and chronic diseases, since DNA markers will be used more and more in epidemiologic research as potential risk factors for a variety of conditions. The human genome project also will lead to major improvements in DNA technology. This development will be most rapid during the first 5 years of the project. From an epidemiologic perspective, advanced DNA technology will provide the necessary tools for the incorporation of molecular testing into large epidemiologic investigations. In addition, the project focuses on mapping and sequencing nonhuman organisms, permitting the identification and classification, at the DNA level, of additional risk factors (such as infectious agents). Finally, the accumulating knowledge on genetic susceptibility to diseases will require epidemiologists to address ethical, social, and legal issues, including genetic screening, discrimination, and confidentiality (Holtzman, 1989). The human genome project undoubtedly will have enormous impact on the practice of medicine and public health.

References

Ahearn, J. M., and Hochberg, M. C. (1988). Epidemiology and genetics of ankylosing spondylitis. *J. Rheumatol.* 15(Suppl. 16), 22–28.

Antonorakis, S. E. (1989). Diagnosis of genetic disorders at the DNA level. *N. Engl. J. Med.* **320**, 153–163.

Antonorakis, S. E. (1990). The mapping and sequencing of the human genome. *South. Med. J.* **83**, 876–878.

Antonorakis, S. E., Lewis, J. G., Adelsburger, P. A., *et al.* (1990). Parental origin of the extra chromosome in trisomy 21 revisited: DNA polymorphism analysis suggests that in 95% of cases the origin is maternal. *Am. J. Hum. Genet.* **47**(Suppl.), A207.

Ardinger, H. H., Buetow, K. H., Bell, G. I., *et al.* (1989). Association of genetic variation of the transforming growth factor alpha gene with cleft lip and palate. *Am. J. Hum. Genet.* **45**, 348–353.

Barbosa, J., Chern, M. M., Anderson, V. E., *et al.* (1980). Linkage analysis between the major histocompatibility system and insulin-dependent diabetes in families with patients in two consecutive generations. *J. Clin. Invest.* **65**, 592–601.

Barnett, A. H., Eff, C., Leslie, R. D. G., *et al.* (1981). Diabetes in identical twins: A study of 200 pairs. *Diabetologia* **20**, 87–93.

Bertram, J., and Baur, M. (1984). Insulin-dependent diabetes mellitus. *In* "Histocompatibility Testing," pp. 348–368. Springer-Verlag, Heidelberg.

Betterle, C., Zanette, F., Pedini, B., *et al.* (1984). Clinical and subclinical organ-specific autoimmune manifestations in type I (insulin-dependent) diabetic patients and their first-degree relatives. *Diabetologia* **26**, 431–436.

Bonifacio, E., Bingley, P. J., Shattock, M., *et al.* (1990). Quantification of islet-cell antibodies and prediction of insulin-dependent diabetes. *Lancet* **335**, 147–149.

Bottazzo, G. F. (1986). Death of a β-cell. Homicide or suicide? *Diabetic Med.* 3, 119–30.

Brenner, S. (1990). The human genome: The nature of the enterprise. *Ciba Found. Symp.* 149, 6–12.

Bruining, G. J., Molenaar, J. L., Grobbee, D. E., *et al.* (1988). Ten year follow-up study of islet cell antibodies and childhood diabetes mellitus. *Lancet* 1, 1100–1002.

Cartwight, R. A., Glashan, R. W., Rogers, H. J., *et al.* (1982). Role of N-acetyltransferase phenotypes in bladder carcinogenesis: A pharmacogenetic-epidemiologic approach to bladder cancer. *Lancet* 2, 842–845.

Cavalli-Sforza, L. L., and Bodmer, W. F. (1971). "The Genetics of Human Populations." Freeman, San Francisco.

Cavender, D. E., Wagener, D. K., Rabib, B. S., *et al.* (1984). The Pittsburgh insulin-dependent diabetes mellitus (IDDM) study: HLA antigens and haplotypes as risk factors for the development of IDDM in IDDM patients and their siblings. *J. Chron. Dis.* 37, 555–568.

Childs, B., and Motulsky, A. G. (1988). Recombinant DNA analysis of multifactorial disease. *Prog. Med. Genet.* 7, 180–194.

Congress of the United States (1988). "Mapping our Genes. Genome Projects: How Big, How Fast?" The Johns Hopkins University Press, Baltimore, Maryland.

Cooper, D. N., and Clayton, J. F. (1988). DNA polymorphisms and the study of disease associations. *Hum. Genet.* 78, 299–312.

Crow, J. F. (1986). "Basic Concepts in Population, Quantitative, and Evolutionary Genetics." Freeman, New York.

Day, N., and Simmons, N. (1976). Disease susceptibility genes—Their identification by multiple case family studies. *Tissue Antigens* 8, 109–119.

Deckert, T., Poulsen, J. E., and Larsen, M. (1978). Prognosis of diabetics with diabetes onset before the age of thirty-one. *Diabetologia* 14, 363–370.

Diabetes Epidemiology Research International Study Group (1988). Geographic patterns of childhood insulin-dependent diabetes mellitus. *Diabetes* 37, 1113–1119.

Diabetes Epidemiology Research International Study Group (1990). The epidemiology and immunogenetics of IDDM in Italian-heritage populations. *Diabetes Metab. Rev.* 6, 63–69.

Donis-Keller, H., and Botstein, D. (1988). Recombinant DNA methods: Applications to human genetics. *Prog. Med. Genet.* 7, 17–42.

Dorman, J. S., LaPorte, R. E., Kuller, L. H., *et al.* (1984). The Pittsburgh insulin-dependent diabetes mellitus (IDDM) morbidity and mortality study. Mortality results. *Diabetes* 33, 271–276.

Dorman, J. S., Trucco, M., LaPorte, R. E., *et al.* (1988). Family studies: The key to understanding the genetic and environmental etiology of chronic diseases? *Genet. Epidemiol.* 5, 305–310.

Dorman, J. S., LaPorte, R. E., Stone, R. E., *et al.* (1990a). Worldwide differences in the incidence of type I diabetes are associated with amino acid variation at position 57 of the HLA-DQ beta chain. *Proc. Natl. Acad. Sci. U.S.A.* 87, 7370–7374.

Dorman, J. S., Lipton, R. B., Trucco, M., Yoon, J. W., and LaPorte, R. E. (1990b). Familial insulin-dependent diabetes mellitus (IDDM): An investigation of host–environmental interactions. *Am. J. Hum. Genet.* 47(Suppl.), A132.

Easton, D. F. (1989). Linkage analysis in non-Mendelian disorders. *Prog. Clin. Biol. Res.* 329, 159–164.

Emanuel, B. S. (1988). Invited editorial: Molecular cytogenetics: Toward dissection of the contiguous gene syndromes. *Am. J. Hum. Genet.* 43, 575–578.

Erickson, J. D., and Bjerkedal, T. (1981). Down syndrome associated with father's age in Norway. *J. Med. Genet.* 18, 22–28.

Erlich, H. A., Bugawan, T. L., Scharf, S., *et al.* (1990). HLA-DQ beta sequence polymorphism and genetic susceptibility to IDDM. *Diabetes* 39, 96–103.

Gardiner, R. M. (1990). The human genome: A prospect for paediatrics. *Arch. Dis. Child* 65, 457–461.

Gibbs, R. A., and Caskey, C. T. (1989). The application of recombinant DNA technology for genetic probing in epidemiology. *Annu. Rev. Publ. Health* **10**, 27–48.

Green, J. R., and Woodrow, J. C. (1977). Sibling method for detecting HLA-linked genes in disease. *Tissue Antigens* **9**, 31–35.

Greenberg, F., Elder, F. F. B., Haffner, P., *et al.* (1988). Cytogenetic findings in a prospective series of patients with DiGeorge anomaly. *Am. J. Hum. Genet.* **43**, 605–611.

Haseman, J. K., and Elston, R. C. (1972). The investigation of linkage between a quantitative trait and a marker locus. *Behav. Genet.* **2**, 3–19.

Hassold, T., and Jacobs, P. A. (1984). Trisomy in man. *Annu. Rev. Genet.* **18**, 69–97.

Haywood, A. M. (1986). Patterns of persistent viral infections. *N. Engl. J. Med.* **315**, 939–948.

Holtzman, N. A. (1989). "Proceed with Caution: Predicting Genetic Risks in the Recombination DNA Era." The Johns Hopkins University Press, Baltimore, Maryland.

Hook, E. B., and Porter, I. H. (eds.). (1981). "Population and Biological Aspects of Human Mutation." Academic Press, New York.

Khoury, M. J., Adams, M. J., and Flanders, W. D. (1988). An epidemiologic approach to eco-genetics. *Am. J. Hum. Genet.* **42**, 89–95.

Khoury, M. J., Beaty, T. H., and Flanders, W. D. (1990). Epidemiologic approaches to the use of DNA markers in the search for disease susceptibility genes. *Epidemiol. Rev.* **12**, 41–55.

Khoury, M. J., Flanders, W. D., Lipton, R. B., and Dorman, J. S. (1992). The affected sib-pair method in the context of an epidemiologic study design. *Genet. Epidemiol.* **8**, 417–423.

Khoury, M. J., Beaty, T. H., and Cohen, B. H. (1993, in press). "Fundamentals of Genetic Epidemiology." Oxford University Press, New York.

Kimbel, P. (1985). Proteolytic damage and emphysema pathogenesis. *In* "Chronic Obstructive Pulmonary Disease" (T. L. Petty, ed.), 2d Ed., pp. 105–128. Marcel Dekker, New York.

King, M. C., Lee, G. M., Spinner, N. B., Thomson, G., and Wrensch, M. R. (1984). Genetic epidemiology. *Annu. Rev. Publ. Health* **5**, 1–52.

Knowler, W. C., Williams, R. C., Pettitt, D. J., and Steinberg, A. G. (1988). Gm (3;5,13,14) and type 2 diabetes mellitus: An association in American Indians with genetic admixture. *Am. J. Hum. Genet.* **43**, 520–526.

Kwok, W. W., Mickelson, E., Masewicz, S., *et al.* (1990). Polymorphic DQ alpha and DQ beta interactions dictate HLA Class II determinants of allo-recognition. *J. Exp. Med.* **171**, 85–95.

LaPorte, R. E., and Cruickshank, K. J. (1985). Incidence and risk factors for insulin-dependent diabetes. *In* "Diabetes in America" (R. F. Hamman and M. I. Harris, eds.), pp. 111–122. NIH Publication No. 85-1468. U.S. Government Printing Office, Washington, D.C.

Leclere, J., and Weryha, G. (1989). Stress and auto-immune endocrine diseases. *Horm. Res.* **31**, 90–93.

Lipton, R. B., and LaPorte, R. E. (1989). Epidemiology of islet cell antibodies. *Epidemiol. Rev.* **11**, 182–203.

Lipton, R. B., Atchison, J. A., and Becker, D. J. (1989). Prediction of IDDM among first-degree relatives of children with IDDM. *Diabetes* **38** (Suppl. 2), 91A.

Lipton, R. B., Atchison, J., Dorman, J. S., *et al.* (1992). Genetic, immunologic and metabolic determinants of risk for insulin-dependent diabetes mellitus in families: The Pittsburgh etiology and epidemiology of diabetes study. *Am. J. Epidemiol.* (*submitted*).

McKusick, V. A. (1988). "Mendelian Inheritance in Man: Catalogs of Autosomal Dominant, Autosomal Recessive, and X-Linked Phenotypes, 8th Ed. The Johns Hopkins University Press, Baltimore, Maryland.

Marrack, P., and Kappler, J. (1987). The T cell receptor. *Science* **238**, 1073–1079.

Mattei, J. F., Mattei, M. G., and Giraud, F. (1985). Prader-Willi syndrome and chromosome 15: A clinical discussion of 20 cases. *Hum. Genet.* **54**, 356–362.

Morel, P. A., Dorman, J. S., Todd, J. A., *et al.* (1988). Aspartic acid at position 57 of the HLA-DQ beta chain protects against type 1 diabetes: A family study. *Proc. Natl. Acad. Sci. U.S.A.* **85**, 8111–8115.

Morton, N. E. (1982). "Outline of Genetic Epidemiology." Karger, Basel, Switzerland.

Morton, N. E. (1984). Linkage and association. *Proc. Clin. Biol. Res.* **147,** 245–265.

Neel, J. V., Satoh, C., Goriki, K., *et al.* (1988). Search for mutations altering protein charge and/ or function in children of atomic bomb survivors: Final report. *Am. J. Hum. Genet.* **42,** 663–676.

Notso, K., Goto, Y., Sakurami, T., *et al.* (1983). A population study of pancreatic islet cell antibodies and antithyroid antibodies. *Tohoku J. Exp. Med.* **141** (Suppl.), 261–264.

Olshan, A. F., Baird, P. A., and Tesckle, K. (1989). Paternal occupational exposures and the risk of Down syndrome. *Am. J. Hum. Genet.* **44,** 646–651.

Orchard, T. J., and Rosenbloom, A. L. (1985). The development of insulin-dependent diabetes mellitus among relatives. *Diabetes Care* **8** (Suppl. 1), 45–50.

Ottman, R. (1990). An epidemiologic approach to gene-environment interaction. *Genet. Epidemiol.* **7,** 177–185.

Pak, C. Y., Eun, H. M., McArthur, R. G., and Yoon, J. W. (1988). Association of cytomegalovirus infection with autoimmune type 1 diabetes. *Lancet* **2,** 1–4.

Penrose, L. S. (1953). The general sib-pair linkage test. *Ann. Eugen.* **18,** 120–144.

Reiss, J., and Cooper, D. N. (1990). Application of the polymerase chain reaction to the diagnosis of human genetic disease. *Hum. Genet.* **85,** 1–8.

Rewers, M., LaPorte, R. E., Walczak, M., *et al.* (1988). Apparent epidemic of insulin-dependent diabetes mellitus in Midwestern Poland. *Diabetes* **39,** 1113–1119.

Risch, N. (1988). Assessing the role of HLA-linked and unlinked determinants of disease. *Am. J. Hum. Genet.* **40,** 1–14.

Risch, N., Claus, E., and Giuffra, L. (1989). Linkage and mode of inheritance in complex traits. *Prog. Clin. Biol. Res.* **329,** 183–188.

Rothman, K. J. (1986). "Modern Epidemiology." Little, Brown, Boston.

Schattner, A., and Rager-Zisman, B. (1990). Virus-induced autoimmunity. *Rev. Infect. Dis.* **12,** 204–222.

Schmickel, R. D. (1986). Contiguous gene syndromes: A component of recognizable syndromes. *J. Pediatr.* **109,** 231–241.

Schwartz, C. E., Johnson, J. P., Holycross, B., *et al.* (1988). Detection of submicroscopic deletions in band 17p13 in patients with Miller–Dieker syndrome. *Am. J. Hum. Genet.* **43,** 597–604.

Scott, F. W. (1990). Cow's milk and insulin-dependent diabetes mellitus: Is there a relationship? *Am. J. Clin. Nutr.* **51,** 489–491.

Scriver, C. R., Kaufman, S., and Woo, S. L. C. (1989). The hyperphenylalaninemias. *In* "The Metabolic Basis for Inherited Disease" (C. R. Scriver, A. L. Beaudet, W. S. Sly, and D. Valle, eds.), 6th Ed., pp. 495–546. McGraw-Hill, New York.

Sherman, S., Takaesu, N., Freeman, S., *et al.* (1990). Trisomy 21: Association between reduced recombination and non-disjunction. *Am. J. Hum. Genet.* **47** (Suppl.), A97.

Srikanta, S., Ganda, O. P., Rabizadeh, A., *et al.* (1985). First-degree relatives of patients with type I diabetes: Islet cell antibodies and abnormal insulin secretion. *N. Engl. J. Med.* **313,** 461–464.

Stephens, J. C., Cavanaugh, M. L., Gardie, M. I., Mador, M. L., and Kidd, K. K. (1990). Mapping the human genome: Current status. *Science* **250,** 237–244.

Suarez, B. K. (1978). The affected sib-pair ibd distribution for HLA-linked disease susceptibility genes. *Tissue Antigens* **12,** 87–93.

Tarn, A. C., Thomas, J. M., Howard, W., *et al.* (1987). Nine year follow-up in a family study: The risk for insulin-dependent diabetes. *Diabetes* **36** (Suppl. 1), 72.

Tarn, A. C., Thomas, J. M., Dean, B. M., *et al.* (1988). Predicting insulin-dependent diabetes. *Lancet* **2,** 845–850.

Thomson, G. (1986). Determining the mode of inheritance of RFLP-associated diseases using the affected sib-pair method. *Am. J. Hum. Genet.* **39,** 207–221.

Tockman, M. S., Khoury, M. J., and Cohen, B. H. (1985). The epidemiology of COPD. *In*

"Chronic Obstructive Pulmonary Disease" (T. L. Petty, ed.), 2d Ed., pp. 43–92. Marcel Dekker, New York.

Todd, J. A., Bell, J. L., and McDevitt, H. O. (1987). HLA-DQ beta gene contributes to susceptibility and resistance to insulin-6dependent diabetes mellitus. *Nature (London)* **329**, 559–604.

Trucco, M., and Dorman, J. S. (1989). Immunogenetics of insulin-dependent diabetes mellitus in humans. *CRC Crit. Rev. Immunol.* **9**, 201–245.

Vogel, F., and Motulsky, A. G. (1986). "Human Genetics: Problems and Approaches," 2d Ed. Springer-Verlag, Berlin.

Vogel, F., and Rathenberg, R. (1975). Spontaneous mutation in man. *Adv. Hum. Genet.* **6**, 223–318.

Watson, J. D. (1990). The human genome project: Past, present, and future. *Science* **248**, 44–49.

Watson, J. D., and Cook-Dugan, R. M. (1991). Origins of the human genome. *FASEB J.* **5**, 8–78.

Weeks, D. E., and Lange, K. (1988). The affected pedigree member method of linkage analysis. *Am. J. Hum. Genet.* **42**, 315–322.

Williamson, R., and Kessling, A. M. (1990). The problem of polygenic disease. *Ciba Found. Symp.* **149**, 63–70.

Wolf, E., Spencer, K. M., and Cudworth, A. G. (1983). The genetic susceptibility to type I (insulin- dependent) diabetes: Analyses of the HLA-DR associations. *Diabetologia* **23**, 224–230.

15

Biologic Markers in the Epidemiology of Reproduction

Grace Kawas Lemasters
and Paul A. Schulte

Although most of the literature in reproductive epidemiology focuses on clinically recognized biological events, such as fetal loss or malformations, current and future studies are likely to involve a range of biologic markers, either as outcomes or to indicate exposure or susceptibility. Literature on effect markers such as semen analysis and early pregnancy loss is growing. Advances in molecular biology are providing new tools with which to investigate poorly understood disorders of reproduction. [See review by the National Research Council (NRC), 1989.] Wide-scale use of molecular genetic diagnostic tests is almost certain to be common in clinical medicine, yielding new insights into the etiology, mechanism, and risk of inherited conditions. Advances in DNA technology have allowed exploration of the previously speculative role of molecular mutation that results in germ line mutations, most likely leading to early pregnancy loss. Defects in genetic coding for products critical for embryonic or fetal development have been hypothesized to play a causative role in euploidic abortion in humans (Butler and Mcdonough, 1989).

The framework for employing biologic markers involves three separate but interlinked systems. A couple's reproductive success depends on a delicate physiochemical balance between and within the paternal, maternal, and fetal systems. Any disruption of this balance can result in a broad range of effects. The consequence of exposure of men and women to mutagens, teratogens, and carcinogens is described in Figure 15.1. Exposure of both genders may result in cancer or chromosomal damage, potentially increasing individual risk of cancer. Germ line mutation may lead to infertility, adverse pregnancy outcomes, or heritable alterations expressed in future generations.

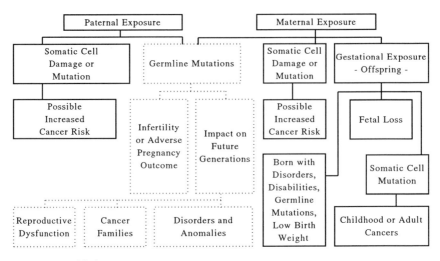

FIGURE 15.1   Consequences of exposures to carcinogens, mutagens, or teratogens.

Gestational exposure of the offspring may lead to such effects as fetal loss, disorders and disabilities, or childhood and adult cancers. To date, limited inquiries have been made to determine the impact of exposure on future generations. The identification of susceptible subgroups or, conversely, resistant groups in epidemiologic studies may lend itself to this backward tracing of genetic differences in the population. Technologies such as radioactively labeled DNA probes and polymerase chain reactions have proved fruitful in identifying genetic disorders and may help in understanding biologic differences in response to toxicants. In this chapter, we identify a framework for using practical and available biologic markers in epidemiologic research of reproduction. Markers of exposure, effect, and susceptibility are discussed for each gender. Figure 15.2 displays the interplay between these categories of markers.

Markers of Female Reproduction and Pregnancy Outcome

Markers of Exposure

In epidemiologic studies of reproduction, the use of biologic markers to assess exposure has been limited, but the need and the potential for such use exists. The need stems from the same weakness that has been recognized for most epidemiologic studies, namely, the classification of exposure or identification of early effects. In females, markers of exposure are not easily acces-

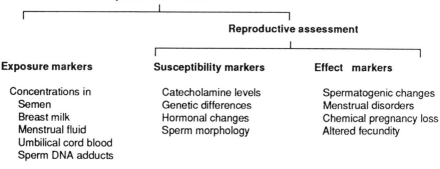

FIGURE 15.2   Exposure and reproductive health assessment.

sible from target tissues. Hence, general exposure-related markers in routinely collected biologic media, such as blood, urine, and breath, are often used. Specimens that are more difficult to collect, such as body fat, or more temporally limited, such as breast milk, umbilical cord blood, or menstrual blood, may be important exposure markers.

Depending on the research question, a unique issue in exposure assessment pertaining to reproduction will be the critical period of exposure. If, for example, fetal loss is the effect of interest, then the determining exposure marker might be collected just prior to conception or during early pregnancy. Exposure markers for teratogens generally are targeted during the period of organogenesis, up to days 55–60 of gestation, but for some anomalies, for example, of the central nervous system, eye, and genitalia, markers may need to be measured throughout the pregnancy. Figure 15.3 approximates a relationship between the time that an exposure marker may be associated with a particular effect. Time frames of exposure also can be represented by protein or DNA adducts. Hemoglobin, for instance, is a very accurate dosimeter of exposures during the preceding 4 months; albumin addresses the preceding month. In both cases, these dosimeters reflect cumulative dose. An example of the efficacy of the approach was described by Everson *et al.* (1988), who evaluated the association between DNA damage in the human placenta and maternal smoking and birth weight. The investigators evaluated the extent of an association between maternal smoking and birth weight assessed by questionnaire data, biochemical measures for smoking exposure, and molecular methods (Table 15.1). They found no association between birth weight and biochemical measures of smoking, but a clearly significant association between birth weight and the adduct levels. These findings were similar for birth length. Although these associations cannot be interpreted as evidence for a causal association between level of adducts and a decrease in

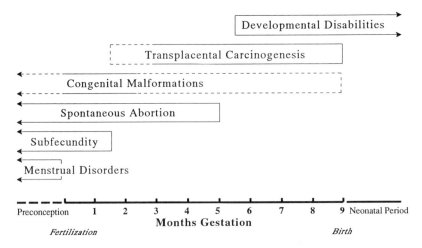

FIGURE 15.3 Reproductive outcomes associated with timing of maternal exposure. Solid lines indicate the most probable timing of exposure for a particular outcome; dashed lines indicate less probable, but still possible, timing of exposure; arrows suggest that a defined cutoff point for a specific outcome is not known. (Adapted from Selevan and Lemasters, 1987.)

birth weight or length, they do serve as a demonstration that the DNA adduct is an accurate dosimeter reflecting exposure to cigarette smoke that causes low birth weight (Everson *et al.,* 1988). The lesson of the study is the need to integrate various approaches in assessing how specific environmental agents might impair development of the fetus. Although widespread application will depend on further development and field testing, these exposure markers may be useful for depicting the time period surrounding conception and gestation.

Markers of Effect

Studies on the reproductive effects of toxic exposure of women are unique because two individuals are potentially at risk—the woman and, if she is pregnant, her developing offspring. Exposure to either may result in a wide range of adverse responses. Maternal exposure to a toxicant may cause infertility, genetic damage, menstrual disorders, illness during pregnancy, chromosomal aberrations, breast milk alteration, early onset of menopause, and suppressed libido. Adverse fetal effects include preterm delivery, fetal loss, perinatal death, lowered birth weight, altered sex ratio, congenital malformation, childhood malignancies, infant or childhood illness, inheritable genetic damage, and developmental disabilities. This section discusses the markers and monitors used in epidemiologic studies of menstrual cycle variability, fetal loss, and birth anomalies of the offspring.

TABLE 15.1 Association between Birth Weight and Smoking Exposure Comparing Results of Questionnaire and Biochemical and Molecular Methods for Assessing Intensity of Exposure to Smoking[a]

Parameter for smoking exposure	Birth weight			Birth length		
	R^2	Significance of model (P value)	Significance of adding smoking parameter to basic model (P value)	R^2	Significance of model (P value)	Significance of adding smoking parameter to basic model (P value)
None	0.41	.009	—	0.31	.05	—
Questionnaire data (Number of cigarettes smoked in 2nd trimester)	0.41	.02	.85	0.34	.06	.28
Biochemical and questionnaire summary of smoking exposure score	0.48	.006	.09	0.43	.01	.03
Level of adduct 1	0.52	.002	.025	0.47	.007	.01

Source: Reprinted with permission from Everson *et al.* (1988).

[a] All models present data for the 30 smokers for whom results of ^{32}P postlabeling assays and questionnaire data were available. Basic model includes terms for maternal age, maternal race, maternal years of education, and gestational age of the newborn. The R^2 indicates the proportion of variability of the dependent variable (birth weight or birth length) that is explained by the independent variables. Statistical significance of the model is increased as the independent variables explain a greater proportion of the variation in the dependent variable, but decreases if a larger number of independent variables are required to explain the same proportion of variability.

Menstrual Cycle Variability

In humans, between 3 and 4 million follicles are present in each ovary at birth; shortly after birth, the primary oocytes are arrested in late prophase of meiosis I. At puberty, gonadotropins stimulate meiosis and the number of follicles reduces to less than 400,000. During each ovarian cycle, follicles are recruited in groups, the leading follicle ovulates, and the remainder undergo atresia. After menopause, few if any follicles are present in the ovary; their absence is accompanied by reduced ovarian steroid synthesis. Reproductive senescence can occur if xenobiotics block oogenesis in the fetus or destroy oocytes, causing premature ovarian failure (Mattison, 1983). Susceptibility also depends on the developmental stage of the individual. In studies of girls receiving combination antineoplastics, prepubertal girls were generally less susceptible to ovarian damage than girls receiving similar therapy after puberty (Siris *et al.*, 1976; Lentz, 1977; Stillman, 1981; Chapman, 1983).

This sensitive and complex process requires the integrated function of the hypothalamus, pituitary, and ovary, representing several potential targets for damage. Primary ovarian failure can be evaluated using hormonal markers that indicate raised gonadotropins or lowered estrogens, followed by menopausal symptoms. One study found urinary hormones to be the most reliable in predicting ovulatory function in a small group of women (Kesner *et al.*, 1992). The use of urinary hormones blunts the effects of hormone pulsatility and may not be informative unless serially collected, preferably over several cycles (Scialli and Lemasters, 1993, in press). Hormonal concentrations in blood are unreliable measures of reproductive impairment because of erratic fluctuation over very short intervals. Less severe damage to the ovulatory process may be expressed in disorders of menstruation which, in turn, may be a surrogate of other events such as a decrease in individual fertility potential or a very early pregnancy loss in which menses may or may not be delayed. Several potential mechanisms associated with toxic exposure and disruption of cyclic ovarian function have been summarized (NRC, 1989). Agents that mimic the action of naturally occurring hormones such as polychlorinated biphenyls (PCBs) or DDT, for example, may alter (1) normal estrogen feedback between the gonad and brain, (2) hormone synthesis and storage, and (3) hormone release and metabolism.

Menstrual changes as effect markers related to exposures are relatively sparse, but studies suggest that menstrual patterns may be sensitive monitors of environmental influences. Although the characteristics of a normal cycle vary among women, variations in individual women are slight. The average age of menarche is 12.5 years with a range of 9 to 16 years (McFarland, 1980). The average duration of menses is between 2 and 7 days; the interval between menses ranges from 23 to 35 days with a mean of 28.1 days (Chiazze *et al.*, 1968; Goldsmith and Weiss, 1986). Regularity of intervals between menses for an individual generally falls within 5 days (Fogel and

Woods, 1981). The average menstrual blood loss is 30 to 100 milliliters (Goldsmith and Weiss, 1986). The estimated mean age at natural menopause is 50.5 years.

Menstrual abnormalities can be divided into three broad categories of (1) cycle length or rhythm, (2) characteristics of bleeding patterns, and (3) the presence of pain. Exposure to solvents such as benzene, toluene, and xylene have shown menstrual disturbances primarily associated with abnormal bleeding (Michon, 1965; Butarewicz *et al.*, 1969; Syrovadko *et al.*, 1973; Beskrovnaja, 1979). Exposure to perchloroethylene and other solvents in dry cleaners was associated with a significant excess of menstrual disorders including cycle length, menorrhagia, dysmenorrhea, and premenstrual syndrome (Zielhuis *et al.*, 1989). Of workers exposed to trinitrotoluene in explosive manufacturing, 83% experienced changes in their menstrual cycle with symptoms diminishing to 65% and 20% after 1 and 2 years postexposure, respectively (Gesev, 1967). The prevalence of menstrual irregularity and amenorrhea was shown to be higher in nurses who handled cytotoxic drugs than in controls and was most pronounced in women over age 30 (Shortridge, 1988).

Using menstrual disorders as monitors of reproductive toxicity presents many challenges. Often clinical signs arise only after the occurrence of major functional changes. Disruption in menses cannot discriminate between ovarian, hormonal, or other central nervous system target sites. Data collection procedures may be subject to recall bias. Studies requiring calendar event recordings are cumbersome, resulting in poor compliance. Biologic samples combined with questionnaire data can, however, be powerful tools for eliciting answers to female reproductive toxicity.

Fetal Loss

Exposure of the conceptus to a toxicant can result in diverse effects, depending on when the exposure occurred during embryonic–fetal development. Transport time of a fertilized ovum before implantation is 2–6 days. During this early stage of embryogenesis, exposures to chemical compounds may occur by penetration of chemicals into the uterine fluids. Absorption of xenobiotics may be accompanied by degenerative changes, alteration in the blastocystic protein profile, or inability to implant. The period of late embryogenesis is characterized by differentiation, mobilization, and organization of cells and tissues into organ rudiments. Embryogenesis is the period of greatest sensitivity to teratogens. Untoward responses during embryogenesis may culminate in loss of the embryo or other structural or developmental defect. The fetal period extends from embryogenesis to birth. The distinction between the embryonic and fetal period is somewhat arbitrary, approximating when the developing organism has a crown rump length of 30 mm or approximately 55–60 days after conception. The fetal period is characterized developmentally by growth, histogenesis, and functional maturation.

Toxicity may be manifested by a reduction in cell size and number. The brain is still sensitive to injury; myelination is incomplete until after birth. Growth retardation, functional defects, disruption in the pregnancy, behavioral effects, transplacental carcinogenesis, or death may result from toxicity during the fetal period.

Epidemiologic studies of fetal wastage are based primarily on relatively few methodologies, that is, biochemical assays, record searches, and interview and survey approaches. Wilcox and others (1987,1988) have described a practical biologic marker of human trophoblast cell development to assess very early "chemical" pregnancy loss. Urinary human chorionic gonadotropin (hCG) is measured with immunoradiometric assay using a detection antibody. It is measured with no cross-reactivity of human leuteinizing hormone (hLH) at almost 100% specificity. The lowest hCG concentration detectable was 0.01 ng/ml. Screening methods involved collection of daily urine specimens within a 15-day portion of each cycle, beginning 10 days before the onset of menstrual bleeding. The definition of early loss was hCG levels of 0.025 ng/ml for 3 consecutive days, a fairly conservative criterion. A decrease in hCG titers signals loss of the pregnancy. The use of hCG as a biomarker, however, does not distinguish if the pregnancy failure is of maternal or trophoblastic origin. To discern this difference, a multiple hormone profile such as estrone conjugates and relaxin can be used (Stewart *et al.,* 1990). Future research will be enhanced by the further development and improvement of hormonal markers in easily obtained biologic specimens, such as urine and saliva (NRC, 1989). These methods are needed for application in epidemiologic studies.

Approximately 15% of fertilized eggs have been estimated to be lost prior to implantation, leaving 85% available for implantation (Leridon, 1977; Schlesselman, 1979; Edmonds *et al.,* 1982; Jones *et al.,* 1983; Little, 1988). In a study using measures of hCG (Wilcox *et al.,* 1988), wastage of fertilized ova occurring postimplantation but subclinically was 22%. This early loss rate translates into 19% of the remaining conceptions (.85 × .22). Recognized losses constituted 9% of remaining conceptions or 8% of the potential total available ova (.85 × .09) (Hertz-Picciotto and Samuels, 1988). Thus, the "true" potential rate of spontaneous abortions among fertilized ova is high, approximately 42% (.15 + .19 + .08) for fetal losses occurring up to 28 weeks of gestation.

Congenital Anomalies

Numerous reviews are available on the etiology, mechanism, and types of malformations. This section highlights data on incidence rates, risk factors, and the use of congenital anomalies as events signaling the introduction of a new teratogen into our environment or serving as a barometer of change that has occurred in the germ line of previous generations.

Malformations may be single or multiple. Chromosomal defects generally lead to multiple defects whereas single gene changes or exposure to environmental agents may cause single defects or a syndrome (Warkany, 1971). A major malformation can be defined as a defect resulting in death, requiring surgery or medical treatment, or causing a substantial physical or psychological handicap.

The known causes of birth anomalies are genetic (10.1%), multifactorial inheritances (23%), uterine factors (2.5%), twinning (0.4%), or teratogens (3.2%) (Nelson and Holmes, 1989). The remaining 43.2% of anomalies are of unknown etiology but are probably multifactorial. Certainly one important mechanism of action of teratogens may be mutation associated with cell death. Many chemical mutagens are not direct acting but require metabolic activation by the cytochrome P450 proteins. Placental tissue, as well as the embryo, is capable of this oxidative metabolism. Studies have shown that the inducibility of the maternal and the embryonic genotype is important in the teratogenicity of polycyclic aromatic hydrocarbons (PAHs) because of the binding of the metabolites to DNA (Manchester, 1991).

Several environmental exposures have been associated with congenital anomalies in the offspring. These include maternal consumption of food contaminated with methyl mercury during gestation in Japan and Iraq, causing morphological, central nervous system, and neurobehavioral abnormalities (Bakir *et al.*, 1973; Amin-Zaki *et al.*, 1974). In Japan, the cluster of cases was linked to the consumption of fish and shellfish contaminated with mercury derived from the effluent of a chemical factory. Maternal ingestion of PCBs from contaminated rice oil in 1968 gave rise to babies with several disorders, including growth retardation, dark brown skin pigmentation, early eruption of teeth, gingival hyperplasia, wide sagittal suture, edematous face, and exophthalmos (Kuratsune *et al.*, 1972).

Several registries in the United States and other countries monitor births with anomalies and link information about exposures that occurred during the pregnancy. One reason these registries are important is that the offspring may be more sensitive to xenobiotics during the period of organogenesis than at any other stage in the life cycle (Wilson, 1973). If rates of a particular defect, syndrome of defects, or unusual anomaly suddenly appear or increase, presumably these monitors will serve as early harbingers of a change in exposure. Registries have provided useful clues about potentially hazardous occupations but have fallen short of providing early warning signals of the introduction of a new teratogen.

The incidence of malformations is dependent on the status of the conceptus: (1) live births, (2) spontaneous abortuses, or (3) stillbirths. Miller and Poland (1970) found 88% abnormal abortuses in 28-day-old embryos (73 of 83). The overall incidence by 20 weeks of gestation was 43% (223 of 498). In the earliest age group, multiple system defects and severe growth disorganization were found and became less frequent with each developmental stage. Birth defect incidence figures for live births are dependent on age of diagnosis. The frequency of major congenital defects reported shortly after birth can triple on re-examination at 9 months of age. The incidence of all major defects combined fluctuates between 1 and 3%, with a frequency of about 2.2% (Nelson and Holmes, 1989). Rates for all minor defects combined ranges between 3 and 15% (Ekelund *et al.,* 1970; Warkany, 1971; Bloom, 1981). This large variation in reported rates for major and minor defects is created by differences in data sources, populations, definitions, and specialty of individuals performing the examinations.

One challenge in using congenital anomalies as effect monitors is deciding how to group for analysis (Källén, 1988; Kline *et al.,* 1989). Often, all malformations are combined or grouped in major and minor categories. The rationale for grouping all these together is that the majority arose during organogenesis. The advantage of this approach is the larger number of cases, thereby increasing statistical power. Assume, for example, a case-control study in which the combined incidence of major and minor defects was hypothesized to be 10%, in contrast to a specific single anomaly with a rate of 1%. With 300 cases and controls in each group, the study has approximately 80% power to detect a doubling effect with a two-sided alpha of 0.05, in contrast to less than 20% power for the single anomaly. If an exposure effect is specific to a particular type of malformation, that is, central nervous system, grouping all major and minor defects together may mask the effect. Alternatively, grouping may be done by organ system. Although this method may be an improvement, certain defects may dominate the class, for example, varus deformities of the feet when examining the musculoskeletal system. The optimal solution, given the limitation of power, is dividing the defects into pathogenically homogeneous groups (Källén, 1988). Other considerations should be given to the exclusion or inclusion of certain malformations, such as those likely to be caused by chromosomal defects, autosomal dominant conditions, or *in utero* positioning. When analyzing congenital anomalies, a balance must be maintained between precision and statistical power.

Markers of Susceptibility

Markers of susceptibility for female reproductive abnormalities are indicators that are inherited or acquired that increase the likelihood of an abnormality. Inherited markers might include maternal inborn errors of metabolism that influence the potential of the fetus to be carried to term, such as those pertaining to blood clotting or the oxygen carrying capacity of blood, that is, those factors necessary for the formation and maintenance of the placenta. Examples of acquired susceptibility markers are catecholamine as a marker of stress to distinguish sensitive subgroups and alterations in hormonal profiles due to an agent that competes for receptor sites. Bernstein

TABLE *15.2*   Adjusted Geometric Mean Estradiol, Sex-Hormone-Binding Globulin-Binding Capacity, and Human Chorionic Gonadotropin Levels in Sera by Smoking Status at Time of Sampling[a]

Variable	Smoking status at sampling				P value[b]
	Nonsmoker		Smoker		
Sample size(n) for E_2	83		64		
E_2 (pg/ml)[c]	158.7		130.7		0.037
By cigarettes per day	None	1–5	6–10	11[c]	
n	83	16	19	29	
	158.7	152.5	125.5	123.5	0.025[d]
Sample size (n) for SHBG-bc	55		48		
SHBG-bc (µg/dl)[c]	5.93		5.20		0.153
By cigarettes per day	None	1–5	6–10	11[c]	
n	55	9	16	23	
	5.93	5.87	4.91	5.17	0.182[d]
Sample size (n) for hCG	66		48		
hCG (iu/ml)[c]	50.5		39.6		0.044
By cigarettes per day	None	1–5	6–10	11[c]	
n	66	13	15	20	
	50.5	47.3	40.9	34.4	0.013[d]

[a] E_2, Estradiol; SHBG-bc, sex-hormone-binding globulin-binding capacity; hCG, human chorionic gonadotropin.
[b] P value associated with analysis of covariance F test, with adjustment for data set and length of gestation.
[c] All geometric means adjusted for data set and length of gestation.
[d] P value for trend with adjustment for data set (because hormone and SHBG-bc levels were determined at separate times for each study) and length of gestation (Bernstein *et al.*, 1989).

et al. (1989) examined the impact of smoking on hCG, estradiol (E2), and sex-hormone binding globulin-binding capacity (SHBG-bc) as shown in Table 15.2. The decreased levels of E2 ($p = 0.025$), SHBG-bc ($p = 0.182$), and hCG ($p = 0.013$) in smokers were evident over the range of gestation studied, that is, 43–113 days. The authors believe that the smoking effect on E2 is explained largely in statistical terms by the associated hCG effect (that is, adjusting the E2 values for the associated hCG values accounted almost completely for the difference in E2 values between smokers and nonsmokers). Hence, the hCG marker measures a direct effect on the placenta, and the between-person differences in level of E2 mirror differences in hCG level. Key to using such susceptibility markers are understanding whether the markers are related to the exposure, to the outcome, or to both, and understanding how to treat the markers in the design and analysis of the study. If research is aimed at assessing the role of genes in a reproductive event, markers of susceptibility that depict preexisting or previously acquired mutation might be useful in assessing whether all study subjects are equally at risk. Hence, the impact of the particular gene being studied can be determined with minimum confounding or effect modification. Little research has been performed thus far on markers of susceptibility for untoward reproductive outcomes. One exception is based on the observations between abnormal fibrogens and spontaneous abortion (Weatherall, 1991). One fibrogen variant that shows a clear association with spontaneous abortions is the Metz variant. The Metz variant is characterized by a defective fibrinopeptide resulting from an arginine-to-cysteine substitution at position 16 in the alpha chain (Weatherall, 1991). Prenatal diagnosis by amniocentesis and chorionic villus sampling can be used to obtain specimens of susceptibility markers for cytogenetic, biochemical, or molecular studies. Once fetal DNA is available, it can be analyzed in several ways for genetic disorders that might influence whether the pregnancy will come to term. Mutations can be identified directly by restriction endonuclease mapping or indirectly using restriction fragment length polymorphisms (RFLPs). Chapter 14 should be reviewed for a more detailed discussion on these genetic markers.

Markers of Male Reproduction

Markers of Exposure

Unlike the female, the male provides greater accessibility to reproductive fluids, allowing the opportunity to evaluate relevant target biologic materials and obtain estimates of biologically effective dose. Markers of biologically effective dose that have been identified include DNA adducts in sperm, xenobiotics in seminal plasma, and various enzymes, metalloproteins, flavoproteins, and mucoproteins in seminal plasma (NRC, 1989). Additionally, markers most commonly thought of as effect markers, such as DNA mutations in sperm, also may be used to indicate exposure when used in studies comparing groups nominally classified as with and without putative exposure (Segg, 1991). In such studies, the best indicator of exposure might involve the combination of a nominal exposure indicator and the presence of altered parameters (that is, markers of biologically effective dose or early biologic effects). In this approach, it is necessary to consider confounding factors that can account for differences among groups. Certain changes in sperm parameters, for example, may be used as exposure or reproductive effect markers. Sperm DNA adducts may be markers of exposure whereas changes in sperm morphology may be markers of susceptibility or effects.

Markers of Effect

Semen Assays

Manifestation of male reproductive toxicity has been defined as alteration in sexual behavior, fertility, pregnancy outcomes of the partner, or modifications in other functions that are dependent on the integrity of the male reproductive system (*Fed Reg.*, 1988). Noticeably absent from this definition is reference to male-mediated effects, for example, childhood cancers, as expressed in the offspring. There is some debate about the proper placement of these events. Because the insult to the human germ cell is a prerequisite for male-mediated expression, we suggest that the definition should be inclusive based on where the biologic insult has occurred, not on where it is expressed.

Assessment of male reproductive toxicity has relied heavily on semen assays because seminal fluid represents a relatively easily collected and analyzed medium in which to observe biologic effects of exposure. Sperm assays provide a direct measure of male reproductive impairment and a possible indirect measure of transmission of genetic damage. Spermatogenesis is a renewing, synchronous process requiring at least 14 intermediate stages wherein the germ cell proceeds through mitotic division for cell proliferation, meiotic divisions to generate genetic diversity and to decrease the chromosome number by half, and differentiation steps prior to the release of immature spermatozoa from the testes. Any of the developing cell types, from spermatogonia, spermatocytes, and spermatids to immature and mature spermatozoa, may be susceptible to toxic exposure. The process of spermatogenesis requires approximately 74 days.

Given the site of insult and the specific sperm parameter affected, different adverse outcomes may be observed. The absence of sperm (azoospermia) or reduced sperm count, less than 20 million per milliliter semen (oligospermia), and sperm with low motility are associated with reduced fertility. For nonmutagenic events, the most likely outcome associated with insult to the spermatogonia (stem cells) may be cell death. Thus, the affected cell would be phagocytized early and would not matriculate through spermatogenesis to appear in the ejaculate. Although cell death also may occur in later stages, the speed and efficiency of phagocytizing processes are uncertain. In addition to decreases in concentration, one may also observe decreased viability, decreased motility, and degenerate cells. Perturbations of the biochemical milieu in which the mature cells are maintained may be reflected as alterations in motility, decreases in viability followed by cellular degeneration, and eventual declines in concentration (Lemasters and Zenick, 1985). Sperm with abnormal forms (teratosperm) or degenerative forms may be associated with basic interferences in cellular processes necessary for maintaining the integrity and viability of the cell. These abnormal forms are more likely to be associated with infertility or early fetal loss. There are large gaps in our knowledge concerning the interrelationships among these various markers. In one of our clearest examples of an occupational exposure affecting male reproduction, workers exposed to a nematocide, 1,2-dibromo-3-chloropropane (DBCP) not only experienced sterility (Whorton *et al.,* 1977) but experienced an increased incidence of spontaneous abortions in their wives (Kharrazi *et al.,* 1980).

The most common biologic markers used in semen studies are sperm count, sperm velocity, percentage motile sperm, percentage viable sperm, and percentage sperm with normal morphology. A longitudinal study of 34 unexposed men was conducted to obtain data on normative values and statistical variation (Schrader *et al.,* 1988); from that study and another (Schrader *et al.,* 1987) these mean values (and standard deviations) are derived. Sperm count is measured as the total number of sperm ejaculated ($\bar{X}$ = 204.8 ± 199.2). Sperm concentration is reported in millions per milliliter ($\bar{X}$ = 47.4 ± 22.9). The percentage of motile sperm measured is defined as the proportion of sperm that fulfills at least minimum movement criteria ($\bar{X}$ = 59.8 ± 20.9). Sperm velocity (in μm/sec) is the swimming speed of sperm either along its swimming path or in a straight line from one set time to the next. Computer instruments are used to evaluate swimming speed and patterns. Percentage viable sperm is expressed as the portion of living sperm measured by a stain exclusion test ($\bar{X}$ = 71.4 ± 7.4) or by hypoosmotic swelling ($\bar{X}$ = 64.1 ± 9.1). The stain exclusion assays measure structural integrity whereas placement in a hypoosmotic solution to measure cell swelling assesses the functional integrity of sperm membranes (i.e., maintenance of ionic gradients). Sperm morphology is an assessment of the shape of the sperm head but also may incorporate the midpiece and tail. Morphology may be reported as the percentage normal ($\bar{X}$ = 80.2 ± 9.5) or described in terms of percentage macro, micro tapered, double head or tail, percentage immature, percentage amorphous, and so on. Alterations in sperm morphology may be considered markers of susceptibility for subfecundity, fetal loss, and heritable genetic abnormalities (Wyrobek *et al.,* 1984). Sperm motility and morphology are considered two of the most sensitive indicators of susceptibility for subfecundity (Jouannet *et al.,* 1988; Grunert *et al.,* 1989).

Although sperm count is the marker most frequently used in field studies, it lacks precision for detecting population differences; the average coefficient of variation within subject sperm count is high: 44% (Schrader *et al.,* 1988). Sperm velocity, on the other hand, has low overall between- and within-person coefficients of variation but the within-person variation is greater, indicating that fluctuations within a subject are almost as great as those among subjects. This finding suggests that multiple samples from the same individual may optimally stabilize the results. Parameters of morphology and viability have relatively good intraclass correlations and low coefficients of variation, allowing good precision for detecting trends and population differences.

These standard tests of sperm function may be of limited value, however, in predicting fecundity. For example, after intrauterine insemination, no significant differences were found in semen parameters among those who did or did not achieve pregnancies (Huszar and DeCherney, 1987). Biochemical markers of sperm quality are needed to identify deficiencies in sperm function. Huszar *et al.* (1990) have attempted to take this next step by evaluating sperm creatine kinase (CK) activity to predict sperm fertilizing potential of oligospermic men. A comparison was made of oligospermic men who were fertile (N=33) with an infertile group (N=66) with identical mean sperm concentrations, 11.9 million sperm/ml, and similar mean motility values of 23.7 and 23.0, respectively. Sperm CK was significantly lower in the fertile oligospermic group. These findings also were supported in normospermic men (Huszar *et al.*, 1988). Certainly future investigations will need to augment the standard tests of sperm parameters with biochemical measures that have the ability to signal defects of sperm development and fertilizing potential.

Survey Methods

Approaches for estimating effects of exposure on reproduction, other than the collection of biologic specimens, include measures of fertility or fecundity as biologic markers. The survey methods described in this section are equally applicable in evaluating subfecundity of an exposed female population. Often in studies of male workers, the spouses are the actual source of the interview data.

Terminology used in defining fertility can be confusing. Fertility statistics are based on the ability to deliver a live born child, whereas fecundity impairments are more inclusive and address the physiologic capacity to conceive. A couple is considered subfecund if they have difficulty conceiving or maintaining a pregnancy over a specified time period. Population statistics indicate that, for married women between the ages of 15 and 44 years, approximately 16% have nonsurgical fecundity impairments (U.S. Department of Health and Human Services, 1982). This rate is considerably higher in African-Americans (23%) than in Caucasians (15%).

In epidemiologic studies, three approaches may be used to assess subfecundity. The first approach asks specific questions that determine if the individual has identified a period of time, usually 1–2 years, during which the couple was trying to conceive. Dates for total time period(s) are obtained. The rate and length of time in days, weeks, or months that the couple was subfecund are compared for exposed and unexposed employees. Requiring a minimum of 1 full year of unprotected intercourse, however, may miss more subtle effects of exposures. This length of time requires the passage of about five sperm cycles and may be inappropriate for exposures that are acute or intermittent (Lemasters *et al.*, 1991).

Another statistical approach was employed by Wong *et al.* (1979) to assess subfecundity of workers exposed to ethylene dibromide. This method compared the observed number of births for exposed person–years to an expected number estimated from maternal birth rates specific to maternal age, parity, race, and year of birth of the woman. Levine *et al.* (1980,1981) and Starr and Levine (1983) extended this method to examine the fertility experience, comparing preexposure to exposure time periods. The disadvantages of this approach are that important covariates such as contraceptive history and functional infertility are not considered. Expected rates are missing for some ages, cohorts, or parities, and there may be a lack of comparability of marital status between the study group and national statistics. Because standardized fertility ratios use only live births as the end point, other important reproductive events are ignored.

Determining the number of noncontracepting cycles that are required for a couple to conceive after complete termination of birth control is a third method (Baird *et al.*, 1986). This "time-to-pregnancy" approach incorporates a wider complement of reproductive experiences since it provides an estimate of the per cycle probability of conceiving a detectable pregnancy. Data are collected that may include information on menstrual periods, contraception, and frequency of sexual intercourse. Baird *et al.* (1986) have shown that the Cox proportional hazard model is quite robust to minor violations of assumptions and provides an estimated relative risk measure; the discrete proportional risk model, however, may be the preferred choice, depending on the specific question being asked.

The choice of approach in analyzing infertility or subfecundity is dependent on the study design. If a referent population is unavailable, the standardized fertility analysis might be considered with an understanding of design restrictions inherent in this approach. If considerable details are known on potential confounders and a referent group is available, the time-to-pregnancy approach is preferred. It is probably prudent to include more than one approach, for example, responses to specific questions about the couple's recognition of an infertility problem and the life-event calendar approaches inherent in the time-to-pregnancy analyses.

Markers of Susceptibility

The chromosomal constitution of human sperm can be used as a biomarker in studies to determine the potential impact of xenobiotics on the human genome and the future conceptus. An approach in analyzing this genetic information involves the fusion of human sperm with eggs of the golden hamster. The sperm chromatin decondenses in the activated egg and undergoes DNA synthesis, replicating the human and hamster genome. Abnormal chromosome spreads are verified by microscope analysis of the region surrounding the egg and by karyotyping the hamster chromosome complement to assess normality (Martin *et al.*, 1987). Structural aberrations are classified sess normality (Martin *et al.*, 1987). Structural aberrations are classified according to an international system (ISCN, 1978). An example of the application of this test was shown with a significant dose-dependent increase in the frequency of sperm chromosomal abnormalities after radiotherapy in cancer patients (Martin and Rademaker, 1987). The range in the testicular radiation dose was 0.4 to 5.0 Gy and the frequency of sperm chromosomal abnormalities was 6 to 67%. Although quite laborious and not currently feasible for large population studies, this method nevertheless is promising for the identification of susceptible subgroups.

Another promising sperm assay currently being used to assess workplace exposures is the sperm chromatin structure assay (SCSA). This assay, originally developed and validated in livestock, uses flow cytometric analysis of acridine orange-stained sperm to prove structural integrity of the sperm chromatin; native DNA fluoresces green and denatured DNA fluoresces red (Evenson *et al.*, 1991). The derived measure is "alpha t," red fluorescence divided by the combination of red and green fluorescence showing good stability and minimal variability. It is speculated that, had the male workers at the Sellafield nuclear plant been followed from initial hire, an earlier "susceptibility" marker on sperm may have been observed prior to the reporting of an excess of childhood leukemia and non-Hodgkin's lymphoma in the offspring of fathers receiving a total preconceptual ionizing radiation dose of 100 mSv or more (Gardner *et al.*, 1990). Without development of some earlier warnings, it will always be too late to prevent disease.

Analyzing the effects of exposures on older or on younger individuals also may elucidate susceptible subgroups. For example, the older male is at greater risk for sperm chromosome structural abnormalities, but there is no relationship between age and numerical abnormalities in sperm (Martin and Rademaker, 1987). In addition, sperm parameters distinguish potentially high-risk groups in studies of environmental exposures. A prospective study of the effect of an exposure on fecundity, which is postulated to be related only to an effect on sperm motility, might use sperm count as a marker of susceptibility. A low sperm count due to normal variability may act as an intervening factor since these individuals also have lower probabilities of successful fertilization than do men with higher counts. The group of men with lower sperm counts might, therefore, constitute the only group with no built-in margin of safety.

The whole issue on margin of safety refers to the definition of subfertility, which is a sperm concentration of <20 million sperm per ml or <50 million sperm per ejaculate, <1.0 ml semen volume, <60% motile sperm, and <60% normal morphologic forms (Cunningham, 1978). Therefore, the group of individuals whose preexposed sperm count is borderline, that is between 60 and 70 million per ejaculate compared to the average of 200 million, has less biologic redundancy before their fertility potential is greatly compromised. In summary, in this situation, an analysis requires examining crude versus adjusted rates by stratifying preexposure low- and high-count individuals and examining group differences.

Hormonal changes may serve as excellent markers of susceptibility but attention must be given to whether these markers correlate with exposure or disease. Spermatogenesis requires an intact hypothalamic–pituitary–testicular axis and hormonal alterations may provide a reading on the successful integration of this system. Luteinizing hormones (LH) stimulates Leydig cells in the testis to secrete testosterone. Testosterone diffuses from the interstitial space in the testis into the spermatogenic tubules. Testosterone in the tubules affects spermatogenesis, either directly or indirectly through the Sertoli cells (Overstreet *et al.*, 1985). Follicle stimulating hormone (FSH) also acts on the Sertoli cells to stimulate spermatogenesis. Repeated measurements of FSH, LH, and testosterone, both pre- and postexposure, may be useful for assessing the temporal effects of exposure that precede a more permanent effect on a couple's fecundity (NRC, 1989).

The possible effects of tobacco smoke on sperm concentration, morphology, and motility provides an interesting case study. Although study results have been mixed, it has been suggested that the effects of smoking on these semen parameters may be caused by the presence of intermediary factors such as increased levels of choline acetyltransferase inhibitors, catecholamines, prolactin, serum estradiol, or testosterone (Albin, 1986a,b; Klaiber and Broverman, 1988). Alterations in any of these measurements might be early indicators of susceptibility that precede direct effects on sperm values.

The challenge in using hormonal markers is to detect them sufficiently early in the exposure–disease process. The goal is to distinguish whether hormonal imbalances are a consequence of a primary effect on the hypothalamus or pituitary axis or are secondary to disruption of the testicular–hypothalamic feedback mechanism. Hence, markers may be useful for distinguishing (stratifying) a population with an environmental exposure to determine if there is effect modification or confounding due to a preexisting susceptibility factor. The ability to distinguish the marker as preexisting, that is, acquired prior to the exposure period of interest or inherited, is of utmost importance. Acquisition will be assessed by prospective studies, and inheritance may be assessed by family studies (i.e., pedigree studies).

References

Albin, R. J. (1986a). Cigarette smoking and quality of sperm. *N.Y. State J. Med.* 86(2), 108.
Albin, R. J. (1986b). Prolactin: A link between smoking and increased fertility? Reply of the author. *Fertil. Steril.* 46(3), 531–532.
Amin-Zaki, L., Elhassani, S., Majeed, M. A., Clarkson, T. W., Doherty, R. A., and Greenwood, M. (1974). Intra-uterine methylmercury poisoning in Iraq. *Pediatrics* 54, 587–595.

Baird, D. D., Wilcox, A. J., and Weinberg, C. R. (1986). Use of time to pregnancy to study environmental exposure. *Am. J. Epidemiol.* **124**(3), 470–480.

Bakir, F., Damluji, S. F., Amin-Zaki, L., Murtadha, M., Khalidi, A., Al-Rawi, N. Y., Tikriti, S., Dhahir, H. I., Clarkson, T. W., Smith, J. C., and Doherty, R. A. (1973). Methylmercury poisoning in Iraq. *Science* **181**, 230–241.

Bernstein, L., Pike, M. L., Lobo, R. A., Depue, R. T., Ross, R. K., and Henderson, B. E. (1989). Cigarette smoking in pregnancy results in marked decrease in maternal hCG and oestradiol levels. *Br. J. Obstet. Gynaecol.* **96**, 92–96.

Beskrovnaja, N. J. (1979). Gynecological morbidity in women workers in the rubber industry. *Gig. Tr. Prof. Zabol.* **8**, 36–38.

Bloom, A. D. (1981). "Guidelines for Studies of Human Populations Exposed to Mutagenic and Reproductive Hazards." March of Dimes Birth Defects Foundation New York.

Butarewicz, L., Gosk, S., and Gluszczopma, M. (1969). Examination of the health of female workers in the leather industry, especially from the gynecological viewpoint. *Med. Pr.* **20**(2), 137–140.

Butler, W. J., and McDonough, P. G. (1989). The new genetics: Molecular technology and reproductive biology. *Fertil. Steril.* **51**(3), 375–386.

Chapman, R. M. (1983). Gonadal injury resulting from chemotherapy. *Am. J. Ind. Med.* **4**, 149.

Chiazze, L., Brayer, F. T., Macisco, J. J., Parker, M. P., and Duffy, B. J. (1968). The length and variability of the human menstrual cycle. *J. Am. Med. Assoc.* **203**(6), 377–380.

Cunningham, G. R. (1978). Medical treatment of the subfertile male. Symposium on Male Infertility. *Urol. Clin. North. Am.* **5**(3), 537–548.

Edmonds, D. K., Lindsay, K. S., Miller, J. F., Williamson, E., and Wood, P. J. (1982). Early embryonic mortality in women. *Fertil. Steril.* **38**, 447–453.

Ekelund, H., Kullawder, S., and Källén, B. (1970). Major and minor malformations in newborns and infants up to one year of age. *Acta Paediatr. Scand.* **59**, 297.

Evenson, D. P., Jost, L. K., Baer, R. K., Turner, T. W., and Schrader, S. M. (1991). Individuality of DNA denaturation patterns in human sperm as measured by the sperm chromatin structure assay. *Reprod. Toxicol.* **5**, 115–125.

Everson, R. B., Randerath, E., Santella, R. M., Avitts, T. A., Weinstein, I. B., and Randerath, K. (1988). Quantitative association between DNA damage in human placenta and maternal smoking and birth weight. *J. Natl. Cancer Inst.* **80**, 567–576.

Federal Register (1988). **126**, 24849–24869.

Fogel, C., and Woods, N. (1981). "Health Care of Women." Mosby, St. Louis, Missouri.

Gardner, M. J., Snee, M. P., Hall, A. J., Powell, C. A., Downes, S., and Terrell, J. D. (1990). Results of case-control study of leukemia and lymphoma among young people nearest Sellafield nuclear plant in West Cumbria. *Br. Med. J.* **30**, 423–429.

Gesev, G. (1967). Changes in the menstrual cycle of women occupationally exposed to trinitrotoluene. *Parva Nac. Konf. Aspir. Med. Fizkult. Sofia* 259–262.

Goldsmith, L., and Weiss, G. (1986). Puberty, adolescence and the clinical aspects of normal menstruation. *In* "Obstetrics and Gynecology" (D. N. Danforth and J. Scott, eds.), pp. 148–162. Lippincott, Philadelphia.

Grunert, J. H., deGeyter, C., Bordt, J., Schneider, H. P. G., and Nieschlag, E. (1989). Does computerized image analysis of sperm movement enhance the predictive value of semen analysis for *in vitro* fertilization results? *Int. J. Androl.* **12**, 329–338.

Hertz-Picciotto, I., and Samuels, S. J. (1988). The incidence of early loss of pregnancy. *N. Engl. J. Med.* **319**(22), 1483–1484.

Huszar, G., and DeCherney, A. (1987). The role of intrauterine insemination in the treatment of infertile couples: The Yale experience. *Semin. Reprod. Endocrinol.* **51**, 11–21.

Huszar, G., Corrales, M., and Vigue, L. (1988). Correlation between sperm creatine phosphokinase activity and sperm concentrations in normospermic and oligospermic men. *Genet. Res.* **19**, 67–75.

Huszar, G., Vigue, L., and Corrales, M. (1990). Sperm creatine kinase activity in fertile and infertile oligospermic men. *J. Androl.* **11**(1), 40–46.

International System for Cytogenetic Nomenclature (1978). An international system for human cytogenetic nomenclature. *Cyt. Cell Genet.* **21**, 309–404.

Jones, H. W., Jr., Acosta, A. A., Andrews, M. C., Garcia, J. E., Jones, G. S., Mantzavinos, T., McDowell, J., Sandow, B. A., Veeck, L., Whibley, T. W., Wilkes, C. A., and Wright, G. L., Jr. (1983). What is a pregnancy? A question for programs of *in vitro* fertilization. *Fertil. Steril.* **40**, 728–733.

Jouannet, P., Ducot, B., Fenleux, D., and Spira, A. (1988). Male factors and the likelihood of pregnancy in infertile couples. I. Study of sperm characteristics. *Int. J. Androl.* **11**, 379–394.

Källén, B. (1988). "Epidemiology of Human Reproduction." CRC Press, Boca Raton, Florida.

Kesner, J. S., Wright, D. M., Schrader, S. M., Chin, N. W., and Krieg, E. F., Jr. (1992). Methods to monitor menstrual function in field studies. *Rep. Toxicol.* **6**, 385–400.

Kharrazi, M., Potashnik, G., and Goldsmith, J. R. (1980). Reproductive effects of dibromochloropropane. *Isr. J. Med. Sci.* **16**, 403–406.

Klaiber, E. L., and Broverman, D. M. (1988). Dynamics of estradiol and testosterone and seminal fluid indexes in smokers and nonsmokers. *Fertil. Steril.* **50**(4), 630–634.

Kline, J., Stein, Z., and Susser, M. (1989). Conception to birth—Epidemiology of prenatal development, pp. 18–30. Oxford University Press, New York.

Kuratsune, M., Yoshimura, T., Matsuzaka, J., and Yamaguchi, A. (1972). Epidemiologic study of Yusho, a poisoning caused by ingestion of rice oil contaminated with a commercial brand of polychlorinated biphenyls. *Environ. Health Perspect.* **1**, 119–128.

Lemasters, G. K., and Zenick, H. (1985). "Assessing Reproductive Effects of Workers Exposed to Hazardous Wastes." Paper presented at the Seventh Annual Rocky Mountain Conference for Occupational and Environmental Health, October 1985, Salt Lake City, Utah.

Lemasters, G. K., Zenick, H., Hertzberg, V., Hansen, K., and Clark, S. (1991). Fertility of workers chronically exposed to chemically contaminated sewer wastes. *Rep. Toxicol.* **5**, 31–37.

Lentz, R. D., Bergstein, J., Steffes, M. W., Brown, D. R., Prem, K., Michael, A. F., and Vernier, R. L. (1977). Postpubertal evaluation of gonadal function following cyclophosphamide therapy before and during puberty. *J. Pediatr.* **91**, 385–94.

Leridon, H. (1977). "Human Fertility: The Basic Component." University of Chicago Press, Chicago.

Levine, R. J., Symons, M. J., Balogh, S. A., Arndt, D. M., Kaswandik, N. T., and Gentile, J. W. (1980). A method for monitoring the fertility of workers. I. Method and pilot studies. *J. Occup. Med.* **22**(12), 781–791.

Levine, R. J., Symons, M. J., Balogh, S. A., Milby, T. H., and Whorton, M. D. (1981). A method for monitoring the fertility of workers. II. Validation of the method among workers exposed to dibromochloropropane. *J. Occup. Med.* **23**(3), 183–188.

Little, A. B. (1988). There's many a slip 'twixt implantation and the crib. *N. Engl. J. Med.* **319**, 241–242.

McFarland, K. F. (1980). Amenorrhea. *Am. Fam. Physician* **22**(6), 95–101.

Manchester, D. K. (1991). Human mutagen exposures and reproductive risks. *Org. Teratol. Inf. Serv. Newsl.* Spring, 1–2.

Martin, R. H., and Rademaker, A. W. (1987). The effect of age on the frequency of sperm chromosomal abnormalities in normal men. *Am. J. Hum. Genet.* **41**, 484–492.

Martin, R. H., Hildebrand, K., Yamamoto, J., Rademaker, A., Barnes, M., Douglas, G., Arthur, K., Ringrose, T., and Brown, S. (1986). An increased frequency of human sperm chromosomal abnormalities after radiotherapy. *Mutat. Res.* **174**, 219–225.

Mattison, D. R. (1983). "Reproductive Toxicology." Liss, New York.

Michon, S. (1965). Connection between aromatic hydrocarbons and menstrual disorders analyzed. *Pol. Tyg. Lek.* **20**, 1648–1649.

Miller, J. R., and Poland, B. J. (1970). The value of human abortuses in the surveillance of developmental anomalies. I. General overview. *Can. Med. Assoc. J.* **103**, 501–502.

National Research Council (1989). "Biologic Markers in Reproductive Toxicology." National Academy Press, Washington, D.C.

Nelson, K., and Holmes, L. B. (1989). Malformations due to presumed spontaneous mutations in newborn infants. *N. Engl. J. Med.* **320**(1), 19–23.

Overstreet, J. W., Sokol, R. Z., and Rajfer, J. (1985). Infertility in the male. *Ann. Intern. Med.* **103**, 906–919.

Schlesselman, J. J. (1979). How does one assess the risk of abnormalities from human *in vitro* infertilization? *Am. J. Obstet. Gynecol.* **135**, 135–148.

Schrader, S. M., Ratcliffe, J. M., Turner, T. W., and Hornung, R. W. (1987). The use of new field methods of semen analysis in the study of occupational hazards to reproduction: The example of ethylene dibromide. *J. Occup. Med.* **29**(12), 963–966.

Schrader, S. M., Turner, T. W., Breitensteien, M. J., and Simon, S. D. (1988). Longitudinal study of semen quality of unexposed workers. *Rep. Toxicol.* **2**, 183–190.

Scialli, A. R., and Lemasters, G. K. (*in press*). Epidemiologic aspects of reproductive toxicology. *In* "Target Organ Toxicology Series: Reproductive Toxicology" (A. W. Hayes, J. A. Thomas, and D. E. Gardner, eds.). Raven Press, New York.

Segg, G. A. (1991). Adducts in sperm protamine vs mutation frequency. *In* "New Horizons in Biological Dosimetry" (B. L. Gledhill and F. Mauro, eds.), pp. 521–530. Wiley–Liss, New York.

Selevan, S. G., and Lemasters, G. K. (1987). The dose-response fallacy in human reproductive studies of toxic exposure. *J. Occup. Med.* **29**(5), 451–454.

Shortridge, L. A. (1988). Assessment of menstrual variability in working populations. *Rep. Toxicol.* **2**, 171–176.

Siris, E. S., Leventhal, B. G., and Vaitukaitis, J. L. (1976). Effects of childhood leukemia and chemotherapy on puberty and reproductive function in girls. *N. Engl. J. Med.* **294**(21), 1143–1146.

Starr, T. B., and Levine, R. J. (1983). Assessing effects of occupational exposure on fertility with indirect standardization. *Am. J. Epidemiol.* **118**(6), 897–904.

Stewart, D., Celniker, A., Taylor, C., Cragun, J., Overstreet, J., and Lasley, B. (1990). Relaxin in the peri-implantation period. *J. Clin. Endocrinol. Metab.* **70**, 1771–1773.

Stillman, R. J. (1981). Ovarian failure in long-term survivors of childhood malignancy. *Am. J. Obstet. Gynecol.* **139**, 62.

Syrovadko, O. N., Skormin, V. F., and Pron'kova, E. N. (1973). Effect of working conditions on the health and some specific functions in female workers exposed to white spirit. *Gig. Tr. Prof. Zabol.* **16**(6), 5–8.

U.S. Department of Health and Human Services (1982). "Reproductive Impairments among Married Couples." DHHS Publication No. (PHS) 83-1987. U.S. Government Printing Office, Hyattsville, Maryland.

Warkany, J. (1971). "Congenital Malformations: Notes and Comments." Year Book Publishers, Chicago.

Weatherall, D. J. (1991). "The New Genetics and Clinical Practice." Oxford University Press, Oxford.

Weinstein, I. B. (1988). Cigarette smoking and its fingerprint on DNA. *J. Natl. Cancer Inst.* **80**, 548–49.

Whorton, D., Krauss, R. M., Marshall, S., and Milby, T. H. (1977). Infertility in male pesticide workers. *Lancet* **2**, 1259–61.

Wilcox, A. J., Baird, D. B., Weinberg, C. R., Armstrong, E. G., Musey, P. I., Wehmann, R. E., and Canfield, R. E. (1987). The use of biochemical assays in epidemiologic studies of reproduction. *Environ. Health Perspect.* **75**, 29–35.

Wilcox, A. J., Weinberg, C. R., O'Connor, J. F., Baird, D. D., Schlatterer, J. P., Canfield, R. E.,

Armstrong, E. G., and Nisula, B. C. (1988). Incidence of early loss of pregnancy. *N. Engl. J. Med.* **319**, 189–194.

Wilson, J. G. (1973). "Environment and Birth Defects." Academic Press, New York.

Wong, O., Utidjian, H. M. D., and Karten, V. S. (1979). Retrospective evaluation of reproductive performance of workers exposed to ethylene dibromide (EDB). *J. Occup. Med.* **21**, 98–102.

Wyrobek, A. J., Watchmaker, G., and Gordon, L. (1984). An evaluation of sperm test as indicators of germ-cell damage in men exposed to chemical or physical agents. *Terat. Carc. Mutagen.* **4**, 83–107.

Zielhuis, G. A., Gijsen, R., and van der Gulden, J. W. J. (1989). Menstrual disorders among dry cleaning workers. *Scand. J. Work Environ. Health* **15**, 238.

16

Immune Markers in Epidemiologic Field Studies

Robert F. Vogt, Jr.,
and Paul A. Schulte

Introduction

The mammalian host defense system is a complex network of cells and mediators with recognition and response functions throughout most tissues of higher organisms. The primary functions of the host defense system are repairing injured tissue, identifying and removing foreign substances, destroying or containing infectious agents, and, in some cases, eradicating cancer cells. These functions are carried out through both nonspecific mechanisms of innate or natural immunity and through specific mechanisms of acquired immunity that develops as the organism encounters environmental agents or antigens. The term "immune system" will be used to refer to all components of nonspecific innate immunity and antigen-specific acquired immunity, since their components and activities are invariably intertwined.

Over the last few decades, the cellular and molecular basis for many host defense functions has been uncovered through the use of emerging laboratory technologies. The use of these laboratory methods to detect biologic markers in epidemiologic studies already has provided critical information in many areas of basic and public health science. Their future potential is even greater. However, this potential can be realized only by applying scientifically valid and logistically feasible markers in field studies. The absence of either characteristic can result in flawed or failed research through the use of markers that are not measured properly or cannot be interpreted with respect to exposure, susceptibility or health effects (NRC, 1992).

This chapter outlines the major considerations involved in using laboratory tests for immune markers in epidemiologic studies. Although it cannot provide a complete description of the host defense system, for which many excellent texts and reviews are available (e.g., Hood *et al.,* 1990; Roitt, 1990; Paul, 1992), it discusses many of the cellular and molecular components of the immune system and their use as biologic markers.

General Aspects of Biologic Markers of the Host Defense System

The benefits and limitations of immune markers in epidemiologic studies may be appreciated best by understanding their relationship to the basic biology of host defense function. Biologic considerations are especially important for the effective use of immune markers in epidemiologic studies because of the dynamic and variable nature of the immune system, its complex organization, the large number of components, and its decentralized location in the body. Since the immune system is designed to change with environmental exposures, markers that are adaptive must be distinguished from those that are pathognomic. Moreover, pathologic alterations of the immune system may result in disease through over-reactivity (autoimmunity, hypersensitivity) or impairment (immune deficiency). Familiarity with this range of normal and abnormal perturbations is required before tests for markers of the immune system can be used effectively. The numerous confounding factors that can influence the immune response also must also be considered.

Almost all markers used as tests of immune status are active participants in protective, regulatory, or pathogenic processes of the immune system. This direct biologic relevance provides special opportunities to learn about the mechanisms of host injury and response through tests for immune components. However, this relevance also can make interpretation more difficult, since physiologic interactions among markers can mask changes or create internal confounders. Moreover, the continual changes that occur as the immune system senses and responds to environmental influences make the normal range of variability for immune constituents very large among individuals and even within individuals over time. Finally, the immune system of each individual continues to evolve throughout life, its course determined by a combination of inherited influences and acquired exposures. Normal ranges of immune cells and mediators therefore may be very broad in the general population, and relatively clustered in genetically or environmentally restricted populations. This characteristic has important implications for the use of immune biomarkers in epidemiologic studies, in which subpopulations may have their own "normal" immunologic values.

A final point for emphasis concerns the biologic material used to test for immune components. Human studies often are limited to sampling peripheral blood, which does provide a convenient source of cells and mediators. However, peripheral blood by no means represents the immune system as a whole. Host defense activities take place primarily in the lymphoid tissues (spleen, lymph nodes, epithelial-associated lymphoid tissues) and in interstitial tissue at local sites of injury and infection. Cell traffic and recirculation through the blood is controlled carefully (Figure 16.1), and activated cells and molecules are removed quickly. In contrast, some cells and mediators persist within or outside the bloodstream for days and even years (Figures 16.1 and 16.2). Therefore, blood samples often may represent an inappropriate surrogate of the actual immune system components being evaluated.

Components of the Host Defense System

Table 16.1 presents a summary of the major molecular and cellular constituents of the host defense system. The following synopsis [adapted from Vogt (1991)] elaborates briefly on these components.

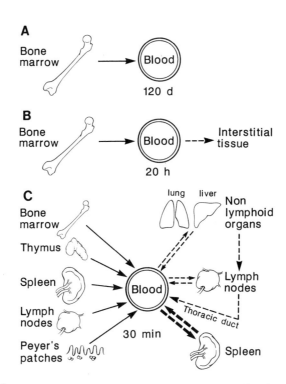

FIGURE 16.1   Comparison of (A) erythrocyte, (B) granulocyte, and (C) lymphocyte kinetics. Lymphocytes differ greatly from erythrocytes and granulocytes because they are produced in many organs. They have a very short mean transit time through the blood and can migrate. Interpretations based on the number of lymphocytes in the blood are much more difficult to make than those based on the number of erythrocytes or granulocytes. The solid arrows indicate the organs releasing newly formed lymphocytes into the blood; the broken arrows indicate migration routes. Note the outstanding role of the spleen in lymphocyte migration. (From Westermann and Pabst, 1990.)

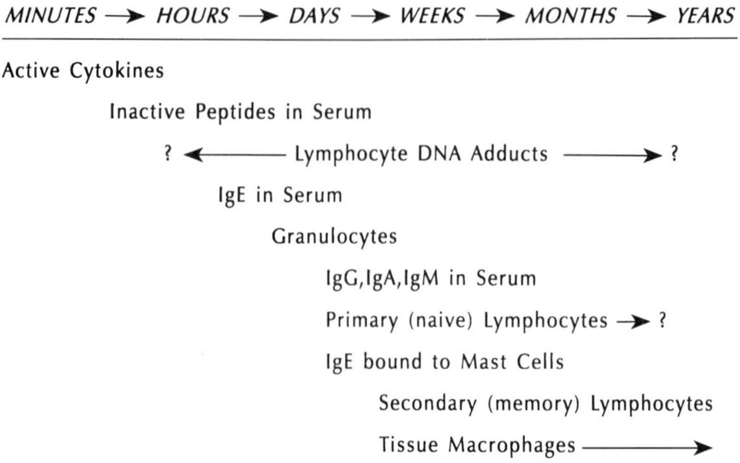

MINUTES ➞ *HOURS* ➞ *DAYS* ➞ *WEEKS* ➞ *MONTHS* ➞ *YEARS*

Active Cytokines

       Inactive Peptides in Serum

           ? ⬅——— Lymphocyte DNA Adducts ———➤ ?

           IgE in Serum

             Granulocytes

               IgG,IgA,IgM in Serum

               Primary (naive) Lymphocytes ➞ ?

               IgE bound to Mast Cells

                   Secondary (memory) Lymphocytes

                   Tissue Macrophages ———————➤

FIGURE 16.2   The persistence of immune markers varies widely among the different humoral mediators and cellular components, and may also depend on conditions within the organism. Highly reactive humoral mediators that act locally (such as cytokines) often are inactivated within minutes of their formation, but their inactivation products may circulate in serum much longer. Serum IgE is cleared more quickly than serum IgG, but IgE bound to mast cells or basophils persists much longer than any serum immoglobulin. Primary lymphocytes with new specificities do not survive long, but if they are stimulated by contact with antigen, their clonal offspring colonize the organism and persist indefinitely.

Chemical Mediators of the Immune System

Many of the defense and regulatory functions of the immune system are conducted by chemical mediators released from its cells. Antibodies (also called immunoglobulins, Igs) are the only antigen-specific mediators; they are secreted by stimulated B lymphocytes and comprise several major classes with different functional capacities. IgM and IgG antibodies, the most general-purpose types, facilitate phagocytosis, antigen clearance, and destruction of parasites. IgA antibodies are secreted at the mucous membranes, where they help prevent attachment and invasion by microbes and parasites that come in contact with these surface tissues. IgE antibodies, bound to the outer membrane of mast cells and basophils, help initiate immune responses and are involved particularly with immunity against worms and mites; they are also the antibodies responsible for allergic reactions such as hay fever.

Cytokines are extremely potent peptide molecules that activate or suppress target cell populations that express the appropriate receptors. More than a dozen immune cytokines (many called interleukins) have been identified as participants in the complex network of immune regulation (Figures 16.3 and 16.4).

Complement is one of several plasma proteins involved with acute nonspecific responses to tissue injury and invasion. Complement is actually a

TABLE 16.1 Major Components of the Host Defense System

Molecular mediators	
Component	**Function**
Proteins	Viral inactivation; antigen clearance;
Immunoglobulins (antibodies)	complement activation; opsonization
Cytokines	Intercellular signaling
Interferons	
Interleukins	
Growth factors	
Complement (interacting with the kinin,	Parasite destruction; chemotactic
fibrin, and plasmin systems)	stimulation; acute inflammatory reactions
Heat shock	Protein binding and preservation; cross-reactive antigenicity
Lipid-derived	
Prostaglandins	
Leukotrienes	Intercellular signaling

Molecular cell surface receptors	
Component	**Function**
Immunoglobulins; T-cell antigen receptor	Specific antigen recognition on lymphocytes
Immunoglobulin E	Specific antigen recognition on mast cells and basophils
Class I histocompatibility proteins	
Class II histocompatibility proteins	
Immunoglobulin-related proteins (CD4, CD8, β_2-Microglobulin)	Cell–cell interactions
Cytokine receptor proteins	Receptors for the various cytokines
Cell adhesion molecules	Cell traffic and migration control

Cell lineages and subsets	
Component	**Function**
Granulocytes	
Neutrophils	Phagocytosis and antigen destruction
Eosinophils	Parasite destruction; regulation
Basophils	Parasite destruction; regulation
Monocytes/macrophages	Phagocytosis and antigen destruction; antigen processing and presentation; regulation
T-Lymphocytes	
Helper (CD4) cells	Activation of antigen-specific responses
Suppressor (CD8) cells	Suppression of antigen-specific responses
Cytotoxic (CD8) cells	Destruction of virus-infected and neoplastic cells
B-Lymphocytes	Antibody production; regulation
Plasma cells	
Natural killer cells (NK)	Destruction of virus-infected and certain neoplastic cells
Dendritic cells	Antigen presentation
Platelets	Blood clotting; activation

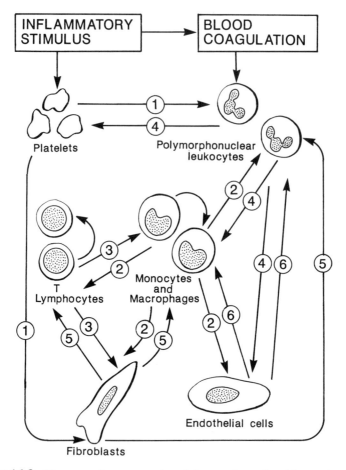

FIGURE 16.3 Diagrammatic representation of the complex series of interactions between numerous cell types, depicted as a series of arrows, involved in such events as inflammation, wound healing, tissue repair and tissue formation. These interactions are made possible through the release and recognition of cytokines acting in autocrine and paracrine fashions. The arrows are numbered and the cytokines involved in the interactions are listed below illustration. (From Pugh-Humphries *et al.*, 1991.) (1) *Platelet-derived factors:* PDGF-A and -B, TFG-β, bFGF; (2) *Monocyte/macrophage-derived factors:* PDGF-A and -B, TFG-α and -β, IL-1, IL-6, IL-8, TNFα, INFα, INFβ, GM-CSF, bFGF, EFG, IGF-1; (3) *T-helper-derived factors:* IL-2, IL-3, IL-4, IL-5, IL-6, GM-CSF, IFNγ, TNF-α and β; (4) *PMN-derived factors:* Many noncytokine factors, including arachidonic acid metabolites; (5) *Fibroblast-derived factors:* IL-1, IL-6, PDGF-A and -B, GM-CSF; (6) *Endothelial-cell-derived factors:* IL-1, IL-6, TNFα, PDGF-A and -B, GM-CSF, bFGF.

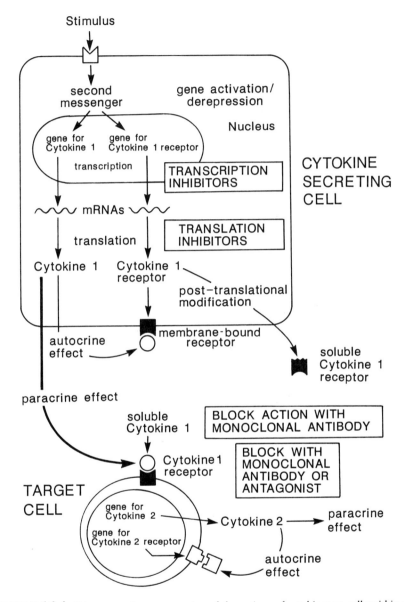

FIGURE 16.4 Diagrammatic representation of the actions of cytokines on cells within the cytokine network. The activation of cytokine and cytokine receptor genes within cytokine-secreting cells, and the sites of interference with gene expression using transcription and translation inhibitors, are indicated. Cytokine-secreting cells not only release cytokines which can have autocrine or paracrine actions, but they can also synthesize cytokine receptors, parts of which can be cleaved extracellularly and released as soluble cytokine receptors. The interactions of cytokines with cells can be blocked using antibodies directed either against the cytokines and/ or against their receptors. Cytokines can induce the release of other cytokines within target cells through activation of the relevant cytokine genes. (Adapted from Pugh-Humphries *et al.,* 1991.)

cascading system of different protein molecules that can be activated by a variety of stimuli, including antigen–antibody complexes, blood clotting proteins, and other mediators. Complement activation products have a number of activities, including chemotaxis, clearance, and destruction of cells. Other such "acute phase" serum proteins include transferrin and plasmin.

Several nonprotein molecules are also important immune mediators. They include different lipid-derived chemicals (such as prostaglandins) that have a wide variety of effects on many different tissues, including the activation or suppression of immune cells and the dilation or constriction of blood vessels and airways. Histamine, which is stored in the granules of mast cells and basophils, causes dilation and leakage in small blood vessels and has effects on immune cells and other tissues; it is responsible for many of the symptoms of allergy.

Several other chemical mediators influence cells of the immune system, although they are not central to its function. These include catecholamines (such as adrenalin), endorphins, and insulin.

Cells of the Immune System

Most of the several different types of cells that constitute the immune system spend at least part of their lifetime in the peripheral blood, where they constitute the white blood cells or leukocytes. The major types of leukocytes are lymphocytes, monocytes, and granulocytes.

Lymphocytes (B cells and T cells) are the specific recognition cells of the immune system. Each family (clone) of lymphocytes has unique recognition molecules on its surface. If the lymphocyte is activated by recognizing a foreign protein (antigen), a specific immune response is initiated. Activated lymphocytes proliferate and engage in a variety of host defense functions, such as producing antibody (B cells) and killing virus-infected cells or regulating immune activities (T cells).

Monocytes are immature cells that differentiate into macrophages after they emigrate from the blood. Macrophages are distributed throughout many tissues including the lung, liver, skin, brain, and bone marrow. Their innate activities of phagocytosis and digestion are nonspecific, but they become part of the specific immune response when they "present" processed fragments of foreign protein to lymphocytes.

Granulocytes are important auxiliary cells with activities that are critical to host defense but also may contribute to disease processes. Neutrophils, like macrophages, are avid phagocytes, but are short-lived and less versatile. Mast cells, basophils, and eosinophils are involved in immunity to larger parasites such as worms, and are the primary participants in the allergic responses to pollens, foods, and other substances. They also appear to be involved in inflammatory reactions to certain toxic and sensitizing chemical exposures (Vogt *et al.*, 1984).

Laboratory Measurements of Immune Cells, Mediators, and Functions

Laboratory tests can be used to measure the concentrations of many immune mediators, the types and numbers of immune cells, their functional capacities, and factors influencing disease susceptibility (Tables 16.2, 16.3). However, measurements for most immune components are not well standardized, so the proper choice and evaluation of analytical methods is critical to the success of epidemiologic studies using such measurements. The number and type of tests employed depend on the purpose of the field study and on indications that an immunologic end point (biomarker) may be involved. Clearly, no single immunologic test can evaluate the entire immune system; rather, a comprehensive panel of assays should be selected carefully. Method evaluation has become even more important since many assays have evolved from simple dichotomy (positive/negative results) to semiquantitative (weak/

TABLE 16.2   **Immune Markers Associated with Health Effects and Exposures**

Effects or Exposures	Marker
Inflammatory disease	
Autoimmune disorders	Antibodies to tissue antigens; histocompatibility genotypes
Allergic (hypersensitivity) reactions	Antibodies to environmental antigens; *in vivo* reactions (e.g., skin tests)
Immunoproliferative disease	
Chronic lymphocytic leukemia	Peripheral blood lymphocyte counts
Multiple myeloma	Monoclonal serum immunoglobulin
Lymphoma	Monoclonal lymphocyte infiltration
Infectious disease	
Parasitic	Antibodies to parasite antigens; peripheral blood eosinophil counts
Bacterial	Antibodies to bacterial antigens; peripheral blood granulocyte counts; *in vivo* reactions (e.g., TB skin tests)
Viral	Antibodies to viral antigens; cellular cytotoxic responses; peripheral blood lymphocyte subset counts
Neoplastic disease	
Solid tumors	Antibodies to tumor-specific antigens
Environmental exposures	
Infectious agents	Antibodies and cellular responses to specific antigens
Other biologic antigens	
Chemical antigens	
Volatile organics (benzene)	Peripheral blood leukocyte counts
Volatile irritants	Leukocyte infiltration of mucous membranes

TABLE 16.3   Susceptibility Factors

Factors	Immune related		Nonimmune
	Antigen specific	Nonspecific	
Genetic	Antibody variable-region genes T-cell receptor variable-region genes	Antibody constant-region genes T-cell receptor constant-region genes Histocompatibility genes	Metabolism related Gender related Other
Environmental	Sensitizing exposures	Adjuvant inflammatory reactions	Behavioral (smoking, etc.) Concurrent illness Neurogenic/psychogenic

strong reactions) and fully quantitative (concentration units). The important parameters are analytical accuracy, precision, sensitivity, and specificity.

Analytical accuracy, the extent to which the measurement gives the "true value" for the analyte, must be evaluated to prevent the false impression of biologic differences actually caused by bias among methods. Accuracy cannot be assessed for most immune markers because no true reference standards exist. However, some indication of accuracy is provided by the degree of consensus between values on reference materials obtained from different laboratories and methods. Consensus evaluation of reference materials is of critical importance to epidemiologic studies, since only unbiased methods can be compared among different studies and different times, or pooled into databases to determine reference ranges and long-term predictive values. Unfortunately, consensus evaluation is available for only a few immune markers, such as serum immunoglobulins and complete blood counts. Even these standard reference values are often subject to method-related biases and changes with time (see Chapter 4). Reference ranges based on consensus evaluation are beginning to emerge for the major lymphocyte subsets measured by flow cytometry (Kidd and Vogt, 1989), although considerable work is still needed in this area. Little, if any, consensus evaluation exists for functional immune assays such as cell proliferation or cytoxicity.

Analytical precision, the reproducibility of a measurement, is a critical aspect of laboratory assays and becomes even more important in the absence of an acceptable accuracy base. Imprecision generally indicates some inherent weakness in the assay methodology and lowers predictive value by blurring the distinction between true differences in results. The smaller the true differences between distributions in the populations tested, the more detrimental is the effect of imprecision on predictive value. The broad distributions of many immune parameters make small differences difficult to detect. Analytical imprecision can add to this problem by concealing biologically significant differences between groups. Unfortunately, many tests for immune markers (particularly cellular and functional measurements) have poor reproducibility. In any event, study designs for any biologic marker should include assessment of analytical imprecision to account for all the sources of variability in the final distributions of marker results.

Analytical sensitivity may be considered the lowest level of an analyte that can be measured with reasonable precision and accuracy. If the laboratory method is not able to detect the analyte at levels important to pathophysiology, the test is likely to have little value.

Sensitivity is especially important for immune markers because most testing is performed on peripheral blood, whereas host defense functions are based primarily in localized tissue reactions. Although active immune mediators and cells may become very concentrated in a microscopic area of tissue, the overall amount of analyte in peripheral blood or serum samples is often too low to measure. In fact, activated cells and molecules are removed from circulation rapidly to prevent harmful systemic reactions. Analytical sensitivity, therefore, becomes a critical issue in developing tests to probe active immune processes.

Analytical specificity is the extent to which other influences alter the result of a biomarker measurement. If the measurement of the biomarker is subject to interference from other substances, the results will not be correct. Interference is a common problem with assays in which antibodies are measured or used as reagents for other analytes because of nonspecific binding and cross-reactivity. For instance, some recipients of a standard influenza vaccine have shown a false-positive serologic reaction for antibodies to HIV. If not detected, such false results obviously can be disconcerting to the individuals tested and also can lead to inappropriate public health decisions. For this reason, assays that measure only antibody binding do not by themselves establish the presence of antigen-specific antibodies. Specificity must be documented by competitive binding assays and, preferably, by purification (or at least concentration) of the specific antibody population. Nonspecific influences also can interfere with *in vitro* and *in vivo* cellular assays. For instance, some skin reactions attributed to nickel hypersensitivity are more likely to be caused by nonspecific irritation (deBoer *et al.*, 1988; Staberg and Serup, 1988). Moreover, because of the potential for immune cross-reactivity between similar antigens (and even some dissimilar antigens), analytical specificity in immunochemical assays still does not prove biologic specificity.

Applications of Immune Markers for Determining Exposure, Health Effects, and Susceptibility

Exposure

Because the immune system responds specifically to a variety of foreign material, immune markers may be used to detect exposures to many environmental substances, including infectious agents, chemical and biologic materials that cause allergic reactions, and foreign tissue proteins from blood transfusions, organ transplants, or developing fetuses. The foreign material is called an antigen. A specific immune response may be demonstrated by (1) chemical assays that detect antibodies specifically bound to the antigen, (2) cellular function assays that show lymphocytes reacting specifically to the antigen, or (3) specific hypersensitivity reactions in tissues exposed to the antigen (e.g., skin tests).

Changes in host defense components may occur without specific immune responses to particular antigens. For instance, many chemicals will cause an irritative inflammatory response characterized by redness, swelling, heat, and pain, although no antigen-specific immune response occurs (Vogt *et al.*, 1984). Such nonspecific acute inflammatory activity might be detected by local tissue reactions or by increases of peripheral blood granulocytes and acute-phase reactant serum proteins. Other exposures (such as antiproliferative agents) may damage host components nonspecifically and suppress the functional capability of the system, as is true for cancer chemotherapeutic drugs. Tests for cellular and molecular events that accompany these nonspecific changes also may serve as markers of exposure.

Health Effects

Immune System Disorders

The most obvious health effects that might be revealed by changes in immune cells and mediators are those involving the immune system itself. Three general types of disorders of the immune system may have adverse health consequences: immune deficiencies, inappropriate immune reactivities, and unregulated immune proliferation.

Immune deficiency disorders are those in which the immune system fails to mount adequate protective responses against infection or certain forms of cancer. Depending on the nature of the deficiency, the health consequences can range from almost undetectable (increases in the incidence of mild infections) to life threatening (overwhelming sepsis). Immune deficiencies may be indicated by low or absent levels of serum immunoglobulins, low or absent numbers of immune cells, or decreased functional responses.

Immune reactive disorders are those in which immune activity damages host tissues because of inappropriate or poorly regulated responses. Again, depending on their cause and nature, such disorders can be very mild or very severe. Common allergies are caused by inappropriate responses that release histamine and lipid-derived mediators. These allergic reactions often are directed against airborne antigens and may contribute to the pathogenesis of asthma. Depending on the causative antigen, *in vitro* tests for IgE serum antibodies may be good markers for allergies. Autoimmune diseases are debilitating immune reactive disorders in which the immune system reacts against its own body's tissues. Autoimmune reactions can damage the skin, liver, kidneys, various glands, joints, and other tissues, leading to diseases such as rheumatoid arthritis, ankylosing spondylitis, systemic lupus erythematosis, thyroiditis, multiple sclerosis, myasthenia gravis, and some types of diabetes. Autoimmune reactions often are associated with antibodies that react to self-proteins in particular tissues or cell components, and can serve as excellent markers of health effects.

Immune proliferative disorders include lymphoma, multiple myeloma, and some types of leukemia. Like other forms of cancer, they involve the relentless expansion of one family (clone) of cells. However, since clonal expansion is normally an essential part of immune function, immune proliferative disorders often have distinctive characteristics that can be detected by their cell-surface protein phenotypes and molecular genotypes (Figure 16.5) (Waldmann, 1987; Fey *et al.*, 1991).

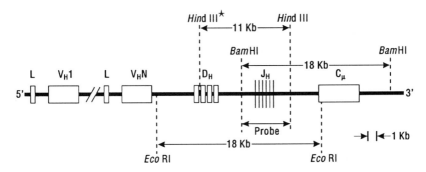

FIGURE 16.5   Genetic restriction map of the immoglobulin heavy chain locus on human chromosome 14. Lymphocytes are the only known cells in which somatic recombination is used to provide a source of diversity within an individual organism. In each lymphocyte family (clone), one of the many variable region genes (V genes to the left of the configuration) has combined with one of the few constant region genes (C genes to the right of the configuration) to form a single gene that encodes the heavy chain of an antibody molecule. The recombination site particularly involves regions designated D (for diversity) and J (for joining). Use of appropriate restriction enzymes and complementary probes can identify the recombinant "signature" unique to each lymphocyte clone. Similar V gene/C gene recombination occurs for the kappa light chain locus (human chromosome 2), the lambda light chain locus (human chromosome 22), the alpha T-cell receptor locus (human chromosome 14), the beta T-cell receptor locus (human chromosome 7), the gamma T-cell receptor locus (human chromosome 7), and the delta T-cell receptor locus (human chromosome 14). (Reprinted with permission from Cossman, *et al.*, 1991. Copyright © 1991 by J. B. Lippincott.)

Health Effects in Other Tissues

Infectious diseases that involve any organ tissues are likely to cause changes in the host defense system. In fact, many of the symptoms associated with infections are caused not by the infectious agents themselves but by cellular and molecular activities of the host response. Some solid tumors that release tumor-specific antigens may elicit autoimmunogenic responses that could serve as markers of the malignancy (Mavligit and Stuckey, 1983). Malnutrition, stress, pregnancy, and a variety of other factors all can influence the immune system (Table 16.3). Immune markers can be used as indicators of such health effects; conversely, these effects can be confounding variables when immune markers are used in attempts to characterize the host defense system itself.

Susceptibility

Susceptibility may be considered the relative propensity of an individual or population to develop dysfunction or disease. Increased susceptibility to infectious diseases and to certain cancers often is associated with deficiency or suppression of the immune system, as discussed earlier. Here we consider the susceptibility of the immune system itself to influences that alter its normal function, as in allergic, autoimmune, and immunoproliferative diseases.

Some well-characterized susceptibility markers have been established for certain immune-related disorders (Nowell and Croce, 1988; Clayton *et al.*, 1989; Wordsworth, 1991). Many of these markers may be detected by direct probes for the gene or by characterization of the gene-related protein products. Most of the genetic markers useful for assessing susceptibility to autoimmunity encode proteins directly involved with the immune system itself, but certain genetic markers for ancillary systems also may be helpful in assessing immune status. Genetic markers concerned directly with the immune system may be related to antigen-specific functions or to nonspecific activities.

In addition to these genetic factors, susceptibility to immune disorders may be influenced by a variety of behavioral and environmental factors such as stress, smoking, and preexisting disease. Thus, susceptibility markers can fall into one of several categories (Table 16.3, on page 416).

Immune-Specific Genetic Factors

The antigen specificity of every immune response depends on two genetically derived systems of recognition proteins: variable genes (V genes) that encode antibodies (Figure 16.5) and V genes that encode T cell receptors. These genes are expressed as the variable regions of antibody proteins or T-cell surface receptor proteins, respectively. The repertoire of these proteins clearly influences immune reactivity and could, therefore, reveal bio-markers of susceptibility. One pathologic process in which susceptibility can be attributed directly to V genes is experimental allergic encephalitis (EAE), an animal disease model with similarities to human multiple sclerosis. EAE is caused by immunization with nerve tissue protein; the pathogenic response depends on a particular V gene in the T-cell receptor repertoire (Clayton *et al.*, 1989). Multiple sclerosis may involve similar interactions between V genes and histocompatibility antigens (Sinha *et al.*, 1991).

Assays for specific V genes or their protein products (idiotypes) are not performed readily at this time, nor is the understanding of V gene biology comprehensive enough to permit focused use of such assays. This is especially true for V genes of the T-cell receptors, which have been characterized only recently (Marrack and Kappler, 1986, 1987, 1990) and already have been implicated in human diseases such as toxic shock syndrome (Choi *et al.*, 1990). Further examples of susceptibility related to the V gene repertoire no doubt will be uncovered as gene probe techniques become more common and our knowledge of idiotypes increases. This area of research should be fruitful ground for development of assays to identify susceptible populations.

Nonspecific Immune Genetic Factors

Two major gene families that also can have strong influence on susceptibility to immune disorders encode immune proteins other than specific antigen recognition structures. One is the family of constant-region genes (C genes), which encode parts of the antibody and T-cell receptor molecules that are not involved in antigen recognition. The second is the family of histocompatibility genes that encode the so-called "transplantation" antigens.

The C genes for antibodies determine the isotypes (class and subclass) of antibodies (IgG, IgA, IgM, IgD, or IgE), which are found in all individuals, and also determine certain allotypes that differ among individuals. The C genes for the T-cell receptor determine whether the T cell expresses receptors of the alpha–beta type or the gamma–delta type. Both alpha–beta and gamma–delta T cells are found in all individuals, but some reports suggest that differences in lymphocyte compartmentation and antigen specificities within individuals are related to the type of T-cell receptor (Davis and Bjorkman, 1988).

The most well-known susceptibility factor related to antibody C genes is the IgE antibody isotype. Individuals who respond to allergens by producing primarily IgE (reaginic) antibodies may become sensitized for allergic reactions, whereas those who produce IgG (blocking) antibodies generally avoid such reactions. The mechanisms of IgE isotype regulation are only partly understood (Ishizaka, 1987, 1988; Vercelli and Geha, 1989; Maggi *et al.*, 1989). At this time, there are no clear tests to determine susceptibility to reaginic sensitization. However, it has been well established that individuals with relatively high IgE levels are likely to be atopic. The impact of atopy on risk of disease in occupational exposures has been demonstrated using a case-control design to shown that atopy was a predisposing factor in the development of work-related hand eczema. Further elucidation of IgE isotype regulation will be required before more informative susceptibility factors can be found.

Other antibody C-gene susceptibility factors may lie among the different isotype subclasses and among different allotypes, particularly in association with certain allotypes of the major histocompatibility complex (MHC). C genes encoding the gamma–delta T-cell receptor proteins may be associated with susceptibility to particular autoimmune disease because of their interactions with heat-shock proteins (Winfield and Jarjour, 1991).

The genes of the MHC are the second family of nonspecific constitutive genetic factors known to influence susceptibility to a variety of immune-mediated diseases. The proteins that these genes encode are involved in presentation of antigen to T cells (Figure 16.6), a critical aspect of immune function. These genes are clearly major influences in the susceptibility to autoimmune disorders (Sinha *et al.*, 1990). Evidence has been reported for MHC regulation in the development of contact dermatitis to nickel (Braathen, 1988; Emtestam *et al.*, 1988) and the development of the scleroderma-like illnesses caused by vinyl chloride (Black *et al.*, 1983) and Spanish toxic oil (Vicaro *et al.*, 1982). Other susceptibilities have been identified among certain combinations of MHC and the C-gene allotypes discussed previously.

Ancillary Genetic Factors

Other genetic factors that are not directly involved with the immune system could act as ancillary markers of susceptibility to immune-mediated disease. The metabolism of xenobiotics may provide ancillary susceptibility markers; for instance, the probability of developing autoimmune illness from procainamide exposure is influenced by the genetically determined rate at which the chemical is acetylated (Reidenberg and Drayer, 1986). The XX karyotype (i.e., female gender) definitely predisposes to autoimmune disease (Bias *et al.*, 1986). The activation of certain oncogenes (Haluska *et al.*, 1986; Nowell and Croce, 1988) also may signal an increased susceptibility, especially to immunoproliferative disorders.

Nongenetic Susceptibility Factors

A number of preexisting conditions as well as behavioral and environmental factors also can influence susceptibility to immune-related health effects. Perhaps the most obvious example of a preexisting condition related to immune effects is previous sensitization by an antigen causing susceptibility to hypersensitivity disorders. Nonspecific inflammation also can influence susceptibility by enhancing immune responsiveness to specific antigens, the so-called "adjuvant effect." Air pollutants, such as ozone and diesel ex-

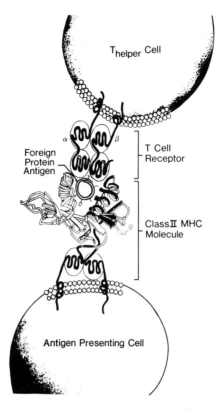

FIGURE 16.6 The three-way (ternary) complex that forms between an antigen-presenting cell (APC) and a T-lymphocyte. The APC binds antigen in its Class II protein of the major histocompatibility complex (MHC), and the T-lymphocyte binds to the antigen–MHC II complex via its specific receptor. This type of receptor-mediated cell–cell interaction is essential for many immune functions. (Reprinted with permission from Sinha, 1990. Copyright 1990 by the AAAS.)

haust particulate matter, can act as adjuvants, increasing the susceptibility to specific hypersensitivity reactions (Muranaka *et al.*, 1986; Koenig, 1987; Bascom *et al.*, 1990).

A number of behavioral factors, include smoking and drug use, can influence immune parameters and susceptibilities. Neurogenic and psychogenic factors also can influence immune function (Figure 16.7). Stress, for instance, can modulate immune reactivity and increase the propensity to develop active autoimmune or allergic disease. Neurogenic factors probably are involved in the pathogenesis of atopic reactions (Greene *et al.*, 1988). Conditioning also can cause changes in immune status and function. Suppressive and reactive immunologic changes have been observed in response to conditioning stimuli such as taste, odor, and audiovisual cues (Kusnecov *et al.*,

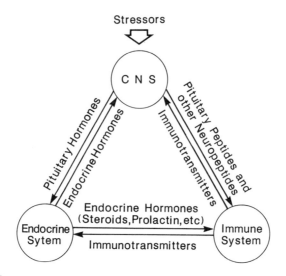

FIGURE 16.7   The direct influence of stress on the immune system may occur via hypothalamic pituitary peptides and the sympathetic branch of the autonomic nervous system. Stress also may influence the immune system through bidirectional communications between the CNS, the endocrine system, and the immune system. Thus, the influence of stressors on the immune response can be viewed as feedback regulatory loops between the CNS, the endocrine system, and the immune system. (Reprinted with permission from Greene *et al.*, 1988.)

1989; MacQueen *et al.*, 1989). Therefore, under some circumstances, the perception of exposure may trigger the appearance of a specific immune reaction (such as allergy) in a conditioned animal without any actual exposure to antigenic material, or it may inhibit an immune response to a sensitizing agent.

Confounding Variables

As the preceding discussion on susceptibility indicates, the cells and mediators of the immune system may be influenced by a wide variety of pathophysiologic processes, some of which are completely unrelated to immune function. For instance, hypergammaglobulinemia may be secondary to dehydration. Malnutrition affects a variety of systems, including the immune response. Other confounding factors that may affect immune parameters include aging, pregnancy, stress, genetic polymorphisms, prior or concurrent disease, previous toxicant exposures, psychoactive drugs (tobacco, alcohol, illicit drugs), therapeutic medications, and neurologic and endocrine influences.

   The multiplicity of confounding factors adds another layer of complexity to the statistical analysis of multiple immune parameters. The complex interactions of the immune system require that a judgment be made about whether immunologic markers are biologically independent, so they can be treated correctly as statistically independent or statistically correlated. Proper control groups (to control for confounders and to assess variation between populations) and longitudinal studies (to assess variation within populations) are the keys to successful evaluation. Confounding factors may be controlled in epidemiologic study designs by restriction or matching and in analyses by stratification or multivariate methods. When genetic factors are confounding, the choice of inheritance models is important for interpreting results.

Host defense activities involve many transitional stages, so the distinction between an intermediate factor and a confounding factor may be difficult to determine. When there is uncertainty about the mechanism, handling a potential confounding factor as both confounding and not confounding in different analyses may be justified (Rothman, 1986). If the variable is actually an intermediate in the process under study, then controlling for it in the analysis as a confounding factor may lead to serious underestimation of effect.

Guidelines for Using Immune Markers in Field Studies

General Approaches

Tests for immune markers may be included when searching for exposure to biologics or chemicals, susceptibility to particular health effects, or biologic effects from exposure or disease processes. Biologic effects range from the earliest stages of pathogenesis or exposure, through stages of subclinical disease, to overt health effects (NRC, 1992).

General applications for immunologic biomarkers are presented in Table 16.4 under two categorical end points: antibodies and immune response. Antibodies can be used in two general ways to effectively estimate internal dose, preclinical biologic responses, or clinical health effects from chemical exposures. The first approach is to exploit antibodies as powerful molecular *probes* to detect a compound of interest, made feasible by scientists' ability to produce purified antibodies with known specificity. For example, specific antibodies commonly are used in forensic toxicology to detect the presence of specific abused drugs, such as morphine, in biologic fluids. Another growing use of antibodies as specific probes is to incorporate them into immunoassays that can detect the presence of biomolecules, such as cytokines or hormones, with exquisite sensitivity. The second approach for using antibodies in biologic monitoring is to measure antibody *titers* (levels of immunoglobulins that are able to react with a specific antigen of interest) in biologic fluids (usually serum). Antibody titers, like probes, also can provide specific and sensitive assessments of exposures to chemicals and the resulting early biologic responses or adverse effects on health. For example, positive

TABLE 16.4   Approaches for Using Immunologic Biomarkers[a]

Exposed individual	Biologic end points		
	Antibodies		Immune response function
	Probes	Titers	
Susceptibility	Genotype, metabolism[b]	Hypogammaglobulinemia[c]	SCID, Atopy[d]
Internal dose	Xenobiotics, toxicants[e]	—	—
Early response	—	Specific serum Ig[f]	Immune profile[g]
Adverse effects	Biologic molecule[h]	Allergy(IgE)[i]	Resistance to disease[j]

[a] Table adapted from presentations developed by Gerry Henningsen, NIOSH, showing immunologic indicators of environmental exposure, internal dose, early preclinical responses, adverse clinical effects, and susceptibility.
[b] Antibody probes used to detect susceptibility to a specific chemical exposure.
[c] Antibody titers used to detect susceptibility to specific pathogenic diseases.
[d] Inherent deficiencies used to detect susceptibility to general pathogenic diseases. SCID, Severe Combined Immunodeficiency Disorder.
[e] Antibody probes used to detect dose from a specific chemical exposure.
[f] Antibody titers used to detect response to a specific chemical exposure.
[g] Selective changes in immune response coincide with certain chemical exposures.
[h] Antibody probes used to detect effects from a specific chemical exposure.
[i] Antibody IgE titers used to detect effects from a specific chemical exposure.
[j] Immunocompetency changes detect effects from a specified chemical exposure.

antibody titers to isocyanates (absent in nonexposed people) denote a prior exposure to these chemicals; if significant specific IgE antibody titers are found, the titers support the ability to attribute adverse effects (allergies) to the specific chemical exposure. The second category of immunologic biomarker is immune response in which the *function* of the immune system is evaluated for selective changes that are, in general, characteristics of a specific chemical exposure. Such immunologic profiles of selective changes in immune response usually are characterized first in animal models or by *in vitro* experiments. Many dose-related selective responses to certain chemical exposure conditions have been described. For example, ethanol selectively interferes with the interleukin 2 receptor, whereas dioxin (TCDD) primarily affects T lymphocytes, and pentachlorophenol has more generalized effects on immune responses. Finally, changes in resistance to disease also can be measured, as a result of known exposure to specified chemicals or certain other environmental immunomodulating agents that can produce potentially harmful effects, including increased susceptibility to infections or cancer and autoimmune disorders. Depending on the design of the field study, any one or a combination of these immunologic end points may be used as an epidemiologic biomarker to assess susceptibility, internal dose, early responses, or adverse effects. Susceptibility to certain environmental agents also potentially can be measured through *in vitro* immunization tests to evaluate the propensity of an individual to develop antibody titers to a specific chemical, or by evaluating baseline immune responses to estimate the impact of immunomodulating agents on the ability of individuals to resist disease.

Choice of Tests

Tests for immune markers used in field studies should be selected to provide the most cost-effective information relevant to the focus of investigation. A serviceable approach to categorizing and selecting immune markers was developed by a subcommittee of the Centers for Disease Control (CDC) and the Agency for Toxic Substances and Disease Registry (ATSDR), convened to develop guidelines for the use of biomarker tests in health assessment studies conducted at Superfund sites (Centers for Disease Control, 1990).

The subcommittee identified three general categories of tests: (1) basic tests that provide a general evaluation of immune status; (2) focused/reflex tests that address particular aspects of immune function as indicated by clinical findings, suspected exposures, or results of prior tests, and (3) research tests that require evaluation in well-defined control populations (Table 16.5).

Tests in both the basic group and the focused/reflex group should have clinical interpretations for disease end points when values lie outside established reference ranges. Tests from the basic group should be included in most studies, since they provide the minimal "core" assessment of immune status. Although they may be omitted in studies addressing very specific concerns, the interpretation of other tests may suffer without the supporting data. Tests from the focused/reflex panel are suggested by particular clinical symptoms, prior laboratory findings, or specific exposures; they may be used individually or be augmented by tests from the basic group. Research tests should be used under the auspices of an investigative protocol with control populations that have known exposure or disease end points. Before a test is considered to have completed the investigative phases, the biochemical or physical abnormalities associated with changes in the marker should be identified, and the nature of any disease associations should be determined. Because of the intrinsic variability of the immune system within and between individuals, longitudinal studies are essential in evaluating research tests for immune markers.

Study Design

In addition to test selection, the overall study design must be orchestrated carefully to insure interpretability of results. The basic goal should be to identify all sources of variability in the tests: analytical (laboratory error), within individuals (over time), among individuals within each group, and among study groups. Analytic variability can be assessed by including a subset of duplicate (split) samples. Variability within individuals can be assessed only by longitudinal studies. Variability among individuals within a study

TABLE 16.5   Classification of Tests for Immune Markers

Test Category	Characteristics	Specific tests
Basic Should be included with general panels	General indicators of immune status Relatively low cost Assay methods are standardized among laboratories Results outside reference ranges are clinically interpretable	Complete blood counts Serum IgG, IgA, IgM levels Surface marker phenotypes for major lymphocyte subsets
Focused/Reflex Should be included when indicated by clinical findings, suspected exposures, or prior test results	Indicators of specific immune functions/events Cost varies Assay methods are standardized among laboratories Results outside reference ranges are clinically interpretable	Histocompatibility genotype Antibodies to infectious agents Total serum IgE Allergen-specific IgE Autoantibodies Skin tests for hypersensitivity Granulocyte oxidative burst Histopathology (tissue biopsy)
Research Should be included only with control populations and careful study design	Indicators of general or specific immune functions/events Cost varies; often expensive Assay methods are usually not standardized among laboratories Results outside reference ranges are often not clinically interpretable	In vitro stimulation assays Cell activation surface markers Cytokine serum concentrations Clonality assays (antibody, cellular, genetic) Cytotoxicity tests

group may be quite high, and may require a large number per group to assess properly. Identification of biologically significant variability among groups, the general goal of controlled studies, is possible only with careful selection of the populations to control for the many differences in susceptibility and the confounding variables that influence the immune system.

Once a significant difference among groups is established for one or more markers, long-term follow-up is required to determine the extent to which such differences are predictive for overt health effects. Identifying predictive immune markers in epidemiologic field studies will lead to a better understanding of immune-related disease mechanisms and enhanced measures for prevention and intervention.

Data Management

Interpretation of many immune markers requires that results be evaluated in the face of evolving analytical methods, shifting calibration values, and a plethora of confounding variables. A properly designed database will allow any result to be related to analytical method, reagents, bias relative to reference calibrators, analytic and biologic variability, and, ultimately, human health outcome. The structure of the database should be an integral part of designing the study and should precede any data collection (see Chapter 4).

Illustrations Using Immune Markers in Health Effect Studies

The following examples have been chosen to illustrate the potential of immune system components as cellular and molecular markers as well as some of the practical considerations involved with such use. Each example is given as a brief summary to point out its main features of interest. Readers should consult the original references for full information and analyses.

Immune Markers in HIV Infection

Infection by the human immunodeficiency virus (HIV) leads to the acquired immune deficiency syndrome (AIDS), usually long after the initial infection. A few years after a surface protein called CD4 was identified on a subpopulation of helper T lymphocytes, this protein was shown to be the site of viral attachment by HIV (McDougall *et al.*, 1985). The CD4 protein is present not only on helper T cells, but also on monocytes, macrophages, and dendritic cells. Consequently, these cells also can be infected by HIV. Subsequent events in the pathogenesis of AIDS are not well understood, but it appears to be more complex than the simple destruction of virus-infected cells (Margolick and Vogt, 1992). Many different immune markers have been critically important in the diagnosis, monitoring, and evaluation of HIV infection, as well as in attempts to understand and mitigate the basic disease processes.

Assays for antibodies to HIV are the mainstay of screening and diagnosing the infection. Generally a two-stage testing protocol is used; the sensitivity and specificity of the combined tests both exceed 99% if properly performed (MMWR, 1990). Antibody tests can be applied to bloodspots taken routinely for neonatal screening; these results provide the most broad-based data on the prevalence of the infection (Quinn *et al.*, 1992). Such bloodspots also can be tested for other factors potentially related to HIV infection, for example, illicit drugs (Henderson *et al.*, 1992), providing important information to help focus efforts of prevention.

Although antibody tests are valuable tools for HIV screening and diagnosis, they do not provide much information about the progress of the infection. Low peripheral blood counts for CD4 lymphocytes are currently the best laboratory indicator of impending opportunistic infections by *Pneumocystis carinii* (Figure 16.8A). CD4 counts also correlate with the response to therapy by AZT (Figure 16.8B); because of the slow progress of the disease and earlier therapeutic intervention, CD4 counts will become increasingly important in the evaluation of new therapeutics and vaccines. A number of technical issues concerning CD4 lymphocyte measurements are still unresolved (Kidd and Vogt, 1989), but are receiving increased attention in light of the public health importance of the test (MMWR, 1992).

Changes in other lymphocyte subset phenotypes also may provide important insight into the pathogenesis of AIDS. A subtle trend in peripheral blood T cells associated with HIV seroconversion was identified when over 1500 results from 268 subjects were compiled from a multisite study (Margolick *et al.*, 1989) (Figure 16.9). Over a 27-month period after the anti-HIV serum antibodies first became detectable, the number of T cells with neither CD4 nor CD8 surface markers increased significantly, suggesting that seroconversion is associated with an expansion of natural killer (NK) or gamma–delta T cells. Although predictive value of this change is low, it provides another clue to the immunologic responses that accompany HIV infection. Such subtle changes would be difficult or impossible to detect in the face of wide normal variation without large longitudinal multisite studies using standardized methods.

Leukocyte Counts in Benzene Exposure

The volatile organic compound benzene is toxic to bone marrow cells. Exposure at high levels can lead to aplastic anemia (a complete loss of all blood-forming stem cells), acute myelogenous leukemia, and possibly immunoproliferative malignancies including lymphomas and multiple myeloma (Young, 1989; Goldstein, 1990). This bone marrow toxicity can cause subclinical decreases in peripheral blood leukocytes in humans exposed to benzene at levels that once were common in occupational settings. Kipen *et al.* (1988) found that the average peripheral blood leukocyte counts monitored from 1940 to 1975 in a large rubber factory cohort increased linearly over an 8-year period during which the exhaust system of the plant was improved and a recommended exposure standard was implemented (Figure 16.10). The investigators used this relationship for a retrospective exposure assessment that seemed warranted, but changes in the blood counts also have been attributed to methodologic differences. Since that time benzene has been regulated to levels at which this effect is no longer detectable, but the acceptable levels for minimal health risk remain controversial (Brett *et al.*, 1989; Nicholson and Landrigan, 1989) Moreover, methodologic biases are still possible, even on very simple markers such as blood cell counts (see Chapter 4).

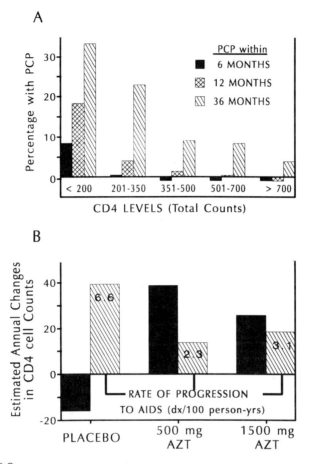

FIGURE 16.8 (A) CD4 lymphocyte counts in peripheral blood indicate susceptibility to pneumocystic carinii pneumonia (PCP) in HIV-infected individuals. When counts are above 200 cells/μl, the risk of PCP within 1 year is negligible; below 200, the risk approaches 20%. The average CD4 in normal individuals is around 1000 cells/μl with a lower limit around 400. (Adapted from Morbidity and Mortality Weekly update, 1989). Guidelines for prophylaxis against pneumocystitis carinii pneumonia for persons infected with Human Immunodeficiency Virus. 38(S-5), 6. (B) CD4 lymphocyte counts also reflect the response to treatment of HIV-infected individuals with AZT. Among those receiving placebo treatment, CD4 counts fell about 15% per year, and the rate of progression to AIDS was 6.6 diagnoses per 100 person-years. Lower-dose AZT treatment resulted in an annual increase in CD4 counts and cut the rate of progression to AIDS by two-thirds. Higher-dose AZT treatment also increased the CD4 count, but not as much as the lower-dose regimen; moreover, the rate of progression to AIDS was higher than that in the lower-dose cohort. (Adapted with permission of the New England Journal of Medicine from Volberding et al., 1990.)

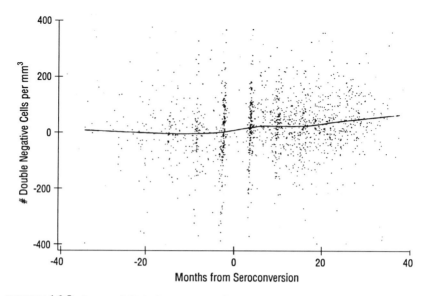

FIGURE 16.9 Increased CD3+CD4−CD8− cells in HIV-1 infection. This longitudinal study showed a subtle but significant increase in an unusual peripheral blood T-lymphocyte subset when participants became infected by HIV-1.

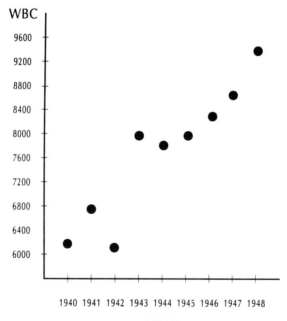

FIGURE 16.10 Annual average white blood cell count (WBC) in a cohort occupationally exposed to benzene. The results showed, on average, a striking progressive increase in WBC over the 8-year period during which benzene exposure levels declined due to revisions of the recommended exposure standard and improvements in the exhaust system of the plant. (Reprinted with permission from Kipen *et al.*, 1988.)

Air Pollution and Acute Inflammation

Complement is a series of proteins activated in a cascading sequence by a variety of nonspecific stimuli as well as by specific antibody–antigen interactions. The most prominent protein in the complement system, designated C3, is one of the acute-phase reactant serum proteins that increase in concentration with ongoing stimulation of acute inflammatory responses (Dowton and Colten, 1988; Katz *et al.*, 1990) However, if large-scale acute inflammation continues long enough, C3 levels actually can decline because depletion surpasses synthesis.

When serum C3 levels were compared in populations residing in areas with varying degrees of air pollution, the results showed small but significant increases as the pollution index rose (Stiller-Winkler *et al.*, 1989) (Figure 16.11). These findings emphasize both the inflammatory effects of air pollution and the potential impact of subtle confounding variables on immune markers.

Infiltration of respiratory mucosal surfaces by neutrophils is another indicator of acute inflammatory reactions to airborne irritants or sensitizers. Controlled exposures of normal subjects to ozone at part per million levels has been shown to cause an increase in the number of neutrophils obtained by nasal lavage (Koren, 1990), another confirmation that air pollution is associated with respiratory tract inflammation.

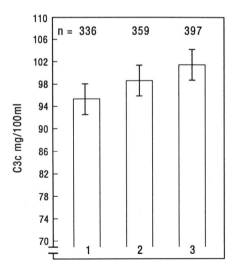

FIGURE *16.11*   Serum levels of complement component C3 in males living in areas characterized by little (1), moderate (2), or heavy (3) air pollution. Complement is an acute-phase reactant that shows a nonspecific increase under a variety of proinflammatory conditions. This effect could represent a marker of exposure, a marker of effect, or a confounding variable, depending on the goal and design of the study. (Reprinted with permission from Stiller-Winkler *et al.*, 1989.)

Chronic Lymphocytic Leukemia and Cell Surface Protein Expression

B-cell chronic lymphocytic leukemia (B-CLL), a disease most prominent in older populations and that sometime runs in families, is a clonal expansion of B cells that may be undetected clinically for many years. In the leukemic form of B-CLL, blood and bone marrow lymphocytosis predominates, whereas in the lymphomatous form, lymphadenopathy is dominant. Mixtures of these forms also occur. Peripheral blood lymphocyte counts can increase more than 10-fold above normal levels before symptoms are noticeable, and the disease generally remains indolent for some time after that. Although B-CLL cells appear very similar to normal resting lymphocytes on stained blood smears, small morphologic differences often can be discerned. The cause of CLL is unknown, but a discrete molecular lesion affecting regulatory function is suspected (Marti and Fleisher, 1987). Gene probe studies in one familial cohort suggest that immunoglobulin heavy-chain genes are rearranged in CLL cells, whereas the T-cell receptor gene is not rearranged; moreover, the rearrangements were different in the affected cases (Shah *et al.*, 1992).

When immunophenotyping by flow cytometry was applied to the study of B-CLL, deviations from the normal expression of several cell-surface markers became apparent through differences in the fluorescence intensity of the stained cells (Marti *et al.*, 1989). In particular, B-CLL cells were stained more strongly than normal cells by antibodies to Class II histocompatibility proteins, and less strongly by antibodies to the B-cell lineage-specific CD20 and CD22 proteins (Figure 16.12). The differences do not appear to be caused by technical factors (Marti et al., 1992); therefore, they represent alterations in the actual amounts of surface protein receptors expressed by the malignant cells, presumably related to the molecular lesion responsible for their neoplastic behavior.

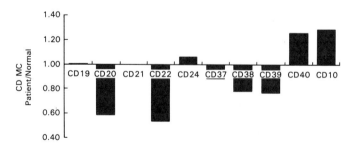

FIGURE *16.12*   Relative differences in the fluorescent intensity of normal lymphocytes and chronic lymphocytic leukemia (CLL) lymphocytes when cells are stained with fluorescent probes for different surface proteins. The CD20 and CD22 surface proteins stain much less strongly in CLL cells than in normal cells, suggesting that the CLL cells express fewer of these proteins on their surfaces. In contrast, the CD40 and CD10 proteins appear to be expressed more abundantly in CLL cells.

The morphologic and surface marker differences in B-CLL cells make immunophenotyping by flow cytometry an ideal technique for detecting subclinical disease (Figure 16.13). Since these tests rarely are applied on a broad population basis, the true prevalence of B-CLL is unknown but is probably much higher than current figures suggest, especially in older populations.

Ankylosing Spondylitis and Histocompatibility Genes

Ankylosing spondylitis (AS) is an arthritis-like autoimmune disease affecting the joints of the spine and pelvis that can lead ultimately to bony fusion of the vertebra. AS afflicts about 1 in 1000 individuals in Caucasian populations; 96% of those individuals possess the HLA-B27 histocompatibility gene, compared with 7% of the general population (Benjamin and Parham, 1990). The specificity of the HLA-B27 relationship is not strong, however, since only about 2% of HLA-B27 positive individuals contract AS. Other genetic factors also influence risk, since the disease runs in families and varies among races in prevalence as well as in association with HLA-B27.

The molecular basis for the connection between AS and HLA-B27, a class I histocompatibility human leukocyte antigen (HLA), is not clear. In general, the class I HLA molecules are cell-surface proteins that restrict the cytotoxic activity of CD8 T cells by binding and presenting antigenic peptides within a molecular "groove" of their three-dimensional structure (Figure 16.14). Some regions of amino acid sequences that compose the groove display considerable variability among different HLA alleles and even between different subtypes of a particular family such as B27, whereas other regions are conserved strongly in all B27 proteins yet differ from the other HLA-B molecules. The epidemiologic observation that AS is associated with the four most common B27 subtypes may, therefore, be translated directly into molecular terms: the disease-associated parts of the molecule must be unique to HLA-B27 but shared by all common subtypes within that family. Protein structure analysis suggests as a likely candidate a five amino acid residue stretch positioned in or near the antigen-binding groove (Figure 16.13). If this association is authentic, it could help explain the tissue specificity of disease activity, because the B27 groove could bind a peptide unique to the target tissue (Benjamin and Parham, 1990). Other theories based on the molecular structure of B27 have been proposed; further epidemiologic and laboratory investigation will be necessary to find the true mechanism(s) of the pathogenic process.

In Vitro *and* in Vivo *Methods for the Evaluation of Inhalent Allergy*

Common allergies to inhalent antigens such as pollens and skin danders are caused by immediate hypersensitivity reactions. Immediate hypersensitivity occurs when IgE molecules bound to the surface of mast cells or basophils become cross-linked through contact with antigen, causing cell degranula-

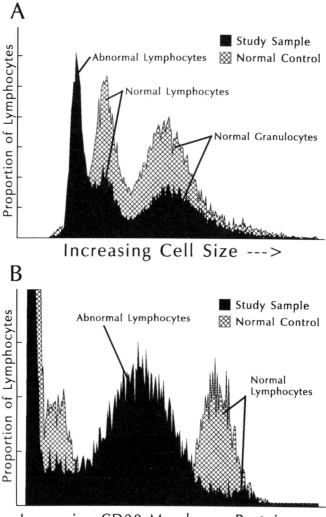

FIGURE 16.13 Fluorescent flow cytograms from an individual with subclinical B-cell chronic lymphocytic leukemia (CLL) compared to those from a normal individual. The light scatter pattern (A) from the CLL sample shows a distinct subpopulation to the lower left of the lymphocyte cluster not seen in the normal sample. The fluorescence pattern (B) shows an increase in cells staining for the B-cell marker CD20, and most of the cells are much dimmer than the CD20 staining in normal B cells. A few normal B-cells are present in the CLL sample, as shown by their brighter staining. The findings are consistent with a monoclonal expansion of B cells expressing low CD20 surface protein, or a preclinical CLL. This condition was easily recognized by flow cytometric analysis, although it was not apparent from the complete blood count.

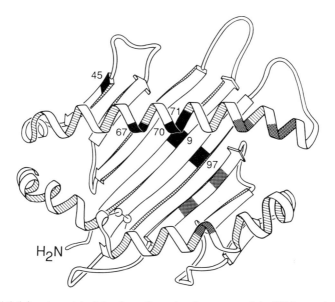

FIGURE 16.14   A model of the three-dimensional structure of the HLA protein molecule showing the conserved (solid) and polymorphic (stippled) sites in the four HLA-B27 subtypes associated with an increased risk of ankylosing spondylitis (AS). The conserved residues (numbered) that line the "groove" of the molecule are unique to the B27 type among HLA-B molecules and may be the specific sites associated with the increased risk of AS. (Reprinted with permission from Benjamin and Parham, 1990.)

tion and the release of active mediators, including histamine. Allergy skin tests detect cell-bound IgE by the rapid onset of wheal-and-flare reaction when antigen causes mast cell degranulation and immediate hypersensitivity inflammatory reaction in the area of the test site. Skin tests often are done by first applying a very small amount of antigen with a "puncture" method, then following weak or negative reactions with a larger amount through intracutaneous injection. The sensitivity and specificity of skin tests depends largely on the person who applies the antigen and interprets the reaction.

Although some IgE is present in serum, the levels are ordinarily very low (less than a million-fold below IgG levels). Serum IgE does not contribute directly to the pathogenesis of allergic reactions. However, serum IgE represents a sampling of total IgE production and therefore can serve as a marker of susceptibility to immediate hypersensitivity reactions. A number of laboratory methods to measure antigen-specific IgE are in common use; they differ in both technical and interpretative aspects.

The optimal approach to using skin tests (*in vivo* testing) and the various serum IgE assays (*in vitro* testing) in the evaluation of allergy has been a continuing source of controversy, particularly in the lower ranges of skin test reactivity and with antigen-specific IgE levels close to the limit of detection. In a careful study comparing skin tests with three *in vitro* methods for evaluating inhalent allergy (Williams *et al.*, 1992), the investigators found generally good agreement among all methods, but ROC analysis (Figure 16.15; see Chapter 3) showed that skin testing by an experienced allergist was the most sensitive and specific approach. A modification of the original *in vitro* test increased sensitivity, but only with a corresponding decrease in specificity, whereas a newer *in vitro* assay showed a marginal improvement in both sensitivity and specificity.

A number of issues impact the relative sensitivity and specificity of these tests. The major biologic considerations are that (1) IgE bound to mast cells persists much longer than IgE in the serum, which has a half-life of less than 1 day (Tada *et al.*, 1975), and (2) the skin test response involves biologic amplification through the immediate hypersensitivity reaction. Both factors contribute to the better sensitivity of allergy skin tests, but the newest technologies for measuring specific IgE approach sensitivity of skin tests for many inhalent allergens. Technical considerations include the subjective nature of reading skin tests compared with objective numerical results of serum

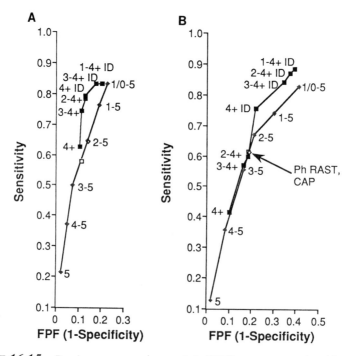

FIGURE 16.15   Receiver–operator characteristic (ROC) curves comparing skin tests and three different *in vitro* assays for the diagnosis of allergy to timothy (A) and cat dander (B). The skin test (■) curves go from a 4+ positive puncture test (the least sensitive) to a negative puncture test (P) with a positive intradermal (ID) (the most sensitive). The ROC curve for skin tests is better at all points than the multipoint curve for the modified radio-allergosorbent test (modified RAST ◆), for the single cut-off point for the original RAST (Phadebas RAST □), or for the newer CAP system (◇). (Reprinted with permission from Williams *et al.*, 1992.)

IgE assays. In less experienced hands, *in vitro* tests may well be superior to *in vivo* tests.

Summary

The discussion and examples presented in this chapter should confer a sense of cautious enthusiasm to those who would use immune cells and mediators as markers of health-related events in epidemiologic studies. The enthusiasm is warranted because of the rapid evolution of technologies that allow extensive characterization of cellular and molecular markers. It is justified further by the success of using immune markers in epidemiologic studies to elucidate health effects and environmental exposures in many infectious, autoimmune, allergic, and neoplastic diseases as the preceding examples demonstrate. However, the rapid pace of technology also should invoke a sense of caution, because it means that changing methodologies will be applied to problems that may have an uncertain biologic basis and an unclear public health significance. The interpretation of results from such studies can be highly problematic.

Methodology is the most critical aspect of immune marker use in epidemiologic field studies. In the field, suitable control groups, appropriate study designs, and proper sample handling are essential elements. In the laboratory, quality control of assay methods through comparisons within and between runs, days, methods, and laboratories is equally essential. The selection of tests and the interpretation of results are other critical issues that are closely intertwined. Investigators should select tests that have known clinical significance or can be interpreted in light of some understanding of the underlying biologic processes. Since most immune markers have not been tested in large groups of normal individuals, the extent to which outlying values or shifts in overall population distributions reflect genuine health effects is often uncertain.

In summary, the great promise of immune markers as tools for exposure and health assessment can be realized only through the application of rigorous epidemiologic and laboratory methods coupled with an understanding of the basic biology of the host defense system. Efforts to this end have already begun through multicenter activities such as the AIDS Clinical Trials Group (Margolick *et al.*, 1989; Paxton *et al.*, 1989) and the Immune Biomarkers Demonstration Project for environmental health studies (Vogt *et al.*, 1990). Future activities should strengthen the role immune markers can play in epidemiologic studies.

Acknowledgments

This work was supported in part with funds from the Comprehensive Environmental Response, Compensation, and Liability Act trust fund by interagency agreement with the Agency for Toxic Substances and Disease Registry, U.S. Public Health Service, Department of Health and Human Services. Use of trade names is for identification only and does not constitute endorsement by the Public Health Service of the U.S. Department of Health and Human Services.

References

Balkwill, F. R., and Burke, F. (1989). The cytokine network. *Immuno. Today* **10**, 299–303.

Bascom, R., Naclerio, R. M., Fitzgerald, T. K., Kagey-Sobotka, A., and Proud, D. (1990). Effect of ozone inhalation on the response to nasal challenge with antigen of allergic subjects. *Am. Rev. Resp. Dis.* **142**, 594–601.

Benjamin, R., and Parham, P. (1990). Guilt by association: HLA-B27 and ankylosing spondylitis. *Immun. Today* **11**, 137–142.

Bias, W. B., Reveille, J. D., Beaty, T. H., Meyers, D. A., and Arnett, F. C. (1986). Evidence that autoimmunity in man is a Mendelian dominant trait. *Am. J. Hum. Genet.* **39**, 584–602.

Black, C. M., Welsh, K. I., Walker, A. E., Bernstein, R. M., Catoggio, L. J., McGregor, A. R., and Jones, J. K. (1983). Genetic susceptibility to scleroderma-like syndrome induced by vinyl chloride. *Lancet* **1**, 53–55.

Brett, S. M., Rodricks, J. V., and Chinchilli, V. M. (1989). Review and update of leukemia risk potentially associated with occupational exposure to benzene. *Environ. Health Perspect.* **82**, 267–281.

Centers for Disease Control (1990). "Internal Report on Biomarkers of Organ Damage and Dysfunction for the Renal, Hepatobiliary, and Immune Systems." Subcommittee on Biomarkers of Organ Damage and Dysfunction of the Centers for Disease Control and Agency for Toxic Substances and Disease Registry, Atlanta, Georgia. [Available from Clinical Biochemistry Branch, Mailstop F-19, CDC, Atlanta, Georgia 30333.]

Choi, Y., Lafferty, J. A., Clements, J. R., Todd, J. K., Gelfand, E. W., Kappler, J., Marrack, P., and Kotzin, B. L. (1990). Selective expansion of T cells expressing V beta 2 in toxic shock syndrome. *J. Exp. Med.* **172**(3), 981–984.

Clayton, J. P., Gammon, G. M., Ando, D. G., Kono, D. H., Hood, L., and Sercarz, E. E. (1989). Peptide-specific prevention of experimental allergic encephalomyelitis. Neonatal tolerance induced to the dominant T cell determinant of myelin basic protein. *J. Exp. Med.* **169**(5), 1681–1691.

Cossman, J., Zehnbauer, B., Garrett, C. T., Smith, L. J., Williams, M., Jaffe, E. S., Hanson, L. O., and Love, J. (1991). Gene rearrangements in the diagnosis of lymphoma/leukemia. Guidelines for use based on a multiinstitutional study. *Am. J. Clin. Pathol.* **95**(3), 347–354.

Davis, M. M., and Bjorkman, P. J. (1988). T-cell antigen receptor genes and T-cell recognition. *Nature (London)* **334**, 395–402.

Dowton, S. B., and Colten, H. R. (1988). Acute phase reactants in inflammation and infection. *Semin. Hematol.* **25**(2), 84–90.

Emtestam, L., Carlsson, B., Marcusson, J. A., Wallin, J., Moller, E. (1989). Specificity of HLA restricting elements for human nickel reactive T cell clones. *Tissue Antigens* **33**(5), 531–541.

Fey, M. F., Kulozik, A. E., Hansen-Hagge, T. E., and Tobler, A. (1991). The polymerase chain reaction: A new tool for the detection of minimal residual disease in haematological malignancies. *Eur. J. Cancer* **27**(1), 89–94.

Goldstein, B. D. (1990). Is exposure to benzene a cause of human multiple myeloma? *Ann. N.Y. Acad. Sci.* **609**, 225–230.

Greene, R., Fowler, D., MacGlashan, D., and Wienreich, D. (1988). IgE-challenged human lung mast cells excite vagal sensory neurons *in vitro*. *J. Appl. Physiol.* **64**, 2249–2253.

Haluska, F. G., Tsujimoto, Y., and Croce, C. M. (1986). Molecular genetics of B- and T-cell neoplasia. *Int. Symp. Princess Takamatsu Cancer Res. Fund* **17**, 151–158.

Henderson, L. O., Powell, M. K., Hannon, W. H., Miller, B. B., Martin, M. L., Hanzlick, R. L.,

Vroon, D., and Sexon, W. R. (1992). Radioimmunoassay screening of dried blood spot materials for benzoylecgonine. *Anal. Toxicol. (in press)*

Hood, L. E., Weissman, I. L., and Wood, W. B. (1978). "Immunology." Benjamin/Cummings, Menlo Park, California

Katz, P. R., Karuza, J., Gutman, S. I., Bartholomew, W., and Richmond, G. (1990). A comparison between erythrocyte sedimentation rate (ESR) and selected acute-phase proteins in the elderly. *Am. J. Clin. Pathol.* **94**(5), 634–640.

Kidd, M., and Vogt, R. F. (1989). Evaluation of T-cell subsets during HIV infection and AIDS: A workshop report. *Clin. Immunol. Immunopathol.* **52**, 3–9.

Kipen, H. M., Cody, R. P., Crump, K. S., Allen, B. C., and Goldstein, B. D. (1988). Hematologic effects of benzene: A thirty-five year longitudinal study of rubber workers. *Toxicol. Ind. Health* **4**, 411–430.

Koenig, J. Q. (1987). Pulmonary reaction to environmental pollutants. *J. Allergy Clin. Immunol.* **79**, 833–843.

Koren, H. S., Hatch, G. E., and Graham, D. E. (1990). Nasal lavage as a tool in assessing acute inflammation in response to inhaled pollutants. *Toxicol.* **60**, 15–25.

Kusnecov, A., King, M. G., and Husband, A. J. (1989). Immunomodulation by behavioural conditioning. *Biol. Psychol.* **28**, 25–39.

McDougall, J. S., Mawle, A., Cort, S. P., Nicholson, J. K. A., Cross, G. D., Scheppler-Campbell, J. A. Hicks, D., and Sligh, J. (1985). Cellular tropism of the human retrovirus HTLV-III/LAV. I. Role of T cell activation and expression of the T4 antigen. *J. Immunol.* **135**, 3151–3162.

MacQueen, G., Marshall, J., Perdue, M., Siegel, S., and Bienenstock, J. (1989). Pavlovian conditioning of rat mucosal mast cells to secrete rat mast cell protease II. *Science* **243**, 83–85.

Maggi, E., Del Prete, G. F., Parronchi, P., Tiri, A., Macchia, D., Biswas, P., Simonelli, C., Ricci, M., and Romagnani, S. (1989). Role of T cells, IL-2 and IL-6 in the IL-4-dependent in vitro human IgE synthesis. *Immunol.* **68**, 300–306.

Margolick, J. B., et al. (1989). Development of antibodies to HIV-1 is associated with an increase in circulating CD3+CD4−CD8− lymphocytes. Clin. Immunol. Immunopathol. **51**(3), 348–361

Margolick, J. B., and Vogt, R. F. (1992). Environmental effects on the human immune system and the risk of cancer: Facts and fears in the era of AIDS. *Environ. Carc. Exotoxicol. Rev.* **C9**(2), 155–206.

Marrack, P., and Kappler, J. (1986). The T cell and its receptor. *Sci. Am.* **254**, 35–45.

Marrack, P., and Kappler, J. (1987). The T cell receptor. *Science* **238**, 1073–1079.

Marrack, P., and Kappler, J. W. (1990). The T cell receptors. *Chem. Immunol.* **49**, 69–81.

Marti, G. E., and Fleisher, T. A. (1987). B Cell CLL Immunophenotypes. *In* "Chronic Lymphocytic Leukemia: Recent Progress, Future Directions." (R. P. Gale and K. Rai, eds.), pp. 195–204. UCLA Symposia on Molecular and Cell Biology, New Series, Volume 59. Alan R. Liss, New York.

Marti, G. E., Zenger, V., Caproaso, N. E., Brown, M., Washington, G. C., Carter, P., Schechter, G., and Noguchi, P. (1989). Antigenic expression of B-cell chronic lymphocytic leukemic lymphocytes. *Anal. Quant. Cytol. Histol.* **11**(5), 315–323.

Marti, G. E., Faguet, G., Bertin, P., Agee, J., Washington, G. C., Ruiz, S., Carter, P., Zenger, V., Vogt, R., and Noguchi, P. (1992). CD20 and CD5 expression in B-chronic lymphocytic leukemia. *Ann. New York Acad. Sci.* **651**, 480–483.

Mavligit, G. M., and Stuckey, S. (1983). Colorectal carcinoma. Evidence for circulating CEA–anti-CEA complexes. *Cancer* **52**(1), 146–9.

Morbidity and Mortality Weekly Report (1990). Update: Serologic testing for HIV-1 antibody—United States, 1988 and 1989. 39(22), 380–383.

Morbidity and Mortality Weekly Report (1992). Guidelines for the performance of CD4+ T-cell determinations in persons with human immunodeficiency virus infection. 47(RR-8), 1–17.

Muranaka, M., Suzuki, K., Koizumi, S., and Takafuji, S. (1986). Adjuvant activity of diesel-exhaust particulates for the production of IgE antibody in mice. *J. Allergy Clin. Immunol.* 77, 616–623.

Nicholson, W. J., and Landrigan, P. J. (1989). Quantitative assessment of lives lost due to delay in the regulation of occupational exposure to benzene. *Environ. Health Perspect.* 82, 185–188.

Nowell, P. C., and Croce, C. M. (1988). Chromosomal approaches to oncogenes and oncogenesis. *FASEB J.* 2(15), 3054–60.

NRC (National Research Council) (1992). Biologic markers in immune toxicology. Washington, D.C. National Academy Press. 206 pp.

Paul, W. (1989). "Fundamental Immunology," 2d Ed. Raven Press, New York.

Pugh-Humphries, R. G. P., Woo, J., and Thomson, A. W. (1991).

Quinn, M., Redus, M. A., Granade, T. C., Hannon, W. H., and George, J. R. (1992). HIV-1 serologic test results for one million newborn dried blood specimens: Assay performance and implications for screening. *J. AIDS (in press).*

Reidenberg, M. M., and Drayer, D. E. (1986). Procainamide, N-acetylprocainamide, antinuclear antibody and systemic lupus erythematosus. *Angiology* 37, 968–971.

Roitt, I. M., Brostoff, J., and Male, D. K. (1987). "Immunology." Mosby, St. Louis, Missouri.

Rothman, K. J. (1986). "Modern Epidemiology." Little, Brown, Boston.

Rystedt, I. (1985). Atopic background in patients with occupational hand eczema. *Contact Derm.* 12, 247–254.

Shah, A. R., Maeda, K., Deegan, M. J., Roth, M. S., and Schnitzer, B. (1992). A clinicopathologic study of familial chronic lymphocytic leukemia. *Am. J. Clin. Pathol.* 97, 184–188.

Sinha, A. A., Lopez, M. T., and McDevitt, H. O. (1990). Autoimmune diseases: The failure of self tolerance. *Science* 248, 1380–1388.

Sinha, A. A., Bell, R. B., Steinman, L., and McDevitt, H. O. (1991). Oligonucleotide dot-blot analysis of HLA-DQ beta alleles associated with multiple sclerosis. *J. Neuroimmunol.* 32, 61–65.

Staberg, B. and Serup, J. (1988) Allergic and irritant skin reactions evaluated by laser Doppler flowmetry. *Contact Derm.* 1, 40–45.

Stiller-Winkler, R., Kramer, U., Fiedler, E., Ewers, U., Dolgner, R. (1989). C3c concentrations in sera of persons living in areas with different levels of air pollution in Northrhine-Westphalia. (Federal Republic of Germany). *Environ. Res.* 49, 7–19.

Tada, T., Okumura, K., Platteau, B., Beckers, A., and Bazin, H. (1975). Half-lives of two types of rat homocytotropic antibodies in circulation and in the skin. *Int. Arch. Allergy Appl. Immunol.* 48, 116–131.

Vicaro, P. J. L., Serrano-Rios, M., Andres, F., *et al.* (1982). HLA-DR3, DR4 increase in chronic stage of Spanish Oil Disease. *Lancet* 1, 276.

Vogt, R. F. (1991). Use of laboratory tests for immune biomarkers in environmental health studies concerned with exposure to indoor air pollutants. *Environ. Health Perspect.* 95, 85–91.

Vogt, R. F., Dannenberg, A. M., Papirmeister, B., and Scofield, B. (1984). Pathogenesis of skin lesions caused by sulfur mustard. *Fund. Appl. Toxicol.* 4, S71–78.

Waldmann, T. A. (1987). The arrangement of immunoglobulin and T cell receptor genes in human lymphoproliferative disorders. *Adv. Immunol.* 40, 247–321.

Westermann, J., and Pabst, R. (1990). Lymphocyte subsets in the blood: a diagnostic window on the lymphoid system? *Immunol. Today* 11(11), 406–410.

Williams, P. B., Dolen, W. K., Koepke, J. W., and Selner, J. C. (1992). Comparison of skin testing and three in vitro assays for specific IgE in the clinical evaluation of immediate hypersensitivity. *Ann. Alergy* 68(1), 35–45.

Winfield, J. B., and Jarjour, W. N. (1991). Stress proteins, autoimmunity, and autoimmune disease. *Curr. Top. Microbiol. Immunol.* 167, 161–89.

Wordsworth, P. (1991). PCR-SSO typing in HLA-disease association studies. *Eur. J. Immunogen.* 18, 139–146.

Young, N. (1989). Benzene and lymphoma. *Am. J. Ind. Med.* 15, 495–498.

17

Biomarkers of Pulmonary Disease

Melvyn S. Tockman,
Prabodh K. Gupta,
Norman J. Pressman,
and James L. Mulshine

Pulmonary Carcinogenesis

For two decades, detection of neoplastic changes in exfoliated sputum epithelial cells has been the standard clinical marker of "early" bronchogenic
carcinoma (Saccomano *et al.*, 1974), yet we have learned that cytomorpho-
logic detection of neoplasia is too insensitive for lung cancer mass screening.
Screening of 30,000 middle-aged male cigarette smokers who participated in
three collaborating National Cancer Institute (NCI)-sponsored clinical trials
(at Johns Hopkins University, Memorial Sloan-Kettering Hospital, and the
Mayo Clinic) has demonstrated that, compared with chest radiography, sputum cytology screening can detect some presymptomatic earlier-stage carcinomas, particularly carcinoma of the squamous cell type (Frost *et al.*,
1984a,b; Stitik *et al.*, 1985; Tockman *et al.*, 1985). Although the study
groups (dual screened by cytology and radiography) enjoyed higher resect-
ability and survival rates than the controls (screened by X-ray alone), enhanced detection of slower growing cancers did not, unfortunately, result in
a study group advantage for lowered (overall) lung cancer mortality. Less
than 10% of lung cancers in the early lung cancer detection trial were added
by morphologic examination of sputum cells. Length-biased sampling, lead-
time bias, and misclassification, in addition to failures of detection and intervention, contributed to the lack of improvement in mortality rates (Frost
et al., 1986; Tockman, 1986; Mulshine *et al.*, 1989).

Progress in tumor biology has provided new tools with great potential as markers of early neoplasia. As summarized in Table 17.1, a variety of biologic features of shed bronchial epithelial cells (or cell products) can be used to probe the progression toward carcinogenesis. Ultimately, perhaps a combination of these targets might serve most effectively to detect the presence of a transformed bronchial epithelium. Our work is now focused on detection of early lung cancer through monoclonal antibody recognition of tumor-associated antigens (Tockman *et al.*, 1988). This class of biomarkers will be used to illustrate this chapter. Careful attention to the biology of lung carcinogenesis will, no doubt, yield other intermediate end points that recognize the onset and progression of neoplasia before its clinical presentation. Appreciation of the complex issues that surround the selection, validation, quantitation, and population application of lung cancer biomarkers has led us to propose the following conceptual organization (Mulshine *et al.*, 1991; Tockman *et al.*, 1991).

The process of bringing markers from the biology laboratory to application in human populations requires new insight at four frontiers, that is, spheres of research that overlap between traditional disciplines (see Table 17.2): (1) biomarker identification and selection, (2) biomarker validation against disease end points, (3) biomarker quantitation, and (4) biomarker application to specimens from representative populations. Expertise across these frontiers seldom is possessed by individual investigators, yet expertise often is required, since it is unwise to assume that solutions appropriate for the application of one biomarker simply can be transferred to another. Clarification of the issues at these frontiers helps underscore the need for collaborative interaction within a new research paradigm. This chapter surveys and evaluates salient points in this process.

TABLE 17.1   Potential Targets in Bronchial Fluids and/or Sputum for Early Lung Cancer Detection

Differentiation markers (e.g., glycolipid expression)
Specific tumor products (e.g., mucins, matrix proteins, surfactant)
DNA ploidy
Polyamines
Nucleosides
Growth factors
Oncogenes or oncogene products
Cytogenetic changes
Specific chromosomal deletions or rearrangements
DNA repair enzymes

Source: Adapted from Mulshine *et al.* (1989).

TABLE 17.2   **Paradigm for Application of Lung Cancer Biomarkers**

A. Biomarker identification and selection
   1. End point definition
      a. Presence of cancer
      b. Susceptibility to cancer
      c. Exposure to carcinogens
   2. Type of specimen
      a. Sample of airway (compartment at risk)
      b. Cells/tissue premalignant
   3. Discrimination from background
B. Validation against acknowledged disease end points
   1. Specimen collected in advance of clinical cancer
   2. Marker presence or absence linked with histologically confirmed disease
C. Quantitative criteria for marker presence or absence
   1. Objective
   2. Reproducible
D. Biomarker confirmation in population specimens
   1. Banked specimens
   2. Populations at different levels of risk

Biomarker Identification and Selection

Biomarker identification is the province of the tumor biologist who, in exploring the role of a potential marker for early detection, is confronted with three fundamental issues: (1) to define clearly the end point for which the putative index is a marker, (2) to identify the type of clinical specimen from which the marker can be measured, and (3) to establish and expected (i.e., normal or background) range of variability from which abnormals may be distinguished.

End Point Definition

Carcinongenesis is believed to progress from initiation, through promotion, to progression following a multistage model that has potential genetic or phenotypic markers at each stage (see Table 17.3). Tumor initiation marks the genetic change from a normal to a transformed cell, a cell bearing an altered genome. Oncogene activation and inactivation of tumor suppressor genes may emerge as markers of this earliest stage of carcinogenesis (although these events also may occur later at the stage of malignant transformation known as tumor progression) (Harris *et al.*, 1990). Activation of six families of oncogenes—*ras, raf, jun, fur, neu,* and *myc*—has been associated with human lung cancer (Harris *et al.*, 1990). The sequence and importance of the activation of individual oncogenes, however, remains poorly understood. Transfection of activated Ha-*ras* oncogenes into human small cell lung cancer (SCC) cells expressing activated c-*myc* has been observed to alter

TABLE 17.3   **Early Markers of Existing Cancer**

A. Markers of initiation or transformation
  1. Oncogene activation (*ras, raf, jun, fur, neu, myc*)
  2. Suppressor gene deletion [heterozygosity at 3p, 11p, 13q14 (rb locus), 17p]
  3. Tumor-associated antigen expression
B. Markers of promotion
  1. Autocrine growth factor expression
      EGF (epidermal/transforming-α)
      FGF (fibroblast)
      GRP (bombesin)
      IGF (insulin-like)
      PDGF (platelet derived)
      TGF (transforming-β)
  2. Transcription factor expression (*fos, jun*)
  3. Tumor differentiation antigen expression (SSEA-I)
C. Markers of progression or malignant invasion
  1. Blood-borne markers (CEA, TPA, CA-125)

the phenotype (from SCC to non-small cell, NSCC) and alter the growth rate and retinoic acid-receptor response of model lung cancer systems (Mabry *et al.*, 1991). Ki-*ras* expression frequently has been detected in bronchial adenocarcinoma tumor tissue (Pulciani *et al.*, 1982; Rodenhuis *et al.*, 1987) and associated with shortened survival for both early stage (Slebos *et al.*, 1990) and advanced disease (Mitsudomi *et al.*, 1991). Elevation of c-*myc* expression has been associated with growth deregulation and loss of terminal differentiation in squamous cell and small cell tumors (Birrer *et al.*, 1989). Heterogeneity for c-*myc* amplification and rearrangement of *myc* family oncogenes suggests that activation of these oncogenes may occur later in carcinogenesis, during tumor progression (Yokota *et al.*, 1988). To date, however, no investigator has described the expression of activated oncogenes in premalignant specimens.

Chromosomal deletions may lead to loss of transcription factors for suppressor genes. Probing for allelic deletion of specific chromosomal regions by restriction fragment length polymorphisms (RFLPs), Yokota and colleagues (1987) have found frequent loss of heterozygosity (expression of only a single allele) in human small cell lung tumors on chromosomes 3p (100%), 13q (91%), and 17p (100%). The tumors of 83% of adenocarcinoma patients also displayed loss of heterozygosity on chromosome 3p. Although the karyotypes in non-small cell lung cancer are very complex, Miura and colleagues found recurrent loss of 17p, 3p, and 11p (in 67%, 57%, and 48% of cases, respectively), suggesting "hot spots" for genetic alteration in lung cancer (Miura *et al.*, 1990). Breakpoints indicating other candidate recessive oncogenes are found on 1q, 3q, 5p, 7p, 16q24, and 21p (Whang-Peng *et al.,* 1991). Thus, specific chromosomal deletions may become markers of human lung cancer. However, as with oncogene activation, evidence of suppressor gene deletions in premalignant specimens is lacking.

The shift in the balance from cell (terminal) differentiation to growth marks the selective clonal expansion characteristic of tumor promotion. Oncogenes may activate other key genes to encode growth deregulation, thus altering the balance between cell growth and differentiation (Harris *et al.,* 1990). The mammalian cell response to peptide growth factors includes expression of two critical "early response" transcription factors, *fos* and *jun* (Schutte *et al.,* 1989; Beardsley, 1991). The neoplastic cellular response to growth factors and expression of growth factors by neoplastic cells is now receiving a great deal of attention. For example, a peptide growth factor, bombesin, released by pulmonary neuroendocrine cells has been shown to induce growth and maturation of human fetal lung in organ culture (Sunday *et al.,* 1990). A functional membrane-associated bombesin receptor has been isolated from human small cell lung carcinoma (NCI-H345) cells (Kane *et al.,* 1991); bombesin-like peptides have been found in the bronchial lavage fluid of asymptomatic cigarette smokers (Aguayo *et al.,* 1989). Thus markers of growth factor expression, insofar as they reflect oncogene activation, also may hold promise for the detection of early (premalignant) lung cancer, perhaps at the stage of promotion. However, growth factor expression, also observed after exposure to inflammatory stimuli, cannot be considered specific for premalignant lung cancer. Techniques to discriminate cells that encode growth factor because of inflammation from cells with neoplastic growth factor expression have not yet been described.

Gene transcription and signal transduction lead to a proliferation of peptide products expressed in the cytoplasm and on the surface of transformed cells. By virtue of more frequent binding sites, these products of activated oncogenes will be far easier to detect on individual cells than would specific allelic polymorphisms. In fact, expression of tumor-associated antigens may be the only detectable signal if cancer is to be recognized before frank tumor tissue becomes available through clinical detection and biopsy.

Immunization of mice and rats against intact cultured tumor cells, followed by fusion of B cells with a non-immunoglobulin-producing murine myeloma cell line, has led to the development of antibody-producing hybridomas, from which monoclonal antibodies can be selected by their preferential reactivity with tumor cells over normal cells (Cottitta *et al.,* 1981; Mulshine *et al.,* 1983). The two murine monoclonal antibodies selected for our analysis exhibited the best reactivity against tumor-associated antigens of SCC and NSCC cultured cell lines and clinical specimens (Tockman *et al.,* 1988).

Although no single tumor-specific chemical structure is responsible for the specificity of tumor-associated antigens (Hakomori, 1989), Hakomori

(1986) and others (Fukushi *et al.*, 1989) have shown that many of these differentiation markers are defined by carbohydrate antigens. For example, the stage-specific embryonal antigen I (SSEA-I) cited by Hakomori (1989) is expressed during fetal development, but uncommonly by mature tissues. These tumor-associated antigens are absent in progenitor cells and show limited expression in other normal tissues, but are again highly expressed at the tumor cell surface. Recognition of this high antigen density or a specific conformation (epitope) induced by the high density is believed to underlie monoclonal antibody tumor specificity (Hakomori, 1989). Neoplastic transformation of the pulmonary epithelial cells might lead to the expression of these epitopes. Activated oncogenes encoding transcription factors may permit the overexpression of fetal antigens such as the carbohydrate SSEA-I. Alternatively, SSEA-I epitope expression might be the result of unspecified posttranslational enzyme modification (Hakomori, 1986). Posttranslational phosphorylation, for example, has been suggested as a reversible mechanism to modify the activity of cellular proteins (Dosaka-Akita *et al.*, 1991). Thus, several possible mechanisms would allow tumor-associated antigens to be expressed as a function of tumor growth and development. The contrasting expression of lacto-type II antigens in normal and malignant epithelial tissues has been mapped (Combs *et al.*, 1984; Hakomori, 1986). Complexity (numbers of fucosyl repeats) of glycosphingolipid antigen expression has been correlated with nuclear differentiation and survival (Fukushi *et al.*, 1989). If cancer is considered a retroversion of ontogeny, then such differentiation markers may provide signposts along the early steps of carcinogenesis. Changed marker expression might be expected with the advancement of carcinogenesis. Differentiation markers also could provide useful prognostic clues of tumor growth and response to therapy.

We have reported the successful recognition of tumor-associated antigens from sputum samples collected prior to lung cancer detection, extending the directions suggested by Saccomanno (Saccomanno *et al.*, 1974) and Hakomori (1986,1989). Tockman, Gupta, and co-workers (Tockman *et al.*, 1988) used two monoclonal antibodies to bind an immunocytochemical label (biotinylated diaminobenzidine, DAB) to sputum specimens collected during the NCI-sponsored Johns Hopkins Lung Project (JHLP) early lung cancer detection trial. During the JHLP, approximately half (5,226) of the 10,386 community-dwelling high-risk individuals (males, age > 45 years, currently smokers of $\geq$ 1 pack/day) randomly were allocated to receive cytologic and radiographic (dual) screening. Moderate atypia was found in one or more specimens of 626 (12%) of these individuals. The first atypical, and all subsequent, specimen of these individuals was preserved in Saccomanno's preservative (50% alcohol and 2% Carbowax 1540) at room temperature. The first morphologic atypical specimens of these 626 JHLP participants were divided into four groups (Table 17.4). Two of the groups consisted of specimens that demonstrated only moderate atypia: most of these partici-

TABLE 17.4   Allocation of JHLP Participants with Stored Sputum by Atypia Grade and Cell Type of Eventual Lung Cancer

Atypia severity/cancer development	N	%
	626	100
Group I atypia < marked (× 2)		
No lung cancer	537	86
Group II atypia < marked (× 2)	40	6.4
Squamous	12	
Small-cell	9	
Adeno-	7	
Large-cell	8	
Other, mixed	4	
Group III atypia > marked (× 2)		
No lung cancer	3	0.5
Group IV atypia > marked (× 2)	46	7.4
Squamous	41	
Adeno-	3	
Large cell	2	

Source: Adapted from Tockman *et al.* (1988).

pants (*n* = 537, 86%) never developed lung cancer (group I). It is important to observe that of the 40 individuals (6.4%) who did progress to lung cancer (group II), all four major lung cancer cell types eventually arose. Groups III and IV consisted of specimens with marked atypia on at least two occasions. Three individuals (group III) never developed lung cancer. The majority of those with marked atypia (group IV) progressed to cancer; all were NSCC, primarily of the epidermoid cell type. These observations suggest that cells exfoliated at the stage of moderate atypia may be a morphologic correlate of a neoplastic stem cell capable of differentiation into all four major cell types of lung cancer.

To determine whether they expressed tumor-associated antigens, the earliest preserved sputum specimens collected during the JHLP from participants with at least moderate atypia were incubated with murine monoclonal antibodies (624H12 and 703D4, which bind to a glycolipid SSEA-I antigen of SCC (a fucosylated ceramide pentasaccharide, lacto-N-fucopentaose III) (Huang *et al.*, 1983; Kyogashima *et al.*, 1989) and a 31-kDa peptide antigen of NSCC (Spitalnik *et al.*, 1986), respectively. Immunostaining was applied using a double-bridge immunoperoxidase technique with DAB chromogen (Gupta *et al.*, 1985). Specimens from individuals who ultimately developed lung cancer stained with a sensitivity of 91% 2 years (on average) before the earliest appearance of neoplasia, whether determined by routine sputum cytology, radiography, or clinical criteria. Specificity was 88% among specimens from individuals who remained free of lung cancer (see Table 17.5).

TABLE 17.5 Result of Double-Bridge Immunoperoxidase Staining[a] of MAb
Surface Markers Applied to First Atypical Sputum Specimen[b]

	Lung cancer	No lung cancer	Total
Satisfactory			
Stain +	20	5	25
Stain −	2	35	37
Subtotal	22	40	62
Unsatisfactory	4	3	7
Total	26	43	69

Source: Adapted from Tockman *et al.* (1988).
[a] Sensitivity, 91%; specificity, 88%; odds ratio, 70–95%; CI, 10.46–297.8; $p < 1 \times 10^{-6}$.
[b] Or atypical specimen approximately 2 yr in advance of clinical Ca.

The emphasis in this section on markers of the early stages of carcinogenesis was intentional. Evidences of a transformed genome by expression of tumor-associated antigens, oncofetal growth factors, or specific chromosomal deletions has clear biologic plausibility as markers of existing preclinical lung cancer. If indices of gene transformation can be validated as markers of existing early lung cancer and the therapeutic roles of micronutrient (Hong *et al.*, 1990) or photodynamic endobronchial laser (Edell and Cortese, 1987) can be confirmed, there will be grounds to revisit the issue of lung cancer mass screening.

In contrast, susceptibility markers act as markers of existing carcinogenesis but as effect modifiers (Hulka and Wilcosky, 1988), by which subgroups of otherwise similar individuals show an enhanced probability of disease (see Table 17.6). Most genetic abnormalities associated with human lung cancer are acquired somatic mutations, yet several lines of evidence are suggestive of an inborn predisposition to lung cancer (Ninna *et al.*, 1989). First-degree relatives of lung cancer patients are found to have a 2.4-fold risk of developing lung cancer, independent of personal cigarette smoking habits (Ooi *et al.*, 1986). Inheritance of a recessive tumor suppressor gene heterozygosity at 11p has been associated with NSCC (Heighway *et al.*, 1986).

Caporaso and colleagues compared the abilities of 245 lung cancer patients and 234 controls (with chronic obstructive pulmonary diseases,

TABLE 17.6 Markers of Lung Cancer Susceptibility

Possibly inherited chromosomal abnormalities
Metabolic oxidation phenotypes (including debrisoquine, p450 isotypes)
DNA repair

COPD) to metabolize debrisoquine, an antihypertensive drug (Caporaso *et al.*, 1989). These investigators found that, after similar doses of the drug, extensive metabolizers excrete up to several hundred times more of the urinary metabolite 4-hydroxydebrisoquine than do poor metabolizers. Debrisoquine phenotypes are defined by the metabolic ratio (MR) of debrisoquine to its chief metabolite recovered in an aliquot of an 8-hr urine sample. Deficient metabolism is inherited as an autosomal recessive. This study demonstrated Hardy–Weinberg equilibrium conditions in the controls, whereas the lung cancer case group contained an excess of extensive metabolizers. Patients who metabolize the drug extensively were at about 8 times greater risk of getting lung cancer than were patients who metabolize the drug poorly or moderately. The key metabolic step is an oxidation performed by P4502D6 cytochrome. Since 90% of the United States population are estimated to be extensive or moderate metabolizers of debrisoquine, an advantage would accrue to the 10% of the United States population that metabolizes debrisoquine poorly and thus would be at lower risk of lung cancer.

An assay for measuring nucleotide excision repair of UV- and chemically induced DNA damage has been reported by Athas *et al.* (1991). These investigators transfected damaged, nonreplicating, recombinant plasmid DNA harboring a chloramphenicol acetyltransferase (*cat*) reporter gene into peripheral blood lymphocytes. Excision repair of the damaged bacterial *cat* gene was monitored as a function of time. Although this assay has been validated by discriminating the reduced DNA repair ability of xeroderma pigmentosum patients from that of controls, there is clear potential to explore DNA repair as a component of lung cancer susceptibility.

The last major group of lung cancer biomarkers consists of the nonspecific markers of gene damage, for example DNA adducts (Hemminki *et al.*, 1988), micronuclei (Lippman *et al.*, 1990a; Stich 1986), and sister chromatid exchanges (Liou *et al.*, 1989). In the absence of evidence of gene transformation, this group might be considered markers of exposure to known or suspected carcinogens (see Table 17.7). These molecular dosimeters might be regarded as indicators of a biologically effective exposure. These markers differ conceptually from the markers discussed earlier because their validation comes not from correlation with the development of cancer (which may take

---

TABLE 17.7 **Biomarkers of Carcinogen Exposure**

DNA adduct detection
   [32]P postlabeling assay
   ELISA assay
Sister-chromatid exchanges
Micronuclei
Biomarkers of smoking (including nicotine adducts/metabolites)

10–15 years) but from correlation with environmental and tissue or fluid levels of the carcinogen.

However, the interpretation of studies that demonstrate elevation of exposure markers in patients with lung cancer may be confusing. The biologic change associated with most exposure markers is of short duration (weeks to months). Current elevation of an exposure marker may or may not be relevant over the latent period of lung cancer development. The distinction between chronic low level and acute high level exposures is not addressed, nor is the contribution of genome repair. Nevertheless, with further refinement, exposure markers may make a substantial contribution to the recognition of individuals at risk of lung cancer because of elevated carcinogen exposure.

Covalent DNA addition products (adducts) are formed when chemical carcinogens react with DNA. Randerath *et al.* use a sensitive ^{32}P-postlabeling assay to determine that aromatic DNA adducts occur in a dose- and time-dependent manner in the lung, bronchus, and larynx of smokers with cancer of these organs (Randerath *et al.*, 1989). Enzyme-linked immunosorbent assays (ELISA) have been used by Poirier *et al.* to measure adducts in the femtomole range (approximately 1 adduct in 10^7 normal nucleotides) (Poirier *et al.*, 1990). Unfortunately, the high degree of sensitivity is accompanied by high variability and high cross-reactivity. Nevertheless, the greater sensitivity of this technique may yet prove useful in providing incontrovertible evidence of human exposure, compared, for example, to sister chromatid exchange (Liou *et al.*, 1989). In this latter study, Liou *et al.* found that a specific indicator of exposure, benzo[*a*]pyrene-diol-epoxide–DNA (adduct) antigenicity, was more informative than a general indicator of chromosomal injury, sister chromatid exchange, in the monitoring of mutagenic exposure among fire fighters.

Micronuclei have been advocated as an intermediate phase (end point) in carcinogenesis (Lippman *et al.*, 1990a). Micronucleated cells are easily obtained and quantified biologic markers thought to reflect chromosomal or genetic damage from a carcinogenic insult such as smoking or irradiation. They are small hyperchromatic bodies in the cytoplasm that stain with features of the main nucleus. Chronic exposure to genotoxic agents, resulting in elevated micronuclei frequencies, may reflect both the level of carcinogenic exposure and the inherent epithelial responsiveness to such carcinogenic or mutational insults. Stich (1986) has documented the rapid disappearance of micronucleated cells from buccal mucosa of irradiated patients following the cessation of radiotherapy. Lippman's finding that bronchial micronuclei counts in prior (ex-)smokers were not significantly different from never-smokers, confirms and extends these results, suggesting that micronuclei expression reflects only active or ongoing genotoxic exposure or susceptibility to pyrolytic carcinogens in tobacco, rather than evidence of cellular transformation (Lippman *et al.*, 1990a).

Arylamines such as 4-aminobiphenyl (ABP) found in cigarette smoke have been recognized as human and animal bladder carcinogens. Bartsch *et al.* (1990) observed that levels of ABP–hemoglobin adduct were approximately five times higher in smokers of black tobacco and three times higher in smokers of blond tobacco than in nonsmokers, and that the magnitude of increased risk of bladder cancer among smokers of the two tobacco types, compared with that among nonsmokers, was proportional to the concentrations of ABP–hemoglobin adducts. These investigators found that a combination of slow acetylator/fast oxidizer phenotype was associated with the highest level of ABP–hemoglobin adduct. ABP–hemoglobin adduct levels correlated better with urinary mutagenicity than did cotinine levels.

Marker selection depends on the hypothesis, of course, and both markers of individual early disease and markers of enhanced group risk have a role. Further, markers of disease and markers of susceptibility can be combined for population screening. This dual-phase strategy in which a simple, highly sensitive, but not necessarily specific, initial screen might be followed by a more specific confirmatory test has been suggested as most suitable for the vast populations at risk (Makuch and Muenz, 1987).

Type of Tissue Specimen

Bronchogenic carcinoma is thought to arise from the dividing cells of the bronchial epithelium (McDowell *et al.*, 1982). Focus on the airway as the site of pulmonary carcinogenesis underscores the importance of Saccomanno's insight that exfoliated epithelial cells recovered in the sputum may provide intermediate end points of developing lung cancer (Saccomanno *et al.*, 1974). Although cytomorphology-based lung cancer screening failed to reduce overall lung cancer mortality in the NCI-sponsored Early Lung Cancer (ELC) detection trial (Tockman, 1986), the original concept of early lung cancer recognition through detection of exfoliated cell markers remains intact. Pursuit of preclinical lung cancer markers from airway cells and fluids is conceptually attractive from a theoretical viewpoint; one would not expect alternative (bloodborne) markers to distinguish the earliest stages of lung cancer. First, bloodborne markers become detectable only after bronchogenic cancer crosses the basement membrane and invades the pulmonary parenchyma and vascular system, that is, becomes a systemic (not a localized) disease. Second, the volume of extracellular fluid traversed by the marker would dilute small concentrations of marker to obscurity. On empirical grounds, a variety of tumor markers in serum have been evaluated for use to detect early cancer, including carcinoembryonic antigen (CEA), tissue polypeptide antigen (TPA), neuron-specific-enolase (NSE), CA-125, and others. None of these has recognized preclinical lung cancer reliably (Gail *et al.*, 1988).

Cellular elements of peripheral blood also have been examined for lung cancer markers. Peripheral blood lymphocyte DNA has been extracted and digested with restriction endonuclease (*Msp*I) to evaluate the association of RFLPs of the P450IA1 gene with lung cancer (Kawajiri *et al.*, 1990). Although the frequency of the homozygous rare allele of P450IA1 gene was 3-fold higher among lung cancer patients than among healthy controls, no information is available about this marker in the early (preclinical) stages of carcinogenesis. Studies of genetic changes in peripheral blood lymphocytes eventually may indicate host exposure or susceptibility but are not expected to detect the presence of early stages of lung carcinogenesis.

Specimens bearing markers of early lung cancer must represent the airway (the compartment at risk) and come from transformed preinvasive cells. Multiple sputum specimens may sample the airway epithelium adequately, particularly for centrally located clinical lesions (Erozan and Frost, 1983), yet it is not known conclusively whether a focus of premalignant cells may be represented adequately in the sputum. Strong suggestive evidence that preneoplastic cell markers are detectable in the sputum was provided by Frost *et al.* (1986) who have shown that the greater the degree of atypical (preneoplastic) sputum cytomorphology, the more likely the subsequent diagnosis of lung cancer. The 12% of the dual-screened participants in the JHLP early detection trial who produced moderate (or more severe) atypia accounted for more than one-third (86/233, 37%) of the lung cancers that developed over the subsequent 5–8 years (Tockman *et al.*, 1988). Studies of the entire bronchial tree (from the surgical margin of resection to the ends of the subsegmental bronchi) from patients with early (*in situ* and microinvasive) lung cancer frequently have shown large areas of preneoplastic epithelium surrounding the neoplastic focus (Eggleston *et al.*, 1982; Frost *et al.*, 1983). Lippman and co-workers (1990b) have adopted Slaughter's concept of "field cancerization" (Slaughter *et al.*, 1953) to describe the widespread multifocal morphologic change observed in the airway epithelium assaulted by inhaled carcinogens (that is, tobacco smoke). Ease of obtaining a sufficient number of cells that express tumor-associated antigen molecules for immunoprobe detection in sputum make this marker/medium combination attractive. With future refinements, molecular evidence of specific genome transformation may come from biologic material provided from sputum or endoscopic biopsy.

Discrimination from Background

When any potential marker is applied repeatedly to the same individual or to similar individuals within a group, each marker will demonstrate an inherent unreliability and inevitable lack of constancy (Fleiss, 1986). For a biologic change to be considered a marker of disease, it must produce a recognizable departure from normal; the change produces a difference from the mean or usual value by an amount greater than is likely due to random or expected variability. Having established that a biologic index may be useful for recognizing the early stages of carcinogenesis, the interaction of the biologist with a statistician or epidemiologist is recommended.

Peto *et al.* have responded to a report that the inheritance of rare hypervariable alleles at the Ha-*ras*-1 locus is associated with a predisposition to human cancer by establishing five general guidelines (Peto *et al.*, 1988). In brief, these investigators recommend that (1) the initial analysis of association between marker frequency and cancer be based on the data of the entire study population, not on subgroup analysis, (2) bias be considered as a possible explanation for a positive association, (3) if bias can be excluded, inspection for a data-derived hypothesis to account for the variation should be done, (4) the data-derived hypothesis be tested on a fresh cohort to support the validity of the marker hypothesis before its publication, and (5) that the biologic plausibility of the data-derived hypothesis be insured.

Biomarker Validation against Acknowledged Disease End Points

After selection of a biomarker, the sensitivity and specificity of label–epitope binding in premalignant specimens must be validated to a known (histology or cytology confirmed) cancer outcome. This domain is the province of the pathologist, with some communication with the statistician or epidemiologist.

The terms "sensitivity" and "specificity" indicate how well a particular biologic change indicates a disease outcome (e.g., cancer). These indices usually are determined by applying the marker to specimens from one group of persons who have (or will develop) the disease and to specimens from another group who do not, and comparing the results. For our discussion, sensitivity refers to the proportion that stains positive among all those with or that develops cancer, whereas specificity is the proportion that stains negative among all those without or do not develop cancer (Lilienfeld and Lilienfeld, 1980). The more sensitive an indicator is for a disease, and the more specific it is for that disease only, the better it functions as a test. In contrast to "predictive value" or "false-positive rate," the sensitivity and specificity are constant for a given test, even when different groups or populations are tested (Goyar and Rogan, 1986). The accuracy of markers based on the peptide products of activated oncogenes has not yet been proved. Although easier to detect than specific genomic alterations, measurement of a product may be expected to result in more frequent false positives and false negatives.

Before testing monoclonal antibody (MA6) in patient sputa, MAbs were selected based on binding in tissue culture and histologic sections of tumor and normal lung (Tockman *et al.*, 1988). The precision of epitope localization that results from optimal MAb selection was enhanced further by modifying the avidin–biotin–peroxidase complex (ABC) immunostain by the method of Gupta *et al.* (1985). This modification entails the addition of a second biotinylated antibody–ABC reagent layer. The species-specific secondary immunoglobulin binds to the layers previously applied and increases the size of the lattice-like bridge between the antigen and the enzyme molecules that catalyze the staining reaction. This method has been shown to enhance the sensitivity of immunostaining in both Saccomanno-fixed cytologic and paraffin-embedded histologic tissue.

The essential element of the validation of an early detection marker is the ability to test the marker on clinical material obtained from subjects monitored in advance of clinical cancer and link those marker results with subsequent histologic confirmation of disease. The experience gained through the NCI early detection trial at Johns Hopkins again is instructive. That study provided an archived bank of sputum specimens with a record of the clinical course and long-term follow-up for the patients from whom the specimens were obtained (Frost *et al.*, 1984b; Stitik *et al.*, 1985; Tockman *et al.*, 1985). Clinical follow-up for an average of 8 years from specimen collection was available for all 626 individuals who showed moderate (or greater) atypia (Tockman *et al.*, 1988). Histologic slides were obtained for almost every case from biopsy or autopsy to confirm the link between the intermediate end point and a standard pathologically confirmed case of lung cancer. A reasonable (5- to 8-year) cancer-free follow-up period was required for each control. Other investigators presently are preserving a bank of endoscopic samples from individuals with high risk of lung cancer for histologic follow-up of subsequent cases and controls to validate their marker studies (Lippman *et al.*, 1990). This irrefutable link between antecedent marker and subsequent acknowledged disease is the essence of a valid intermediate end point.

Quantitative Criteria for Marker Presence or Absence

The lack of a unique chemical structure for tumor-associated antigens signifies that a qualitative presence or absence criterion of marker binding is not sufficiently specific. These circumstances require the development of rigorous quantitative criteria for positive or negative marker binding based on the number of probe adherence sites per cell and the frequency of labeled cells per specimen. We have been engaged in studies to quantify immunolabeled cell detection by characterizing the source and magnitude of the optical or electronic probe signal over all other sources of variation (noise) (Tockman, 1991). Noise may arise from technical variation in the specimen collection or preparation, from variation in the assay, from biologic variation in the host (e.g., the degree of cytologic atypia), and in the quantitation of marker uptake.

Automated cytology systems are able to augment human ability to detect and interpret biologically significant cellular and tissue changes (Bartels *et al.*, 1974; Olson *et al.*, 1980; Frost *et al.*, 1986). Automated instruments are capable of determining spectral characteristics of stained cellular proteins, DNA content, and ploidy (Tyrer *et al.*, 1980; Frost *et al.*, 1983b; Bacus and Grace, 1987). Commercially available, integrated optical microscope and computer systems are available for the pathologist to recognize morphologic and cytochemical markers for solid tumors of the bladder, breast, colon, lung, pancreas, prostate, and thyroid and for non-Hodgkin's lymphoma

(Broers *et al.*, 1988; Cohen *et al.*, 1988; Carruba *et al.*, 1989; Gavoille *et al.*, 1989; Kommoss *et al.*, 1989; Lesty *et al.*, 1989; Oud *et al.*, 1989; Norazmi *et al.*, 1990). However, recognition of biomarkers for preneoplastic lesions represents a new departure for this technology. Greenberg *et al.* have focused their initial studies on automation of traditional morphometry of atypical cells in sputum specimens, leading these investigators to the development of a "cell atypia profile" that may prove useful as a marker of carcinogenesis if validated in clinical specimens (Greenberg *et al.*, 1987).

Our preliminary analyses have focused on quantitation of three image-derived properties (spectral signature, optical texture, and morphology) of labeled and unlabeled malignant cells. Probe characterization (i.e., marker recognition studies) have been accomplished initially by performing spectral analyses on neoplastic (SCC and NSCC) cell lines, prepared using the standard Saccomanno technique for sputum cytology, followed by DAB immunostaining and methylene blue counterstaining of specimens that have been incubated with and without the primary antibody (for positive and negative controls, respectively). The morphometric (e.g., size and shape parameters) and photometric (e.g., texture and DNA content/distribution) features were analyzed using conventional univariate statistical techniques.

Spectral signature   Transmission spectra were obtained over the visible spectrum (i.e., 400 nm to 700 nm) for multiple cells per slide from each population. Log-absorbance (i.e., log optical density) spectra were computed and used to estimate the variability of the spectral characteristics of the probe within slides (i.e., intraslide variability) and between slides (i.e., interslide variability). These spectral studies were performed to determine the optimal wavelength(s) at which to collect morphometric and densitometric data. The spectra of probe-positive and probe-negative specimens were measured on the Zeiss Axiomat microscope in the Frost Center Laboratory at the Johns Hopkins School of Hygiene. The log optical density spectra of probe-negative and probe-positive cells were compared. Results show that positive and negative cells are maximally different at approximately 510 nm and 600 nm. These wavelengths maximally discriminate DAB-labeled cells from unlabeled (but counterstained) cells, given the estimated variability associated with both the control specimens and their preparation methods. A spectral-ratio parameter was tested successfully with respect to its discriminatory potential to separate control-negative cells from control-positive and sputum-derived cells. Narrow-band dual-wavelength optical scanning appears to be a powerful approach to discriminate probe from counter stain, according to this study.

Optical texture   Optical texture was determined quantitatively by measuring statistical moments of the frequency distribution of optical densities within the cytoplasm of individual cells. The frequency distributions of cytoplasmic optical densities may be useful for cell-class discrimination when the histograms are normalized for cytoplasmic area. Measurements of the variation of absorbance measurements within cells and cytoplasm show that the cells under investigation have variable optical densities (i.e., are textured). A texture parameter (i.e., short run length emphasis, SRLE) was measured for cells from each of the six classes under investigation. This single feature, when applied to the measurement of cytoplasmic texture, accurately discriminated each of the 6 control-negative cells from the total sample of 18 measured cells including an additional 6 control-positive cells and an additional 6 cells from the sputum of patients.

Morphology   Morphometric studies of cell areas and shapes were conducted by measurement of nuclear area and total cell area and by analysis of Fourier coefficients of closed linear contours. Area measurements were evaluated after normalization to μ^2 units. Shape was quantified using simple parameters such as perimeter^2/area, and by the Fourier coefficients of the closed linear contours that define the cell boundaries. Results showed that manually traced cell boundaries were reproducible compared with intercellular or internuclear area measurements, although the shape of manually traced nuclei varied more significantly from tracing to replicate tracing. Results also show that the shape of the cells from the sputum cases sampled were more irregular than those of either positive or negative control cells. Fourier analysis was tested on these data, and demonstrated a potential for discriminating irregularly shaped nuclei from those with more round shapes.

Features such as these with demonstrated discriminatory potential are compared by correlation matrix analyses. Multivariate (e.g., stepwise linear discrimination analysis) statistical techniques are employed to produce discriminant functions that combine relatively independent (i.e., orthogonal) parameters. Training sets are used to generate these discriminant functions and the functions are tested on test data sets to determine their prognostic value in differentiating atypical cells from patients who subsequently developed lung cancer from those atypical cells from patients who did not.

Biomarker Confirmation in Population Specimens

The positive predictive value of a test declines as the risk of disease in a population falls. Therefore, population application of even a valid test will be justified only after the predictive value of an early detection marker has been balanced against the population incidence of lung cancer. The development of such a strategy is the domain of the epidemiologist, who, through an ongoing dialogue with the biologist, may sequentially combine validated early detection markers to greatly enhance the accuracy of marker-based population screening.

Preliminary evidence of biomarker benefit came from marker application to banked sputum specimens for which the clinical outcome was recorded and histologically documented. We are currently engaged in a study
to determine the validity (sensitivity, specificity, predictive value) of these
monoclonal antibodies as markers of a new/continuing process of lung carcinogenesis in a population sufficiently large to provide for cell-type (subgroup) analyses. Patients who have undergone successful resection for post-
surgically staged-(Stage I) lung cancer have a 5% annual incidence of
developing a second primary lung cancer (Thomas and Piantadosi, 1987).
Over a 1-year period, approximately 900 of these patients from 11 collaborating university-affiliated oncology programs will be recruited to provide
informed consent and undergo questionnaire interview, forced expiration,
and sputum induction. After the accrual period 3 years of observation will
provide individual patient follow-up periods of 1–3 years. Diagnosis and
treatment of second primary lung cancer will follow standard clinical practice. All lung cancer diagnoses and all causes of death will be confirmed with
pathology review. The sputum specimens will be stained by routine (Papanicolaou) and immunologic methods to determine the validity of morphology
and tumor-associated antigen detection independently and together for recognition of second primary lung cancer. Specimens will be preserved in the
sputum bank for the continuing exploration of new and refined markers of
the earliest stages of carcinogenesis.

Summary

Developments in tumor biology in general, and monoclonal antibody recognition of tumor-associated antigens in particular, hold great promise for detection of the stages of carcinogenesis well in advance of clinical cancer.
However, prior to our application of emerging biomarkers to population-
based early lung cancer detection, we face a series of collaborative research
challenges. These challenges might be considered frontiers to be bridged between established biomedical disciplines, often requiring expertise beyond
the range of individual investigators. Before the successful application of
newly described markers can occur, further cross-disciplinary research must

1. refine the selection of biologically appropriate markers. Selection of
   valid markers of the transformed cell that appear in clinically accessible material is fundamental.
2. validate such markers against acknowledged disease end points. The
   use of specimen banks to validate existing and anticipated markers
   will be essential to rapid progress in the development of intermediate end points.
3. establish quantitative criteria for marker presence or absence. The
   absence of tumor-specific end points for tumor-associated antigens

requires, at least for this class of markers, that quantitative criteria be established.

4. confirm marker predictive value in prospective population trials. Marker predictive value must be demonstrated in populations trials, allowing for the anticipated effect of population disease risk on test validation.

Cystic Fibrosis and α_1-Antitrypsin

Two of the most common nonmalignant genetic diseases of Caucasians result in pulmonary damage: cystic fibrosis and α_1-antitrypsin deficiency. As the biology of these conditions unfolds in the laboratory, the possibility of population-based screening for these conditions is discussed with increasing frequency (Silverman *et al.*, 1989; Lemna *et al.*, 1990; Workshop on Population Screening, 1990; Hammond *et al.*, 1991). The research paradigm for proceeding from biology laboratory to population application discussed in detail for lung cancer may be generalized to these nonmalignant conditions.

Cystic Fibrosis

Biomarker Identification and Selection

Cystic fibrosis, characterized by chronic pulmonary and gastrointestinal dysfunction, is reported to be the most prevalent lethal hereditary disease in the white population (Hammond *et al.*, 1991). Of persons of European ancestry, 1 in 25 is a carrier, having one normal and one recessive cystic fibrosis gene, whereas 1 in 2500 persons of European ancestry is affected clinically (Workshop on Population Screening, 1990). Clinical evidence of improved prognosis with early detection, nutritional supplementation, and infection control has led to considerations of population screening. Within 15 years of the recognition of cystic fibrosis as a distinct clinical entity, the sweat test was described in 1953 to be an inexpensive specific test that allowed accurate diagnosis in symptomatic persons (Colten, 1990). However, the normal electrolyte content of sweat in heterozygotes precludes sweat test detection of asymptomatic carriers. The identification of the cystic fibrosis gene in 1989 by investigators in Toronto and at the University of Michigan led to the possibility of gene-based screening of populations (Kerem *et al.*, 1989; Rommens *et al.*, 1989), yet although a specific deletion in the codon for phenylalanine at position 508 (ΔF508) accounts for 70% of the mutations in cystic fibrosis (Colten, 1990), this carrier frequency allows identification of only about half the couples at risk (Workshop on Population Screening, 1990). Up to 20 additional, individually rare mutations also produce cystic fibrosis. Further, mutation and disease frequency vary according to racial and ethnic background. The heterogeneity of the underlying genetic defect in cystic fibrosis, therefore, makes gene screening for cystic fibrosis less likely in the near future (Workshop on Population Screening, 1990; Hammond *et al.*, 1991). A third approach to neonatal cystic fibrosis screening depends on detection of the persistent elevation of immunoreactive trypsinogen, a pancreatic enzyme precursor, in the dried bloodspots that are collected for mandated phenylketonuria screening of infants (Hammond *et al.*, 1991).

Validation against Acknowledged Disease End Points

The validity of trypsinogen measurements strongly depends on the cutoff level for repeat testing and the age-related decline in trypsinogen concentrations (Workshop on Population Screening, 1990). To address these issues, Hammond *et al.* (1991) measured immunoreactive trypsinogen in 279,399 Colorado newborns. The diagnosis of cystic fibrosis was based on the elevation of sweat chloride concentrations (>60 mmol per liter) according to Gibson–Cooke quantitative pilocarpine iontophoresis (Gibson, 1973).

Quantitative Criteria for Marker Presence or Absence

For most of their study, Hammond *et al.* (1991) set the initial cut off level at ≥ 140 µg trypsinogen per liter. At an average age of 38 days, infants with an elevated initial level were retested. A lower cut off (80 µg per liter) was used to allow for the expected decline in trypsinogen levels with age. A specimen was considered positive if the average of duplicate assay values exceeded the cut off value. If the two values differed by more than 15%, the specimen was reassayed.

Biomarker Confirmation in Population Specimens

Screening of all infants born in Colorado between April 1, 1982, and March 31, 1984, and between October 1, 1984, and September 30, 1987, revealed an incidence of cystic fibrosis among white infants of 1 in 2521, close to the expected incidence. Overall, 95.2% of the infants with cystic fibrosis (95% confidence interval, 85–99%) who did not have meconium ileus were identified by a trypsinogen cut off level of 140 µg per liter on initial testing and 80 µg per liter on repeat testing (Hammond *et al.*, 1991).

Based on these data, screening with trypsinogen, but not gene screening, seems to be a valid method for detection of asymptomatic cystic fibrosis carriers. However, the public health policy implications that derive from these results are complex. In an accompanying editorial, Holtzman points out that, for every 100,000 infants tested, 300 had false-positive results (Holtzman, 1991). Further, although 88% of all infants with cystic fibrosis (CF) in the study were detected by the trypsinogen assay, 12% of CF infants were missed. Measurement of a peptide product rather than the altered genome has resulted in a test with imperfect accuracy. Screening costs beyond those of additional evaluation are to be considered, including lingering concern for the health of falsely positive infants and the possible delay in the diagnosis of infants that test falsely negative. The context of screening costs must be defined by the net benefit that results from screening. Controlled trials are now in progress to determine whether the good clinical status of infants screened for CF might be because of early detection rather than inherently mild disease.

α_1-*Antitrypsin Deficiency*

Biomarker Identification and Selection

The major pulmonary protection against the inflammatory destruction of neutrophil elastase is α_1-antitrypsin, a 52-kDa protein encoded by the *PI* (proteinase inhibitor) locus located on the long arm of chromosome 14 at position q31–31.2 (Lai *et al.*, 1983; Kalsheker and Morgan, 1990). The major source of plasma α_1-antitrypsin is the liver, although small amounts produced by monocytes may be significant at sites of local inflammation (Kalsheker and Morgan, 1990). Although more than 75 different *PI* alleles have been identified, only a few, including the *Z, S,* and null alleles, have been associated with reduced plasma levels of α_1-antitrypsin compared with the common *M* allele (Silverman *et al.*, 1989). Increased susceptibility to emphysema occurs when the mean serum concentration falls below 35% (Wewers *et al.*, 1987), whereas *PiZ* phenotypes (*ZZ* homozygotes or *Z*null genotypes) typically have approximately 15% of normal plasma antitrypsin levels (Silverman *et al.*, 1989). One in 3000 individuals in the United Kingdom are reported to be *ZZ* homozygotes with a frequency of heterozygote carriers (*MZ*) thought to be similar to that of individuals heterozygous for cystic fibrosis (Kalsheker and Morgan, 1990). However, there may be significant method-associated bias in estimating *PiZ* prevalence from allele frequency data (Silverman *et al.*, 1989).

Replacement infusion therapy with α_1-antitrypsin derived from plasma has raised *PiZ* patient plasma levels successfully without producing immunologic side effects (Weweb *et al.*, 1987). Although the long-term benefit of replacement therapy in slowing the progression of emphysema is currently the subject of clinical trials, methods of screening are being considered. The advent of the polymerase chain reaction and sequencing of the amplified products has allowed characterization of many of the α_1-antitrypsin gene mutations. Synthetic oligonucleotide libraries could be used to identify DNA that contains a single point mutation (Kalsheker and Morgan, 1990). Although this technique has been used clinically, screening of asymptomatic populations for α_1-antitrypsin variants using DNA technology has not yet been reported. An automated immunoassay has been used to detect reduced α_1-antitrypsin levels to plasma samples from 20,000 blood donors from the St. Louis area (Silverman *et al.*, 1989).

Validation against Acknowledged Disease End Points

The St. Louis survey used the traditional method of isoelectric focusing in polyacrylamide gels to determine the *PI* type of the plasma samples (Silverman *et al.*, 1989). However, although screening for lowered α_1-antitrypsin levels may be effective in detecting individuals with the *PiZ* phenotype, detection of the *PI* phenotype may not be the relevant clinical end point. Only about 2% of the cases of chronic airflow obstruction demonstrate deficient levels of α_1-antitrypsin (Mittman *et al.*, 1973). Further, surveys in the United Kingdom have shown that asymptomatic *PiZ* individuals, identified through family studies as relatives of symptomatic cases, remained free of pulmonary disease (Tobin *et al.*, 1983). Finally, using Southern blot and hybridization techniques, the *Taq*I polymorphism has been found in approximately 20% of unrelated patients with chronic airflow obstruction and occurs in the absence of plasma α_1-antitrypsin deficiency (Kalsheker and Morgan, 1990). The pathophysiology that underlies the association of the *Taq*I polymorphism with a 13-fold risk for developing chronic airflow obstruction has not yet been described.

Summary

Rapid progress in identification biomarkers of preclinical disease is not limited to tumor biology. Similar cross-disciplinary frontiers of (1) biomarker identification and selection, (2) biomarker validation against disease end points, (3) biomarker quantitation, and (4) biomarker application to specimens from representative populations are encountered when proceeding from biology laboratory to population application of markers for nonmalignant pulmonary disease. The approach described in this chapter may clarify the types of research questions likely to be encountered in the application of biomarkers to the early detection of lung disease.

Acknowledgments

This research was supported in part by a Collaborative Research and Development Agreement between the National Cancer Institute, the Johns Hopkins Medical Institutions, the University of Pennsylvania, and Abbott Laboratories.

References

Aguayo, S. M., Kane, M. A., King, T. E., Jr., Schwarz, M. I., Grauer, L., and Miller, Y. E. (1989). Increased levels of bombesin-like peptides in the lower respiratory tract of asymptomatic cigarette smokers. *J. Clin. Invest.* **84,** 1105–1113.
Athas, W. F., Hedayati, M. A., Matanoski, G. M., Farmer, E. R., and Grossman, L. (1991).

Development and field-test validation of an assay for DNA repair in circulating human lymphocytes. *Cancer Res.* 51, 5786–5793.

Bacus, J. W., and Grace, L. J. (1987). Optical microscope system for standardized cell measurements and analyses. *Appl. Optics* 26, 3280–3293.

Bartels, P. H., Bahr, G. F., Jeter, W. S., Olson, G. B., Taylor, J., and Wied, G. L. (1974). Evaluation of correlational information in digitized cell images. *J. Histochem. Cytochem.* 22, 69–79.

Bartsch, H., Caporaso, N., Coda, M., Kadlubar, F., Malaveille, C., Skipper, P., Talaska, G., Tannenbaum, S. R., and Vineis, P. (1990). Carcinogen hemoglobin adducts, urinary mutagenicity, and metabolic Phenotype in active and passive cigarette smokers. *J. Natl. Cancer. Inst.* 82, 1826–1831.

Beardsley, T. (1991). Smart genes. *Sci. Am.* 265, 87–95.

Birrer, M. J., Raveh, L., Dosaka, H., and Segal, S. (1989). A transfected L-*myc* gene can substitute for c-*myc* in blocking murine erythroleukemia differentiation. *Mol. Cell Biol.* 9, 2734–2737.

Broers, J. L. V., Pahlplatz, M. M. M., Katzko, M. W., Oud, P. S., Ramaekers, F. C. S., Carney, D. N., and Vooijs, G. P. (1988). Quantitative description of classic and variant small cell lung cancer cell lines by nuclear image cytometry. *Cytometry* 9, 426–431.

Caporaso, N., Pickle, L. W., Bale, S., Ayesh, R., Hetzel, M., and Idle, J. (1989). *Genet. Epidemiol.* 6, 517–524.

Carruba, G., Pavone, C., Pavone-Macaluso, M., Mesiti, M., d'Aquino, A., Vita, G., Sica, G., and Castagnetta, L. (1989). Morphometry of *in vitro* systems. An image analysis of two human prostate cancer cell lines (PC3 and DU-145). *Pathol. Res. Pract.* 185, 704–708.

Cohen, O., Brugal, G., Seigneurin, D., and Demongeot, J. (1988). Image cytometry of estrogen receptors in breast carcinomas. *Cytometry* 9, 579–587.

Combs, S. G., Marder, R. J., Minna, J. D., Mulshine, J. L., Polovina, M. R., and Rosen, S. T. (1984). Immunohistochemical localization of the immunodominant differentiation antigen lacto-N-fucopentaose III in normal adult and fetal tissues. *J. Histochem. Cytochem.* 32, 982–984.

Cuttitta, F., Rosen, S., Gazdar, A. F., and Minna, J. D. (1981). Monoclonal antibodies that demonstrate specificity for several types of human lung cancer. *Proc. Natl. Acad. Sci. U.S.A.* 78, 4591–4595.

Dosaka-Akita, H., Rosenberg, R. K., Minna, J. D., and Birrer, M. J. (1991). A complex pattern of translational initiation and phosphorylation in L-*myc* proteins. *Oncogene* 6, 371–378.

Edell, E. S., and Cortese, D. A. (1987). Bronchoscopic phototherapy with hematoporphyrin derivative for treatment of localized bronchogenic carcinoma: A 5-year experience. *Mayo Clin. Proc.* 62, 8–14.

Eggleston, J. C., Tockman, M. S., Baker, R. R., Erozan, Y. S., Marsh, B. R., Ball, W. C., Jr., and Frost, J. K. (1982). *In situ* and microinvasive squamous cell carcinoma of the lung. *Clin. Oncol.* 1, 499–512.

Erozan, Y. S., and Frost, J. K. (1983). Cytopathologic diagnosis of lung cancer. *In* "Lung Cancer, Clinical Diagnosis, and Treatment" (M. J. Straus, ed.) 2d Ed., pp. 113–125. Grune & Stratton, New York.

Fleiss, J. L. (1986). Statistical factors in the early detection of health effects. *In* "New and Sensitive Indicators of Health Impacts of Environmental Agents" (D. W. Underhill and E. P. Radford, eds.), pp. 9–16. University of Pittsburgh, Pittsburgh, Pennsylvania.

Frost, J. K., Erozan, Y. S., Gupta, P. K., and Carter D. (1983a) Cytopathology. *In* "Atlas of Early Lung Cancer," pp. 39–70. Igaku-Shoin Medical Publishers, New York.

Frost, J. K., Pressman, N. J., Gill, G. W., Showers, R. L., Bitman, W. R., Frost, J. K. Jr., Albright, C. D., Erozan, Y. S., and Gupta, P. K. (1983b) Quantitative nuclear analysis of developing squamous cell carcinoma of the human lung. *Anal. Quant. Cytol.* 5, 207.

Frost, J. K., Ball, W. C., Jr., Levin, M. L., Tockman, M. S., Baker, R. R., Carter, D., Eggleston,

J. C., Erozan, Y. S., Gupta, P. K., Khouri, N. F., Marsh, B. R., and Stitik, F. P. *et al.* (1984a). "Early lung cancer detection: Results of the initial (prevalence) radiologic and cytologic screening in the Johns Hopkins study. *Am. Rev. Resp. Dis.* **130**, 549–554.

Frost, J. K., Ball, W. C., Jr., Levin, M. L., and Tockman, M. S. (1984b). "Final Report: Lung Cancer Control, Detection and Therapy, Phase II." NCI Contract N01-CN-45037. National Cancer Institute, Washington, D.C.

Frost, J. K., Ball, W. C., Jr., Levin, M. L., Tockman, M. S., Erozan, Y. S., Gupta, P. K., Eggleston, J. C., Pressman, N. J., Donithan, M. P., and Kimball, A. W. (1986). Sputum cytopathology: Use and potential in monitoring the workplace environment by screening for biological effects of exposure. *J. Occup. Med.* **28**, 692–703.

Fukushi, Y., Ohtani, H., and Orikasa, S. (1989). Expression of lacto series type 2 antigens in human renal cell carcinoma and its clinical significance. *J. Natl. Cancer Inst.* **81**, 352–358.

Gail, M. H., Muenz, L., McIntire, K. R., Radovich, B., Braunstein, G., Brown, P. R., Deftos, L., Dnistrian, A., Dunsmore, M., Elashoff, R., Geller, N., Go, V. L. W., Hirji, K., Klauber, M. R., Pee, D., Petroni, G., Schwartz, M., Wolfsen, A. R. (1988). Multiple markers for lung cancer diagnosis: Validation of models for localized lung cancer. *J. Natl. Cancer Inst.* **80**, 97–101.

Gavoille, A., Kahn, E., Bosq, J., and Malki, M. B. (1989). A user-oriented software for cytological image analysis application to automatic DNA content measurement of thyroid cells. *Pathol. Res. Pract.* **185**, 821–824.

Gibson, L. E. (1973). The decline of the sweat test: Comments on pitfalls and reliability. *Clin. Pediatr.* **12**, 450–453.

Goyer, R. A., and Rogan, W. J. (1986). When is biologic change an indicator of disease? *In* "New and Sensitive Indicators of Health Impacts of Environmental Agents." (D. W. Underhill and E. P. Radford, eds.), pp. 17–25. University of Pittsburgh, Pittsburgh, Pennsylvania.

Greenberg, S. D., Spjut, H. J., Estrada, R. G., Hunter, N. R., and Grenia, C. (1987). Morphometric markers for the evaluation of preneoplastic lesions in the lung. Diagnostic evaluation by high-resolution image analysis of atypical cells in sputum specimens. *Anal. Quant. Cytol. Histol.* **9**, 49–54.

Gupta, P. K., Myers, J. D., Baylin, S. B., Mulshine, J. L., Cuttitta, F., and Gazdar, A. F. (1985). Improved antigen detection in ethanol-fixed cytologic specimens. A modified avidin-biotin-peroxidase complex (ABC) method. *Diagn. Cytopathol.* **1**, 133–136.

Hakomori, S. (1986). Glycosphingolipids. *Sci. Am.* **54**, 44–53.

Hakomori, S. (1989). Biochemical basis and clinical application of tumor-associated carbohydrate antigens: Current trends and future perspectives. *Jpn. J. Cancer Chemother.* **16**, 715–731.

Hammond, K. B., Abman, S. H., Sokol, R. J., and Accurso, F. J. (1991). Efficacy of statewide neonatal screening for cystic fibrosis by assay of trypsinogen concentrations. *N. Engl. J. Med.* **325**, 769–774.

Harris, C. C., Reddel, R., Modali, R., Lehman, T. A., Iman, D., McMenamin, M., Sugimura, H., Weston, A., and Pfeifer, A. (1990). Oncogenes and tumor suppressor genes involved in human lung carcinogenesis. *Basic Life Sci.* **53**, 363–379.

Heighway, J., Thatcher, N., Cerny, T., and Hasleton, P. S. (1986). Genetic predisposition to human lung cancer. *Br. J. Cancer* **53**, 453–457.

Hemminki, K., Perera, F. P., Phillips, D. H., Randerath, K., Reddy, V., and Santella, R. M. (1988). Aromatic deoxyribonucleic acid adducts in white blood cells of foundry and coke oven workers. *Scand. J. Work Environ. Hlth.* **14**, 55–56.

Holtzman, N. A. (1991). What drives neonatal screening programs? *N. Engl. J. Med.* **325**, 802–804.

Hong, W. K., Lippman, S. M., Itri, L. M., Karp, D. D., Lee, J. S., Byers, R. M., Schantz, S. P., Kramer, A. M., Lotan, R., Peters, L. J., Dimery, I. W., Brown, B. W., and Goepfert, H.

(1990). Prevention of second primary tumors with isotretinoin in squamous-cell carcinoma of the head and neck. *N. Engl. J. Med.* **323**, 795–801.

Huang, L. C., Brockhaus, M., Magnani, J. L., Cuttitta, F., Rosen, S., Minna, J. D., and Ginsburg, V. (1983). Many monoclonal antibodies with an apparent specificity for certain lung cancers are directed against a sugar sequence found in lacto-N-fucopentaose III. *Arch. Biochem. Biophys.* **220**, 318–320.

Hulka, B. S., and Wilcosky, T. (1988). Biological markers in epidemiologic research. *Arch. Env. Hlth.* **43**, 83–89.

Kalsheker, N., and Morgan, K. (1990). The α_1-antitrypsin gene and chronic lung disease. *Thorax* **45**, 759–764.

Kane, M. A., Aguayo, S. M., Portanova, L. B., Ross, S. E., Holley, M., Kelley, K., and Miller, Y. E. (1991). Isolation of the bombesin/gastrin-releasing peptide receptor from human small cell lung carcinoma NCI-H345 cells. *J. Biol. Chem.* **266**, 9486–9493.

Kawajiri, K., Nakachi, K., Imai, K., Yoshii, A., Shinoda, N., and Watanabe, J. (1990). Identification of genetically high risk individuals to lung cancer by DNA polymorphisms of the cytochrome P450IA1 gene. *FEBS Lett.* **263**, 131–133.

Kerem, B., Rommens, J. M., Buchanan, J. A., Markiewicz, D., Cox, T. K., Chakravarti, A., Buchwald, M., and Tsui, L. (1989). Identification of the cystic fibrosis gene: Genetic analysis. *Science* **245**, 1073–1080.

Kommoss, F., Bibbo, M., Colley, M., Dytch, H. E., Franklin, W. A., Holt, J. A., and Wied, G. L. (1989). Assessment of hormone receptors in breast carcinoma by immunocytochemistry and image analysis. I. Progesterone receptors. *Anal. Quant. Cytol. Histol.* **11**, 298–306.

Kyogashima, M., Mulshine, J., Linnoila, R., Jensen, S., Magnani, J. L., Nudelman, E., Hakomori, S., and Ginsburg, V. (1989). Antibody 624H12, which detects lung cancer at early stages, recognizes a sugar sequence in the glycosphingolipid difucosylneolactonorhexaosylceramide. *Arch. Biochem. Biophys.* **275**, 309–314.

Lai, E. C., Kao, F. T., Law, M. L., and Woo, S. L. C. (1983). Assignment of the alpha$_1$-antitrypsin gene and a sequence-related gene to human chromosome 14 by molecular hybridisation. *Am. J. Hum. Genet.* **35**, 385–392.

Lemna, W. K., Feldman, G. L., Kerem, B., Fernbach, S. D., Zevkovich, E. P., O'Brien, W. E., Riordan, J. R., Collins, F. S., Tsui, L. C., and Beaudet, A. L. (1990). Mutation analysis for heterozygote detection and the prenatal diagnosis of cystic fibrosis. *N.Engl.J Med.* **322**, 291–296.

Lesty, C., Raphael, M., Nonnenmacher, L., and Binet, J. L. (1989). Two statistical approaches to nuclear shape and size in a morphometric description of lymph node sections in non-Hodgkin's lymphoma. *Cytometry* **10**, 28–36.

Lilienfeld, A. M., and Lilienfeld, D. E. (1980). "Foundations of Epidemiology," 2d Ed. Oxford University Press, New York.

Liou, S. H., Jacobson-Kram, D., Poirier, M. C., Nguyen, D., Strickland, P. T., and Tockman, M. S. (1989). Biological monitoring of fire fighters: Sister chromatid exchange and polycyclic aromatic hydrocarbon-DNA adducts in peripheral blood cells. *Cancer Res.* **49**, 4929–4935.

Lippman, S. M., Peters, E. J., Wargovich, M. J., Dixon, D. O., Dekmezian, R. H., Cunningham, J. E., Loewy, J. W., Morice, R. C., and Hong, W. K. (1990a). The evaluation of micronuclei as an intermediate endpoint of bronchial carcinogenesis. *Prog. Clin. Biol. Res.* **339**, 165–177.

Lippman, S. M., Lee, J. S., Lotan, R., Hittelman, W., Wargovich, M. J., and Hong, W. K. (1990b) Biomarkers as intermediate end points in chemoprevention trials. *J. Natl. Cancer Inst.* **82**, 555–560.

Mabry, M., Nelkin, B. D., Falco, J. P., Barr, L. F., and Baylin, S. B. (1991). Transitions between lung cancer phenotypes—Implications for tumor progression. *Cancer Cells* **3**, 53–58.

McDowell, E. M., Harris, C. C., and Trump, B. F. (1982). Histogenesis and morphogenesis of bronchial neoplasms. *In* "Morphogenesis of Lung Cancer" (M. Melamed, Y. Shimosato, and P. Nettesheim, eds.) pp. 1–36. CRC Press, Boca Raton, Florida.

Makuch, R. W., and Muenz, L. R. (1987). Evaluating the adequacy of tumor markers to discriminate among distinct populations. *Semin. Oncol.* **14,** 89–101.

Minna, J. D., Pass, H., Glatstein, E., and Ihde, D. C. (1989). Cancer of the lung. In "Cancer: Principles & Practice of Oncology," (V. T. DeVita Jr, S. Hellman, and S. H. Rosenberg, eds.) 3d Ed., pp. 591–705. Lippincott, Philadelphia.

Mitsudomi, T., Steinberg, S. M., Oie, H. K., Mulshine, J. L., Phelps, R., Viallet, J., Pass, H., Minna, J. D., and Gazdar, A. F. (1991). *ras* gene mutations in non-small cell lung cancers are associated with shortened survival irrespective of treatment intent. *Cancer Res. (in press).* **51,** 4999–5002.

Mittman, C., Teevee, B., and Liberman, J. (1973). Alpha$_1$-antitrypsin deficiency as an indicator of susceptibility to pulmonary disease. *J. Occup. Med.* **15,** 33–38.

Miura, I., Siegfried, J. M., Resau, J., Keller, S. M., Zhou, J. Y., and Testa, J. R. (1990). Chromosome alterations in 21 non-small cell lung carcinomas. *Genes Chrom. Cancer* **2,** 328–338.

Mulshine, J. L., Cuttitta, F., Bibro, M., Fedorko, J., Fargion, S., Little, C., Carney, D. N., Gazdar, A. F., and Minna, J. D. (1983). Monoclonal antibodies that distinguish non-small cell from small cell lung cancer. *J. Immunol.* **131,** 497–502.

Mulshine, J. L., Tockman, M. S., and Smart, C. R. (1989). Considerations in the development of lung cancer screening tools. *J. Natl. Cancer Inst.* **81,** 900–906.

Mulshine, J. L., Linnoila, R. I., Jensen, S. M., Magnani, J. L., Tockman, M. S., Gupta, P. K., Scott, F. S., Avis, I., Quinn, K., Birrer, J. J., Treston, A. M., and Cuttitta, F. (1992) Rational targets for the early detection of lung cancer. *J. Natl. Cancer Inst. (in press).* **13,** 183–190

Norazmi, M. N., Hohmann, A. W., Skinner, J. M., Jarvis, L. R., and Bradley, J. (1990). Density and phenotype of tumour-associated mononuclear cells in colonic carcinomas determined by computer-assisted video image analysis. *Immunol.* **69,** 282–286.

Olson, G. B., Donovan, R. M., Bartels, P. H., Pressman, N. J., and Frost, J. K. (1980). Microphotometric differentiation of human T and B cells tagged with monospecific immunoadsorbent beads. *Anal. Quant. Cytol. J.* **2,** 144–152.

Ooi, W. L., Elston, R. C., Chen, V. W., Bailey-Wilson, J. E., and Rothschild, H. (1986). Increased familial risk for lung cancer. *J. Natl. Cancer Inst.* **76,** 217–222.

Oud, P. S., Pahlplatz, M. M. M., Beck, J. L. M., Wiersma-van Tilburg, A., Wagenaar, S. J., and Vooijs, G. P. (1989). Image and flow DNA cytometry of small cell carcinoma of the lung. *Cancer* **64,** 1304–1309.

Peto, T. E., Thein, S. L., and Wainscoat, J. S. (1988). Statistical methodology in the analysis of relationships between DNA polymorphisms and disease: Putative association of Ha-ras-I hypervariable alleles and cancer. *Am. J. Hum. Genet.* **42,** 615–617.

Poirier, M. C., Weston, A., Gupta-Burt, S., and Reed, E. (1990). Measurement of DNA adducts by immunoassays. *In* "DNA Damage and Repair in Human Tissues" (B. M. Sutherland and A. D. Woodhead, eds.), pp. 1–11. Plenum Press, New York.

Pulciani, S., Santos, E., Lauver, A. V., Long, L. K., Aaronson, S. A., and Barbacid, M. (1982). Oncogenes in solid human tumours. *Nature (London)* **300,** 539–542.

Randerath, E., Miller, R. H., Mittal, D., Avitts, T. A., Dunsford, H. A., and Randerath, K. (1989). *J. Natl. Cancer Inst.* **81,** 341–347.

Rodenhuis, S., van de Wetering, M. L., Mooi, W. J., Evers, S. G., van Zandwijk, N., and Bos, J. L. (1987). Mutational activation of the K-*ras* oncogene. A possible pathogenetic factor in adenocarcinoma of the lung. *N. Engl. J. Med.* **317,** 929–935.

Rommens, J. M., Iannuzzi, M. C., Kerem, B., Drumm, M. L., Melmer, G., Dean, M., Rozmahel, R., Cole, J. L., Kennedy, D., Hidaka, N., Zsiga, M., Buchwald, M., Riordan, J. R., Tsui, L., and Collins, F. S. (1989). Identification of the cystic fibrosis gene: Chromosome walking and jumping. *Science* **245,** 1059–1065.

Saccomanno, G., Archer, V. E., Auerbach, O., Saunders, R. P., and Brennan, L. M. (1974). Development of carcinoma of the lung as reflected in exfoliated cells. *Cancer* 33, 256–270.

Schutte, J., Minna, J. D., and Birrer, M. J. (1989). Deregulated expression of human c-*jun* transforms primary rat embryo cells in cooperation with an activated c-Ha-*ras* gene and transforms rat-1a cells as a single gene. *Proc. Natl. Acad. Sci. U.S.A.* 86, 2257–2261.

Silverman, E. K., Miletich, J. P., Pierce, J. A., Sherman, L. A., Endicott, S. K., Broze, G. J., Jr., and Campbell, E. J. (1989). Alpha-1-antitrypsin deficiency. High prevalence in the St. Louis area determined by direct population screening. *Am. Rev. Resp. Dis.* 140, 961–966.

Slaughter, D. P., Southwick, H. W., and Smejkal, W. (1953). Field cancerization in oral stratified squamous epithelium: Clinical implications of multicentric origin. *Cancer* 6, 963–968.

Slebos, R. J., Kibbelaar, R. E., Dalesio, O., Kooistra, A., Stam, J., Meijer, C. J., Wagenaar, S. S., Vanderschueren, R. G., van Zandwijk, N., Moot, W. J., Bos, J. L, and Rodenhuis, S. (1990). K-*ras* oncogene activation as a prognostic marker in adenocarcinoma of the lung. *N. Engl. J. Med.* 323, 561–565.

Spitalnik, S. L., Spitalnik, P. F., Dubois, C., Mulshine, J., Magnani, J. L., Cuttitta, F., Civin, C. I., Minna, J. D., and Ginsburg, V. (1986). Glycolipid antigen expression in human lung cancer. *Cancer Res.* 46, 4751–4755.

Stich, H. F. (1986). The use of micronuclei in tracing the genotoxic damage in the oral mucosa of tobacco users. *In* "Mechanisms in Tobacco Carcinogenesis" (H. Hoffman and C. Harris, eds.), pp. 99–109. Cold Spring Harbor Laboratory, Cold Spring Harbor, New York.

Stitik, F. P., Tockman, M. S., and Khoury, N. F. (1985). Chest radiology. In "Screening for Cancer" (A. B. Miller, ed.), pp. 163–199. Academic Press, San Diego.

Sunday, M. E., Hua, J., Dai, H. B., Nusrat, A., and Torday, J. S. (1990). Bombesin increases fetal lung growth and maturation *in utero* and in organ culture. *Am. J. Respir. Cell. Mol. Biol.* 3, 199–205.

Thomas, P. A., and Piantadosi, S. (1987). Postoperative T1 N0 non-small cell lung cancer. *J. Thorac. Cardiovasc. Surg.* 94, 349–354.

Tobin, M. J., Cook, P. J. L. and Hutchison, D. C. S. (1983). Alpha 1-antitrypsin deficiency: The clinical and physiological features of pulmonary emphysema in subjects homozygous for Pi Type Z. A survey by the British Thoracic Association. *Br. J. Dis. Chest* 77, 14–27.

Tockman, M. S. (1986). Survival and mortality from lung cancer in a screened population—The Johns Hopkins Study. *Chest* 89, 324S–325S.

Tockman, M. S. (1991). Development of labels of early lung cancer at the John K. Frost Center for Imaging of Cells and Molecular Markers. *Lung Cancer Res. Q.* 1, 4–6.

Tockman, M. S., Levin, M. L., Frost, J. K., Ball, W. C., Jr., Stitik, F. P., and Marsh, B. R. (1985). Screening and detection of lung cancer. *In* "Lung Cancer" (J. Aisner, ed.), pp. 25–40. Churchill Livingstone, New York.

Tockman, M. S., Gupta, P. K., Myers, J. D., Frost, J. K., Baylin, S. B., Gold, E. B., Chase, A. M., Wilkinson, P. H., and Mulshine, J. (1988). Sensitive and specific monoclonal antibody recognition of human lung cancer antigen on preserved sputum cells: A new approach to early lung cancer detection. *J. Clin. Oncol.* 6, 1685–1693.

Tockman, M. S., Gupta, P. K., Pressman, N. J., and Mulshine, J. L. (1992). Considerations in bringing a cancer biomarker to clinical application. *Cancer Res.* 52, 2711S–2718S.

18

Biologic Markers in the Genitourinary System

George P. Hemstreet III,
Robert E. Hurst,
and Nabih R. Asal

Molecular Epidemiology of the Genitourinary System

Introduction

Throughout medical history, new treatments for diseases of the genitourinary (GU) system have reflected advances in medicine, epidemiology, and, most recently, molecular biology. Collectively these advances form the basis for an exciting new era in the study of molecular epidemiology. The purpose of this chapter is to relate special clinical urologic information to factors that will influence the design of epidemiologic studies involving biomarkers in the GU tract. The pathophysiology of disease and concepts for defining useful biomarkers of susceptibility, exposure, intermediate effects, and disease will be considered. The chapter is not intended to be all-inclusive but to provide a framework for integrating current and new information emanating from basic research to assist in individual risk assessment of GU diseases.

Special Features of the Urinary Tract

Several anatomic and physiologic features of the GU system make it a target for xenobiotic substances. The most important of these features are the high rate of blood flow to the kidney (20–25% of cardiac output) and the specialized filtration system of the kidney, which may concentrate the xenobiotic substances with potentially toxic effect. The glomerulus and Bowman's capsule constitute a filtration system that salvages larger proteins and cells while permitting the passage of smaller ions and molecules. Compounds important

Molecular Epidemiology: Principles and Practices
Copyright © 1993 by Academic Press, Inc. All rights of reproduction in any form reserved.

to homeostasis are selectively reabsorbed by the tubular system and toxic wastes are eliminated in the urine. The glomerular filter trap of the kidney is susceptible to immunologic deposits and subsequent development of auto-immune diseases such as glomerulonephritis. In Goodpasture's disease, which may result from acute hydrocarbon exposure, xenobiotic substances are toxic to the lung and kidney filtration systems (Atult *et al.,* 1991).

The glomerulus filters the urine into the proximal tubule collecting system where the P_1–P_3 segments, the sites of most kidney tumors, are particularly susceptible to xenobiotics (Trump *et al.,* 1984). When damaged, the proximal tubular cells may release specific enzymes or protein antigen markers, or may be released themselves. Damage to the tubular cells also may alter the concentrating ability of the kidney and reduce absorption of smaller proteins, thereby allowing their escape into the urine. Damage to the glomerulus could permit leakage of red blood cells into the urine. Quantitation of cytologic and phenotypic biomarkers in urine cells may be a sensitive means of detecting subclinical kidney damage. For example, the quantitation of tubular cells has been used to detect early transplant rejection and kidney tubule disease. The soluble proteins (e.g., β_2-microglobulin, growth factors) and cells with their associated biomarkers are potential resources for the molecular epidemiologist performing population studies.

The urinary bladder, lined with transitional epithelium, serves as a reservoir for human liquid excreta, but this "pool" may become stagnant because of obstruction. Calculi, chronic infection, or ulceration can damage the bladder and circumvent its normal defenses against urine and xenobiotic substances. These agents tend to produce squamous cell cancer, which is less common than transitional cell carcinoma. (Chronic schistosomiasis, commonly found in populations along the Nile River, also produces squamous cell carcinoma.) Chemotherapeutic alkylating agents such as cytoxan can result in increased rates of transitional cell malignancy (Catalona, 1987).

Diagnostic and Experimental Accessibility

Medical history records numerous references to the accessibility of the urinary tract, including removal of stones from the bladder and early methods for looking directly into the bladder. In fact, the Hippocratic Oath mentions stone removal as a specialized task.

Knowledge of the clearance properties of the kidney and the early development of cystoscopic equipment to visualize the bladder facilitated confirmation of suspected pathologic conditions in the GU system. Modern ureteroscopic and nephroscopic instruments now make it is possible to visualize the ureters and renal pelvis directly as well, so areas of abnormal epithelium can be localized both visually and cytologically. Moreover, with laparoscopic equipment, it is now possible to perform aggressive lymph node dissections through small abdominal ports to determine the extent of disease.

Xenobiotic and toxic substances may enter the bladder either antegrade, through the bloodstream, or retrograde, through the urinary tract. The bladder, prostate, and kidney parenchyma are reservoirs for infectious processes, but only limited information about the concentration of xenobiotic substances in these organs is available. The mucinous layer of the bladder is composed primarily of glycosaminoglycans, which form a defense against xenobiotics and infectious agents. In contrast, the glandular nature of the prostate makes it a reservoir for infection. Xenobiotic substances also may be important in the etiology of diseases with vague symptom complexes such as interstitial cystitis and prostatitis, both of which are poorly understood.

Marker Localization in the Urinary Tract

The exfoliative properties of the urinary tract, the unique funnel trap nature of urine, and the reservoir of cells in prostatic secretions simplify biomarker testing on the GU system. When abnormal changes are identified, localization to determine the source of cells can be accomplished with Stamey's 3-glass test for sample collection (Sobel and Kaye, 1987). Upper tract localization is possible by differential catheterization of both ureters retrograde through a cystoscope. Careful placement of a catheter at the ureteral orifice permits collection of urine from both the ureter and the kidney. A catheter passed retrograde to the kidney may miss ureteral pathology, but may be introduced percutaneously through the kidney in selected circumstances to sample small lesions with biopsy forceps or a brush.

Biopsy of identified lesions in almost any organ system provides direct access to cells. Fine-needle aspiration (FNA) with a #22 gauge needle facilitates biopsy, since penetration of any organ is possible with a needle of this size, provided the patient has normal clotting function. FNA biopsy of the prostate is commonplace in Europe and is gaining increasing acceptance in North America as pathologists become skilled in the art of FNA interpretation. Repeated daily FNA sampling of transplanted kidneys is not unusual during the acute stages of transplantation or for monitoring the cytotoxic effects of the drug cyclosporin that is used to control transplant rejection (Guiraudor *et al.*, 1989). FNA sampling has not become as routine as blood sampling, but improved sampling methods and controlled fixation will enhance cellular yield and quality for biomarker analysis in molecular epidemiology studies.

Interpretation of Biomarker Results in Diseases

Identification of multiple biomarkers associated with susceptibility, exposure, intermediate effects, and disease has created an interesting challenge concerning the reporting of biomarker results. The clinical usefulness of a biomarker is related directly to the intended intervention. For instance, genetic markers of susceptibility may have only limited utility for defining disease, but may be highly useful for ameliorating risk of exposure. In contrast, a positive cellular biomarker for disease is the standard for diagnosis and is evaluated most frequently in terms of sensitivity and specificity for disease. Other markers such as those for exposure, early biologic effect, or altered structure and function represent intermediate end points in the development of disease and may or may not persist until clinical manifestation of the disease. The importance of the marker will depend on how many people with the marker actually develop the disease or on the strategy for disease intervention. The "Bethesda System" for reporting cervical or vaginal cytologic diagnoses exemplifies standardized reporting in terms of clinical relevance (National Cancer Institute Workshop, 1990). A similar system, combining DNA ploidy determinations with conventional cytology, has been developed for urinary tract diagnoses and risk stratification. Table 18.1 illustrates the clinical usefulness of the two systems. These approaches facilitate the addition of other biomarkers as their relative values become available.

Intermediate end point markers, almost by definition, have poor specificity, particularly if they are abnormal early in the disease process and not all marker-positive individuals develop disease. More important, they do not always correlate with a positive disease outcome, especially if biopsy-proven disease is the standard end point. Consequently, intermediate end point markers should be evaluated in terms of individual risk assessment. The use of multiple marker profiles is expected to improve the specificity and risk assessment potential of early markers and the specificity of late markers associated with disease.

Abnormal DNA ploidy as indicated by exfoliated urinary cells with >5C DNA is an example of an intermediate end point marker that correlates with

TABLE 18.1   Quantitative Fluorescence Image Analysis (QFIA) Risk Compared with Bethesda System

Bethesda system for Pap smears	QFIA risk group for urinary samples
Normal	V-Negative
repeat 2–3 years	follow as clinically indicated
Atypia	IV-Low risk for malignancy
course indicated by cause	follow as clinically indicated
Low-grade SIL	III-Moderate risk for low-grade CA
repeat 3–6 months	repeat 3–6 months
High-grade SIL	II-High risk for low-grade CA
colposcopy and biopsy	cystoscopy and biopsy
Carcinoma	I-High risk for high-grade CA
colposcopy and biopsy	cystoscopy and biopsy

both disease and exposure (Hemstreet *et al.*, 1988). Not all patients with urinary cells with >5C DNA content develop bladder cancer, but the marker definitely is associated with occupational exposure and cigarette smoking; a subset of these patients eventually will develop disease. Most false-positive results with this test are caused by smoking; the clastogens in cigarette smoke apparently induce failed cell divisions that lead to cells with >5C DNA.

The new challenge, which can best be addressed by a multidisciplinary scientific team, is to develop a reliable method for individual risk assessment, that is, the ability to predict different individual responses to the same exposure. Epidemiologists have identified numerous groups at risk. When etiologic agents have not been identified, mapping the human genome and insight into the pathogenesis of disease provide keys for future research. Unfortunately, because of the long latency period from exposure to GU disease, years of longitudinal monitoring are required to establish the value of biomarkers for individual risk assessment. However, clinical approaches are available to determine the markers that are most appropriate for study. The identification and choice of biomarkers can be understood by associating the marker with the tumor, tumor-associated field changes, clinical symptoms, genetic changes, and the monitoring of therapy for tumor recurrence or regression.

Strategy for Identifying Markers of Susceptibility, Exposure, Intermediate Effects, and Disease

Because most diseases develop through a sequence of events, bladder cancer is a useful prototype for understanding biomarkers in the GU system. Changes in biomarkers of malignancy may follow a precise or varied order and involve fundamental alterations, resulting in uncontrolled growth and metastasis. The effects usually include dedifferentiation, cellular proliferation, and other biochemical changes that promote invasion and metastasis (Kerbel, 1989).

Clues to biomarkers for early risk assessment in bladder cancer may be found in the primary tumor, but in undifferentiated tumors chromosomal aberration may occur as a result of genetic instability not necessarily linked to the oncogenic process. Thus, karyotyping may be of limited value in highly undifferentiated tumors. "Hot spots" for gene mutations also may be associated with early translocations or gene deletions, so the study of well-differentiated tumors or the adjacent cellular field associated with high-grade tumor may be more fruitful for identifying the significant genetic changes.

In their study of colon cancer, Vogelstein and associates (1988) introduced the concept that the study of disease in relation to the field can help identify when changes occur in the oncogenic process. Biochemical changes occur in the area adjacent to the tumor as well as in the normal-appearing epithelium; biomarkers associated with the tumor and the field probably

TABLE 18.2 Percentage Abnormal Biopsy Specimens in Bladder Cancer

Marker	Normal tissue	Distant field	Adjacent field	Tumor
Morphology	0	15	33	73
DNA ploidy (cells > 5c)	0	11	46	77
p185 (Her-2/neu)	0	25	46	63
EGFR[a]	0	36	64	79
p300 (M344 antibody)	17	33	67	87
G-actin	0	58	73	100

[a] Epidermal growth factor receptor.

are involved in the sequence of oncogenic events. An example of this approach to studying bladder cancer is shown in Table 18.2; one can appreciate the diversity of marker expression from presence of G-actin to visual confirmation.

A third strategy is identification of biomarkers that persist as abnormal in patients who are in remission following therapy. Once again, bladder cancer is an excellent study model. Of patients in remission after treatment for recurrent bladder tumors with BCG (bacillus Calmette-Guerin), a bacterial vaccine, 30–40% have persistently abnormal levels of markers of dedifferentiation such as G- and F-actin. As the elapsed time after therapy increases, markers that occur late in oncogenesis reappear, as was observed when the *neu* oncogene protein persisted in patients who had been treated for bladder cancer (Bi *et al.*, 1993).

A fourth option is the study of markers in patients with a spectrum of risk factors (e.g., overt disease, previous history, or symptoms associated with bladder cancer). After this clinical risk stratification (Table 18.3), various markers are evaluated.

Finally, we can study diseases that have a clearly inherited genetic state, for example, von Hippel–Lindau disease, in which there is an autosomal recessive gene on the short arm of chromosome 3 (Gladys *et al.*, 1990). These patients have a higher incidence of kidney cancer and may have higher numbers of deletions and translocations on the chromosomes. In family studies, adjacent chromosomal markers linked to the unknown gene may be mapped. When the gene is sequenced, specific gene probes with point mutations may be evaluated by polymerase chain reaction (PCR) followed by gene sequencing.

Markers of Susceptibility

Markers of susceptibility primarily are inherited and involve a constellation of genetic changes. They may include chromosomes directly related to protooncogenes or suppressor genes, or they could be genetically linked to

TABLE 18.3   **Stratification of F Actin with Risk**

Stratification of risk		F actin results			
Classification	Group	Normal	(%)	Abnormal	(%)[a]
QFIA cytotoxicity/DNA ploidy					
Positive or suspected/Abnormal	A	5	(10)	46	(90)
Atypical/Intermediate	B	6	(25)	18	(75)
Negative/Negative					
Hematuria/Previous history of					
bladder cancer					
Positive/Positive	C	18	(34)	34	(66)
Positive/Negative	D	23	(64)	13	(36)
Asymptomatic controls					
(also all neg)	E	38	(93)	3	(7)

[a] F actin defined by flow cytometry. Abnormal if mean F actin < MCN 95 or > 55% of cells had MCN < 100. MCN (mean channel number) is a unit of fluorescence intensity.

the genes controlling activation of the xenobiotic compound (Hanke and Krajewska, 1990). The types of markers for genetic susceptibility are listed in Table 18.4. Susceptibility is probably the single most important factor relative to initiation of xenobiotic-induced disease, as evidenced by the relatively small percentage of people in an exposed cohort that develops disease. For example, epidemiologic studies by Carter and co-workers (1991) showed the importance of genetic factors in younger patients with prostate cancer, especially those with a family history of prostate cancer. Other examples of genetic susceptibility can be found in patients with von Hippel–Lindau disease or kidney cancer.

The biomarkers of susceptibility that are easiest to identify are those for which a definite inherited pattern is established and the linkage can be correlated with a specific chromosome. These markers can be identified in tumor tissue. However, another approach studies the normal cells from cancer patients and assesses point mutations when chromosomal probes are available. The study of mutations in normal cells has been infrequent in GU disease; the best available data come from von Hippel–Lindau disease associated

TABLE 18.4   **Markers of Genetic Susceptibility**

Suppressor oncogenes
Recessive oncogenes
Metabolic pathways activating xenobiotics
Metabolic pathways inactivating xenobiotics
Binding proteins

with chromosome 3 (Glady *et al.*, 1990), Wilms' tumor (Pritchard-Jones and Fleming, 1991), and the genes associated with acetylation and bladder cancer (Hanke and Krajewska, 1990). As the human genome project evolves, the genes will be sequenced and point mutations will be delineated; when this happens, family linkage studies will not be necessary.

Markers of Intermediate Effects

Markers of intermediate effects may be identified in cells, serum, or other body fluids. Virchow recognized that the basis for disease resides in the human cell. Therefore, it is probable that markers that occur in the cell are more sensitive than serum markers because of the absence of dilutional effects.

Markers of intermediate effects may be transient or persistent. If they persist, they also may be markers of disease. If they are transient, they also may be markers of exposure. Markers that occur prior to the health effect have been termed "intermediate end point markers." Their value in epidemiologic studies will depend on their ability to predict the disease process and whether they are early or late in the sequence of the disease process.

Markers of Disease

In most disease processes, the marker for the disease is the direct identification of a defined pathogen (i.e., bacteria in the case of infection) or of discrete histopathologic features. For example, high-grade tumor cells present in the urine function as a marker for bladder cancer. Biomarkers may define disease but frequently are associated with a premalignant condition; the phenotypic markers, when identified in combination, may constitute a more powerful tool for defining disease.

Bladder Cancer

Etiology

Bladder cancer is associated with exogenous occupational exposure to xenobiotic substances. Rehn (1895), in an autopsy series, first observed an increased incidence of bladder cancer among a group of workers exposed to aromatic amines used in the dye industry. Today it is estimated that 20% of bladder cancers are caused by occupational exposure (Cole *et al.*, 1972). An extensive number of compounds is known to be associated with human bladder cancer; many others are considered to be potential bladder carcinogens based on the results of animal studies. Many workers exposed to these chemicals in the past are at risk and are now approaching the known latency period for bladder cancer development. A second etiologic agent associated with bladder cancer is smoking which, in conjunction with occupational exposure, may be contributing to the increased incidence of bladder cancer in women. The mean age for bladder cancer diagnosis is 69 years. However, as life expectancy of the population increases, mortality from bladder cancer will increase also.

The urinary bladder is lined with transitional epithelium and covered by a layer of glycosaminoglycans (GAG), and has special properties to protect it from the toxic effects of urine or xenobiotic substances. Transitional cells are the origin of 90% of bladder cancers. Transitional epithelium also lines the renal pelvis and the ureters; cells from these sites and from tumors present there are shed into the urine, from which they are easily sampled. Two other primary types of cancer—squamous cell carcinoma and adenocarcinoma—also arise in the bladder. Squamous cell carcinoma is associated with bladder stones and has a high incidence among paraplegic patients. Adenocarcinoma is associated with a urachal cyst, an anomaly formed from a remnant of the umbilical cord, and with the congenital condition exotrophy of the bladder, which results from a failure of the abdominal wall and bladder to separate (Catalona, 1987).

Bladder cancer frequently is associated with field changes, that is, other areas of cellular dedifferentiation, in the epithelium (Catalona, 1987). Evidence of field changes, which can be recognized cytologically or histologically, is a major risk factor for bladder cancer. Histopathology studies have shown clearly the field changes that occur in smokers; in one study, 30% of smokers had significant histopathologic field changes (Auerbach and Garfinkel, 1989). Other classic studies by the National Bladder Cancer Project showed the importance of histopathologic field changes. Mounting evidence suggests that an array of biomarkers also occurs in patients with bladder cancer who have a condition of risk such as smoking. A molecular biology study by Sidransky *et al.* (1992) found that many bladder tumors may be of clonal origin, presumably spreading from a single site. This is an important finding that could help delineate the significance of markers of susceptibility, if there is a common gene involved in bladder cancer.

Transitional cell carcinoma (TCC) may be either high- or low-grade disease. About 80% of bladder cancers manifest as low-grade disease; 20% of those recur and progress to a high-grade invasive form (Catalona, 1987). High-grade disease, theoretically, is detected easily on routine cytologic examination because of the overt cytologic features of the cells. However, screening programs with Papanicolaou cytology have failed to detect half of the high-grade malignancies in stages early enough for effective cancer control (Cartwright, 1986). Some investigators argued that early detection is not worthwhile because of ineffective therapy, but most urologists recognize that early operative intervention is curative in most cases. Lead-time bias, the argument that effective therapies are not available, and the presumed high cost of occupational screening programs have slowed the implementation of widespread bladder cancer screening.

Advances in both systemic chemotherapy for invasive bladder cancer and intravesical therapy for noninvasive disease are compelling reasons to further investigate the important role of biomarkers in occupational bladder cancer screening. This disease provides an excellent model for understanding molecular biology and effective vaccine therapy, and has served as a model for notification programs. The ability to monitor the bladder epithelium noninvasively establishes a rich resource for epidemiologic studies and testing of biomarkers in human disease.

Special Considerations for Biomarkers

The toxic properties of urine, the inconsistent quantity of cells shed into the urine and the small size of many tumors, as well as the lack of quality control in sample fixation and processing, have limited the value of routine cytology for bladder cancer screening. Using this traditional method, nearly 50% of the tumors, many of them high grade, are missed during screening of high-risk cohorts (Cartwright, 1986).

Urine, by nature, is degenerative and toxic to cells so, to avoid false-positive results caused by cellular degeneration, markers should be those that are considered positive when they are higher rather than lower than normal. For example, F-actin, a polymerized cytoskeletal component of the cell, is lowered in early dysplastic changes associated with malignancy, whereas G-actin, the monomeric precursor of F-actin, is elevated (Pollard and Cooper, 1986). Degeneration also decreases F-actin, so G-actin is more useful for study of exfoliated cells found in the urine than F-actin, which is applicable to bladder wash samples.

Biomarkers in Bladder Cancer

Biomarkers may be classified as genotypic or phenotypic, or by their function as markers of susceptibility, exposure, intermediate effects, or disease. The various markers may be identified by the coding DNA, the messenger RNA, or the protein gene product. Since the protein is the functional unit, its measurement may provide the most useful information with respect to function, but because it lacks stability it may not be useful for retrospective studies of tissue blocks because some antigens are unstable under routine fixation. Many markers in cancer are not qualitatively different, but may be expressed differently quantitatively. Since some tumor-associated markers may be expressed, quantitative fluorescence image analysis (QFIA) and flow cytometry are two useful techniques for monitoring.

Markers of Susceptibility

Genetic linkage studies, karyotyping, and classic epidemiologic studies are the major approaches that provide information about chromosomal changes related to markers of susceptibility. Family studies of young patients with bladder cancer with or without occupational exposure are limited. Considerable information is available about karyotyping of bladder cancer cells. Deletions and genetic instability associated with inherited genetic mutations provide clues about which chromosomes may be important in the pathogenesis of this disease. Several investigators have found that chromosomes 9,11,17, and 6 express genetic changes associated with bladder cancer (Gibas *et al.*, 1984; Atkin and Baker, 1985; Fearon *et al.*, Smeets *et al.*, 1987; Tsai *et al.*, 1990; Sidransky *et al.*, 1991). Restriction fragment length polymorphism (RFLP) analysis has helped localize the areas of genetic defect. As tumors become markedly undifferentiated, many express chromosomal rearrangements that are totally unrelated to initiation of the disease process. Identification of specific mutants before marked chromosomal instability occurs requires gene sequencing and is a major area of research focus.

Sidransky *et al.* (1991) reported an association of the *p53* suppressor oncogene with bladder cancer. The marker is expressed primarily in high-grade tumors, indicating a late event in the carcinogenic process associated with marked ploidy changes. Work by Tsai and co-workers (1990) indicated a genetic defect on chromosome 9 in low-grade tumors. This chromosome may carry a suppressor gene that is linked to the *abl* gene, which is closer to the centromere on chromosome 9.

Another marker that may be associated with bladder cancer susceptibility is the fast- or slow-acetylator status of the subject (Hanke and Krajewska, 1990). Subjects with slow acetylation phenotype have been reported to have an increased incidence of bladder cancer after exposure to beta-naphthylamine. Acetylation status now can be determined by genotypic analysis. The genotypic change may exist as a recessive gene expressed in the homozygous or the heterozygous state. Heterozygous gene expression results in an intermediate quantitative phenotypic change. Preliminary results in a study of Chinese workers have not confirmed an association between the genetic phenotype and bladder cancer in benzidine-exposed workers, because the diamines are acetylated only partially. Nevertheless, this result does emphasize a mechanism by which inactivation of xenotoxic substances can be important in determining genetic susceptibility to disease.

Markers of Intermediate Effects and Disease

Bladder cancer is an excellent model for studying markers of preclinical effect and disease, although the two must be defined clearly. Because they represent a continuum, they are not always distinguishable, which may be particularly relevant in the oncogenic process, in which there is a progression of markers from normal to the final carcinogenic event that converts premalignant cells with many abnormal markers to bladder cancer. Some biomarkers include the spectrum of changes that occur in the various oncogenic events. During or immediately after exposure, a marker of effect may be expressed transiently, in which case it would be a marker of exposure. If the marker persists, it may become a marker of intermediate effects and, when the final step in oncogenesis occurs, the biomarker may represent disease. This contrasts with the relative simplicity of diagnosing a urinary tract infection, when the marker is an infectious agent. Certain markers may be present; when they occur we consider them markers of disease.

Histopathologic confirmation is the hallmark of a positive diagnosis of bladder cancer; both visual appearance of the cells and cellular orientation are important in the diagnosis. For example, when low-grade tumor cells are present in urine, the diagnosis of bladder cancer may be more difficult because cellular orientation is lost. The addition of biomarkers to visual cytology may improve early detection (Hemstreet *et al.*, 1990). Using quantitative fluorescence imaging techniques and >5C DNA as a positive marker, we have shown increased sensitivity for bladder cancer detection. An important observation is that the added value of another marker such as DNA may depend on the population being screened and on whether the tumors are multiple and recurrent. The fact that low-grade tumors containing >5C DNA are more likely to recur supports this observation (Hemstreet *et al.*, 1991).

Hematuria as a Bladder Cancer Marker in Asymptomatic Patients

In recent years, there has been progressive interest in developing biomarkers for bladder cancer screening. One of the simplest tests is for hematuria, but a single dipstick method for detecting blood in the urine has sensitivity of only approximately 37%. To improve sensitivity, Messing and co-workers (1989, 1992) asked screenees to use the dipstick for 14 consecutive days. Only 45% of the patients recruited complied with the regimen; 283 (21%) of those tested positive and 191 underwent follow-up evaluation. In rescreening of the participants who initially tested negative, 6% had a positive test and 1% was found to have bladder cancer. The relatively low participation rate for multiday initial testing suggests that two-tier screening could provide a clinical strategy that would be cost effective and, at the same time, reduce the need for unnecessary clinical evaluation.

Sequential events in the oncogenic process may serve as a guide for combining routine markers with cytology. Ideally, markers with improved sensitivity and specificity would be identified and would reduce the overall cost of bladder cancer screening, but a strategy is needed to avoid "overevaluation" of patients who test positive. In the past, patients who tested positive for hematuria all have undergone cystoscopy and intravenous pyelogram (IVP). However, the value of IVP has been questioned. Ultrasound would be a less invasive alternative and, if performed only on patients with persistent hematuria, would reduce the total cost of evaluation significantly.

Interstitial Cystitis

Possible Etiologies

Interstitial cystitis (IC) is a urologic symptom complex consisting of frequency, urgency, nocturia, and pain on urination. In some patients with long-term disease, bladder capacity decreases and there is a loss of bladder compliance. The etiology of IC has not been elucidated, but xenobiotic substances could be involved. Few epidemiologic studies have been performed and familial tendencies for development of the condition at an early age have not been explored. No association with defined infectious agents has been identified, but evidence suggests that the infectious process may be a triggering event. Suggested autoimmune, viral, and neurogenic etiologies have not stood up to further testing, but no mechanism has as yet been proved.

As deliberation continues on the pathogenesis of the disease, an estimated 15,000 women are plagued with this symptom complex each year. Many become emotionally distraught, some require bladder augmentation procedures, and others are relieved of their symptoms only by bladder removal (cystectomy) and urinary diversion. Conservative treatments for the disease include hydraulic dilation and instillation of dimethylsulfoxide (DMSO) or chlorpactin. DMSO is a known differentiation agent, but preliminary studies indicate that the F-actin content of the transitional cells in patients with IC is normal, so the mechanism of action of DMSO remains unknown. Urine is definitely a causal agent, because diversion immediately produces symptomatic relief. One theory is that the protective layers of proteoglycan and glycoprotein on the bladder surface, and other mechanisms that prevent penetration of urinary components through the bladder, have failed. The bladders of IC patients are abnormally permeable to small molecules. Moreover, protamine, an agent that neutralizes the charge of sulfated glycosaminoglycans, thereby eliminating much of their ability to trap water instilled into the normal bladder, induces a transient permeability and pain–urgency symptoms in the normal individual. In addition, administration of heparin seems to ameliorate some of the symptoms in patients. However, whether this loss of impermeability is produced by exogenous agents acting from urine or by changes in the epithelium produced from the stromal side is not clear.

Biomarkers of Interstitial Cystitis

Because of the multifactorial etiology, the difficulty in making accurate diagnoses, and the probable existence of several pathogenic processes, a search for biomarkers continues. One marker that has been observed on biopsy is the presence of increased numbers of mast cells. However, in a large study

Johansson and Fall (1990) found no statistically significant elevation in smooth muscle mast cell content; Parsons and Hurst (1990) have suggested that the increased numbers of mast cells seen in earlier reports may have been due to a tendency to biopsy near previous biopsy scars that resemble ulcers. Hurst and co-workers (1993) assayed the GAG content in urine from patients with interstitial cystitis and reported that the GAG content was about half that of control samples, suggesting that the deficit represents binding of endogenous GAG by the damaged urothelial surface. Because of the multiple factors involved in this complex disease, the final resolution is likely to be found in an approach to understand the pathogenesis of the disease itself.

Kidney Diseases

Kidney Cancer

The incidence of kidney cancer is 1 in 40,000, with about 10,000 deaths per year. Knowledge of the classifications and pathologic types of kidney cancer is critical to understanding the available epidemiologic data (Williams, 1987). Development of methods for individual risk assessment in renal cell carcinoma (RCC) is important because of weak associations with occupational exposure and mounting evidence that a specific chromosome, 3, is important in the pathogenesis of the disease. Markers on this chromosome may serve as markers of susceptibility (Zbar *et al.*, 1987).

Biomarkers of the proximal kidney tubular cell also are found in the majority of RCC, supporting the hypothesis that it is the cell of origin (Bander, 1989). The infrequent shedding of cells into the urine and the retroperitoneal location of the tumor make it difficult to detect. Blood and soluble urine markers or imaging techniques are candidates for the early detection of this disease.

Aliphatic vs Cyclic Hydrocarbons as Etiologic Agents

Attempts have been made to link gasoline exposure and aliphatic hydrocarbons to RCC (Kadamani and Asal, 1989), based on observations in animals that gasoline inhalation promoted the incidence of RCC in male rats. However, extrapolation of the animal data to humans has been difficult; only a single epidemiologic study supports the possibility, and the pathologic mechanisms involving α_2-microglobulin have not been substantiated. Neither a pathologic review of hydrocarbon-exposed cases (Pitha *et al.*, 1987) nor a clinical case study confirmed any pathologic changes in the normal kidney adjacent to RCC in exposed patients.

At present, few biologic markers are available to identify individuals at risk. Males are at higher risk than females, suggesting that occupational exposure is a factor. Hormones or smoking may be primarily etiologic components; smokers are at twice the risk of nonsmokers. Epidemiologic studies of dry cleaners have yielded some useful information, but studies indicate that the association with exposure is only weak, further emphasizing the need for individual risk assessment. Schulte and Kaye (1988) have compiled a list of the known etiologic agents for kidney cancer.

Genetic Markers of Susceptibility and Exposure

Familial clustering of RCC suggests that genetic susceptibility may be of primary importance. In family studies by Cohen and co-workers (1979), reciprocal translocations with chromosome 3 were observed; loss of heterozygosity was shown by RFLP analysis (Zbar *et al.*, 1987) to confirm the importance of the short arm of chromosome 3 in the pathogenesis of this disease. The data strongly suggest that a recessive suppressor gene exists on chromosome 3 and that cancer-prone individuals inherit one defective copy, requiring only a single mutational event on the remaining copy to initiate development of bladder cancer. Gene typing for the heterozygous carrier state may be of benefit in the future. Approximately 70% of karyotypes of papillary carcinomas report trisomy of chromosome 7 and a partial or complete trisomy of chromosome 17, in addition to changes such as trisomy of chromosomes 12, 16, or 20 (which are considered the most significant). This result is particularly important because RCC is located in the retroperitoneum and frequently is large in size and has metastasized or extended to the lymph nodes at the time of discovery (Williams, 1987). Few cells are shed in the urine unless the tumor has broken through the renal pelvis, at which time the patient frequently has microscopic or gross hematuria, which could be considered a biomarker for the disease.

Asal and co-workers (1988) reported that obesity was a significant risk factor for kidney cancer in a case-control study. This finding is important since hydrocarbons have a propensity to localize to fatty tissues. Sampling of the fat of nephrectomized specimens and exposed individuals with spectroscopic analysis could provide a future marker for exposure.

RCC also exists in a variant papillary form in which the cells are derived from the loop of Henle or, in some cases, the convoluted tubular cells. Genetic analysis of these tumors is different and suggests an altered mechanism of oncogenic development from the renal cell to the clear cell type. Both tumor types have been associated with the cystic disease of dialysis, the etiology and pathogenesis of which is unknown, but some evidence exists that tumor promoters or environmental factors associated with the dialysis process may be involved.

TCC of the renal pelvis is associated with increased phenacetin abuse, renal calculi, and pyelonephritis. These tumors in the kidney are not to be confused with adenocarcinoma from the kidney parenchyma. The latter two are associated with the presence of nitrosourea compounds related to gram-negative organisms and may stimulate cellular proliferation, a key factor in the current concepts of carcinogenesis. Markers related to TCC of the renal pelvis are considered similar to those for TCC of the lower urinary tract.

Markers of Intermediate Effects and Disease in the Renal Pelvis

Biomarkers associated with disease include increased DNA in high-grade tumors with associated poor prognosis and monoclonal antibodies against tumor-derived antigens from the proximal tubular epithelium (Karlssun-Parra *et al.*, 1989). The presence of tumor-associated antigens may explain an increased rate of spontaneous tumor regression in response to attack by autologous lymphocytes or lymphocytes stimulated *in vitro*, which has become important as a therapeutic modality. In addition, antigens to kidney tubular cells have been detected by hemagglutination tests. The opportunity exists for further identification of markers in this area. Development of new biomarkers in RCC will focus on serum-detectable factors and markers of susceptibility that are associated with inherited genetic changes. A third possibility is the presence of specific antigens in urine. The potential exists for detection by imaging techniques with agents that localize to the kidney lesion.

Nephrotoxicity

The kidney is prone to nephrotoxicity and, potentially, end-stage renal disease through a variety of mechanisms that are not totally understood. However, 20% of kidney failure is estimated to be attributable to environmental exposure to toxic compounds or to iatrogenic induction by various drugs or procedures. On a per-patient basis, end-stage renal disease is the most expensive kidney or urologic disease. In 1987, the Medicare End-Stage Renal Disease Program spent $3.4 billion for the cost of dialysis and transplantation alone, benefitting 157,944 patients; in the year 2000, the patient population is estimated to exceed 250,000 and the cost to be over $6.5 billion (National Kidney and Urologic Diseases Advisory Board, 1990).

Kidney damage may be prerenal, renal, or postrenal. Prerenal disease is attributed primarily to a decrease in blood flow and perfusion of the kidney. Consequently, diseases or exogenous substances that interfere with the renal vasculature may be extremely important. An example is the drug cyclosporin, which is prescribed to transplant patients to control graft rejection. The majority of kidney damage is renal in origin. Mechanisms of renal toxicity may involve the glomerulus, the kidney tubular cells, or the interstitium. The glomerulus is affected most frequently by autoimmune processes subsequent to streptococcal infections or other autoimmune diseases, but other factors such as hydrocarbon exposure may cause nephrotoxicity through the rare Goodpasture's disease. The majority of nephrotoxicity involves the kid-

TABLE 18.5   Common Nephrotoxins

Industrial and Environmental
Glycols
Heavy metals
Organic solvents
Insecticides, herbicides, fungicides
Drugs
Prescription
Antibiotics
Antibacterial agents
Antiviral agents
Antifungal agents
Immunosuppressive agents
Antineoplastic agents
Nonprescription
Nonsteroidal anti-inflammatory agents
Illicit (recreational)
Heroin
Cocaine

ney tubules. Table 18.5 lists compounds in various categories of substances that have been associated with tubular cell damage and compounds that may affect primarily the interstitium.

Historically, nephrotoxicity has been diagnosed largely by an increase in serum creatinine or a decrease in glomerular filtration rate. However, to observe an increase in serum creatinine, 40–50% of the renal parenchyma must be diseased. Creatinine clearance is more sensitive but requires determining the blood creatinine and the urine creatinine clearance over a prescribed time interval, which is not practical in large-scale epidemiologic studies. Further, subclinical renal damage may occur, followed by repair, making it difficult to determine the immediate effect of nephrotoxic agents in relation to the toxin or to a toxic metabolite.

Principles of Biomarkers for Monitoring Nephrotoxicity

The major goal of monitoring nephrotoxicity is to detect subclinical disease in a quantitative manner specific for the site of toxicity. Urine contains both soluble and insoluble substances for biochemical and immunological analysis. Insoluble substances include various types of cells, cell products, or an accumulation of cellular products such as myeloid bodies and various types of renal cell casts. Soluble markers are diluted by urine and frequently require increased concentrations to be assayed. Many substances are dependent on creatinine clearance, are unstable, and often adhere to glassware. Soluble proteins and insoluble cell casts have been quantitated in urine, but there is little evidence that they have been assessed quantitatively together.

Determination of Cellular Products

Light microscopic evaluations of red blood cells, white blood cells, tubular cells, and casts have been, historically, the mainstay for determining glomerular and tubular injury. Human urinary sediments have been evaluated by electron microscopy to quantitate and characterize tubular cells from patients receiving aminoglycosides, comparing the relative toxicity of tobramycin and gentamycin. In other studies, tubular cells have been quantitated serially to monitor transplant rejection.

Determination of soluble markers of exposure and disease has been refined during recent years. Damage to the glomerular filtration system allows passage of albumin, β_2-microglobulins, and red blood cells. Enzymes associated with the proximal tubular cell and substances such as NAD (nicotinamide adenine dinucleotide), sialic acid, or Tamm–Horsfall protein also have been assayed.

A potentially powerful quantitative approach may be to use specific reagents, such as monoclonal antibodies, to determine cell fragments or cell types in the urine. Monoclonal antibodies have been developed against various segments of the nephron, separating the identification of each of the cellular types. These reagents, combined with imaging techniques and multiple color fluorescence probes, may provide the resolution needed to detect subclinical toxicity. The reagents available for detecting the various cells in urine have been used in the past with enzyme-linked immunosorbent assay (ELISA), primarily to quantitate soluble substances. Further research is needed in this area.

Markers of Susceptibility

Increased susceptibility in the old and the very young has been attributed to differences in metabolic enzyme development in the urinary collecting system. The primary cause, however, for increased susceptibility is multiple insults to the kidney that occur simultaneously in many cases. A concrete example in clinical medicine is the diabetic patient with underlying hypertension and glomerular disease who receives aminoglycosides (a nephrotoxic antibiotic), cisplatin (a nephrotoxic chemotherapeutic agent), and toxicity from radioopaque contrasts.

Specific genetic markers of susceptibility for nephrotoxic effects probably do exist, but have not yet been defined. Such markers are likely to be related to drug transport or activation or inactivation of toxic metabolites. An example is provided by the male rat, in which sex is the predominant factor controlling susceptibility. When the animal is exposed to 12-chain branched hydrocarbons, the liver increases production of α_2-microglobulin, which then collects in the proximal tubule cells and results in tubular cell damage. No such counterpart has been observed in humans. In fact, Pitha *et al.* (1987), in a study using light microscopy, could find no kidney damage in tissue samples from individuals who had known hydrocarbon exposure.

More subtle examples are the acetylation of a drug such as Loradine or the genetic inheritance of the activity of P450 enzymes, which may be involved in drug activation. Much further study is needed to determine markers of susceptibility to nephrotoxins.

Prostate

Background

The prostate is the primary accessory sex gland in the male. It is located between the bladder neck and the external sphincter comprising the pelvic floor. Urine passes from the bladder through the prostatic urethra into the membranous urethra which joins the bulbous urethra. During sexual function and with ejaculation, the bladder neck closes and the semen enters the prostate through the ejaculatory ducts. Prostatic secretions constitute 15% of semen. McNeal and Bostwick (1990) have provided detailed anatomical descriptions of the lobes of the prostate and the orientation of these lobes with respect to each other, information which is invaluable to the use of ultrasound of the prostate.

Three major urologic problems—infection, benign growth, and malignant growth—arise in the prostate. Infectious processes are complicated by the glandular nature of the organ; limited drainage routes are complicated further by an unusual cellular pH milieu that does not facilitate antibiotic delivery to the gland. One disease process that is poorly understood is nonspecific prostatitis. Two other diseases, benign prostatic hypertrophy (BPH) and cancer, are related to growth control of epithelial and stromal cells.

Like the female breast, the prostate is under hormonal growth control that directs its development during embryogenesis and influences the formation and maturation of the glandular structure. Serum testosterone, derived primarily from the testicle under the direction of the hypothalamic and pituitary axis, is critical to this control. Testosterone is converted to dihydrotestosterone by an enzyme, 5-alpha reductase. Dihydrotestosterone binds to a receptor and is transferred to the nucleus where it influences cellular proliferation. Drugs to control the growth of cells at various stages in the biochemical control are being developed. The saga of growth control is complicated further by the influence of stromal cells on the epithelial cells that form the gland. Some researchers have suggested that the balance between estrogen and testosterone or dihydrotestosterone is the critical regulatory mechanism. An understanding of biomarkers in the prostate undoubtedly will be related to understanding subtle regulatory growth control mechanisms and developing assays for these quantitative alterations in the cells or the serum.

Prostate Cancer

Prostate cancer is a disease of the aging population. Adenocarcinoma of the prostate clinically affects 1 in 9 men. It is the third most common cause of cancer death in men and the primary cause in men over the age of 70. Prostate cancer is unusual because serial sections of glands obtained at autopsy of men who died of other causes have shown that 50% of 80-year-old men had prostate cancer that could be detected by histology, yet only 10% of men will ever have a clinically detectable prostate cancer, and only 1–2% will die of the disease.

Use of the serum marker prostate-specific antigen (PSA), especially in conjunction with digital rectal examination (DRE) and ultrasound-directed needle biopsy, has increased the number of prostate cancer diagnoses significantly. However, some of the cases detected may represent occult disease that may never be active clinically, creating a dilemma for both physician and patient. In an effort to reduce the necessity for surgery and consequent patient morbidity, an aggressive search for definitive prognostic markers is underway.

The etiology of prostate cancer is unknown, but is probably multifactorial in origin. Several features of the disease are intriguing. Elevated serum testosterone has been associated with an increased incidence of the disease. Although the incidence of clinical, or symptomatic, prostate cancer is much lower in Asian populations, there is no difference between the prevalence of histologically defined diseases in the United States and Asia, indicating that some specific environmental factor may trigger the clinically significant form of prostate cancer. Schwartz and Hulka (1990) have suggested that a deficiency of vitamin D may be a contributing factor.

Only recently have case-control studies clearly demonstrated an increased risk of prostate cancer in family studies. The proband whose father or uncle and grandfather had prostate cancer has a relative risk 8–24 times that of age-matched family controls. This result pinpoints a unique subgroup for studies to identify markers of genetic susceptibility (Carter *et al.*, 1991).

Occupational etiologies are less well documented, but increased incidence has been reported in cadmium workers, farmers, and workers in the rubber, dye, and hydrocarbon industries. Cadmium exposure is of particular interest because of the high concentration of zinc normally found in prostate tissues. One mechanism of transcription of DNA relies on the incorporation of zinc into proteins that initiate the transcription of the RNA message—the so-called "zinc fingers." Substitution of cadmium for zinc could conceivably influence the transcription. Although early reports indicated increased prostate cancer risk among vasectomized men, this finding has not been confirmed. Unlike many other malignancies, the influence of cigarette smoking is controversial.

Prostate cancer detection has relied primarily on DRE, which detects nodules in the posterior lobe of the prostate where the majority of prostate cancers arise in the peripheral glands. Approximately 50% of prostate nodules prove to be malignant on biopsy and approximately 75% of prostate malignancies are detected by this route of examination. In recent years, ultrasound has been employed as a guide to biopsy to detect occult disease in patients with the elevated biomarker PSA (Cooner *et al.*, 1990). The relationship of PSA to ultrasound-guided biopsy and DRE will be discussed in more detail later.

Biomarkers in Prostate Cancer

For initial prostate cancer screening, serum biomarkers appear to be most useful because of the inability to access cells without invasive needle biopsy procedures. Once initial risk is defined, follow-up FNA may obtain cells for quantitative biomarker analysis and individual risk assessment. Cellular markers will reflect biochemical changes associated with differentiation, proliferation, invasion, and metastasis. A major concern in prostate cancer management will be to determine which tumors will become clinically active, since the prevalence of occult disease is at least five times that of clinically manifest disease.

Transrectal ultrasound (TRUS) was viewed initially as another potentially useful screening tool, but its high false-positive rate has been shown to limit its usefulness as an independent screening tool. However, improved results were achieved when ultrasound was used as a follow-up to PSA and DRE (Table 18.6; Cooner *et al.*, 1990; Cooner, 1991). TRUS was positive in 2.5% of patients with negative PSA and DRE and in 55.3% of patients with above-normal PSA and positive DRE. How many of the tumors are actually in the hypoechoic area that is biopsied and in other random sites biopsied at the time of ultrasound is unknown. Determining the usefulness of PSA as a biomarker has been a challenge because of the perceived high false-positive rate secondary to undefined occult disease and BPH. Operative decisions are made more difficult because of these factors. PSA and TRUS have identified large numbers of patients with occult disease that may never become clinically significant, particularly in relation to expected survival. The combination of multiserum biomarkers in the sequential monitoring of low-level elevation should assess risk more accurately.

Biomarkers of Susceptibility

The finding of increased risk of prostate cancer in probands of family members with prostate cancer suggests the importance of genetic factors in disease susceptibility. The lack of information concerning exogenous factors has thwarted a search for enzymes or other markers important in the activation of compounds, but chromosomal changes provide some clues. Karyotypic studies indicate translocations in some cells that may be related to early inherited mutations in some cases.

TABLE 18.6   Cancer Detection Rate of Transrectal Ultrasound Related to Levels of Both Serum PSA and DRE in 2634 Patients

| PSA[a] (ng/ml) | Number of cancers/ number of patients (%) | |
	DRE +	DRE −
≤ 4	46/446 (10.3)	31/1265 (2.5)
> 4	242/438 (55.3)	64/ 48 (13.2)
4.1–10	74/194 (38.1)	19/ 343 (5.5)
>10	168/256 (65.6)	45/ 144 (31.3)

Source: Cooner (1991).
[a] PSA from Hybritech.

Based on animal studies and allelotype analyses of 30 human prostate cancers, an allelic deletion occurs on chromosome 16 which encodes E-cadherin, a molecule important in cell adhesion. The E-cadherin molecule belongs to a superclass of genes that make cell adhesion molecules (CAM). These molecules are important in gap junctions between cells and could influence the histopathologic pattern of prostate cancer cells markedly, as reflected in the Gleason Staging System.

Subtle alteration in the quantitative hormonal regulation of prostate cell growth may be another important marker. Dihydrotestosterone controls growth of the normal and benign prostatic tissue as well as the hormonally sensitive prostate cancer clones. Since serum testosterone is not the active compound, other factors such as level of dihydrotestosterone receptor of 5-alpha reductase enzyme could be important regulating factors. Their quantitation could provide insight into the carcinogenic mechanisms. Specific genetic alterations now can be identified in small amounts of tissue using the PCR reaction. Individual proteins can be quantified at the single cell level with QFIA. Viral markers of exposure have been investigated, but a firm linkage with viral agents has not been forthcoming. An opportunity exists to gather information on markers of exposure by studying patients with presumed exposure and elevated prostate cancer markers such as PSA. These individuals require prostate biopsy, which could facilitate a search for markers of exposure. Large numbers of patients in various geographic areas undergo biopsy, with and without disease, as a result of the high false-positive rates of TRUS and PSA.

Biomarkers of Disease

Until recently, DRE of the prostate was the accepted method for disease detection and biopsy was the standard for diagnosis. This approach was viable because tumors are located primarily in the posterior lobe of the prostate. The serum marker prostate acid phosphatase is released in increased quantities from prostate cancer cells, but its lack of sensitivity for detecting early disease precluded its use as a screening test. Acid phosphatase is elevated in approximately 70% of cases with overt metastatic disease. Although the marker is highly specific and is useful for monitoring patients with disease, it shows considerable daily variation, making it less useful than PSA, which is found only in prostate tissue. Like acid phosphatase, PSA is released in increased amounts by neoplastic prostate cells. Many of the positive PSA results originally perceived as false positive are proving to be associated with occult disease, but elevated PSA also is associated with BPH. Thus, the test may be positive in patients with benign growth.

The combination of DRE followed by TRUS and PSA is proving to be highly effective in early prostate cancer detection. Catalona et al. (1991) reported finding a high incidence of disease when screening 1653 asymptomatic patients over 50 years of age using PSA followed by DRE and prostate biopsy. Ultrasound is particularly useful for guiding the biopsy needle into the target sites of hypoechoic areas or more peripheral lesions within the gland to insure that the tumor actually is biopsied and not bypassed by the needle.

New prostate markers are being identified but their clinical significance is not yet known. One marker, PD41, is elevated in 80–90% of patients with high-grade prostate cancer (Beckett et al., 1991). However, in contrast to PSA and acid phosphatase, its cellular staining characteristics are patchy. Staining also is associated with PIN (prostatic intraepithelial neoplasia); PD41 is positive in about 10% of cases of benign prostatic hyperplasia that also have normal PSA. The significance of this is unknown. It is not unrealistic to expect that at least 10% of the patients with BPH may eventually develop prostate cancer, even though they may have normal PSA at the time the tissue sections are taken, so longitudinal follow-up of these patients is a requisite. On the horizon are new markers that may be more sensitive and specific for disease detection, particularly if they are administered as profiles analogous to the approaches adopted by Mulshine et al. (1991) for lung cancer and Rao et al. (1991) for bladder cancer.

Other biomarkers for disease are associated with differentiation, proliferation, and invasion with metastasis. From the perspective of molecular biology, the prostate is of considerable interest because of the interrelationship of stromal and epithelial components. Mounting evidence suggests a paracrine relationship between stromal cells and the epithelial cells composing the gland. Although neoplastic processes are typically perceived as epithelial in origin, stromal cells send regulatory signals to the epithelium. The stromal cells are under estrogen growth control and can be regulated by aromatase inhibitors. Factors regulating cell growth, fibroblastic growth factors, and epidermal growth factors are known to be increased in epithelial cells of some prostate cancers (Thompson, 1990).

Some experiments indicate the presence of growth factors associated with bone that promote prostate cancer growth. Hence, the frequency of blastic bony metastasis of prostate cancer cells in response to the growth factors may represent one molecular mechanism important in metastasis. Quantitation of growth factor receptors on prostate cells may further define individual risk in patients with slightly elevated serum markers for whom biopsy is indicated.

The primary challenge is to identify risk factors of susceptibility, and specifically design modalities to prevent disease. At present, limited biomarkers are available to predict biologically active disease with progression to cancer death. DNA ploidy, however, has provided some clues.

DNA Ploidy as a Marker of Disease and Prognosis

DNA ploidy in prostate cancer can be determined by flow cytometry and image analysis as an independent variable for tumor progression. Each method has its strengths and weaknesses. A general theme has evolved from the study of DNA ploidy in many tumors. Aneuploid stem line is an independent variable of stage and grade, and portends a poor prognosis. Ploidy studies on tumor cells from patients at various stages of prostate cancer have shown that the percentage of tumors with abnormal ploidy is strongly correlated with stage and grade. At any stage or grade, diploid tumors have a generally favorable prognosis and tetraploid or aneuploid tumors portend a poor prognosis by criteria such as survival or recurrence. The time until death in patients with diffuse prostate cancer is unaffected by tumor ploidy. Thus, ploidy is associated with a poor prognosis but is not causative of that prognosis.

Other Markers

Preliminary results using various methods during the early stages of reagent refinement implicate the importance of dominant oncogenes in prostate cancer. Care must be taken not to implicate the observed changes as causative events in the pathogenesis of the disease. The significance of the *ras* oncogene indicates it may have a minimal role. Perhaps more important is the *neu* oncogene, *erb2*, related to the epidermal growth factor receptor. The *ras* and the *neu* oncogenes have been shown to be significant in other hormonally dependent tumors such as breast cancer.

A unique opportunity exists for further defining markers of susceptibility and exposure. Molecular biology and the sequencing of the human genome probably will be of primary importance in the advances. The large number of patients who undergo prostate biopsy constitutes a key resource for studying markers of exposure. Biochemical analysis of semen has provided initial clues, but human ejaculate, which contains a rich source of prostatic epithelial cells, has not been investigated fully at the cellular level. Even if it proves not to be useful in cancer detection, an interesting and potentially exciting opportunity is the isotopic imaging of quantitative changes at the molecular level to determine biologically active disease.

Summary

The GU system provides the main disposal route for water-soluble waste compounds and is therefore at risk for environmentally caused or triggered diseases, including cancer, kidney failure, and others. A number of potential markers of susceptibility, exposure, effect, and disease have been recognized. Current research is identifying and validating additional ones. The combination of molecular markers with epidemiologic investigations promises to advance our knowledge concerning etiologies of disease in the GU tract and to help devise effective means to control GU disease.

References

Asal, N. R., Geyer, J. R., Risser, D. R. Lee, E. T., Kadamani, S., and Cherng, N. (1988). Risk factors in renal cell carcinoma: I. Methodology, demographics, tobacco, beverages use and obesity. *Cancer Detect. Prev.* **11**, 359–377.

Atkin, N., and Baker, M. (1985). Cytogenic study of ten carcinomas of the bladder: Involvement of chromosomes 1 and 11. *Cancer Genet. Cytogenet.* **15**, 253–268.

Atul, T. R. Braurbar, N. D., and Lee, D. B. N. (1991). Hydrocarbons and renal failure. *Nephron* **58**, 385–392.

Auerbach, O., and Garfinkel, L. (1989). Histologic changes in the urinary bladder in relation to the cigarette smoking and use of artificial sweetners. *Cancer* **64**(5), 983–987.

Bander, N. H. (1989). Monoclonal antibodies to renal cancer antigens. *Sem. Urol.* **7**(4), 264–270.

Beckett, M. L., Lipford, G. B., Haley, C. Y., Schellhammer, P. F., and Wright, G. L., Jr. (1991). Monoclonal antibody PD_{41} recognizes an antigen restricted to prostate adenocarcinomas. *Cancer Res.* **51**(4), 1326–1333.

Bi, W. F., Rao, J. Y., Hemstreet, G. P., Yin, S., Asal, N. R., Zang, M., Min, K. W., Ma, Z., Fang, P., Lee, E., Li, G., Hurst, R. E., Wu, W., Bonner, R. B., Wang, Y., and Fradet, Y. (1993). Field molecular epidemiology: Feasibility of monitoring for the malignant bladder cell phenotype in a benzidine-exposed occupational cohort. *J. Occup. Med.* (in press).

Carter, B., Steinberg, G., Beaty, T., and Walsh, P. (1991). Evidence for mendelian inheritance in the pathogenesis of prostate cancer. *J. Urol.* **145**(4), 213A.

Cartwright, R. (1986). Screening workers exposed to suspect bladder carcinogens. *J. Occup. Med.* **28**, 1017–1019.

Catalona, W. J. (1987). Bladder cancer. *In* "Adult and Pediatric Urology" (J. Gillenwater, J. Grayhack, S. Howards, and J. Duckett, eds.), Vol. 1, pp. 1000–1043. Year Book Medical, Chicago.

Catalona, W. H., Smith, D. S., Ratliff, T. L., Kodds, K. M., Coplen, D. E., Yuan, J. J., Petros, J. A., and Andriole, G. L. (1991). Measurement of prostate-specific antigen in serum as a screening test for prostate cancer. *N. Engl. J. Med.* **324**(17), 1156–1161.

Cohen, A. J., Li, F. P., Berg, S., Marchetto, D. J., Tsai, S., Jacobs, S. C., and Brown, R. S. (1979). Hereditary renal cell carcinoma associated with a chromosomal translocation. *N. Engl. J. Med.* **301**, 592–595.

Cole, P., Hoover, R., and Friedell, G. H. (1972). Occupational and cancer of the lower urinary tract. *Cancer* **29**, 1250.

Cooner, W. H. (1991). Prostate-specific antigen, digital rectal examination, and transrectal ultrasonic examination of the prostate in prostate cancer detection. *Mongr. Urol.* **12**, 3–14.

Cooner, W. H., Mosley, B. R., Rutherford, C. L., Jr., Beard, J. H. Pond, H. S., Terry, W. J., Igel, T. C., and Kidd, D. D. (1990). Prostate cancer detection in a clinical urological practice by ultrasonography, digital rectal examination and prostate specific antigen. *J. Urol.* **143**(6), 1146–1152.

Fearon, E., Feinberg, A., Hamilton, S., and Vogelstein, B. (1985). Loss of genes on the short arm of chromosome 11 in bladder cancer. *Nature (London)* **318**, 377–380.

Gibas, Z., Prout, G. R., Connolly, J. G., Pontes, J. E., and Sandberg, A. A. (1984). Nonrandom chromosomal changes in transitional cell carcinoma of the bladder. *Cancer Res.* **44**, 1257–1264.

Glady, G., Choylee, P., Berton, Z., and Linehan, M. (1990). von Hippel Lindau disease. *Probl. Urol.* **4**(2), 312–330.

Guiraudor, C., Muirhead, N., and Wallace, A. C. (1989). Expression of HLA-DR on tubular cells: An aid in diagnosis of renal allograft rejection with FNAB. *Transpl. Proc.* **21**(4), 3602.

Hanke, J., and Krajewska, B. (1990). Acetylation phenotypes and bladder cancer. *J. Occup. Med.* **32**(9), 917–918.

Hemstreet, G. P., Schulte, P. A., Ringen, K., Stringer, W., and Altekruse, E. B. (1988). DNA hyperploidy as a marker for biological response to bladder carcinogen exposure. *Intl. J. Cancer* **42**, 817–820.

Hemstreet, G. P., Hurst, R. E., Bass, R. A., and Rao, J. Y. (1990). Quantitative fluorescence image analysis in bladder cancer screening. *J. Occup. Med.* **32**(9), 822–828.

Hemstreet, G. P., Rollins, S., Jones, P., Rao, J. Y., Hurst, R. E., Bonner, R. B., Hewett, T., and Smith, B. G. (1991). Identification of a high risk subgroup of Grade I transitional cell carcinoma using image analysis based DNA ploidy analysis of tumor tissue. *J. Urol.* **146**, 1525–1529.

Hurst, R. E., Parsons, C. L., Roy, J. B., and Young J. L. (1993). Urinary glycosaminoglycan excretion as a laboratory marker. *J. Urol.* **149**(1), 31–35.

Johansson, S. L., and Fall, M. (1990). Clinical features and spectrum of light microscopic changes in interstitial cystitis. *J. Urol.* **143**(6), 1118–11124.

Kadamani, S., Asal, N. R., and Nelson, R. Y. (1989). Occupational hydrocarbon exposure and risk of renal cell carcinoma. *Am. J. Ind. Med.* **15**, 131–141.

Karlssun-Parra, A., Dimery, E., and Juhlin, C. (1989). The anti-Leu 4 (CD_3) monoclonal antibody reacts with proximal tubular cells of the human kidney. *J. Immunol. (Scand.)* **30**(6), 719–722.

Kerbel, R. (1989). Towards an understanding of the molecular basis of the metastatic phenotype. *J. Cancer Evol. Tumor Cell Heterogen.* **9**, 329–337.

McNeal, J. E., and Bostwick, D. G. (1990). Anatomy of the prostate. *In* "Pathology of the Prostate" (D. Bostwick, ed.), pp. 1–14. Churchill Livingstone, New York.

Messing, E. M., Young, T. B., Hunt, V. B., Webbie, J. M., and Rust, P. (1989). Urinary tract cancers found by home screening with hematuria dipsticks in healthy men over 50 years of age. *Cancer* **64**, 2361–2367.

Messing, E. M., Vaillancourt, A. M., Hunt, V. B., Young, T. B., Roecke, E. B., Wegenke, J., Hisgen, W., Kuglitsch, E., and Greenberg, B. (1992). Hematuria home screening: Repeat testing 9 months after negative tests. *J. Urol.* **147**(4), 423A.

Mulshine, J. L., Treston, A. M., Scott, F. M., Avis, I, Boland, C., Phelps, R., Kaspryzyk, P. G., Kakanishi, Y., and Cuttitta, F. (1991). Lung cancer: Rational strategies for early detection and intervention. *Oncology* **5**(5), 25–32.

National Cancer Institute Workshop (1990). The 1988 Bethesda System for reporting cervical/vaginal cytologic diagnoses. *Hum. Pathol.* **21**(7), 704–708.

National Kidney and Urologic Diseases Advisory Board (1990). "1990 Long-Range Plan." NIH Publication No. 90-583. U.S. Government Printing Office, Washington, D.C.

Parsons, C. L., and Hurst, R. E. (1990). Decreased urinary uremic acid levels in individuals with interstitial cystitis. *J. Urol.* **143**(4), 690–693.

Pitha, J. V., Hemstreet, G. P., Asal, N. R., Trump, B. F., Silva, F. G., and Petrone, R. L. (1987). Occupational hydrocarbon exposure and renal histopathology. *Toxicol. Ind. Health* **3**(4), 491–506.

Pollard, T., and Cooper, J. (1986). Actin and actin-binding proteins: A critical evaluation of mechanisms and functions. *Ann. Rev. Biochem.* **55**, 987–1035.

Pritchard-Jones, K., and Fleming, S. (1991). Cell types expressing the Wilms' tumor gene (WT1) in Wilms' tumours: Implications for tumour histogenesis. *Oncogene* **6**(12), 2211–2220.

Rao, J. Y., Hemstreet, G. P., Hurst, R. E., Bass, R. A., Min, K. W., and Jones, P. L. (1991). Cellular F-actin levels as a marker for cellular transformation: Correlation with bladder cancer risk. *Cancer Res.* **51**, 2762–2767.

Rehn, L. (1895). Ueber Blasentumoren bei Fuchsinarbeitern. *Arch. Kind. Chir.* **50**, 588.

Schulte, P. A., and Kaye, W. E. (1988). Exposure registries. *Arch. Environ. Health* **43**(2), 55–61.

Schwartz, G. G., and Hulka, B. S. (1990). Is vitamin D deficiency a risk factor for prostate cancer? (Hypothesis). *Anticancer Res.* **10**(5A), 1307–1311.

Sidransky, D., Frost, P., von Eschenbach, A., Oyasu, R., Preisinger, A. C., and Vogelstein, B. (1992). Clonal origin of bladder cancer. *N. Engl. J. Med.* **326**(11), 737–740.

Sidransky, D., von Eschenbach, A. V., Tsai, Y. C., Jones, P., Summerhayes, I., Marshall, F., Paul, M., Green, P., Hamilton, S. R., Frost, P., and Vogelstein, B. (1991). Identification of *p53* gene mutations in bladder cancers and urine specimens. *Science* **252**, 706–709.

Smeets, W., Pauwels, R., Laarakkers, L., Debruyne, F., and Geraedts, J. (1987). Chromosomal analysis of bladder cancer. III. Nonrandom alterations. *Cancer Genet. Cytogenet.* **29**, 29–41.

Sobel, J., and Kaye D. (1987). Urinary tract infections. *In* "Adult and Pediatric Urology" (J. Gillenwater, J. Grayhack, S. Howards, and J. Duckett, eds.), Vol. 1, pp. 246–302. Year Book Medical, Chicago.

Thompson, T. C. (1990). Growth factors and oncogenes in prostate cancer. *Cancer Cells* **2**, 345–354.

Trump, B. F., Jones, T. W., and Heatfield B. M. (1984). The biology of the kidney. *In* "Advances in Modern Environmental Toxicology" (M. Mehlman, G. Hemstreet, J. Thorpe, and N. Weaver, eds.), Vol. VII, Renal Effects of Petroleum Hydrocarbons, pp. 27–49. Princeton Scientific, Princeton, New Jersey.

Tsai, Y. C., Nichols, P. W., Hiti, A. L., Williams, Z., Skinner, D. G., and Jones, P. A. (1990). Allelic loss of chromosomes 9, 11, and 17 in human bladder cancer. *Cancer Res.* **50**, 44–47.

Vogelstein, B., Fearson, E. R., Hamilton, S., Kern, S., Preisinger, A. (1988). Genetic alterations during colorectal tumor development. *N. Engl. J. Med.* **319** (9), 525–523.

Williams, R. D. (1987). Renal, perirenal, and ureteral neoplasms. *In* "Adult and Pediatric Urology" (J. Gillenwater, J. Grayhack, S. Howards, and J. Duckett, eds.), Vol. 1, pp. 513–554. Year Book Medical, Chicago.

Zbar, B., Brauch, H., Talmadge, C., and Linehan, W. M. (1987). Loss of alleles of loci on the short arm of chromosome 3 in renal cell carcinoma. *Nature (London)* **327**, 721–727.

19

Neurologic Disease

Jack D. Griffith
and Vincent F. Garry

Introduction

Although the nervous system constitutes only about 2% of the tissues of the human body, it forms a network that encompasses all body structures. The nervous system is composed of nerve cells and fibers, brain, brain stem, and spinal cord (Figure 19.1). It responds to stimuli, aids in acclimating the organism to its environment, and protects against pathogenic insult. In other words, the nervous system is an extremely complex network for input, retrieval, transmission, storage, and integration of information.

Information processed by the nervous system determines the physiologic function, behavior, and cognitive activities of the human organism. It is a marvel of hierarchical organization based on well-defined anatomic structure. The organization of the nervous system consists of multiple arrays of nerve fibers and neurons extending from the central nervous system (CNS), beginning with the brain and its components, to the peripheral nervous system. The peripheral system forms a final common pathway of nerve fibers for control of the physiologic functions of every organ and tissue in the organism. Input from the external and internal environment of the organism begins processing in the peripheral nervous system. Final interpretation is made by the central nervous system at some hierarchical control level. From minute-to-minute routine blood pressure maintenance to a highly organized fight or flight response, all are parts of the levels of integration and hierarchical control exerted by the central nervous system.

Pathologically, dysfunction of the nervous system can be described as a process of destruction, release, or irritation of nerve tissue. Destruction results from a transient or permanent loss of function. For example, if the cortex of the brain is damaged by disease or trauma, specific functions may be

Molecular Epidemiology: Principles and Practices
Copyright © 1993 by Academic Press, Inc. All rights of reproduction in any form reserved.

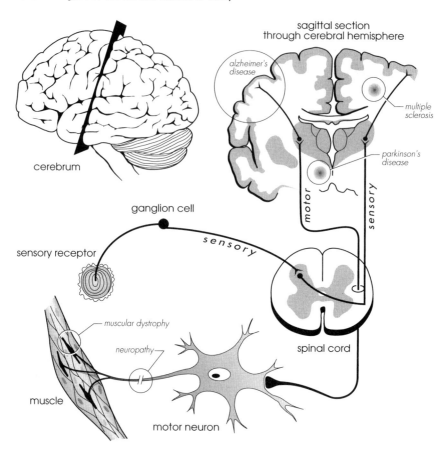

FIGURE 19.1   Organization of the nervous system.

lost, resulting in paresis, hypesthesia, blindness, and even a decrease in cognitive reasoning. Release involves the inability of the nervous system to perform specific inhibitory functions. This occurs when the pyramidal system is compromised, resulting in an increase in tone and reflexes, causing an exaggerated response due to the loss of inhibitory function and the release of intact centers from higher control.

Dysfunction from irritation of the nervous system results from increased stimulation of the tissue. For example, when the peripheral nerve is damaged, pain and muscle spasm follow. With damage or irritation of the motor cortex, jacksonian seizures can occur. As the seizure progresses, more and more motor functions are involved, affecting muscle groups on the side of the body opposite the CNS lesion. Perturbation of the nervous system by disease at any level, that is, peripheral or central, affects the integrated behavioral and cognitive function of the organism.

For this reason, public health practitioners attach considerable importance to early detection and prevention of diseases of the nervous system. However, many of the routine detection systems currently available—for example, the EEG, neuroradiology CT, pneumoencephalography, cerebral angiography, plain skull X-ray, and tests of motor functions, reflexes, and sensibilities—focus on confirmation of signs and symptoms of clinical neurologic disease reported by patients to their attending physicians and do not lend them selves easily to early detection. In neurologic epidemiology, there is an evident need for sensitive quantitative tests to supplement the existing conventional and mostly qualitative tests used in the clinical neurologic examination. Systems with the level of sensitivity needed for early detection tend to be highly sophisticated technologies whose use for large scale public health investigations is limited.

In this chapter, we focus our exploration on the link of epidemiology to biomarkers that seem to promise detection and prevention of human diseases of the nervous system.

Perhaps the most exciting area for the application of these technologies to epidemiologic investigation of CNS disorders lies in the study of neuromuscular disorders. Among the more common neuromuscular disorders are muscular dystrophy and myasthenia gravis. Much more prevalent are non-neuromuscular diseases, including multiple sclerosis, Down's syndrome, convulsive disorders, and Parkinson's disease. All these diseases are chronic and constitute a long-term societal burden.

Although genetic and familial studies have uncovered the etiologic sources of a limited number of neurologic diseases, the study of the distribution of the disease in specific populations, and the factors that determine that distribution, have also played an important role in understanding the etiology of selected diseases.

Previously, the epidemiology of chronic neuromuscular diseases has been based on an evaluation of disease incidence, prevalence, or mortality. The epidemiologists use populations characterized by age, sex, race, and type of condition and selected risk factors, such as geographic distribution, heredity, nativity, marital status, migration, latency, evaluation of disease incidence, prevalence, and mortality to focus the search for disease etiology. For example, multiple sclerosis (MS) is rare in the indigenous population of the tropics, increases in the subtropics, and is of increasing frequency from the lower to the higher temperate regions (Kurland and Kurtzke, 1972). At present, the etiology of this disease remains unclear but research now focuses on identification of susceptible populations at risk.

Genetic factors and family history also seem to play a dominant role in the etiology of several neurologic diseases, including some forms of epilepsy, Parkinson's disease, muscular dystrophy (MD), and Down's syndrome. Race apparently plays a role in selected diseases. For example, nonwhites are more likely than whites to develop myasthenia gravis, while whites are more likely to develop Parkinson's disease, Down's syndrome, and multiple sclerosis.

Muscular Dystrophy

Muscular dystrophy (MD) is a hereditary disorder characterized by progressive weakness and wasting of striated muscle. Although there may be some degenerative changes in the ventral horn cells, or a small reduction in their number, the central and peripheral nervous systems are normal. The several forms of the disease differ in the muscle groups affected, the age of onset, and the mode of inheritance.

The overall prevalence rate for MD in adult males and females in the United States is 4 in 100,000. Limb–girdle type MD (incidence rate for males and females is 7 per 100,000 live births) is primarily an autosomal recessive trait that results in MD in both males and females during the first 30 years of life, although it is known to occur later. The Duchenne form of MD is the most common (14 per 100,000 live born males). The disease is first observed early in childhood and is characterized clinically by a waddling gait produced by progressive hip girdle weakness. Mild mental retardation and cardiomyopathy are frequent associated findings. Death in late childhood commonly occurs. Generally, the disease is inherited as a sex-linked recessive trait (i.e., the gene for this disease is carried on the X chromosome), males are affected selectively. In genetics, linkage alludes to gene placement on the chromosome. If genes or DNA segments are very close together on the chromosome, they are more likely to be transferred to offspring. In fact, if two genes are extremely close together, the probability of genetic transmission can be as great as 99%. If genes are not linked, or are separated by great distance on the chromosome, the possibility of transmission becomes a random process. Genetic markers usually fall between the two extremes; linkage estimates can be used to estimate distances between two DNA segments.

Approximately 70% of carriers show significantly elevated serum creatine phosphokinase, an enzyme marker for muscle damage (Greenberg *et al.,* 1988). Studies of patients with this birth defect suggest that the disease involves deletion of small segments of the short arm of the X chromosome (Xp21), detectable at the molecular level. This detection scheme can be employed to identify the carrier state. For this form of muscular dystrophy, it is possible, therefore, to screen populations for at-risk individuals (carriers) using relatively simple laboratory technology (creatine phosphokinase) and to identify the carrier state using molecular probes to identify deletions of DNA segments containing the gene (Mulley *et al.,* 1988). It is also possible to detect the defect *in utero* using molecular probes for specific DNA segments of interest (Ward *et al.,* 1989). A relatively new marker, dystrophin (the protein product of the Duchenne muscular dystrophy locus), also has been used successfully in the differential diagnosis of Duchenne and limb–girdle dystrophies. Masanori *et al.* (1991), through molecular and cellular techniques on muscle tissue and blood cells, have used dystrophin to estimate the prevalence and incidence of progressive muscular dystrophy among the residents of Okinawa.

The separately classified myotonic dystrophy (DM) is the most common form of adult muscular dystrophy and is an inherited disorder (autosomal dominant) characterized by marked variability in expression (prevalence rate of 3 per 100,000 population). Onset of the disease occurs in the early teens and begins clinically with facial muscle weakness and drooping eyelids. Prefrontal balding and cataracts are also characteristic. Clinical examination of patients with this disease demonstrates prolonged contraction of muscle groups on reflex stimulus. The condition is inherited as autosomal dominant; males and females are at equal risk for DM. The prevalence of this disease in any population is affected by the degree to which the gene is penetrant, that is, the number of persons carrying the gene and the degree of expression of the gene which translates to the number of persons with the disease and the severity of the disease.

The fertility rate of persons carrying the DM gene is the final determinant of prevalence. Regional differences in the prevalence of DM also occur. For example, the population in the Saguenay region of Quebec experience an unusually high prevalence rate of 1/475 inhabitants (Bouchard *et al.*, 1988). The gene involved in DM has been linked to the gene of the third component of complement (C3), an important immunologic protein, and one with a well-characterized DNA sequence. The C3 gene has been localized to the long arm of chromosome 19, close to the centromere, by somatic cell genetic and cytogenetic techniques. The exact location of this gene within this chromosome segment and the DNA sequences involved have not been identified. Genetic risk and the carrier state can be determined by locating genes with known function that lie in close proximity to the DM gene. Three genetic markers have been identified: apolipoproteins E and C2 (Roses *et al.*, 1988; Thibault *et al.*, 1989) and creatine kinase (Brunner *et al.*, 1989). The genes for these proteins are linked closely to the MD gene. Analysis of alteration (i.e., polymorphism—natural heterogeneity in the human genome that produces different blood types, eye colors, and different forms of the same enzyme) of the DNA and segregation of subtypes (isoproteins) of the gene products (apolipoprotein E or C2 and creatine kinase) are used to determine genetic risk and the carrier state for myotonic muscular dystrophy.

Myasthenia Gravis

Myasthenia gravis (MG) is a rare disorder of neuromuscular transmission. MG is classically an autoimmune disorder (Pachner, 1988). The combined sporadic and familial prevalence is approximately 5 in 100,000 persons. The peak age of onset for females and males is in the 3rd and 7th decades of life, respectively. In this disease, antibodies are formed against acetylcholine receptor (AChR) sites, blocking normal transmission of nerve impulses. Acetylcholine, the major neurotransmitter substance released by nerve endings, cannot effectively cross the nerve–muscle junction (synaptic site) to bind to the ACh receptors at the postsynaptic site located on the exterior membrane of muscle. This scenario illustrates that the primary anticholinergic activity of the autoantibodies in myasthenia gravis is directed at the ACh receptor sites at the myoneural junction (Tzartos, 1988). Multiple forms of these autoantibodies (epitopes) attack different parts of the receptor (Heidenreich *et al.*, 1988).

Occasionally, autoantibodies also are directed against acetylcholine itself (Souan *et al.*, 1987). Frequently, MG is associated with diseases of the thymus leading to the speculation that MG might be a thymic disorder (Williams *et al.*, 1987). On occasion, MG is associated with other auto-immune diseases, including systemic lupus erythematosus. Clinically, patients with MG note the onset of increasing muscle fatigue relieved by prolonged periods of rest. Facial and neck muscle tone is lost with gradual extension to the limbs over time. In addition to autoantibodies to cholinesterase receptors, pharmacologic biomarkers for this disease are under consideration. Anticholinesterase agents with a reversible action, for example, edrophonium, will increase muscle weakness due to MG transiently and may provide confirmatory identification of this disease.

Multiple Sclerosis

Multiple sclerosis (MS) is a disease of young adults that affects the central nervous system. It is the most common disease of the nervous system among young adults in the United States. Although the pathogenesis of MS is unknown, exogenous factors, genetic predisposition, and immunologic disturbances seem to play a role in the pathogenesis of MS (McFarland *et al.*, 1989). However, MS usually is considered an immunologic disease. A few retrospective studies have reported an increased coincidence of MS with myasthenia gravis, ankylosing spondylitis, ulcerative colitis, iridocyclitis, and other immunologic disturbances (Seyfert *et al.*, 1990).

The age-specific incidence rate is 6–7 per 100,000 for young males and 7–9 per 100,000 for young females. MS rarely occurs before age 15 and seldom occurs after 40 years of age. In this disease process, progressive patchwork destruction and inflammation of the myelin sheath surrounding nerve fibers in the central nervous system occurs. The nerve fibers are spared. Importantly, myelin with its high lipid content (Deber and Reynolds, 1991) allows for very rapid transmission of nerve impulses within the inner cortex of the brain. As a consequence of MS, the speed of neurotransmission in the CNS is slowed and the message transmitted can become garbled. The chronology of the disease, in general, is episodic with periods of remission and progression over time. The disease may begin with periods of blurred vision and vestibular dysfunction and progress with motor and sensory involvement of limbs. Finally, more and more vital functions carried out by the central nervous system are affected.

Geographically, MS is a disease of temperate and cold climates (Kurland and Kurtzke, 1972). It is possible that environmental factors play an as yet unknown role in the etiology and pathogenesis of MS. For example, the prevalence of MS increases with increasing latitude in both the northern and southern hemisphere. Zones delineating high-, intermediate-, and low-risk groups have been defined (Milonas *et al.*, 1990). In high-risk temperate areas, 30 or more cases per 100,000 occur. For example, in Minnesota the prevalence rate is as high as 160 per 100,000 (Wynn *et al.*, 1990). In the tropics the disease is practically unknown (0–4 cases/100,000). As expected, the risk of MS increases with time early in life spent in temperate or cold climates.

The search for biomarkers relating to etiology or risk has been long and arduous. Three primary investigatory areas are being explored to establish clues to the etiology and associated risk of MS. These are genetic, immunologic, and infectious.

Genetic

The risk of MS in persons with a family history of the disease is approximately 15 times greater than in the general population. Certain genetic phenotypes of the HLA haplotype series of lymphocyte antigens are at increased risk for MS. The DR2 antigen of the HLA series (Barroche *et al.*, 1986) is most commonly found in patients with MS.

Immunologic

Inflammation of the myelin sheath of nerves of the CNS is by far the most common anatomic finding in the disease. Examination of spinal fluid for immunoglobins, inflammatory cells, and certain proteins (Filbin and Poduslo, 1986; Tourelotte *et al.*, 1988) has been used to mark prognosis in this disease. Elevated immunoglobulins, altered lymphocyte response to mitogens, and shifts in T and B cell subsets (D'Andrea *et al.*, 1989) suggest an autoimmune component for MS.

Infectious Agents

Persons with a prior history of vaccination and measles (Rikonen, 1989) may be at increased risk for MS. The role of retroviruses has been explored because of the known affinity of members of this viral group (HTLV-1) and other viral agents to produce pathologic effects on myelin (Gessain *et al.*, 1988). To date, the work has been inconclusive.

Down's Syndrome

Down's syndrome (DS) is a major cause of mental retardation. DS usually results from sporadic chromosomal aberrations, occurring equally in males and females. The condition usually is caused by an extra chromosome 21. However, it may be caused by duplication of only the distal half of chromosome 21, band q22. This band is known as the Down's syndrome region and is defined by several characteristics of the syndrome, for example, mental retardation and typical physical features associated with this disease, including a brachycephalic skull, changes in hands and feet with a simian palmar crease and a crease between the hallux and second toe, epicanthal folds, deformed earlobes, narrowed palpebral fissures, large tongue, and depression at the bridge of the nose. Cytogenetic evidence suggests that a duplication of only the subband 21q22.1 and perhaps 21q22.2 may be responsible for most abnormalities (Korenberg *et al.*, 1990).

The incidence rate for Down's syndrome is 1.2 per 1000 live births, increasing with age of the mother to 17 per 1000 live births for women 45 years of age and older. Two markers in maternal serum, estriol and human chorionic gonadotropin (hCG), combined with the level of maternal serum alpha-fetoprotein and maternal age considerably improve the expected results of a screening strategy for Down's syndrome. A study by Norgaard-Pedersen *et al.* (1990) has shown an improvement in the detection rate of DS from 53.0% to 57.6% but, more importantly, the false-positive rate has been shown to decrease from 9.4% to 7.3%. It is possible that, in women less than 35 years of age, a screening strategy based on a combination of maternal age and biochemical markers would improve antenatal care. For older women, the results of such a maternal serum test may refine counseling for genetic amniocentesis.

Parkinson's Disease

Parkinson's disease (PD) is characterized by muscular rigidity, poverty of movement (bradykinesia), and tremor that is typically regular and rhythmic, maximal at rest, and mostly distal in the limbs. The primary pathology involves degenerative changes in the basal ganglia, especially the substantia nigra and corpus striatum, and alterations in the cerebral cortex (Kurland *et al.*, 1973).

Parkinson's disease appears to occur more frequently in whites than in blacks, with an overall mortality rate slightly higher for males (1.8 per 100,000) than for females (1.5 per 100,000). The overall incidence rate is about 20 per 100,000, with a prevalence rate of nearly 200 per 100,000.

Although the etiology of Parkinson's is not understood, there appears to be an increased familial frequency for parkinsonism. This familial aspect suggests that Parkinson's disease may have a genetic component. Exposure to environmental toxicants resulting in damage to the CNS also has been known to cause parkinson-like symptoms. For example, exposure to manganese causes functional disabilities in humans that closely resemble the extrapyramidal signs and symptoms of parkinsonism. Exposure to carbon monoxide produces both cytotoxic and ischemic damage to the CNS. Repeated anoxic episodes tend to damage the blood–brain barrier, resulting in damage to the white matter of the brain. If the damage to the blood–brain barrier is sufficiently great, parkinson-like symptoms may occur.

Other environmental factors, including work in orchards and wood pulp processing, have been related to the development of Parkinson's disease (Tanner *et al.*, 1987; Hertzman *et al.*, 1990) again raising the possibility of environmental chemical induction of this disease. Perhaps the clearest documented relationship between chemical exposure and development of Parkinson's syndrome comes from studies of chemical abuse of the designer drug MPTP (1-methyl-4-phenyl-1,2,3,6-tetrahydropyridine)(Heikkila *et al.*, 1984; Burns *et al.*, 1985). In MPTP-induced parkinsonism and in Parkinson's disease, the biogenic amine homovanillic acid is reduced.

Analyses of monoamine metabolites and neuropeptides in the cerebrospinal fluid (CSF) offer an important complement for early diagnosis and treatment of patients and provide biomarkers to distinguish chemically induced parkinsonism from Parkinson's disease. The monoamines (dopamine, norepinephrine, serotonin) are present in variable concentrations in different parts of the brain and function as neurotransmitters. Thus, analysis of monoamine metabolites and neuropeptides in the CSF will make it possible to grade the disease in different stages and to identify subtypes of the disease (Olsson *et al.*, 1990). Repeated analyses can be of value in guiding drug therapy.

Convulsive Disorders

The factors that initiate or underlie convulsive disorders are not well understood, but trauma, infectious disease, and genetics have been implicated. This condition is characterized by sudden, brief, repetitive, and stereotyped changes in behavior that are thought to be due to a paroxysmal discharge of cortical or subcortical neurons (Leppert *et al.*, 1989). Clinically, seizures have been classified as grand mal (the generalized tonic clonic seizure) or petit mal (frequent but brief episodes of impairment of consciousness, with or without myoclonic jerks or postural alterations, and always with an aura of postictal features). The prevalence rate for recurrent seizures, commonly known as epilepsies, is 5 per 1000 persons. Less than 2% of the general population is affected by age 40, with an overall incidence rate of 40 per 100,000 persons. Males appear to have an excess mortality, incidence, and prevalence of convulsive disorders, compared with females. A familial relationship also appears to be involved in the etiology of the disease, since clinical studies often show a higher frequency in families with a history of the disease.

Molecular markers may be used to shed some light on the etiology of convulsive disorders. For example, in benign familial neonatal convulsions, DNA markers are being used to map a specific gene to a specific chromosomal region, thereby linking the markers to the area on the chromosome that has been associated with this autosomal dominant trait. Also it may be possible to use amino acid markers (e.g., GABA, alanine, glutamate, glycine, methionine, and phenylalanine) and restriction fragment length polymorphisms (RFLPs) to determine the chromosomal location of juvenile myoclonic epilepsy (Delgado-Escueta et al., 1989).

Alzheimer's Disease

Alois Alzheimer (1907) originally described the neurofibrillary tangles and neuritic plaques present in the cerebral cortex of patients suffering from dementia. A few years later, Perusini (1910) corroborated these findings in other patients and named the disease after Alzheimer. Years later, as a result of increased awareness on the part of the public about the magnitude and destructive nature of dementia, increased scientific interest has begun to focus on this terrible disease.

Alzheimer's disease, or senile dementia of the Alzheimer type (SDAT) is becoming one of the most common afflictions of an aging society. Although the etiology of Alzheimer's disease is not well understood, most observers characterize the condition as an unusual form of presenile dementia with onset before 65 years of age. Among persons over 65, approximately 15% have some form of dementia and 55% of this group have dementia of the Alzheimer type (Wisniewski and Kozlowski, 1982). SDAT is believed to be the most common form of brain failure in the elderly (Katzman, 1976; Armbrustmacher, 1979; Plum, 1979). Approximately one-third of the nursing home beds are occupied by Alzheimer patients (Sinex and Merril, 1982). Although SDAT seldom occurs before 40 years of age, it appears that there is increased reporting of the disease among persons under 65 under of age.

The incidence of SDAT is believed to be increasing as the United States population ages, although definitive diagnosis is almost always made at autopsy. Alzheimer's disease has been characterized as the most prevalent degenerative disorder causing dementia (Massimo et al., 1991). The prevalence of SDAT has been reported to range from 2 to 6% among those persons 65 years of age or older in the United States (Thienhaus et al., 1985). Interestingly, at autopsy, 30% of persons who are free of dementia are found to demonstrate morphological changes similar to those found in Alzheimer patients. SDAT mortality is ranked fourth or fifth for all causes of death among the elderly in the United States (Katzman, 1976). Although the prevalence of SDAT appears to be greater among women, this may reflect the fact that more women are in the primary age groups for this disease.

Pathologically, Alzheimer's disease is an insidious progressive form of dementia (i.e., symptoms are difficult to distinguish from other types of brain failure or from the normal aging process). Symptomatically, loss of cognitive functions is characteristic, associated with reduced mental capacity for remembering past events or for assimilating new information. As the disease progresses, memory disorder becomes so severe that orientation and communication are grossly impaired. Finally, persons with SDAT lose their ability for abstract reasoning; concentration lapses, as does the ability to integrate visual and spatial information (Thienhaus *et al.*, 1985).

With the loss of cognitive function and neurologic deterioration, personality changes occur that often result in deterioration of social relationships to the point of collapse (e.g., frequent mood changes and lapses in judgment). Speech and communication begin to deteriorate and depression ensues. Apathy, urinary incontinence, difficulty in swallowing, and seizures are characteristic symptoms. Finally, the patient assumes a childlike dependent state. In Alzheimer patients, death by inanition (infection) brought on by malnutrition is a common occurrence.

At autopsy, there is atrophy of the cortex, particularly in the frontal lobe. Neuropathologic changes are reflected by neuronal depletion and general cell atrophy of the brain, with reduction of the dendritic trees, changes in the cell nuclei, neurofibrillary degeneration, accumulation of senile plaques, amyloid angiopathy, granulovacuolar degeneration, and Lewy and Hirano bodies. Microscopically, there is degeneration and loss of nerve cells of the CNS, primarily those serving cholinergic functions. The neurologic damage occurs throughout the CNS, but is most pronounced in the hippocampus and in parts of the neocortex. Cell loss involves the largest neurons. Characteristically, the most recognizable neuropathologic signs (neurofibrillary tangles and plaques) of Alzheimer's disease are observed.

Although the etiology of Alzheimer's disease is unknown, it has been hypothesized that the disease may be caused by hereditary factors, slow acting viruses, toxic environmental agents (e.g., aluminum), or immune defects or autoimmunity. Although the pathologic mechanism for Alzheimer's disease is not understood, one hypothesis suggests that the intracellular transport of proteins is compromised in some way. Interference in protein transport results in an accumulation of protein near the cell nucleus, rather than in the periphery of the cell. The accumulation near the nucleus deprives the neurofibrils and membranes of the dendrites and axons of necessary proteins, resulting in the degeneration and atrophy of the cell, loss of synaptic function, and, finally, destruction of the neuron itself. Neuronal damage is believed to result in neurotransmitter dysfunction, thus playing a role in the pathogenic nature of the disease.

Although the exact pathologic mechanism is unclear, it is known that neurochemical changes involving cholinergic molecules such as choline acetyltransferase (AChT), acetylcholine (ACh), and acetylcholinesterase (AChE) take place in Alzheimer patients. For example, investigators have noted a marked decrease in ChAT an AChE in the neocortical and hippocampal areas of the brain of Alzheimer patients (Henke and Lang, 1983; Perry and Perry, 1980). The cholinergic enzyme ChAT catalyzes the conversion of acetyl-coenzyme A and choline to ACh. ACh is the chemical transmitter of nerve impulses at endings of postganglionic parasympathetic nerve fibers, somatic motor nerves to skeletal muscle, preganglionic fibers of both parasympathetic and sympathetic nerves, and certain synapses in the CNS (Murphy, 1980). Since ChAT is found primarily, if not exclusively, in the cholinergic neurons of the brain (Thienhaus *et al.*, 1985), it is used as a marker for cholinergic neurons. In SDAT, when cholinergic neurons are damaged or lost, there is also a reduction in ChAT activity.

In the normal aging processes, AChE levels are largely unaffected. However, in SDAT patients, there is a marked reduction in AChE activity that may be due to chemical alterations brought about by damage to cholinergic neurons. Over time, neuronal damage results in the loss of neurons, which is then responsible for a reduction of ChAT and AChE. Reduced AChE levels result in an accumulation of ACh at the nerve synapse. Accumulated ACh in the CNS is thought to be responsible for several neurologic symptoms associated with organophosphate poisoning (e.g., tension, anxiety, restlessness, insomnia, emotional instability and neurosis, apathy, and confusion). Organophosphates are powerful cholinesterase inhibitors. These symptoms are not unlike those encountered in the treatment of patients with Alzheimer's disease. Since AChE levels in the human brain generally are not affected by the normal aging process, reduced levels in Alzheimer patients eventually may serve as a marker for the disease. The cholinergic hypothesis is that SDAT-related defects in learning and memory are caused by the defective synthesis of ACh, a neurotransmitter involved in neural activities in the hypothalamus and in the neurons of the isodendritic core that project widely to the cortex (Perry, 1979).

Features of the earlier onset of Alzheimer's disease, that is, left handed predominance (Seltzer *et al.*, 1987), rate of progression (Huff *et al.*, 1987), and rate of familial occurrence (Ferini *et al.*, 1990), suggest the possibility of different pathogenesis. Both types of SDAT appear to have a genetic component, but this effect is more apparent in the younger age of onset disease. Work by Larsson *et al.* (1963) suggests that the genetic form of the disease is a dominant trait with incomplete penetrance in some families. Further, there is a familial association with patients with Down's syndrome (trisomy 21) who also suffer a similar form of progressive dementia. This singular insight led to a molecular genetic search for markers of Alzheimer's disease in chromosome 21.

The DNA marked D21S13 localized to the pericentric region (21q11) has shown linkage to familial Alzheimer's disease (Stinissen *et al.*, 1990). Other marker studies provide evidence that the familial form of the disease (St. George *et al.*, 1990) is not a single entity. Hardy (1990) suggests that the Alzheimer's disease locus is:

centromere (AD locus: D21S16/13)-D21S52-D21S1/S11-amyloid gene

A family of biomarkers is now being employed to define the genetic heterogeneity of this disease.

Studies in monozygotic twins who show discordance for SDAT (Kumar *et al.*, 1991) indicate that nongenetic or environmental factors play a role in this disease. Thus, it is interesting to note that high levels of aluminum are found in the brains of patients with Alzheimer's disease (Ganrot, 1986). Whether bioaccumulation of aluminum is due to enhanced blood–brain permeability to this element in Alzheimer's disease (Wisniewki and Kozlowski, 1982) or to some environmental risk factor is not well understood.

Matsubara *et al.* (1990) have attempted to screen for SDAT by measuring increased levels of 1-antichymotrypsin in cerebrospinal fluid. Bucht *et al.* (1983) hypothesized that a reduced supply of glucose would create a problem in acetylcholine synthesis, resulting in neurotransmitter dysfunction. Significantly lower blood glucose levels were found in Alzheimer patients than in age-matched controls. Abalan (1984) found low serum albumin levels in Alzheimer patients. Deary and Hendrickson (1986) hypothesized that low calcium levels might be responsible for the increase in the formation of senile plaques and neurofibrillary tangles.

Another potential marker for SDAT is the peptide somatostatin. Wood *et al.* (1982) looked at levels of the peptide somatostatin in CSF of patients with SDAT and found that levels were only 42% of control levels. Somatostatin is believed to act on specific cortical receptors, causing neuronal excitation in the hippocampus and neocortex. Thus, it is possible that somatostatin has either a neuromodulator or neurotransmitter function. Another peptide, somatomedin, is known to play a role in cell growth. Serum levels of somatomedin subtype A (SM-A) are known to decline with age. However, Sara *et al.* (1982) reported mean serum SM-A levels among SDAT patients to be more than 65% above values found in age-matched controls. Psychometric studies suggest that the WAIS-R subtest may be a useful marker for diagnosis of Alzheimer's dementia (Satz *et al.*, 1990).

The markers just described, as well as other potential genetic and nongenetic markers, show promise as predictors of risk or effect. However, reliable markers to identify Alzheimer's dementia have yet to be identified.

Brain and Central Nervous System Tumors

Tumors of the brain and CNS can be classified as primary (intrinsic) or secondary (metastatic; tumors infiltrating the CNS from the lung, breast, kidney, and skin). As described by Rutka *et al.* (1990), primary tumors develop from neoplastic transformation among the glia (astrocytes, oligodendrocytes, and ependymal cells), the pituitary cells, and the leptomeningeal cells. Brain tumors arising in the very young are extremely rare and may signal heightened familial susceptibility to tumorigenesis (Giuffre, 1989; Li *et al.,* 1986).

CNS tumors can be placed into subgroups that include the brain and spinal cord. The morphology and pathobiology of these tumors are diverse, varying from the benign and slow growing tumors (e.g., meningiomas, neurinomas, and hemangioblastomas) to the highly aggressive forms of primary CNS gliomas (i.e., anaplastic astrocytomas and glioblastomas). Although the gliomas cannot be separated easily into specific subgroups, they represent a continuum ranging from benign to highly malignant. Subgroups are determined by the irregular structure of the cell, molecular genetic damage, an abnormal increase in endothelial cells, and the presence of dead tissue.

Although estimates suggest that brain and CNS tumors are relatively rare among all age groups (less than 1% of all cancers), among children brain tumors account for approximately 20% of all cancers, second only to leukemia as the most frequent cancer. Survival rates over 5-years of brain tumor patients are approximately 25% for adults (over 35 years of age) and 45% for children 15 years of age or younger. Treatment has not improved survival in recent years; patients often are physically impaired if they do survive.

Brain and CNS tumor mortality and incidence rates appear to be on the rise in the United States (Ahlbom, 1990; Davis *et al.,* 1990; Greig *et al.,* 1990). Brain and CNS tumor mortality rates for all race groups experienced a statistically significant increase of 11% from 1973 (3.7/100,000) to 1988 (4.1/100,000) (Ries *et al.,* 1991). Almost 40,000 Americans are diagnosed with a primary or secondary brain tumor each year. In 1991, 16,700 new cases were reported and 11,500 deaths resulted from primary brain and CNS tumors (American Cancer Society, 1991).

At this time, relatively little is known about risk factors other than age and sex (female for meningioma) or potential causative agents associated with the development of brain and CNS tumors, although genetic and environmental factors are believed to play a role in their etiology. Several researchers have postulated that the observed increase in mortality is simply a response to better diagnosis and case finding. Others (Davis and Schwartz, 1988) have postulated that the increased rates are due to increasing exposure to occupational and environmental pollutants.

For example, Thomas and Waxweiler (1986) have suggested that exposure to chemicals such as formaldehyde, vinyl chloride, acrylonitrile, polycyclic aromatic hydrocarbons, organic solvents, and phenolic compounds may be responsible for the increasing mortality. Others (Schoenberg, 1982; Ahlbom, 1988; Ahlbom and Rodvall, 1989) have suggested that exposure to such disparate factors as electromagnetic radiation from power lines, dental amalgam, X-rays, and nitrosamine compounds may be responsible for the increased rates. Selected hereditary diseases, such as neurofibromatosis, tuberous sclerosis (Bourneville's disease), Li–Fraumeni's or SBLA cancer syndrome (breast cancer-sarcoma syndrome) Gorlin syndrome (naevoid-basal-cell-carcinoma) NBCC, Gorlin syndrome, and glioma-polyposis (Turcot syndrome) are risk factors for neuroepithelial brain tumors.

Biomarkers of brain tumor risk are not yet available and clinical markers of tumor type are based primarily on tumor morphology. Work at the molecular and cellular level has begun to characterize some of these tumors further in terms of chromosomal or DNA deletions (Fults et al., 1989, 1990; Mashiyama et al., 1991) and oncogene expression (MacGregor and Ziff, 1990). Aside from routine clinical and radiologic diagnostic and prognostic markers of tumor identification and progression, there are few molecular biomarkers of primary brain tumors. Leukocyte alkaline phosphatase activity has been used to assist in the differentiation of brain tumors from cerebrovascular disease (Walach et al., 1991). Neuron specific enolase, a maker of neuron damage (Ko et al., 1990), and levels of IgG in spinal fluid (Onodera et al., 1987) may have an ancillary role in diagnosis and prognosis of brain tumors. van Zanten et al. (1985), in searching for CSF tumor markers, noted that β-glucuronidase was the most useful CSF marker for meningeal spread from solid tumors.

These biochemical measures of tumorigenesis and brain tumor progression indicate a clear need for further refinement of molecular biomarker technology to define primary brain cancer risk and tumorigenesis.

Conclusions

Molecular markers doubtlessly will play an important role in the epidemiology of neurologic diseases in the 1990s. However, at the present time, considerable developmental research is required before expectations can be realized fully. Such research must focus on the types of laboratory assays available, their sensitivity and specificity in study populations, and the important aspects of inter- and intraperson variability when attempting to interpret the use of highly sophisticated methods applied in the study of selected diseases in large populations.

References

Abalan, F. (1984). Alzheimer's disease and malnutrition: A new etiological hypothesis. *Med. Hypotheses* **15**, 385–383.

Ahlbom, A. (1988). A review of the epidemiologic literature on magnetic fields and cancer. *Scand. J. Work Environ. Hlth.* **14**, 337–343.

Ahlbom, A. (1990). Some notes on brain tumor epidemiology. *Ann. N.Y. Acad. Sci.* **609**, 179–185.

Ahlbom, A., and Rodvall, Y. (1989). Brain tumour trends. *Lancet* ii, 1272.

Alzheimer, A. (1907). Ueber eine eigenartige Erkrankung der Hirnrinde. *Allg. Z. Psychiatr.* **64**, 146.

American Cancer Society (1991). "Cancer Facts and Figures." ACS, Atlanta, GA.

Armbrustmacher, V. W. (1979). Pathology of dementia. *Pathol. Ann.* **14**, 145–173.

Barroche, G., Perrier, P., Raffoux, C., Gehin, P., Streiff, F., and Weber, M. (1986). HLA and familial multiple sclerosis. *Rev. Neurol.* **142**, 738–745.

Bouchard, G., Roy, R., and Declos, M. (1988). Reproduction and gene transmission of myotonic dystrophy in the Saguenay region (Quebec). *J. Genet. Hum.* **36**, 221–237.

Brunner, H. G., Korneluk, R. G., and Coerwinkel-Driessen, M. (1989). Myotonic dystrophy is closely linked to the gene for muscle type creatine kinase(CKMM). *Hum. Genet.* **81**, 308–310.

Bucht, G., Adolfsson, R., Lithner, F., *et al.* (1983). Change in blood glucose and insulin secretion in patients with senile dementia of Alzheimer type. *Acta Med. Scand.* **213**, 387–391.

Burns, R. S., LeWitt, P. A., Ebert, M. H., Pakkenberg, H., and Kopin, I. J. (1989). The clinical syndrome of striatal dopamine deficiency. Parkinsonism induced by 1-methyl-4-phenyl-1,2,3,6-tetrahydropyridine (MPTP). *N. Engl. J. Med.* **12**, 1418–1421.

D'Andrea, V., Meco, G., Corvese, F., Baselice, P.F., and Ambrogi, V. (1989). The role of the thymus in multiple sclerosis. *Ital. J. Neurol. Sci.* **10**, 43–48.

Davis, D. L., and Schwartz, J. (1988). Trends in cancer mortality: U.S. white males and females, 1968–83. *Lancet* i, 633–636.

Davis, D. L., Hoel, D., Percy, C., Ahlbom, A., and Schwartz, J. (1990). Is brain cancer mortality increasing in industrial countries? *Ann. N.Y. Acad. Sci.* **609**, 192–204.

Deary, I. J., and Hendrickson, A. E. (1986). Calcium and Alzheimer's disease. *Lancet* i, 1219.

Deber, C. M., and Reynolds, S. J. (1991). Central nervous system myelin: Structure, function and pathology. *Clin. Biochem.* **24**, 113–134.

Delgado-Escueta, A. V., Greenberg, D. A., Treiman, L., Liu, A., Sparkes, R. S., Barbetti, A., Park, M. S., and Terasaki, P. I. (1989). Mapping the gene for juvenile myclonic epilepsy. *Epilepsia* **30**(Suppl. 4), S8–18.

Ferini, S. L., Smirne, S., Garancini, S., Pinto, P., and Franceschi, M. (1990). Clinical and epidemiologic aspects of Alzheimer's disease with presenile onset: A case control study. *Neuroepidemiol.* **9**, 39–49.

Filbin, M. T., and Poduslo, S. E. (1986). A comparison of the glycoproteins and proteins from multiple sclerosis and normal brain tissue. *Neurochem. Res.* **11**, 1151–1166.

Fults, D., Pedone, C. A., Thomas, G. A., and White, R. (1990). Allelotype of human malignant astrocytoma. *Cancer Res.* **50**, 5784–5789.

Ganrot, P. O. (1986). Metabolism and possible health effects of aluminum. *Environ. Health Perspect.* **65**, 363–441.

Giuffre, R. (1989). Biological aspects of brain tumors in infancy and childhood. *Childs. Nerv. Syst.* **5**, 55–59.

Gessain, A., Saal, F., Gout, O., De-The, G., and Peries, J. (1988). HTLV-1 virus and associated chronic neuromyelopathies. Current data and hypotheses. *Nouv. Rev. Fr. Hematol.* **30**, 15–20.

Greenberg, C. R., Jacobs, H. K., and Nylen, E. (1988). Gene studies in newborn males with Duchenne muscular dystrophy detected by neonatal screening. *Lancet* **2**, 425–427.

Greig, N. H., Ries, L. G., Yancik, R., Rapoport, S. I. (1990). Increasing annual incidence of primary malignant brain tumors in the elderly. *Proc. Am. Assoc. Cancer Res.* **31**, 229.

Hardy, J. (1990). Molecular genetics of Alzheimer's disease. *Acta Neurol. Scand.* **129**(S), 31.

Heidenreich, F., Vincent, A., Roberts, A., and Newsom-Davis, J. (1988). Epitopes on human acetylcholine receptor defined by monoclonal antibodies and myasthenia gravis sera. *Autoimmunity* **1**, 285–297.

Heikkila, R. E., Hess, A., and Duvoisin, R. C. (1984). Dopaminergic neurotoxicity of 1-methyl-4-phenyl-1,2,5,6-tetrahydropyridine in mice. *Science* 1451–1453.

Henke, H., and Lang, W. (1983). Cholinergic enzymes in neocortex, hippocampus and basal forebrain and non-neurological and senile dementia of Alzheimer type patients. *Brain Res.* **267**, 281.

Hertzman, C., Weins, M., Bowering, D., Snow, B., and Caine, D. (1990). Parkinson's disease: A case-control study of occupational and environmental risk factors. *Am. J. Ind. Med.* **17**, 349–355.

Heston, L. L., and Master, A. R. (1977). The genetics of Alzheimer's disease: Associations with hematologic malignancy and Down's syndrome. *Arch. Gen. Psychiat.* **34**, 976–981.

Huff, F. J., Growdon, J. H., Corkin, S., and Rosen, T. J. (1987). Age at onset and rate of progression of Alzheimer's disease. *J. Am. Geriatr. Soc.* **35**, 27–30.

Katzman, R. (1976). The prevalence and malignancy of Alzheimer disease. A major killer. *Arch. Neurol.* **33**, 217–218.

Ko, F. J., Chiang, C. H., Wu C. C., and Wu. L. (1990). Studies of neuron-specific enolase levels in serum and cerebrospinal fluid of children with neurological diseases. *Kao Hsiung I Hsueh Ko Tsa Chih* **6**, 137–143.

Korenberg, J. R., Kawashima, H., Pulst, S., Ikeuchi, T., Ogasawara, N., Yamamoto, K. Schonberg, S. A., West R., Allen, L., Magenis, E., Ikawa, K., Taniguchi, N., and Epstein, C. J. (1990). Molecular definition of a region of chromosome 21 that causes features of the Down Syndrome phenotype. *Am. J. Hum. Genet.* **47**, 236–246.

Kumar, A., Schapiro, M. B., Grady, C. L., *et al.* (1991). Anatomic, metabolic, neuropsychological, and molecular genetic studies of three pairs of identical twins discordant for dementia of the Alzheimer's type. *Arch. Neurol.* **48**, 160–168.

Kurland, L. T., and Kurtzke, J. F. (1972). Geographic neuropathology. *In* "Pathology of the Nervous System" (J. Minckler, ed.), Vol. 3, pp. 2771–2803. McGraw-Hill, New York.

Kurland, L. T., Kurtzke, J. F., and Goldberg, I.D. (1973). Parkinsonism. *In* "Epidemiology of Neurologic and Sense Disorders," pp. 41. Harvard University Press, Cambridge.

Larsson, T., Sjogren, T., and Jacobson, G. (1963). A clinical, sociomedical and genetic study. *Acta Psychiat. Scand. Suppl.* **167**, 1–159.

Leppert, M., Anderson, V. E., Quattlebaum, T., Stauffer, D., O'Connell, P., Nakamura, Y., Lalouel, J. M., and White, R. (1989). Benign familial neonatal convulsions linked to genetic markers on chromosome 20. *Nature* (London) **337**, 647–648.

Li, F. P., Jamison, D. S., and Meadows, A. T. (1986). Questionnaire study of cancer etiology in 503 children. *J. Natl. Cancer Inst.* **76**, 31–36.

McFarland, H. F., and Dhib-Jalbut, S. (1989). Multiple sclerosis: Possible immunologic mechanisms. *Clin. Immunol. Immunopathol.* **50**, S96–105.

Masanori, N., Keiichi, N., Hiroaki, Y., *et al.* (1991). Epidemiology of progressive muscular dystrophy in Okinawa, Japan. *Neuroepidemiol.* **10**, 185–191.

Mashiyama, S., Murakami, Y., Yoshimoto, T., Sekiya, T., and Hayashi, K. (1991). Detection of *p53* gene mutations in human brain tumors by single-strand conformation polymorphism analysis of polymerase chain reaction products. *Oncogene* **8**, 1313–1318.

Massimo, F., Airaghi, L., Gramigna, C., *et al.* (1991). ACTH and cortisol secretion in patients with Alzheimer's disease. *J. Neurol. Neurosurg. Psych.* **54**, 836–837.

Matsubara, E., Hiral, S., Amari, M., Shoji, M., Yamaguchi, H., Okamoto, K., Ishiguro, K., Harigaya, Y., and Wakabayashi, K. (1990). Alpha 1-antichymotrypsin as a possible marker for Alzheimer-type dementia. *Ann. Neurol.* **28**, 561–567.

Milonas, I., Tsounis, S., and Logothetis, I. (1990). Epidemiology of multiple sclerosis in northern Greece. *Acta Neurol. Scand.* **81**, 43–47.

Mulley, J. C., Gedeon, A. K., and Haan, E. A. (1988). Application of DNA probes to carrier detection and prenatal diagnosis of Duchenne (and Becker) muscular dystrophy. *Aust. Paediatr. J.* 24, *Suppl. 1*, 92–97.

Murphy, S. (1980). Pesticides. *In* "Casarett and Doull's Toxicology: The Basic Science of Poisons," pp. 367. Macmillan, New York.

Norgaard-Pedersen, B., Larsen, S. O., Arends, J., Svenstrup, B., and Tabor, A. (1990). Maternal serum markers in screening for Down syndrome. *Clin. Genet.* 37, 35–43.

Olsson, J. E., Kaugesaar, T., and Lindvall, B. (1990). CSF monoamine metabolites and neuropeptides in neurodegenerative disorders. *Acta Neurol. Scand. Suppl.* 128, 24.

Onodera, Y., Saitoh, Y., and Nagai, K. (1987). The diagnostic value of immunoglobulin G in the cerebrospinal fluid of brain tumor patients, particularly in malignant tumors. *Japanese Journal of Cancer Clinics* 33, 1402–1406.

Pachner, A. R. (1988). Myasthenia gravis. *Immunol. Allergy Clin. North Am.* 8, 227–293.

Perry, E. K. (1979). Acetylcholine and Alzheimer's disease. *Lancet* i, 42.

Perry, E. K., and Perry, R. H. (1980). The cholinergic system in Alzheimer's disease. *In* "Biochemistry of Dementia" (P. J. Roberts, ed. Wiley, New York.

Perusini, F. (1910). Ueber klinische und histologische eigenartige psychische Erkrankungen des spaeteren Lebensalters. *In* "Histologische und Histopathologische Arbeiten ueber die Grosshirnrinde mit besonderer Beruecksichtigung der Pathologischen Anatomie der Geisteskrankheiten" (F. Nissl and A. Alzheimer, eds.), Vol. 3. Fischer, Jena.

Plum, F. (1979). Dementia: An approaching epidemic. *Nature (London)* 279, 372–373.

Ries, L. A. G., Hankey, B. F., Miller, B. A., Hartman, A. M., and Edwards, B. K. (1991). "Cancer Statistics Review 1973–88." NIH Publ. No. 91-2789. U.S. Government Printing Office, Washington, D.C.

Rikonen, R. (1989). The role of infection and vaccination in the genesis of optic neuritis and multiple sclerosis in children. *Acta Neurol. Scand.* 80, 425–31.

Roses, A. D., Pericak-Vance, M. A., and Bartlett, R. J. (1988). Dystrophia myotonica: Childhood neuromuscular problems. Myotonic dystrophy: Update on progress to define the gene. *Aust. Paediatr. J.* 24, *Suppl. 1*, 66–69.

Rutka, J. T., Trent, J. M., and Rosenblum, M. L. Molecular Probes In Neuro-Oncology: A Review. *Cancer-Invest*, 8(3–4), 425–438.

St. George, H. P. H., Haines, J. L., Farrer, L. A., Polinsky, R., Van Broeckhoven, C., Goate, A., McLachian, D. R., Orr, H., Bruni, A. C., and Sorbi, S. (1990). Genetic linkage studies suggest that Alzheimer's disease is not a single homogenous disorder. FAD Collaborative Study Group. *Nature (London)* 347, 194–197.

Sara, V. R., Hall, K., Enzell, K., *et al.* (1982). Somatomedins in aging and dementia disorders of the Alzheimer type. *Neurobiol. Aging* 3, 117.

Satz, P., Hynd, G. W., D'Elia, I., Daniel, M. H., Van Gorp, W., and Connnor, R. (1990). A WAIS-R marker for accelerated aging and dementia, Alzheimer's type? Base rates of the Fuld formula in the WAIS-R standardization sample. *J. Clin. Exp. Neuropsychol.* 12, 759–765.

Schoenberg, B. S. (1982). Nervous system. *In* "Cancer Epidemiology and Prevention" (D. Schottenfeld and J. F. Fraumeni, eds.). Saunders, Philadelphia.

Seltzer, B., Burres, M. J., and Sherwin, I. (1987). Left handedness in early and late onset and rate of progression of Alzheimer's Disease. *J. Am. Geriatr. Soc.* 35, 27–30.

Seyfert, S., Klapps, P., Meisel, C., Fischer, T., and Jungham, U. (1990). Multiple sclerosis and other immunologic diseases. *Acta Neurol. Scand. Suppl.* 81, 37–42.

Sinex, M. F., and Merril, C. R. 91982). *Ann. N.Y. Acad. Sci.* 396, 1.

Souan, M. L., Geffard, M., and Viellemaringe, J. (1987). Anti-acetylcholine antibodies and the pathogenesis of myasthenia gravis. *Ann. N.Y. Acad. Sci.* 505, 423–438.

Stinissen, P., Van Hul, W., Van Camp, G., Backhovens, H., Wehnert, A., Vandenberghe, A., and

Van Broeckhoven, C. (1990). The pericentric 21 DNA marker pGSM21 (D2s13) contains an expressed HTF Island. *Genomics* 7, 119–122.

Tanner, C. M., Chen, B., Wang, W. Z., Peng, M. L., Lui Z. L., Liang, X. L., Kao, L. C., Gilley, D. W., and Schoenberg, B. S. (1987). Environmental factors in the etiology of Parkinson's disease. *Can. J. Neurol. Sci.* 14(3), 419–423.

Thibault, M. C., Mathieu, J., and Moorjani, S. (1989). Myotonic dystrophy: Linkage with apolipoprotein E and estimation of the gene carrier status with genetic markers. *Can. J. Neurol. Sci.* 16, 134–140.

Thienhaus, O. J., Hartford, J. T., Skelly, M. F., and Bosmann, H. B. (1985). Biologic markers in Alzheimer's disease. *J. Am. Geriat. Soc.* 33, 715–726.

Thomas, T. L., and Waxweiler, R. J. (1986). Brain tumors and occupational risk factors. *Scand. J. Work Environ. Hlth.* 12, 1–15.

Tourelotte, W. W., Baumhefner, R. W., Syndulko, K., Shapshak, P., Osborne, M., Rubinshtein, G., Newton, L., Ellison, G., Myers, L., and Rosario, I. (1988). The long march of the cerebrospinal fluid profile indicative of clinical definite multiple sclerosis and still marching. *J. Neuroimmunol.* 20, 217–227.

Tzartos, S. J. (1988). Myasthenia gravis studied by monoclonal antibodies to the acetylcholine receptor. *In Vivo* 2, 105–110.

van Zanten, A. P., Twijnstra, A., van Benthem, V., Hart, A. A. M., and Ongerboer de Visser, B. W. (1985). Cerebrospinal fluid β-glucuronidase activities in patients with central nervous system metastases. *Clin. Chim. Acta* 147, 127–134.

Walach, N., Guterman, A., Rabinowitz, H., Kaufman, S., and Habot B. (1991). Leukocyte alkaline phosphatase score in patients with cerebrovascular disease and in patients with primary and metastatic brain tumors. *J. Surg. Oncol.* 46, 37–39.

Ward, P. A., Hejtmancik, J. F., and Witowski, J. A. (1989). Prenatal diagnosis of Duchenne muscular dystrophy: Prospective linkage analysis and retrospective dystrophin cDNA analysis. *Am.J.Hum. Genet.* 44, 270–281.

Williams, C. L., Lennon, V. A., Momoi, M. Y., and Howard, F. M. (1987). Serum antibodies and monoclonal antibodies secreted by thymic B-cell clones from patients with myasthenia gravis define striatal antigens. *Ann. N.Y. Acad. Sci.* 505, 168–179.

Wisnieki, H. M., and Kozlowski, P. B. (1982). Evidence for blood-brain barrier changes in senile dementia of the Alzheimer type (SDAT). *Ann. N.Y. Acad. Sci.* 396, 119–129.

Wood, P. L., Etineene, P., Lal, S., *et al.* (1982). Reduced lumbar CSF somatostatin levels in Alzheimer's disease. *Life Sci.* 31, 2073.

Wynn, D. R., Rodriguez, M., O'Fallon, W. M., and Kurland, L. T. (1990). A reappraisal of the epidemiology of multiple sclerosis in Olmstead County, Minnesota. *Neurology* 40, 780–786.

20

Practical Applications of Biomarkers in the Study of Environmental Liver Disease

Carlo H. Tamburro
John L. Wong

Essential Background

Introduction

The concern over adverse health effects caused by exposure to environmental toxins and unique role of the liver in the metabolism of these toxicants makes biomarkers of liver metabolism and disease clinically very important. Hepatic biomarkers are needed to assess exposure, identify subclinical hepatic injury, monitor for chronic disease, assess long-term risk, and allow for preventive intervention before liver injury progresses to an irreversible stage. Examination of these hepatic biomarkers in the clinical, occupational, and environmental settings will verify their ability to detect specific exposures or adverse health effects. Also, under proper conditions, they can provide the means to reassure exposed individuals of no future adverse health risk.

Liver

Structure

Hepatic biomarkers, especially enzymatic ones, are biologically related to the anatomical architecture of the liver. The *architectural unit* of the liver has been described classically as subunits of hexagonal lobules, 1–2 mm in diameter, situated about a central vein. The boundaries of these lobules are demarcated by portal tracts, composed of two blood supplies (arterial and venous) and the biliary excretory system. These portal tracts approximately

follow the angles of the hexagons (Figure 20.1A). Blood flows to the parenchymal tissue from the portal triads at the periphery of the classic lobule and
exits via the central veins. About one-third of the blood supply to the lobules
is provided by the hepatic artery; the remainder is supplied by the portal vein.
Because hepatic blood flow is such an essential component of hepatic function, the portal triads are considered the center of the *functional unit,* called
the asinus. The parenchymal cells surrounding this vascular distribution are
divided into three zones: Zone 1, closest to the arterial and portal blood supplies; Zone 2, between Zones 1 and 3 in the center of the parenchyma; Zone
3, surrounding the central vein region (Figure 20.1B).

Function

The liver has multiple functions. The primary role is metabolic, involving uptake of substrates for storage, metabolism, and distribution via the
blood and bile. Its second major role is conversion of xenobiotic agents and
endogenous materials into excretable compounds. This second metabolic
function can sometimes convert otherwise harmless compounds into toxic
ones. Third, the liver is a major site for clearance of bacterial and other ma-

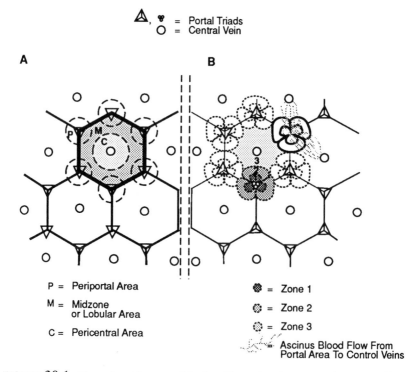

FIGURE *20.1*   Normal architecture of the liver illustrating the structural concept of hepatic
lobule (A) and the functional concept of hepatic ascinus (B).

TABLE 20.1   Useful Biochemical Markers of Liver Function and Injury[a]

Enzymes
Alanine aminotransferase (ALT)
Aspartate aminotransferase (AST)
Gammaglutamyl transpeptidase (GGT)
Lactic acid dehydrogenase (LDH)
Alkaline phosphatase (AP)
Proteins
Albumin (Alb)
Prothrombin (PT)
Bile acids (BA)
Cholyglycine
Total bile acids
Bilirubin (TB)
Conjugated/direct (DB)
Unconjugated/indirect (IB)
Metabolic/Physiologic
Aminopyrine breath test (ABT)
Indocyanine green clearance (ICG)

[a] Referred to as biochemical liver tests (BLTs).

terials by its phagocytic activity via the reticular endothelial system (RES), which also encompasses parts of the immune system. The metabolic role of the liver makes it vulnerable to toxic injury from exposure to a variety of metabolic and xenobiotic insults. Hepatic biomarkers can identify this injury and, in many cases, characterize the location, severity, and nature of the damage. Toxic exposure may manifest itself in three forms: enzyme induction, hepatocellular damage, and cholestasis. Currently useful biochemical markers (clinical tests) that identify these liver responses to toxins are shown in Table 20.1.

Biomarkers Currently in Use or under Consideration

Exposure Detection

Cellular Enzymes

Xenobiotics are known to undergo hepatic biotransformation and produce bioactive, rather than detoxified, metabolites. Although most bioactivation has been demonstrated in animals, acetaminophen, aflatoxin B_1, arsenic, carbon tetrachloride, halothane, isoniazide, and vinyl chloride have been shown to produce acute or chronic disease and malignant transformation in humans. The striking characteristic of biotransformation is the enhancement of cellular enzyme activity. This activity may be used to determine acute and chronic exposure to xenobiotics that exceed the background enzyme induction caused by natural products in the diet and inhaled air. These cellular enzyme markers, as presently characterized, are generic in response to xenobiotic exposure.

Oxidative induction: MFO-P450   Cytochrome P450, the monooxygenase system, is a family of mixed-function oxidase (MFO) enzymes with unusual versatility because of the multiplicity of forms. In humans, Wang *et al.* (1983) have identified six P450s shown to metabolize different compounds. As shown in Table 20.2, these human P450s have different but overlapping broad substrate specificities. Chemical induction of P450 has been divided arbitrarily into two classes, the PB type (phenobarbitol-induced) and the 3MC type (3-methylcholanthrene-induced), on the basis of the induction of characteristic P450 isozymes and the mechanism of induction. For example, a dioxin derivative, TCDD, belongs to the 3MC class (Le Provost *et al.*, 1983). This type of cellular enzyme biomarker may be used to identify specific xenobiotic injury for which organ tissue is available. Indirect measurement of these types of enzyme biomarkers can be done by metabolic clearance tests that are surrogate measures of oxidation.

Surrogate measure of oxidation: aminopyrine breath test   More applicable means of indirect measurements of the hepatic P450 oxidation system are metabolic clearance tests. There are a number of such tests, such as the aminopyrine breath test (ABT) or caffeine clearance (Baker *et al.*, 1983). ABT has had the most extensive use in xenotoxic assessment. Aminopyrine is administered orally and oxidized primarily in the liver, liberating formaldehyde, which undergoes subsequent metabolism to CO_2. This carbon atom is labeled with either ^{14}C or ^{13}C and recovered in the breath, allowing an indirect measurement of P450 activity as a reflection of the functional liver mass.

TABLE 20.2   **Human Hepatic Microsomal P450 Induction in Xenobiotic Metabolism**[a]

Xenobiotic	P450 −2	P450 −3	P450 −4	P450 −5	P450 −7	P450 −8
Acetanilide	L	M	L	L	L	L
Benzo(*a*)pyrene	L	M	M	M	M	M
d-Benzphetamine	M	H	H	H	H	H
Trichloroethylene	L	L	L	L	L	L
1-Naphthylamine	L	—	L/M	—	—	L/M
2-Naphthylamine	L	—	L	—	—	L

Source: Adapted from Wang *et al.* (1983).
[a] Relative rates of metabolism (nmol/min/nmol P450): L, low (≤ 0.09); M, medium (0.1–0.99); H, high (≥ 1.0).

Lidocaine clearance Like ABT, lidocaine clearance is an induced measurement of P450 activity which, in turn, reflects the functional hepatic mass. Lidocaine is aminoethylacetanilide, which undergoes rapid N-deacylation via the hepatic cytochrome P450 system to yield several metabolites, principally monoethyglycinexylidide (MEGX). The concentration of MEGX in serum before and 15 min after iv administration of 1 mg/kg lidocaine can be determined by a fluorescence polarization immunoassay system.

Detoxification induction: glutathione Glutathione (GSH) is the major endogenous protective substance that participates in covalent binding of reactive electrophilic metabolites and in reducing peroxides. The enzymes (transferase) involved in catalyzing the glutathione detoxification effects are also potential biomarkers. These transferases comprise a family of enzymes with overlapping but distinct substrate specificities (Vander Jagt *et al.*, 1985). For example, glutathione S-transferase (GST) is an enzyme involved in catalyzing the detoxification of potential diol-epoxide carcinogenic or mutagenic metabolites of polycyclic aromatic hydrocarbons (Glatt *et al.*, 1983).

Metabolic and Physiologic Tests

Clearance tests: indocyanine green clearance Hepatic function and reserve are related to the ability of the liver to clear substances from the blood. Hepatic extraction (intrinsic clearance) is a determinant of bioavailability that can be measured by the systemic clearance of liver specific substances, for example, iodocyanine green (ICG), galactose, or bile acids. There is good correlation between systemic clearance of ICG and early xenotoxic liver injury. The measurement of hepatic extraction is highly correlated to hepatic blood flow. These substances are given intravenously and their clearance rate is determined by small serial blood samples over 10–15 min (Tamburro and Liss, 1986). Lower clearance of these substances results from lower extraction due to intrahepatic blood flow changes secondary to toxic liver injury.

Bile acids Bile acids are naturally produced hepatic substrates, cleared solely by the hepatocytes, and have shown good correlation with early xenobiotic liver injury. The advantage of this study is that no injection of any substance is required. Bile acids are measurable in serum by radioimmunoassays.

Proteins: Antigens and Antibodies

Some environmental hazards are biologic, for example, viruses that can occur concomitantly with xenobiotic exposure and act as confounders or cotoxins to the liver. Specific antigen–antibody markers are available for identification of exposure and active hepatocellular injury due to a major hepatic virus (Tamburro, 1991). The human virus (e.g., hepatitis B and C) plays a very important role in the causation of liver cancers associated with natural and synthetic xenobiotics (e.g., aflatoxin).

Hepatic fibrosis is the alternative repair mechanism (as opposed to regeneration) for hepatic injury. It is the key indicator of serious hepatic injury after xenobiotic exposure. The detection of hepatic fibrosis is especially important in low-level chronic and subclinical exposure. Noninvasive serum markers of hepatic fibrosis showing early promise include N-terminal propeptides of Type III procollagen and Type IV collagen fragments. These markers of serum concentration correlate well with gene expression (messenger RNA levels) in dimethylnitrosamine- and carbon tetrachloride-induced hepatic fibrosis (Hayasaka *et al.*, 1988; Salvolainen *et al.*, 1988). Further studies are needed to establish baseline variations and to characterize their course in various forms of human liver injury.

Adducts

Assays for xenobiotic binding to GSH, various proteins, and DNA are under development and field application. Antibodies, polyclonal and monoclonal, have been in development for a number of hepatic xenobiotics, for example, aflatoxin B_1 (Sabbioni *et al.*, 1990) and acrylonitrile (Wong *et al.*, 1990). Immunoassays for these xenobiotic antibodies are being developed to detect adducts to GSH, albumin, hemoglobin, and DNA. Clinical trials for each type of adduct provides different information with respect to degree, duration, and dose of exposure. Monoclonal antibody assay for specific hepatic xenobiotics can be applied in many ways, as shown in Table 20.3 (Perera and Weinstein, 1982; Perera *et al.*, 1986).

Genetic Markers

Restriction fragment length polymorphism   Gene susceptibility for the development of alcohol liver injury is suggested, since only a minority of alcoholics develop cirrhosis. Using restriction fragment length polymorphism (RFLP) testing of the Type I collagen gene, the collagen type most prevalent

TABLE 20.3   Useful Molecular Monitoring of Hepatic Xenobiotics by Adduct Formation

Biologic site	Adduct	Half-life
Metabolic	GSH Conjugates	Hours
Protein	Albumin	Days
Cell	Hemoglobin (RBC)	Weeks
Nucleus	DNA	Months/Years *(if no repair)*

in human cirrhosis (Winer *et al.*, 1988), specific agent identification can be made. A RFLP exists when two different but normal nucleotide patterns exist at the same site in the genomic DNA. This difference can be recognized by digestion with a bacterial restriction endonuclease. White blood cell DNA is obtained from exposed individuals, digested with two restriction enzymes, and hydridized with two Type I collagen DNA probes to reveal an RFLP. Six haplotypes or patterns of polymorphisms have been found in alcohol-exposed individuals, based on the presence or absence of these two poly-morphisms. One haplotype is found more frequently in cirrhotic alcoholics than in alcoholics without cirrhosis or in controls. If family studies confirm such a linkage, this type of identified polymorphism could provide a means of identifying individuals at risk of developing liver disease (cirrhosis) when exposed to alcohol. The ability to identify various capabilities of metabolism of xenobiotics, as well as the propensity for formation of collagen after in-jury, can be used as a generic molecular marker for low-dose xenobiotic hepatic exposure or injury.

Activated proto-oncogenes and inactivated tumor suppressor genes Acti-vated transforming genes (oncogenes) have been found in a number of hu-man tumors by use of assays in which transformed foci result from transfec-tion of tumor DNA into NIH3T3 cells. One striking fact that has emerged from screening transfecting DNA is that, for both human and rodent tumor DNA, the transforming genes are virtually all related to the *ras* oncogene family. Activation of *ras* proto-oncogenes by a carcinogenic agent often in-volves base substitutions at codons 12 and 61. Some examples of activated *ras* oncogenes found in liver tumors are given in a review by Harris (1991): aflatoxin B_1-induced $G^{34} \rightarrow T$ and $G^{35} \rightarrow A$ mutations in Ki-*ras* of rat; benzidine-induced $C^{181} \rightarrow A$ mutation in Ha-*ras* of mouse; and urethane-induced $A^{182} \rightarrow T$ mutation in Ha-*ras* of mouse. In addition, Wogan and co-workers (McMahon *et al.*, 1990) reported the presence of Ki-*ras* oncogenes ($G \cdot C \rightarrow A \cdot T$ or $G \cdot C \rightarrow T \cdot A$ in codon 12) in the liver of flounders with hepato-cellular carcinomas that were taken from a contaminated site in Boston Har-bor. DNA samples from histologically normal liver of flounder from a less polluted site showed only wild-type DNA sequences at codon 12 of Ki-*ras*. Since the *ras* oncogene is involved in early, late, and metastatic stages of car-cinogenesis, determination of *ras* mutation in a liver biopsy sample may be used as a biomarker of susceptibility (in the absence of liver impairment) or of effect (in conjunction with liver tumor).

In contrast to proto-oncogenes, tumor suppressor genes are cellular genes that regulate cell growth, induce apoptosis (programmed cell death), and maintain genomic stability. Inactivating the normal allele will cause dys-regulation of growth and differentiation pathways, enhancing cell transformation. In this sense, it is far more likely to disable a gene than to activate a proto-oncogene by point mutation. For example, the *ras* gene is activated by mutation in a few specific codons only. Further, some mutated forms of *p53* are transforming oncogenes. In sum, the *p53* tumor suppressor gene has shown the best association with human liver cancers (Harris, 1991). The majority of the mutations in human tumors occur in exons 5–8 of the *p53* gene, where the hot spots are grouped in the coding region 248–282. A *p53* hot spot mutation in hepatocellular carcinoma has been linked to aflatoxin exposure and hepatitis B virus. Codon 249 mutation of the *p53* gene is strongly associated with high aflatoxin exposure and identifies an endemic form of hepatocellular cancer (Ozturk *et al.*, 1991). Detections of mutations in *ras* and *p53* in small tissue samples are now made possible by polymerase chain reaction (PCR) technology. PCR rapidly is becoming the preeminent area of diagnostic hepatology with respect to environmental hazards, including viruses. Often, exposure to environmental xenobiotics is complicated by latent hepatotoxic agents, especially hepatitis types B, C, and D. Especially relevant is the ability of hepatitis B to become integrated into human DNA and no longer be identifiable by standard immunologic markers. PCR application of specific hepatic viral RNA and DNA species allows detection of such latent confounders, provides more accurate classification of the exposed individual, and has the ability to determine whether any synergistic effect may occur due to the dual hepatotoxin exposure. PCR allows vast amplification of specific DNA species of interest (10^6-fold increases are routine). DNA can be detected from nucleotide samples of less than 10 pg; therefore, the technique is applicable to human liver biopsy samples.

Analytical Techniques

Metabolites in body fluids   Oxygenated derivatives of environmental carcinogens, such as benzo[a]pyrenes and aflatoxin in blood and urine, are direct exposure markers. Analytical techniques involving gas chromatography, mass spectrometry (GC-MS), high-performance liquid chromatography (HPLC), and thin-layer chromatography (TLC) with fluorescence detection are sensitive to nanogram levels. However, blood and urinary metabolites of volatile chemicals (e.g., vinyl chloride and acrylonitrile yielding water-soluble thio acids) are not readily quantitated under low-level exposure conditions. Direct analysis of hepatotoxin metabolites is most useful in heavy metal (e.g., iron, arsenic, copper) dose–response study. Although atomic absorption spectroscopy is widely used to determine metals at ppb concentration in biologic specimens, this method does not provide species information because the analyte is determined at the atomic state. To this end, absorptive stripping voltammetry is being developed for metal speciation, for example, speciation of nickel(II)-histidine as a biomarker for nickel exposure (Wu and Wong, 1991).

Effect/Diagnostic

Application of Biomarkers of Hepatic Effects Caused by Xenobiotic Exposure

Hepatic biomarkers of environmental exposure can identify three major outcomes: acute, chronic, and latent. Chronic and latent diseases (e.g., cancer) are complex multistepped processes. Therefore, any single biomarker is likely to identify only one or a few of the various steps. However, two major steps are essential to all chronic and carcinogenic processes: tissue injury and cellular repair or regeneration. In acute or chronic hepatic exposure, only incomplete cellular repair of toxic injury allows identification of the exposure. Without detectable injury, there is no clinically meaningful exposure. In the carcinogenic process, malignant transformation cannot occur without both tissue injury and cellular regeneration (i.e., dead cells or cells unable to replicate cannot become malignant). Therefore, the most useful hepatic biomarkers assess injury or identify cellular replication.

Presently, the most frequently used markers of hepatic injury are enzymatic or biochemical ones (Table 20.1). These markers are relatively nonspecific with respect to etiology, but are the clinical standard for the absence or presence of hepatic injury. Depending on the degree (level) of hepatic injury, these markers have relative diagnostic usefulness in the detection of hepatotoxic exposure (Table 20.4).

Level 1: adaptive response   At this level, clinical exposure is followed by metabolic or biologic changes that result in no injury, for example, gamma glutamyl transpeptidase (GGT) enzyme induction after alcohol exposure or P450 induction (via abnormal ABT) after synthetic hydrocarbon exposure. Enzyme induction is a physiologic or structural adaptation. There is no cellular damage or death. All other biochemical liver tests (BLTs) and tests of synthetic function are normal.

Level 2: acute injury, mild   This level of clinical exposure causes cellular changes that are nonprogressive, reversible, without disruption of cellular

TABLE 20.4   **Diagnostic Effect: Biomarkers for Various Outcomes**[a]

Adaptive response: Biological change, no injury (P450s, GGT, ABT)
Acute injury: Mild (AST/SGOT, ALT/SGPT, ICG)
Acute injury: Severe (total bilirubin, albumin, PT, transferrin)
Chronic injury (cholyglycine, alkaline phosphatase, procollagen III)
Disease
Nonmalignant: Cirrhosis (albumin, PT, cholyglycine, alkaline phosphatase)
Malignant (α-fetoprotein, lactic dehydrogenase)

[a] Markers in parentheses represent those that are more characteristic of a condition.

function, and without evidence of residual injury, for example, alcohol-induced fatty liver with cellular enzyme leakage (shown by increased alanine aminotransferase (ALT)/aspartate aminotransferase (AST), indocyanine green (ICG) clearance). All other BLTs are normal. At this level, there is structural adaptation without functional impairment or permanent architectural damage, even though some histologic changes are identifiable.

Level 3: acute injury, severe   Clinical exposure to this degree causes disruption of cellular function and leaves residual evidence of liver injury, for example, carbon tetrachloride exposure causing cellular necrosis (shown by increased ALT/AST), disrupted function (by elevated bilirubin), and synthesis (by lowered albumin). There are specific histologic changes (by pericentral necrosis) and later fibrosis and scarring (residual injury shown by elevated alkaline phosphatase). True cellular injury has occurred, with repair. Even with repair, there is residual evidence of damage without major architectural changes. Genetic injury may have occurred but is unlikely to be clinically significant or permanent.

Level 4: chronic injury   Clinical exposure under these circumstances causes cellular disruption and architectural changes that reduce functional hepatic capacity, for example, vinyl chloride-induced fibrosis (shown by procollagen III, IV) and portal hypertension (evidenced by elevated cholylglycine and alkaline phosphatase with decreased ICG clearance). The clinically significant injury is permanent, often with characteristic structural changes. Genetic injury can occur with risk of cancer development.

Level 5: disease   At this stage, clinical exposure has caused permanent structural damage, impaired organ function, and reduced capacity. Genetic injury can cause disruption of cellular control and, with active regeneration, may ultimately lead to malignant transformation, for example, chronic viral hepatitis with cirrhosis (shown by HBV–DNA, HBsAg), primary hepatocellular carcinoma, vinyl chloride fibrosis, peliosis hepatis, and hepatic angiosarcoma. Changes in these molecular and BLT markers correlate with the degree and type of hepatic tissue response to various environmental hepatotoxins (Liss *et al.*, 1985).

Susceptibility

Tests of susceptibility to hepatic xenobiotic injury are governed by the ability of the liver to metabolize and detoxify reactive metabolites. Therefore, markers that identify the oxidative pathway of a xenobiotic or the degree of punitive metabolite detoxification are the best ones to use to assess individual susceptibility. Although some markers (e.g., P450 levels) may indicate increased oxidation and others (e.g., GSH) indicate detoxification, none of the presently available hepatic markers are sufficiently specific or sensitive to identify the metabolic capabilities of an individual. Monoclonal antibodies or adducts with GSH, albumin, or hemoglobin are able to identify exposure or reactive metabolites of specific hepatotoxins. However, tests of future risk must be able to identify specific hepatic changes that will make an adverse outcome more likely. Such hepatic changes include scar formation (fibrosis, i.e., incomplete repair), active regeneration, DNA adduct formation, and oncogene activation. These changes are all related to increased risk of cancer development.

In exposed individuals, for example, identification of *p53* and *ras* gene activity requires concurrent assessment of the putative metabolite (e.g., by adduct occurrence) with the histologic and clinical markers (BLT and serologic) of hepatotoxic injury. Without this form of combined assessment, the differentiation of an exposed individual with subclinical hepatic injury and competent reparative capability from a susceptible individual with genetic injury and high-risk outcomes cannot be accomplished effectively.

The detection of viral confounders (hepatitis B, C, and D), which enhance susceptibility, has improved vastly with the use of PCR. These biomarkers provide proper classification of individuals with chronic liver disease who have exposures to various hepatotoxic chemicals and allow causal differentiation [e.g., Vietnam veterans with viral hepatitis B and dioxin exposure (Tamburro, 1992) and alcoholics with viral hepatitis C (Mendenhall *et al.*, 1991)].

Case Studies

Aflatoxin (Hepatocellular Carcinoma): Natural Environmental Toxin

Human hepatocellular carcinoma (HCC) has been causally associated with chronic active hepatitis (CAH), secondary to hepatitis B virus (HBV) and moldy food grain contaminated by aflatoxins (AF), mycotoxin metabolites of the *Aspergillus* fungus (Harris, 1990). HCC is prevalent in certain regions of Africa and Asia, where HBV carriers and dietary AF, typically AFB_1, are common. AFB_1 has been shown to be a potent carcinogen; its activity depends on the balance of AFB_1 metabolism between oxidation by specific cytochrome P450 phase I isozymes, that produce either the less toxic hydroxylated AFB_1 products or the carcinogenic 2,3-epoxide, and conjugation by glutathione. This balance has been shown (Schrager *et al.*, 1990) to shift with nutritional modulation and chemical intervention, both of which may enhance or diminish liver cancer induced by AFB_1 in rats. The AFB_1 epoxide covalently binds to DNA at the N^7-guanine site, as well as to proteins, for example, via lysine ϵ-amino groups (Figure 20.2). Such chemical reactions

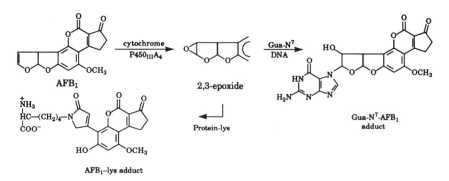

FIGURE 20.2 AFB$_1$ epoxide covalently bonding to DNA, which might lead to hepatocellular transformation.

may lead to transformation of hepatocytes whose clones may expand during the regenerative phase of CAH. CAH behaves as a "viral partial hepatectomy," liberating endogenous proliferative factors.

In addition to environmental monitoring of food contaminants by AFB$_1$ using TLC and HPLC, noninvasive biologic screening of populations to determine the "internal dose" of AFB$_1$ in HCC etiology have been carried out. Immunoassays of the major AFB$_1$–serum albumin adduct, aflatoxin–lysine, have been applied to human populations (Sabbioni *et al.*, 1990). Quantification of this adduct in human serum is achieved by combined immunoaffinity chromatography and HPLC with fluorescence detection. For this method, serum is digested by pronase and the adducts are purified by monoclonal antibody (MAb). The MAb was obtained from a hybridoma of mouse SP-2 myeloma cells with spleen cells of mice immunized with a synthetic antigen of AFB$_1$ epoxide covalently bound to bovine gamma globulin (Sabbioni *et al.*, 1990). One MAb isolated (2B11) was found to be a high IgM antibody with an affinity constant for AFB$_1$ and derivatives of about 1×10^9 liter/mol. A significant correlation coefficient of 0.82 was obtained between the aflatoxin–lysine adduct levels and AFB$_1$ consumption for an epidemiologic study in China. The human data revealed an average aflatoxin–lysine adduct level of 0.38 ng adduct/µg AFB$_1$ from the diet, or a daily albumin adduct burden of 2.9% of the AFB$_1$ daily intake.

The MAb 2B11 also showed significantly cross-reactivity for the major aflatoxin–DNA adducts, the N^7-guanosyl, and the corresponding imidazole-ring opened derivative, suggesting that these adducts share a common antigenic determinant. The antibody was applied by Groopman *et al.* (1985) to quantify AFB$_1$–N^7-G in urine. The MAb first was bound covalently to Sepharose 4B, which made a reusable preparative column for isolating aflatoxin derivatives from human urine. As a measure of MAb sensitivity, a competitive radioimmunoassay (RIA) showed a 50% inhibition value of approximately 300 fmol for AFB_1. When this methodology was applied to human urine samples, the aflatoxin metabolites detected were AFB_1-N^7-G and the hydroxylated aflatoxins M_1 and P_1 in individuals exposed to AFB_1 through dietary contamination at levels of 10–250 ppb.

Although antibody technology facilitates isolation and detection of urinary metabolites of aflatoxins, some studies of aflatoxin exposure may be subject to criticisms. In a cross-sectional ecological survey in China of possible risk factors for primary liver cancer (PLC; Campbell *et al.*, 1990), multiple regression analyses for various combinations of risk factors were attempted that showed that aflatoxin exposure consistently remained unassociated with PLC mortality. In contrast, HBsAg and plasma cholesterol were associated. This unique comprehensive survey included 48 county sites, approximately 600-fold aflatoxin exposure range, a 39-fold range of PLC mortality rates, a 28-fold range of HBsAg carrier prevalence, and estimation of other life-style features. The aflatoxin exposure was determined from 4-hr urine samples, which were analyzed by isolating oxidative aflatoxin metabolites such as AFM_1 (excluding nucleic acid adducts) on an antiaflatoxin MAb affinity column and quantifying them by a competitive ^{3}H–base RIA. This analysis procedure, however, was faulted (Wild and Montesano, 1991) for not being representative of aflatoxin intake; the aflatoxin–albumin adduct was suggested as the proper biomarker for determining recent past exposure to AF. The counterargument (Campbell *et al.*, 1990) is the strong correlation between the intake of AFB_1 and the urinary excretion of AFM_1, as well as the correlation between serum aflatoxin–albumin adduct levels and urinary AFM_1. Since the null effect of aflatoxin in this study contrasts sharply with other surveys, the new provocative conclusion makes it imperative to confirm that the aflatoxin exposure measured during the survey period can represent past intakes when PLC was forming.

The larger question is how to relate aflatoxin exposure to oncogene activation in the etiology of liver cancers. Evidence of such a relationship has appeared. McMahon *et al.* (1987) showed that AFB_1-N^7-G adducts were distributed nonrandomly in tumor-derived DNA of aflatoxin-induced HCC in rats. Such liver tumors also were found to contain activated c-Ki-*ras* oncogenes as identified in NIH3T3 mouse transformants. A single G·C to A·T base mutation in codon 12 was found to activate the *ras* gene. In view of an accumulating body of evidence concerning single base mutations in codons 12, 13, or 61 that arise in cellular *ras* genes after administration of chemical carcinogens, this AF activation of a *ras* gene may not serve the purpose of an exposure biomarker for aflatoxins. However, a combination of positive immunoassay of AFB_1-N^7-G in a dose–response manner, with the presence of multiple c-Ki-*ras* oncogene alleles, will make a compelling case for carcinogenesis induced by aflatoxins. Further, studies have elucidated a significant mutation in the *p53* gene during the development of liver tumors. The *p53* nuclear phosphoprotein appears to function as a cell cycle regulatory molecule, controlling cell proliferation. The wild-type *p53* gene is a tumor suppressor gene and has been mapped to chromosome 17p, a region often reduced to homozygosity in common cancers. It is the most frequently altered gene in human cancers (Jones *et al.*, 1991). In analysis for mutations of *p53* in HCC, in patients from China (Hsu *et al.*, 1991) and from Africa (Bressac *et al.*, 1991), 11 of 13 mutations have resulted in an arginine to serine substitution in codon 249 (AGG) of *p53*. Additionally, 12 of 13 point mutations found in these patients were G → T transversions. Aflatoxin-N^7-G is the most likely cause of mutation. The specific mutant *p53* acts as a dominant oncogene and may interact further with a hepatitis B protein to provide a growth advantage in hepatomas. Other types of mutations, including frameshift and deletion, also may enhance clonal expansions. It appears that *p53* mutations in colon cancer, leukemias, and sarcomas are not induced by carcinogen–DNA adducts (Jones *et al.*, 1991); therefore, patterns of base changes in *p53* induced by aflatoxins may be considered footprints of their activities on DNA.

Vinyl Chloride (Angiosarcoma): Synthetic Environmental Toxin

The original association of vinyl chloride (VC) with angiosarcoma of the liver (ASL) in humans was made at a Louisville plastics and synthetic rubber plant in 1973. Since that initial discovery, the University of Louisville and B. F. Goodrich Company have been involved in a 17-year cooperative prospective medical surveillance study involving 600–1200 active and 150–200 retired employees of the Louisville plant. The biologic data include annual historical, physical, radiologic, physiologic, pathologic, and biochemical data obtained on each employee. The environmental data include rank-ordered exposure estimates to 22 toxic chemicals and yearly individual job and area monitoring for specific vinyl monomers.

The prospective human study of VC-associated ASL illustrates the following points. First, the initial discovery of ASL, a very rare liver tumor, was not linked specifically to VC. Polyvinyl chloride (PVC) and acrylonitrile (AN), as well as other chemicals, were also initially suspect. Medical examinations did not identify the causal agent(s) and, in only a few cases, the existence of liver disease. Basic biochemical screening tests identified one or more abnormalities in 30–35% of the work force. Federally required specific liver tests found abnormalities in 10–20% of the work force. Definitive investigation confirmed only 10% of the work force as having persistent or significant liver dysfunction; 0.4% (four) had pre- or malignant disease.

Individual rank-ordered retrospective or prospective work histories for 22 major work-related chemicals (Table 20.5) were used to identify which of these chemicals' cumulative exposure ranked months (CERMs) correlated with liver disease and angiosarcoma (Figure 20.3A,B). Only four chemicals were associated with ASL cases: VC, hexane, dimethyl maleate (DMM), and

TABLE 20.5   Selected Chemicals for Exposure Indices

Chemical code	Chemical name
01	Acrylic acid
02	Acrylamides—acrylamide, methyl, *n*-octyl
03	Acrylonitrile
04	Acetylene
05	Acrylates—ethyl, methyl, methyl-meth, 2-ethyl hexyl, *N*-butyl
06	Bisphenol A
07	Butadiene
08	Caprylyl chloride
09	Chlorinated solvents—carbon tetrachloride, chloroform, trichloroethylene
10	Chloroethyl vinyl ether
11	Diethyl maleate
12	Mercuric chloride
13	Methanol
14	Phenol
15	Toluene
16	Vinyl chloride
17	Vinylidene chloride
18	Vinyl acetate
19	PVC dust
20	Catalysts
21	Styrene
22	Hexane

catalysts. All other plastics-related chemicals and all the synthetic rubber chemicals showed no relationship. The catalyst group was used for VC products only and hexane was the major solvent for the VC catalyst, therefore both were always present when VC was used. DMM was a specific catalyst for a specialized PVC product and was used only periodically. This chemical is used by toxicologists to deplete GSH in animals in order to potentiate the toxicologic effect of the agent under study. Among the ASL cases, individuals with DMM exposure have shorter latency periods (Tamburro *et al.*, 1984).

VC also causes characteristic histologic liver injury (Tamburro, 1984). These histologic characteristics correlate very well with total (CERMs) relative VC exposure job rank, as shown in Figure 20.4. Study of biochemical and metabolic liver markers in detecting chemical injury, using CERMs and liver histology for specific lesions, revealed that ICG clearance provided the best combination of sensitivity and specificity (Figure 20.5A); GGT provided the highest sensitivity but also had the lowest specificity (highest false positivity; Figure 20.5B); and AP had the highest specificity (Figure 20.5C). Finally, individual job and area monitoring of VC and AN were shown to be

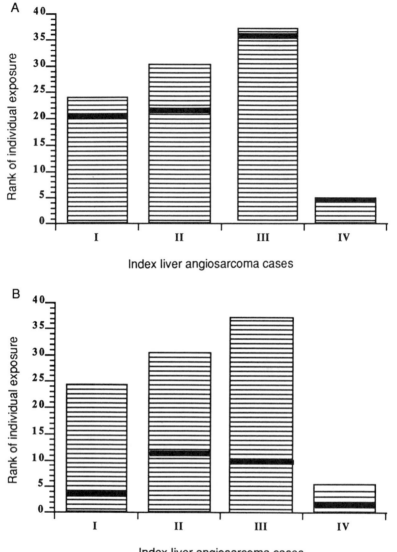

FIGURE 20.3   Relative vinyl chloride (A) and acrylonitrile (B) exposure rankings of index cases of hepatic angiosarcoma relative to their controls (individuals who worked the same years and number of years as index cases). , Angiosarcomas; ▭, matched controls.

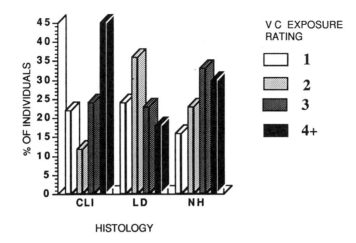

FIGURE 20.4 Correlation of hepatic injury and chemical exposure illustrated by the significantly larger percentage of markers whose liver histology showed evidence of chemical liver injury (CLI) that had vinyl chloride (VC) exposure ranking of 4 or greater. LD, Liver disease, nonchemical; NH, normal histology.

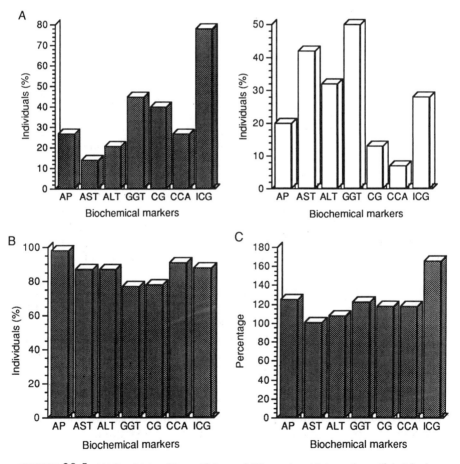

FIGURE 20.5 (A) Sensitivity, (B) specificity, and (C) sum (sensitivity and specificity) for hepatic biochemical biomarkers in chemical (■) and nonchemical (□) liver injury.

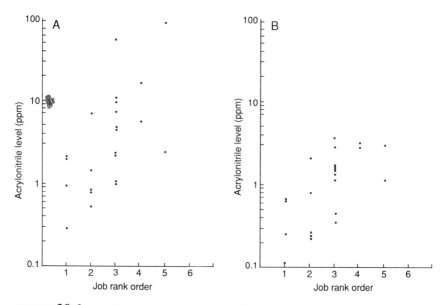

FIGURE 20.6   Correlation of job-specific acrylonitrile environmental exposure with job exposure ranking, at maximum level (A) and at mean/average level (B).

highly correlated to CERMs, to verify the relative exposure estimates of the CERMs, and to provide a ppm value for the CERMs (Figure 20.6).

This study identified and verified the cellular toxicity and carcinogenicity level of VC and established its biologic threshold level for humans. These data now can be used to estimate human risk to past and future exposure accurately (Tamburro, 1984). Finally, similar analysis of the other chemicals, via the relational database system, provided strong evidence that no association or relationship existed between VC exposure and other malignancies in the cohort. A 1982 report summarized the initial multidisciplinary research developments on techniques and methods for the detection and prevention of carcinogenesis in this cohort of industrial workers (Tamburro *et al.*, 1982).

As illustrated in the ASL case, because of the multiple metabolic and synthetic roles of the liver, no single marker, biologic or analytical, is sufficient for molecular epidemiologic purposes. The essential requirements for effective epidemiologic study in the occupational surveillance of hepatic injuries are listed in Table 20.6. Guidelines for detection of hepatotoxicity due to chemical exposure were outlined by Davidson *et al.* (1979).

A comprehensive review of the VC epidemiologic studies (Doll, 1988) confirms that the association between VC and cancer is confined to ASL. However, concern remains regarding individual human variation and prior VC animal exposure studies showing other cancers (Maltoni *et al.*, 1981).

TABLE 20.6 **Essential Elements for Prospective Surveillance of Occupational Environments**

1. Medical history
2. Physical examination
3. Basic biochemical screening tests
4. Specific biochemical markers of liver (target organ)
5. Medical protocol for evaluation of positive finding
6. Defined investigation (radiologic, physiologic, pathologic) for hepatic evaluation
7. Biologic storage bank (blood, tissue)
8. Individual work history with rank-ordered cumulative exposure to key chemicals in work environment
9. Individual job and area analytical monitoring of key chemicals (agents) in the work environment
10. Computerized relational database for storage of all data

Therefore, a dosimetry method based on molecular markers of VC metabolites is needed to evaluate the current regulatory exposure limit of 1 ppm. Potential biomarkers for the metabolites shown in the scheme in Figure 20.7 are under current development (Tamburro *et al.*, 1982; IARC, 1986; Joseph *et al.*, 1990).

Among the DNA adducts detected after exposure of experimental animals to VC, the major product, N^7-(2-oxoethyl)guanine (OEG), is derived from guanine N^7 alkylation by chloroethylene oxide (CEO). The detection limit was 10 pmol OEG/μmol unmodified guanine (Fedtke *et al.*, 1990). Rat tissue DNA was depurinated using mild acid hydrolysis. The hydrolysates

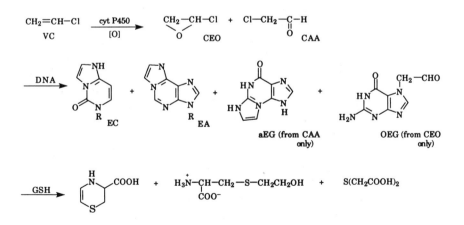

R = H; a ribose

FIGURE 20.7 Potential biomarkers for vinyl chloride metabolites.

were analyzed by HPLC on a reversed-phase strong cation exchange column with a fluorescence detector (excitation at 225 nm with a 340-nm emission cut off filter). This method appears to give more reproducible results than reduction of the N^7-oxoethyl group with tritiated sodium borohydride or derivatization of the oxo group with O-methylhydroxylamine for GC–MS analysis.

Other DNA adducts are the cyclic etheno derivatives whose formation occurs in the order of EG > EC > EA (E for etheno; G, C, and A are DNA bases). The concentration of EG was determined in mild acid DNA hydrolysates by MS. The EG chromatography fraction was electrophore-labeled using pentafluorobenzyl bromide. The dipentafluorobenzyl derivative was quantified relative to an internal standard $^{13}C_4$-EG using GC–MS with negative ion chemical ionization by measuring the m/z ion ratio of 354/358. The limit of detection was 60 fmol EG/μmol guanine (Fedtke et al., 1990), an improvement over the previous HPLC method with fluorescence detection. Thus, the ratio between EG and OEG was found to be approximately 1 : 100 in all tissues of rats immediately after VC exposure (600 ppm by inhalation, 4 hr per day, 5 days per week). This ratio in the liver increased to 1 : 14 1 week after exposure. It was calculated that the half-life of OEG was 62 hr, but that of EG was greater than 30 days, showing a greater persistence of the ethenoguanine adduct. Whether any of these two major VC adducts is required for cell transformations is not known.

Analysis of the two minor cyclic nucleosides EC and EA can be achieved by ^{32}P-postlabeling or by radioimmunoassay. The former procedure reported by Watson and Crane (1989) was preceded by separating the adducts as 3′-monophosphates by ion-pair reversed-phase HPLC. Then the molecules were postlabeled using [γ-^{32}P]ATP, and the mixture was treated with a nuclease for 3′ dephosphorylation. The etheno[5′-^{32}P]monophosphates, collected from HPLC, were quantified by liquid scintillation counting, yielding detection limits of 3–6 fmol EA and EC/μg DNA. Monoclonal antibodies that specifically recognize ethenoadenosine or ethenocytidine at approximately 200 fmol have been developed as an alternative detection method (Young and Santella, 1988). A radioimmunoassay for their presence in exposed rat tissues was reported (Ciroussel et al., 1990). The concentrations measured were 0.49 pmol EC/μmol deoxycytidine and 0.13 pmol EA/μmol deoxyadenosine in the liver DNA of rats exposed to 500 ppm VC (7 hr per day for 14 days). These values were an order of magnitude lower than the EG value from a similar exposure experiment described earlier.

Analysis of these four VC–DNA adducts from animal tissues can be adapted to human dosimetry. The major DNA adduct formed in livers of rats acutely exposed to VC was OEG, whereas the etheno derivatives (but not OEG) were found in liver DNA of rats chronically exposed to VC. In addition to its role as a biomarker of exposure to VC-type chemicals, each adduct should be considered for its role in genotoxicity. The predominant adduct

OEG, derived from the putative metabolite CEO, has a short half-life and lacks miscoding properties (IARC, 1986). It probably contributes only indirectly to the mutagenic effects of VC via depurination and mispairing opposite the apurinic sites. In contrast, the three minor etheno adducts have been reported to be efficient in causing mispairing during DNA replication (Jacobsen *et al.*, 1989; Singer *et al.*, 1987), although conflicting data point to low miscoding efficiency of EA and EC (Bartsch and Singer, 1985). All three cyclic adducts can be attributed to the other putative metabolite chloroacetaldehyde (CAA), thereby suggesting CAA to be responsible for VC genotoxicity. However, bacterial mutagenesis assays showed CEO to be much more potent than CAA (Perrard, 1985). Also, under comparable conditions when CEO was found to produce skin tumors in mice, CAA produced no increase in benign or malignant tumors (Zajdela *et al.*, 1980). Thus, the detoxification of CEO and CAA must be considered in assessing individual risk to VC exposure.

Strengths and Limitations

Enzymes

Microsomal P450

The induction enzymes, such as cytochrome P450, can be measured directly from liver samples obtained by needle or surgical biopsies (McPherson *et al.*, 1982). Such measurements have limitations based on differences in regional distribution of P450 and other enzymes, and on the different forms of the groups of enzymes, the levels of which may be reduced or induced by the xenobiotics themselves. In addition, these methods are limited by overlap (Table 20.2), variable xenobiotic induction, and background induction caused by high natural diet exposure, air pollutants, and life-style factors (Watkins, 1990). Knowledge of the "usual" background level (steady state) of induction is required also to identify any changes attributable to the suspect xenobiotic(s). Indirect measurement by ABT is the alternative to using tissue for these enzyme determinations.

Aminopyrene Breath Test

ABT itself has several limitations. It cannot distinguish among the various levels of liver disease (Hepner and Vesell, 1975). The overlap between individuals with adaptive or mild liver dysfunction makes the test less useful for those in most need of such evaluation, that is, individuals with subclinical disease. ABT has been found to be more reliable in predicting short-term changes, clinical improvement, and the histologic severity of chemical liver disease (e.g., alcohol related) than the more conventional liver tests. The predictive value of ABT for steatonecrosis, pericentral fibrosis, and cirrhosis (inactive) is less than the standard predicted value of a BLT. At the moment, there is no evidence that one breath test has anything to offer over another. Such "surrogate" methods (ABT) with high sensitivity are desirable. However, without concurrent high specificity and high disease occurrence, such surrogate markers can be potentially more psychologically or socioeconomically harmful because of high false-positive and false-negative rates.

Glutathione S-Transferase

The clinical usefulness of GST as a potential biomarker is uncertain. For example, although glutathione S-transferase is involved in catalyzing the detoxification of diol-epoxide carcinogenic metabolites of polycyclic aromatic hydrocarbons (Vander Jagt et al., 1985), these transferases have distinct but overlapping substrate specificities bordering on the complexities of the cytochrome P450s.

Metabolic and Physiologic Tests

ICG and other clearance tests mainly reflect hepatocellular injury or physiologic dysfunction. They provide only indirect evidence of xenobiotic injury. They are effective markers when the agent(s) and its exposure level are known and when other toxic associations can be excluded by epidemiologic or statistical analysis. Their major strength is their selectiveness for the liver.

Proteins

The major limitation of tests of antigen or antibody induced by xenobiotics is their sensitivity and specificity for the chemical agent. Exposure to the chemical agents acting as antigens may not be followed by antibody induction due to inadequate antigen production or structure derangement caused by its hepatic metabolism. Hepatic biomarkers of this type can be enhanced by PCR amplification for better detection.

Adducts

Monoclonal antibodies for specific chemical agents or their metabolites provide the most promising biomarker methods. A major limitation, at present, is that many adduct markers are not hepatically selective. Adducts with GSH, albumin, and hemoglobin reflect highly sensitive methods of identifying hepatic exposure over various periods of time. Monoclonal antibodies to various hepatic metabolites allow identification of different routes of metabolism (albumin and hemoglobin adducts) and the effectiveness of detoxification (GSH adducts). DNA adducts are best used to identify high-risk effects of exposure and provide a potential method of assessing the reparative capability of individual DNA. This methodology can be applied indirectly (circulating tissue: white blood cells) and directly (hepatic tissue: via biopsy).

The data suggesting strong xenobiotic associations or even causations in "group" data are insufficient for use on an individual basis. Such adduct markers still require confirmation of their specificity and sensitivity in individuals whose exposure and hepatic disease has been well characterized. The adduct surrogate (e.g., hemoglobin adduct for a hepatotoxic xenobiotic also must be shown to reflect the target organ (the liver) under surveillance correctly (i.e., selectivity).

Genetic Markers

Polymorphism

Use of gene mutation in the clinical setting of hepatic disease may be limited because (1) gene mutation may be present only in the end stages of the carcinogenic processes, (2) gene mutation may require multiple "hits" before becoming established, and (3) gene mutation may be seen only in liver tissue and not in the more accessible body tissues.

Oncogene Markers

In human cancer, the *ras* oncogene was found in 90% of adenocarcinoma in the pancreas, 50% in the colon, 30% in the lung, 50% in the thyroid, and in 30% of myeloid leukemia (Bos, 1989). The NIH3T3 transfection–transformation assay may not be sensitive enough to select *ras* activation in all the liver tumor DNA. Until a more sensitive assay is used, interpretation of the detection percentages of activated *ras* gene as a biomarker cannot be made with confidence. However, one should note the potential of the *ras* oncogene as a specific disease marker for the causative agent. Increasing evidence suggests that mutational spectra are highly correlated with each chemical carcinogen and reflect the predicted base substitution, that is, $G \cdot C \rightarrow T \cdot A$ transversion for benzo[a]pyrene, which forms predominantly the N^2-BPDE–deoxyguanosine adduct, and $A:T \rightarrow T:A$ resulting from N^6-deoxyadenosine bonded to the diol-epoxide of 7,12-dimethylbenzanthracene (Singer and Grunberger, 1983). It is plausible that molecular analysis of mutationally activated *ras* genes (a feat readily achievable with PCR) will reflect promutagenic DNA adduct formation and the mutagenic activities elicited by specific environmental carcinogens. With more complete molecular information, such structure–function correlations between *ras* DNA adducts and *ras* activities may be made, even in the presence of confounding factors such as spontaneous mutations producing $G \cdot C \rightarrow T \cdot A$ transversions.

The *p53* tumor suppressor gene appears to be more suited as a hepatic biomarker. The *p53* gene is mutated in diverse types of human cancers (Hollstein *et al.*, 1991); germ line mutations in *p53* predispose to cancers of the breast, soft tissues, and brain. Mutant *p53* has been found in hepatocellular carcinoma in connection with aflatoxin and hepatitis B. However, it is not certain which of these two agents has caused the *p53* mutations in the China and southern Africa studies. Analysis of liver tumors from regions in which either aflatoxin or the hepatitis B virus is the predominant agent will be instructive. At present, mutant *p53* is associated with driving selective clonal growth, a critical step in the neoplastic process. Its detection in liver and other tissues means a risky prognosis.

Clinical Field Application of Biomarkers

Further application of these methods in the early detection of hepatic injury in multiple exposure environments, such as the workplace, is needed. Their application for determining hepatic cancer risk, however, will have strong socioeconomic and ethical impacts. Their field application is vital in showing their ability to: (1) identify high-risk individuals, (2) identify an individual's specific hepatic metabolism for xenobiotics (e.g., degree of oxidation/detoxification, DNA adduct/repair), (3) confirm the safety of work environments containing potential carcinogens (e.g., acrylonitrile, TCDD, PCB), (4) show levels of individual exposure not associated with hepatic functional or structural changes beyond the adaptive response (Level 1), and (5) differentiate the cause of hepatic injury in multiple agent involvement (e.g., acrylonitrile and vinyl chloride).

Due to the multiple and complex functions of the liver, molecular epidemiologic investigations require joint disciplinary research between the basic molecular biologist and the clinical hepatologist. By nature, human investigation must be conducted in well-characterized environments, in a prospective surveillance-type system containing and applying the essential elements set forth in Table 20.6. This control is especially relevant to hepatotoxin exposure and hepatogenetic markers. More than any other hepatic biomarkers, the genetic markers raise serious ethical and social questions. In industrially developed countries, the incidence of hepatic cancer is low, whereas in underdeveloped countries it is high. Having DNA damage or a cancer-susceptibility gene does not necessarily lead to cancer, although it may identify an individual as high risk. Determining whether such individual information outweighs the benefits requires continuous reassessment. In the low HCC-incidence populations, high risk identification is associated with increased anxiety, discrimination, depression, decreased job security, or uninsurability. In high HCC-incidence populations, such information often impairs personal economic growth or opportunity.

Research Needs

In the hepatic organ system, no single molecular marker will provide adequate information about exposure, effect, or susceptibility (risk) nor will it answer all clinically relevant needs. Research in hepatic molecular markers is needed in four distinct but interdependent areas.

The first need deals with exposure identification of causative agent(s) when liver disease is found in a new occupational or environmental setting. Because of the variety of chemicals customarily present in such situations, it is often neither practical nor feasible to use specific markers such as MAbs to assess exposures. Under these circumstances, screening approaches are needed to identify the environmental chemicals or their hepatic metabolites. Mature analytical techniques for analysis of body fluids, such as GC-MS, are well developed for stable and volatile compounds such as dioxin (TCDD) and polychlorinated biphenyls (PCB). However, gaseous or gas-like compounds such as formaldehyde, methyl chloride, vinyl chloride, acrylonitrile, or butadiene escape easily from the aqueous samples to be so determined. Better techniques are needed for detection of their hepatic metabolites or conjugates in body fluids or tissue. In addition, more technological development is needed for nonvolatile agents or by-products. Intensive research is on-going to develop mass spectrometry for trace analysis of highly polar and nonvolatile compounds of complex mixtures; none, however, has been directed to liver-specific assays. Under development are derivatizations of adducts or adduct hydrolysates to increase volatility followed by GC-MS; liquid chromatography-mass spectrometry (LC-MS); tandem mass spectrometry (TMS or MS-MS) with desorption ionization (DI) via fast atom bombardment (FAB) or laser microprobe (LAM); LC-MS-MS; and so forth. Particularly promising for liver tissue analysis is the TMS technique. Here, the key innovation is the DI technique to produce ions from nonvolatile surfaces for mass analysis. Ions are formed from sputtered molecules after irradiating samples with a high-energy particle beam (FAB) or a focused laser beam (LAM). TMS is a nonchromatographic method for direct-mixture analysis that can yield molecular weight and structural information. A TMS experiment is performed as the name implies: two mass spectrometers (MS-1 and MS-2) are connected together so that MS-1 separates a particular ion Ma^+ (molecular weight information), formed by direct ionization of the sample, and the fragment ions formed by dissociation of Ma^+ are mass-analyzed by MS-2 (structure identification). Thus, TMS performs both the separation and the analysis step with sensitivity of detection reaching to sub-picogram levels. The technique has been applied to biomolecules such as vitamin B_{12}, chlorophyll, and bradykinin (Burlingame *et al.*, 1984). Burlingame and co-workers reported assignment of specific residues in human hemoglobin modified by styrene oxide using TMS (Kaur *et al.*, 1989).

In the second area, for the more defined exposures under which routine screening of biological samples is required, MAbs to metabolites of commodity or known high-risk chemicals such as vinyl chloride, acrylonitrile, and the environmental toxin aflatoxin are very much wanted. However, the problem of false positives must be addressed more keenly, since cross-reactivity of MAbs that is not identified during laboratory development may become significant in its field applications.

To date, liver histology (biopsy) remains the standard of environmental liver injury. However, specific hepatic tissue biomarker assays are needed for hepatocytes, biliary ductal cells, reticuloendothelial cells, and macrophages to help identify organ tissue target sites. Along this line, liver specific tissue antigen biomarkers could help identify specific liver function impairment (effect) without the need for liver tissue. Further, hepatic biomarkers with high specificity are more important socioeconomically than those with high sensitivity. These markers are needed to avoid the high false-positive findings that often outweigh the true positive benefits.

Especially needed for chronic exposure are quantitative markers of hepatic reserve, both anatomical and functional. Special attention needs to be directed to methods that can quantitate hepatic collagen content, decreased protein substrate, or synthesis. Quantitative biomarkers of hepatic collagen (effect of injury) or substrate content or synthesis (effect of function impairment) would have very high clinical diagnostic value.

Susceptibility research, the third area of hepatic biomarkers of risk, should be directed at two major categories. Methods that will identify putative metabolic pathways of exposure in individuals (individual susceptibility), for example, MAbs and MS, are needed. Another area of susceptibility involves tests to detect gene injury. In liver tissue, susceptibility of groups or populations will depend greatly on the history of exposure (dose, duration, and recentness) and the evidence of hepatic injury. Tests directed at gene injury or activation, for example, analysis of *p53* and the *ras* gene using PCR and RLFP techniques, offer promise for hepatic biomarker development. This development should start with verification of the effect of each gene presently implicated in hepatic tumors (e.g., hepatocellular carcinoma and HBV-aflatoxin or angiosarcoma and vinyl chloride). Retrospective and prospective applications should be in a defined population, for example, first-generation Asian immigrants to the United States and vinyl monomer workers. Both are valid populations in which to apply the outcomes of susceptibility research to demonstrate its benefits or limitations.

Finally, the fourth area, the most time consuming and the most vital, is research in the further development and maintenance of well-characterized prospective surveillance programs of at-risk populations. These studies are absolutely essential to validate the application and interpretation of these evolving techniques. Long-term cooperative nonadversarial efforts between industry, its workforces, and the scientific community already have shown that such programs can be financially, ethically, and scientifically feasible. These programs need to be enhanced by incorporating the standardization of hepatic nomenclature and criteria for hepatic function, injury, and disease (Leevy *et al.*, 1976) into environmental surveillance. Issues of background marker frequency, determination of relative risk assessment, and disease frequency (M, P, and R, respectively) in the workplace and in noncommercial exposed environments cannot be addressed properly without such population or group studies.

References

Baker, A. L., Kotake, A. N., and Schoeller, D. A. (1983). Clinical utility of breath tests for the assessment of hepatic function. *Sem. Liver Dis.* 3, 318–329.

Bartsch, H., and Singer, B. (1985). International meeting on the role of cyclic nucleic acid adducts in carcinogenesis and mutagenesis. *Cancer Res.* 45, 5205–5209.

Bos, J. L. (1989). Ras oncogenes in human cancer: A review. *Cancer Res.* 49, 4682–4689.

Bressac, B., Kew, M., Wands, J., and Ozturk, M. (1991). Selective G to T mutations of p53 gene in hepatocellular carcinoma from southern Africa. *Nature (London)* 350, 429–431.

Burke, M. D., Murray, G. I., and Lees, G. M. (1983). Fluorescence-microscopic measurement of intracellular cytochrome P-450 enzyme activity (ethoxyresorufin O-deethylation) in unfixed liver sections. *Biochem. J.* 212, 15–24.

Burlingame, A. L., Whitney, J. O., and Russell, D. H. (1984). Mass spectrometry: Fundamental reviews. *Anal. Chem.* 56, 417R.

Campbell, T. C., Chen, J., Liu, C., Li, J., and Parpia, B. (1990). Nonassociation of aflatoxin with primary liver cancer in a cross-sectional ecological survey in the People's Republic of China. *China Res.* 50, 6882–6893.

Ciroussel, F., Barbin, A., Eberle, G., and Bartsch, H. (1990). Investigations on the relationship between DNA ethenobase adduct levels in several organs of vinyl chloride-exposed rats and cancer susceptibility. *Biochem. Pharmacol.* 39, 1109–1113.

Colli, A., Buccino, G., Cocciolo, M., Parravicini, R., and Scaltrini, G. (1988). Disposition of a flow-limited drug (lidocaine) and a metabolic capacity-limited drug (theophylline) in liver cirrhosis. *Clin. Pharmacol. Ther.* 44, 642–649.

Davidson, C. S., Leevy, C. M., and Chamberlayne, E. C. (1979). "Guidelines for Detection of Hepatotoxicity Due to Drugs and Chemicals." DHEW (NIH) Publ. No. 79-313. U.S. Government Printing Office, Washington, D.C.

Doll, R. (1988). Effects of exposure to vinyl chloride. *Scand. J. Work Environ. Health* 14, 61–78.

Fedtke, N., Boucheron, J. A., Walker, V. E., and Swenberg, J. A. (1990). Vinyl chloride-induced DNA adducts. II. Formation and persistence of γ-(2′-oxoethyl)guanine and N-2,3-ethenoguanine in rat tissue DNA. *Carcinogenesis* 11, 1287–1292.

Glatt, H., Friedberg, T., Grover, P. L., Sims, P., and Oesch, F. (1983). Inactivation of a diol-epoxide and a K-region epoxide with high efficiency by glutathione transferase X. *Cancer Res.* 43, 5713–5717.

Groopman, J. D., Donahue, P. R., Zhu, J., Chen, J., and Wogan, G. N. (1985). Aflatoxin metabolism in humans: Detection of metabolites and nucleic acid adducts in urine by affinity chromatography. *Proc. Natl. Acad. Sci. U.S.A.* 82, 6492–6496.

Harris, C. C. (1990). Review on hepatocellular carcinoma: Recent advances and speculations. *Cancer Cells* 2, 146–148.

Harris, C. C. (1991). Chemical and physical carcinogenesis: Advances and perspectives for the 1990s. *Cancer Res. Suppl.* 51, 5023s–5044s.

Hayasaka, A., Maddrey, W. C., and Hahne, E. G. (1988). Serum concentrations of amino terminal propeptide of collagen. III. [PIIP] Study state levels of messenger RNA of alpha-1 chain of procollagen III in carbon tetrachloride-induced fibrotic rat livers. *Hepatology* 8, 1304–1309.

Hepner, G. W., and Vesell, E. S. (1975). Quantitative assessment of hepatic function by breath analysis after oral administration of C-aminopyrine. *Ann. Int. Med.* 83, 632–638.

Hollstein, M., Sidransky, D., Vogelstein, B., and Harris, C. C. (1991). p53 mutations in human cancers. *Science* **253**, 49–53.

Hsu, I. C., Metcalf, R. A., Sun, T., Welsh, J. A., Wang, N. J., and Harris, C. C. (1991). Mutational hotspots in the p53 gene in human hepatocellular carcinomas. *Nature (London)* **350**, 427–428.

International Agency for Research on Cancer (1986). "The Role of Cyclic Nucleic Acid Adducts in Carcinogenesis and Mutagensis," (B. Singer, and H. Bartsch, eds.). IARC, Lyon, France.

Jacobsen, J. S., Perkins, C. P., Callahan, J. T., Sembimupti, K., and Humayun, M. Z. (1989). Mechanism of mutagenesis by chloroacetaldehyde. *Genetics* **121**, 213–222.

Jones, P. A., Burckley, J. D., Henderson, B. E., Ross, R. K., and Pike, M. C. (1991). From gene to carcinogen: A rapidly evolving field in molecular epidemiology. *Cancer Res.* **51**, 3617–3620.

Joseph, J. T., Elmore, J. D., and Wong, J. L. (1990). Comparative sulfhydryl reaction pathways of chlorooxirane and chloroacetaldehyde. *J. Org. Chem.* **55**, 471–474.

Kaur, S., Hollander, D., Haas, R., and Burlingame, A. L. (1989). Characterization of structural xenobiotic modifications in proteins by high sensitivity tandem mass spectrometry. *J. Biol. Chem.* **264**, 16981–16984.

Leevy, C. M., Popper, H., and Sherlock, S. (1976). "Disease of the Liver and Biliary Tract: Standardization of Nomenclature, Diagnostic Criteria, and Diagnostic Methodology." Fogarty International Center Proceedings No. 22, DHEW (NIH) Publ. No. 76-725. U.S. Government Printing Office, Washington, D.C.

Le Provost, E., Cresteil, T., Codumelli, S., and Leroux, J. P. (1983). Immunological and enzymatic comparison of hepatic cytochrome P-450 fractions from phenobarbital-, 3-methylcholanthrene-, beta-naphthoflavone-, and 2,3,7,8-tetrachlorodibenzo-*p*-dioxin-treated rats. *Biochem. Pharmacol.* **32**, 1673–1682.

Liss, G. M., Greenberg, R. A., and Tamburro, C. H. (1985). The use of serum bile acids in the identification of vinyl chloride hepatotoxicity. *Am. J. Med.* **78**, 68–76.

McMahon, G., Davis, E., and Wogan, G. N. (1987). Characterization of c-Ki-*ras* oncogene alleles by direct sequencing of enzymatically amplified DNA from carcinogen-induced tumors. *Proc. Nat. Acad. Sci. U.S.A.* **84**, 4974–4978.

McMahon, G., Huber, L. J., Moore, M. J., Stegeman, J. J., and Wogan, G. N. (1990). Mutations in c-Ki-*ras* oncogenes in diseased livers of winter flounder from Boston Harbor. *Proc. Nat. Acad. Sci. U.S.A.* **87**, 841–845.

McPherson, G. A. D., Benjamin, I. S., Boobis, A. R., Brodie, M. J., Hampden, C., and Blumgart, L. H. (1982). Antipyrine elimination as a dynamic test of hepatic functional integrity in obstructive jaundice. *Gut* **23**, 734–738.

Maltoni, C., Lefemine, C., Ciliberti, A., Cotti, G., and Carretti, D. (1981). Carcinogenicity bioassays of vinyl chloride monomer: A model of risk assessment on an experimental basis. *Environ. Health Perspect.* **41**, 3–29.

Mendenhall, C. L., Seeff, L., Diehl, A. M., Ghosn, S. J., French, S. W., Gartside, P. S., Rouster, S. D., Buskell-Bales, Z., Grossman, C. J., Roselle, G. A., Weesner, R. E., Garcia-Pont, P., Goldberg, S. J., Kiernan, T. W., Tamburro, C. H., Zetterman, R., Chedid, A., Chen, T., Rabin, L., and the VA Cooperative Study Group (No. 119) (1991). Antibodies to hepatitis B virus and hepatitis C virus in alcoholic hepatitis and cirrhosis: Their prevalence and clinical relevance. *Hepatol.* **14**, 581–589.

Ozturk, M., Bressaci, B., Puisieux, A., Kew, M., Volkmann, M., Bozcali, S., Bella Mura, J., de la Monte, S., Carlson, R., Blum, H., Wands, J., Takahashi, H., von Weizsacker, F., Galun, E., Kar, S., Carr, B., Schroder, C., Erken, E., Varinli, S., Rustgi, V., Prat, J., Toda, G., Koch, H., Lian, X., Tang, Z., Shouval, D., Lee, H., Vyas, G., and Sarosi, I. (1991). p53 hotspot mutation in hepatocellular carcinoma is linked to aflatoxin exposure. *Hepatol.* **14**, 94A.

Perera, F., and Weinstein, I. B. (1982). Molecular epidemiology and carcinogen–DNA adduct detection: New approaches to studies of human cancer causation. *J. Chron. Dis.* **35**, 581–600.

Perera, F., Santella, R., and Poirier, M. (1986). Biomonitoring of workers exposed to carcinogens: Immunoassays to benzo(*a*)pyrene-DNA adducts as a prototype. *J. Occup. Med.* 28, 1117–1123.

Perrard, M. H. (1985). Mutagenicity and toxicity of chloroethylene oxide and chloroacetaldehyde. *Experientia* 41, 676–677.

Sabbioni, G., Ambs, S., Wogan, G. N., and Groopman, J. D. (1990). The aflatoxin–lysine adduct quantified by high performance liquid chromatography from human serum albumin samples. *Carcinogenesis* 11, 2063–2066.

Santella, R. M., Weston, A., Perera, F. P., Trivers, G. T., Harris, C. C., Young, T. L., Nguyen, D., Lee, B. M., and Poirier, M. C. (1988). Interlaboratory comparison of antisera and immunoassays for benzo[*a*]pyrene-diol-epoxide-I-modified DNA. *Carcinogenesis* 9, 1265–1269.

Savolainen, E. R., Brocks, D., Ala-Kokko, L., and Kivirikko, K. I. (1988). Serum concentrations of the N-terminal propeptide of type III procollagen and two type IV collagen fragments and gene expression of the respective collagen types in liver in rats with dimethylnitrosamine-induced hepatic fibrosis. *Biochem. J.* 249, 753–757.

Schrager, T. F., Nerberne, P. M., Pikul, A. H., and Groopman, J. D. (1990). Aflatoxin-DNA adduct formation in chronically dosed rats fed a choline-deficient diet. *Carcinogenesis* 11, 177–180.

Singer, B., and Grunberger, D. (1983). "Molecular Biology of Mutagens and Carcinogens." Plenum Press, New York.

Singer, B., Spengler, S. J., Chavez, F., and Kusmierek, J. T. (1987). The vinyl chloride-derived nucleoside, ^{2}N-3-ethenoguanosine is a highly efficient mutagen in transcription. *Carcinogenesis* 8, 745—747.

Tamburro, C. H. (1984). Relationship of vinyl monomers and liver cancer; angiosarcoma and hepatocellular carcinoma. *Sem. Liver Dis.* 4, 159–169.

Tamburro, C. H. (1991). Laboratory evaluation of liver disease. In "Diseases of the Liver and Biliary Tract" (G. Gitnick, D. R. LaBrecque, F. G. Moody, eds.), pp. 158–162. Mosby, St. Louis, Missouri.

Tamburro, C. H. (1992). Chronic liver injury in phenoxy herbicide exposed Vietnam veterans. *Environ. Res.* 59, 175–188.

Tamburro, C. H., and Liss, G. M. (1986). Tests of hepatotoxicity: Usefulness in screening workers. *J. Occup. Med.* 28, 1034–1044.

Tamburro, C. H., Kupchella, C. E., Wong, J. L., Barrows, G. H., Du, J. T., Espinosa, E., Feldhoff, R. C., Fortwengler, H. P., Schrodt, G. R., Sonnenfeld, G., Streips, U. N., Tseng, M. T., and Waddell, W. J. (1982). Report on research techniques and methods for the detection and prevention of carcinogenesis in the industrial worker. *Rep. Chem. Manufact. Assoc.* vc7.0, pp. 1–104.

Tamburro, C. H., Makk, L., and Popper, H. (1984). Early hepatic histologic alterations among chemical (vinyl monomer) workers. *Hepatology* 4, 413–418.

Vander Jagt, D. L., Hunsaker, L. A., and Royer, R. E. (1985). Glutathione *S*-transferase in the human liver; 13 forms of GST have been identified. Isolation of the multiple glutathione *S*-transferases from human liver. *J. Biol. Chem.* 260, 1163–1166.

Wang, P. P., Beaune, P., Kaminsky, L. S., Dannan, G. A., Kadlubar, F. F., Larrey, D., and Gengerich, F. P. (1983). Purification and characterization of six cytochrome P450 isoenzymes from human microsomes. *Biochem.* 22, 5375–5383.

Watkins, P. B. (1990). Role of cytochromes P450 in drug metabolism and hepatotoxicity. *Sem. Liver Dis.* 10, 235–350.

Watson, W. P., and Crane, A. E. (1989). HPLC-^{32}P-postlabelling analysis of 1,N^6-ethenodeoxyadenosine and 3,N^4-ethenodeoxycytidine. *Mutagenesis* 4, 75–77.

Wild, C. P., and Montesano, R. (1991). Correspondence re: T. C. Campbell *et al.* and Reply. *Cancer Res.* 51, 3825–3827.

Winer, F. R., Eskreis, D. S., Compton, K. V., Orrego, H., and Zern, M. A. (1988). Haplotype analysis of a type-I collagen gene and its association with alcoholic cirrhosis in man. *Mol. Asp. Med.* **10**, 159–168.

Wong, J. L., Ma, F. F., and Zhang, Y. (1990). Antibodies to acrylonitrile–glutathione conjugate. *Antibody Immunoconj. Radiopharmaceu.* **3**, 194.

Wu, T. G., and Wong, J. L. (1991). Adsorptive stripping voltammetric speciation of nickel(II)-histidine in aqueous ammonia. *Anal. Chim. Acta* **246**, 301–308.

Young, T. L., and Santella, R. M. (1988). Development of techniques to monitor for exposure to vinyl chloride: Monoclonal antibodies to ethenoadenosine and ethenocytidine. *Carcinogenesis* **9**, 589–592.

Zajdela, F., Croisy, A., Barbin, A., Malaveille, C., Tomatis, L., and Bartsch, H. (1980). Carcinogenicity of chloroethylene oxide, an ultimate reactive metabolite of vinyl chloride and bis(chloromethyl)ether after subcutaneous administration and in initiation-promotion experiments in mice. *Cancer Res.* **40**, 352–356.

21

Biomarkers of Musculoskeletal Disorders

J. Patrick Mastin,
Gerry M. Henningsen,
and Lawrence J. Fine

Background

It is estimated that 20 million Americans have impairments, or limitations in the ability to perform various normal daily functions, related to musculo-skeletal (MS) conditions. These conditions include disorders of bone (e.g., osteoporosis), muscles (e.g., muscular dystrophy and idiopathic myositis), joints (e.g., osteoarthritis, rheumatoid arthritis, and some forms of back pain), tendons and tendon sheaths (tendinitis and tenosynovitis), and certain peripheral nerves (e.g., carpal tunnel syndrome). The ability to identify pre-clinical and reversible musculoskeletal insults and injury is highly desirable for potential prevention of, and intervention into, these disease processes. The use of early biologic changes as indicators, or biomarkers, of disease resulting from musculoskeletal trauma offers the prospect of monitoring workers who are at risk for the progressive development of arthritis, myosi-tis, and other musculoskeletal disorders.

Until recently, the repertoire of musculoskeletal biomarkers consisted mainly of clinical diagnostic indicators of irreversible inflammatory changes in joints or other tissues. However, advances in the knowledge of molecular and cellular processes of the physiologic and pathologic changes in musculo-skeletal tissues, coupled with recent developments in biotechnology, are now providing the opportunity to relate biochemical changes with the onset and progression of musculoskeletal diseases. Modern methods, used in con-junction with classical techniques, adapted to evaluate these potential biomarkers include commercial and customized immunochemistry, automated high-resolution chromatography and electrophoresis, and other bioanalytical procedures enhanced by use of advanced optics, lasers, and computers. Defined biomarkers have the potential to screen, diagnose or prognose, monitor, and identify quantitatively predispositions of individuals and populations for musculoskeletal disorders.

The effect of these musculoskeletal disorders on the working population is immense (Kelsey and Cunningham, 1984). Cross-sectional and case-control studies have demonstrated striking associations between occupation and musculoskeletal disorders, including disorders involving the lower extremities, back, upper extremities, and neck. The disorders can involve nerve compression (carpal tunnel syndrome, median nerve), mixed peripheral neurovascular damage in hand–arm vibration syndrome, inflammation of specific tendons (DeQuervain's Disease, stenosing tenosynovitis at the radial styloid involving abductor pollicis longus and extensor pollicis brevis), osteoarthritis, and possibly muscles (tension neck syndrome). For example, substantial occupational risks have been reported for cervical spondylosis (degenerative changes in the cervical spine) in dentists, rotator cuff tendinitis of the shoulder in industrial workers working with their hands above shoulder height, osteoarthritis of the hip in farmers, and knee joint inflammation in carpet installers (Tanaka *et al.*, 1982; Hagberg and Wegman, 1987; Axmacher and Lindberg, 1988).

Musculoskeletal disorders can be caused by nonoccupational factors or exposures as well. For instance, although epidemiologic studies of low back pain in the general population have in general failed to identify factors that are strong predictors or correlates of low back pain, there is limited evidence that cigarette smoking, number of births or pregnancies, and distance traveled to work are related to the risk of developing low back pain. Some occupational factors such as lifting, twisting, prolonged sitting, and driving, of course, occur during recreational activities and may be a cause of low back pain. Similarly, nonoccupational activities such as sports activities certainly can cause a variety of tendinitides of the upper extremity. Finally, the nonoccupational causes of carpal tunnel syndrome, which is the best studied of the upper extremity disorders, include coexisting medical conditions such as rheumatoid arthritis, acute trauma, and probably some recreational activities that involve forceful and repetitive hand movements.

It is logical to hypothesize that some personal characteristics of an individual, such as the strength of the muscles of the trunk or the size of the carpal tunnel, might be associated with the risk of developing a musculoskeletal disorder or injury, although studies have not demonstrated consistently that there are any reliable or strong personal predictors of upper extremity disorders or injuries.

Epidemiologic Studies

Relatively little work has been done in trying to identify and use biomarkers for diagnosis of musculoskeletal disorders. Most of this research has focused on rheumatoid arthritis (RA) and osteoarthritis (OA). Although potential nonspecific markers, such as C-reactive protein, and more specific ones, such as cartilage matrix glycoprotein (see "Types of Biomarkers of Musculoskeletal Disorders"), have been identified, these studies of patients from the general population have not identified biomarkers of early disease. Part of the challenge is that the subjects in these studies have been from relatively heterogeneous populations, with respect to risk factors for degenerative joint disease such as OA. In biomarker studies of occupational musculoskeletal disease, however, the subjects would be from a more homogeneous population, that is, reasonably healthy individuals of working age who have a known risk for musculoskeletal disorders due to their occupations.

Several aspects of many epidemiologic studies potentially could be improved by use of validated biomarkers. For most work-related musculoskeletal disorders, several occupational risk factors have been identified. For example, likely risk factors for carpal tunnel syndrome are repetitive and forceful movement of the wrist. Direct observational measurement of these risk factors for every worker in a study is very expensive and, in some cases, not technologically feasible. Development of biomarkers of the biologically effective dose (the amount of trauma needed to induce early musculoskeletal disease) would allow more accurate studies of the relationship between complex exposures and health end points. In addition, biomarkers of dose might allow the study of interaction between common complex exposures such as local segmental vibration and repetitive work. The mechanisms of damage from chronic exposure to these risk factors often are not clear. Biomarkers of early biologic effect or altered structure could provide needed insight into possible mechanisms. Many work-related musculoskeletal disorders such as low back pain cannot be classified accurately into more specific or homogeneous diagnostic categories with current epidemiologic methods. Biomarkers with prognostic or diagnostic potential would be desirable. A brief review of the current state of knowledge will underline the need for better characterization of occupational exposures and the other causes of these disorders.

Epidemiologic studies that attempt to identify the etiologic factors are at an early stage and have tended to focus more on low back pain than on disorders of the neck or lower and upper extremities. Generally, these studies have not used biologic markers. Results of surveys done on active workers in high-risk industries may differ from those acquired from the general population. The former tend to identify most clearly the role of the work-related factors whereas the latter tend to identify the nonoccupational factors. In both workplace and community studies, preexisting medical conditions rarely explain the majority of new cases of musculoskeletal disorders, such as low back or shoulder disorders. Lower extremity disorders generally have not been studied. The potential for biomarkers in epidemiologic studies of work-related disorders is substantial. Once validated biomarkers have been developed, they will be useful in quantifying exposure, untangling the mechanisms of injury or disorder, improving the detection of mild or intermittent disorders, and identifying more homogeneous health outcomes in epidemiologic studies.

A number of challenges face individuals trying to identify biomarkers of MS disorders, for example, lack of information on pathogenesis and etiology of MS disorders and presence of confounding conditions in patients with MS disease.

Considerations for Using Biomarkers in Epidemiology Studies

As with all studies using biomarkers, consideration must be given to the type of sample to be collected when investigating MS diseases. For instance, analysis of synovial fluid probably would give the best indication of local changes in affected joints. Similarly, specimens of muscle or bone tissue should contain indicators of local changes. However, these specimens generally cannot be obtained easily. Although potential biomarkers produced locally are diluted in the systemic circulation, blood still offers the best option as a specimen because of ease of collection.

Although some biomarkers are disease specific, most are not specific for a particular disorder and can be related to conditions other than the one under study. For instance, biomolecules that indicate an inflammatory response might not be site specific. Similarly, cartilage breakdown products found in serum might result from catabolic processes in knee, back, shoulder, or other joints. The nonspecific biomarkers can, however, be compatible with and aid in supporting the presence and extent of disease when used as part of a profile of appropriate tests. Occasionally, more specific musculoskeletal biomarkers can be found for certain disorders, for example, for autoimmune disorders.

Investigation of animal models of MS disorders can yield useful information in identifying biomarkers of human disorders. Potentially confounding factors, such as genetic or life-style differences, can be controlled in animal studies. Changes in levels of potential biomarkers also can be correlated with pathologic changes.

Types of Biomarkers of Musculoskeletal Disorders

As discussed previously in this book, biomarkers can be categorized as biomarkers of susceptibility, exposure, or effect. Biomarkers of exposure generally are related to chemical exposures and therefore are not relevant to trauma-induced disorders.

Susceptibilities to RA and systemic lupus erythematosus (SLE), both of which are autoimmune diseases that affect the MS system, have been linked to the major histocompatibility antigens. Both syndromes are associated with the HLA-D antigens (Puttick *et al.*, 1989; Fronek *et al.*, 1990). Similarly, ankylosing spondylitis is associated with the HLA-B antigens, specifically HLA-B27 (Tertti and Toivanen 1991). A link between premature OA and a cartilage-specific gene has been reported in a family with inherited polyarthritis (Palotie *et al.*, 1989). Use of these susceptibility markers is problematic at this point, since these diseases have multifactorial causes involving environmental factors as well as genetic ones. To our knowledge, no susceptibility markers for occupational MS disorders have been identified.

Discussions of several MS disorders, as they pertain to biomarker research, follows. Our discussions will center on three representative MS disorders and their potential biomarkers: osteoarthritis, low back pain, and cumulative trauma disorders. All these are important because of the large number of affected individuals. Potential biomarkers have been extensively studied only for OA; hence, these will be given the most attention here. In addition, three other MS disorders that are of interest from a biomarker perspective will be discussed. Table 21.1 provides a list of a number of other types of MS disorders, along with biomolecules that have been or have the potential for being used as biomarkers. Information concerning the fundamental physiologic and pathologic characteristics of MS tissues is included to assist in understanding the rationale and concepts of applying biomarkers.

In general, the types of molecules that have been investigated as potential biomarkers of effect in MS disease are of three types: (1) proteins that are constituents of musculoskeletal tissue and are released through catabolic processes or produced as part of the repair process (e.g., cartilage proteins from joints or myoglobin from muscles), (2) proteins associated with the inflammatory response (e.g., acute-phase proteins) or preinflammatory response (occurring before the inflammatory response and often inducing it), and (3) indicators of immune or autoimmune processes. As discussed earlier, many MS biomarkers are not specific for a particular disorder, but can nevertheless aid in detection of disease when used in combination with other information, such as medical history.

Osteoarthrosis

Syndesmology, or arthrology, is the study of synovial joints. Degeneration of joints is a natural aging process called osteoarthrosis or osteoarthritis, a term that is highly synonymous with osteoarthrosis but denotes inflammation along with degeneration. Figure 21.1 is a schematic representation of the structure of a typical synovial joint, which normally functions to provide

TABLE 21.1 Examples of Potential Musculoskeletal Biomarkers

Disease or condition	Marker
Muscle	
Duchenne muscular dystrophy (DMD)	Haptoglobin
DMD carrier	Myoglobin (after exercise); creatinine kinase
Idiopathic myositis	Lymphocyte activation markers
Inflammatory muscle disease (Myositis)	Antibody to 56-kDa nuclear protein
Joint	
Rheumatoid arthritis (RA)	N-Terminal type III procollagen; antikeratan antibodies; lactoferrin
Active RA	Elastase–proteinase inhibitor
Osteoarthritis	Keratan sulfate; hyaluronan; β-D-xylosyltransferase, pyrilinolin, autoantibodies to cartilage matrix and cell components; cartilage-directed T-cell reactions; proteoglycans and glycosaminoglycans
Systemic lupus erthyematosus (active)	C-Reactive protein (CRP)
Rheumatoid arthritis, psoriatic arthritis, and Reiter's syndrome	Mitogenic response in mononuclear cells
Ankylosing spondylitis	IgA; CRP; α_1-antitrypsin; α_1-acid glycoprotein; haptoglobin
Bone	
Bone turnover	Osteocalcin; alkaline phosphatase; procollagen fragments; pyridinium cross-links
Bone resorption (in the elderly)	Urinary calcium/creatinine; calcium albumin; 1,25-dihydroxyvitamin D
Bone turnover in rheumatoid arthritis and osteoarthritis	Osteocalcin
Osteoporosis (corticosteroid-induced in humans)	Osteocalcin
Osteoporosis (secondary to RA in women)	4-Androstenedione
Bone turnover (hypothyroidism)	Alkaline phosphatase; osteocalcin; urinary hydroxyproline/creatinine
Bone resorption (secondary to RA)	Urinary calcium/creatinine; urinary hydroxyproline/creatinine

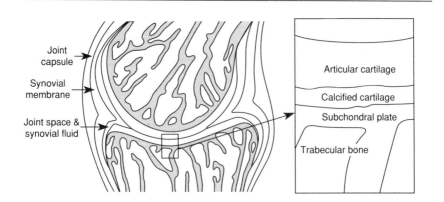

FIGURE 21.1 Schematic cross-section of an articular joint. (Adapted from Radin, 1984.)

smooth and stable hinge-like movement between two bones. Cartilage degeneration is a common, age-related condition in asymptomatic adults; however, such degeneration varies in its potential to progress into clinically recognized degenerative joint disease (DJD). Symptoms of DJD can include varying levels of joint pain, swelling, lameness or dysfunction, heat, and deeper tissue (bone) involvement. Because of the yet uncertain steps in transition between early asymptomatic joint degeneration and clinical osteoarthritis, the pathogenesis is still speculative but has a firm hypothetical basis. Initial lesions may be in the cartilage framework of the joint; such changes can be identified microscopically. Other early changes (offering potential biomarker opportunities) include changes in the synthesis of cartilage proteoglycans by chondrocytes, enhanced enzymatic degradation of cartilage, increased deposition of hydroxyapatite crystals in cartilage, changes in subchondral remodeling at the cartilage–bone interface, reduced subchondral bone resilience, and changes in synovial fluid. Some intermediate changes in osteoarthritis include fibrillation of cartilage, horizontal shearing of cartilage (which detaches it from underlying calcified zones), and abrasive thinning of the cartilage. Later clinical changes usually involve some calcification on the joint surface progressing to osteoarthritic bone formation, osteosclerosis of underlying bone, synovitis, fibrosis, and necrosis. The debris formed by DJD can incite an inflammatory reaction with lymphocyte infiltrations into the joint capsule. Depending on the extent of damage, endogenous repair eventually is overwhelmed by the destructive processes in osteoarthritis (Meachim and Brooke, 1984).

Increasing knowledge of the biochemistry and metabolism of cartilage in normal and osteoarthritic individuals offers the prospect of developing early preclinical biomarkers to detect susceptible persons, exposure to injurious joint trauma, and effects of joint injury that can proceed to DJD and osteoarthritis. The major changes in osteoarthritis occur in the hyaline articular cartilage; additional changes are observed in the capsule, synovium, and subchondral bone depending on the disease severity. It should be understood that mature cartilage is totally avascular, aneural, and alymphatic, deriving its nutrition from the synovial fluid by diffusion. It is also a superhydrated tissue; water constitutes 60–80% of its wet weight. Collagen and proteoglycans make up most of the dry weight; traces of other biomolecules such as a matrix glycoprotein are found also. The proteoglycans consist of a linear protein molecule surrounded by numerous polysaccharides known as glycosaminoglycans. These glycosaminoglycans include chondroitin sulfate, keratan sulfate, and hyaluronic acid. Hyaluronate, along with link proteins, binds the proteoglycan molecules together. Cartilaginous collagen is Type II, which differs from other body collagen in its amino acid content, hydroxylation, and carbohydrate content. Other normal biomolecules include chondronectin, lipids, lysozyme, and matrix proteins, whereas the presence of proteolytic enzymes, acute-phase reactive proteins, cartilage breakdown products, and immunologic or inflammatory molecules may be present at varying levels in DJD. Although much is known about the biochemistry and metabolism of cartilage, there is also tremendous variation in these areas among species, ages, joint sites, and individuals, making it difficult to establish reproducible pathogenic steps. However, improvements in biotechnology make the identification of biomarkers for osteoarthritis more feasible in well-controlled field studies of populations at risk for developing DJD as a result of repeated trauma to joints. These biologic molecules offer ample biochemical opportunities to explore possible biomarkers for DJD (Mankin and Brandt, 1984).

Figure 21.2 shows one possible paradigm for the development of degenerative joint disease as a result of trauma. Investigators have identified the chemicals enclosed in the ellipses as ones that are produced as a result of degradative or inflammatory processes. These molecules have been or could be investigated as possible biomarkers of trauma-induced degenerative joint disease. Obviously, the earlier in the disease process these biomolecules can be detected, the better the chances for successful intervention.

Most investigations of potential biomarkers of osteoarthritis have focused on the presence of cartilage components in serum, primarily proteoglycans and glycosaminoglycans. Average serum levels of keratan sulfate, for instance, have been shown to be increased in patients with OA and in laboratory animals with experimentally induced OA (Brandt *et al.*, 1989b; Mehraban and Moskowitz, 1989; Campion *et al.*, 1991; Mehraban *et al.*, 1991), although there is considerable overlap between patients and controls. Hyaluronic acid also has been reported to be elevated in serum from patients with OA and RA (Goldberg *et al.*, 1991). Although these cartilage components may be released from damaged cartilage, Brandt (1989a) has suggested that they may also result from attempts by the cartilage to repair itself, arguing that if the cartilage components are released slowly over time, as would be the case for osteoarthritis, the concentration in serum would never reach very high levels. Brandt also pointed out that many of the potential markers could come from anatomical sites other than the arthritic joints, for example, intervertebral disks. Myllylä *et al.* (1989) also found elevated levels of procollagen type III peptides (precursors of collagen) and galactosylhydroxylysyl glucosyltransferase (an enzyme involved in the cross-linking of collagen) in serum of patients with RA.

Fife (1988; Fife and Brandt, 1989) has identified a 550-kDa noncollagenous, nonproteoglycan glycoprotein in articular cartilage, as well as in hyaline cartilage and vitreous that was increased in serum from 8 of 12 dogs after cruciate ligament section over presurgical levels. This protein, known as cartilage matrix protein, also was found in synovial fluid from patients with OA, but not from patients with other arthritides (Fife, 1988). This biomolecule might prove to be a useful biomarker in the future.

Researchers also have looked at changes in serum proteins that might

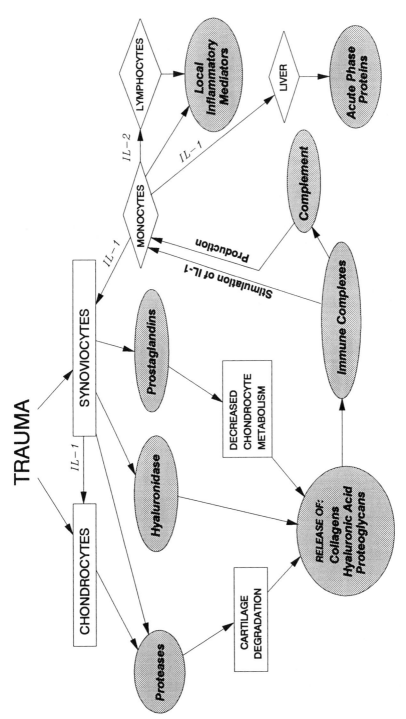

FIGURE 21.2 Paradigm for development of degenerative joint disease resulting from trauma. Ellipses (shaded) denote potential biomarkers. IL-1, Interleukin-1; IL-2, Interleukin-2. (Adapted from Pujol and Loyau, 1987, and McIlwraith, 1982.)

occur as a result of inflammation that accompanies the degenerative processes in patients with OA (see Figure 21.2). Increases in serum levels of ceruloplasmin, α_1-acid glycoprotein, α_1-antitrypsin (Denko and Gabriel, 1979), serum amyloid A, and C-reactive protein (Sukenik *et al.*, 1988) in OA patients compared with controls have been reported, although there was overlap between the two groups. Changes in the glycosylation of IgG have been reported in serum from OA and RA patients (Parekh *et al.*, 1985). It should be noted that changes in these inflammatory proteins can result from nonarticular inflammatory processes.

The role of the immune system in the pathogenesis of osteoarthritis has been gaining recognition. Although cartilage is weakly antigenic, it is immunogenic through the antigenic structures found in proteoglycans, collagen, and chondrocytes. These "sequestered" antigens that normally have been tolerated by the body actually may become recognized by the body's immune system during osteoarthritis and act as autoantigens that evoke an immunologic response in the joint. Many cases of osteoarthritis have both an inflammatory and an immunologic component, which can be recognized in the synovial fluid by increases in mononuclear cells, immunoglobulins, cytokines, other proteins, and complement. Actual proof of autoimmunity in osteoarthritis is weak and may be masked by the accompanying relatively severe biochemical changes that occur in a degenerating inflamed joint. There are some reports of significant cell-mediated immune responses in rabbit models, in which increased blastogenesis to proteoglycans correlated highly with severity of progressive osteoarthritis. The immunologically induced arthritic reactions resemble hypersensitivity, specifically type III hypersensitivity (Arthus or immune complex disease), type II hypersensitivity (cytotoxicity mediated by specific IgG and IgM antibodies), or type IV (delayed reactions mediated by sensitized T lymphocytes) (Dean and Murray, 1991). Although not equal to the severe autoimmune reactions seen in rheumatoid arthritis patients, the possible immunologic contributions to the progressive degeneration of joints in osteoarthritis may be pathogenically important (Goldberg, 1984).

A number of researchers have looked for antibodies against cartilage components in serum from OA patients. Antibodies against collagen have been reported in patients with OA (Stuart *et al.*, 1983). However, Hartmann *et al.* (1989) did not detect increased levels of anticollagen antibodies in rabbits with experimentally induced joint disease. Austin *et al.* (1988) reported that antibodies against link protein (a component of cartilage) were no more common in patients with OA than in healthy people. Antibodies to chondrocyte membrane antigens were found to be elevated in 86 of 86 clinical OA patients compared with controls (Nemeth-Csoka *et al.*, 1988). Cell-mediated immunity to types I, II, and III collagen was demonstrated in 10 of 20 RA patients, but in 0 of 10 OA patients (Stuart *et al.*, 1980).

Low Back Pain

Vertebral myalgia (pain) and degeneration of joints can have similar pathogeneses as osteoarthritis, but is somewhat more complex because there can also be muscular, neurologic, and intervertebral disk involvement. The vertebral column contains the vertebral canal which protects the spinal cord, intervertebral foramina through which spinal nerves pass, and bony processes that serve as the origins and insertions for many muscles and ligaments. The vertebrae have synovial joints between superior and inferior processes and symphyses (nonsynovial joints) that connect the bodies of the vertebrae through the compressible intervertebral disks. The vertebral disks contain an immunologically "privileged" material, the nucleus pulposus, that is antigenic to the body's immune system should it ever leave its isolated site (as in a disk rupture). The lumbar area is one of the more common sites of back injury resulting from physical trauma. Several sequelae can occur from trauma to the back, including muscular tears and inflammation (myositis), strained ligaments and tendons, ruptured disks with immune-mediated inflammation evoked by the discharge of nucleus pulposus, pinched nerves, and osteoarthritis and degeneration of the vertebrae.

Although back pain is among the most common MS disorders in the United States, there has been very little research on biomarkers of early back disease. Parris *et al.* (1990) have reported decreased plasma and saliva levels of substance P, a neurochemical with inflammatory properties, in patients with chronic back pain compared with controls. However, biomarkers potentially might be developed to identify early mild injury to the back that results in biomolecular alterations, such as those described for DJD and autoimmune reactions to "foreign" antigenic disk material. Specific serum autoantibodies could be surveyed as likely biomarkers using immunoassays or other new technologies.

Cumulative Trauma Disorders

Cumulative trauma disorders (CTDs) constitute a group of MS disorders that result from chronic exposure to excessive force, repetitive motion, abnormal postures, or combinations of these that often occur as work-related injuries. The types of disorders in this category include disorders of tendons (tendinitis), disorders of nerves (nerve entrapment, e.g., carpal tunnel syndrome), or neurovascular disorders (thoracic outlet syndrome) (Putz-Anderson, 1988). To our knowledge, no research in the area of biomarkers of cumulative trauma disorders has been published. However, cumulative trauma disorders are chronic progressive disorders, and intervention procedures could be very effective if the conditions were detected early, for example, by detection of early biomarkers. Amyloid has been detected in the tenosynovium of the wrist in some patients with carpal tunnel syndrome (Bjerrum *et al.*, 1984; Kyle *et al.*, 1989). If amyloid protein or precursor proteins could be detected in serum, they might serve as biomarkers for carpal tunnel syndrome of certain etiologies. In addition, since CTDs are inflammatory disorders, it is likely that changes in serum inflammatory proteins (e.g., acute-phase proteins) occur. As the biomolecular pathogeneses of these disorders are eventually determined, the opportunity should arise to develop potential biomarkers of susceptibility, exposure, and early effects.

Others

Investigations of several other MS disorders have led to the possibility of developing biomarkers for the conditions. For instance, several markers of muscle disease have been investigated. Arad-Dann *et al.* (1989) identified autoantibodies against a 56-kDa protein in sera from patients with inflammatory muscle disease. Studies have implicated HTLV II (human T-lymphotropic virus, a relative of HIV) as a possible etiologic factor in the so-called and vaguely defined chronic fatigue syndrome, which is characterized by generalized myalgia, myositis, and fibrositis.

Certain connective tissue disorders, some involving MS tissue, have been linked to certain chemical exposures, for example, silica and vinyl chloride, or drugs, for example, hydrazine, methyldopa, hydralazine, procainamide, isoniazid, chlorpromazine, and penicillamine (Dean *et al.*, 1989). As discussed earlier, some of these conditions are associated with certain alleles of the major histocompatibility antigen (MHC) genes. It is possible that these alleles might be linked to, and hence might act as biomarkers of, susceptibility to xenobiotically induced connective tissue disorders. In addition, these connective tissue disorders might result from autoimmune or hypersensitivity mechanisms, raising the possibility of using these antibodies to autoantigens or cell immunity to the xenobiotics as biomarkers.

Methods for Measuring Biomarkers

Development of new methodologies for the detection and measurement of biomolecules has increased the potential utility of biomarkers for early diagnosis of musculoskeletal, as well as other, disorders greatly. Several technological advances have facilitated the development of these methodologies. In addition to improvements in conventional methods, for example, better liquid chromatography bed packings, the use of lasers and high-efficiency computers and increased availability of well-characterized monoclonal antibodies have resulted in more sensitive and specific methods.

Table 21.2 lists some techniques that are used or have potential use in

TABLE 21.2   **Bioanalytical Techniques for Biomarker Detection**

Technique	Description
Electrophoretic	
Native PAGE	Separates proteins according to size and surface charge
Denaturing (SDS) PAGE	Separates peptides according to molecular weight
Isoelectric focusing	Separates proteins according to isoelectric point (pI)
Two-dimensional PAGE	Separates proteins according to both pI and molecular weight
Immunoelectrophoresis (IEP)	
Crossed IEP and Rocket IEP	Separates proteins according to electrophoretic mobility and capacity to bind and precipitate specific antisera
Immunoblotting (e.g., Western blotting)	Detects proteins, separated by native or SDS-PAGE, according to ability to bind specific antisera or monoclonal antibodies
Chromatography	
Liquid	Several techniques that separate and partially purify proteins based on surface charge, molecular weight, pI, hydrophobicity, and/ or ability to bind specific immobilized ligands
High performance liquid chromatography (HPLC)	Liquid chromatographic separations done under increased pressure; increases resolution and efficiency
Immunochemical	
Enzyme-labeled immunosorbent assay (ELISA); radioimmunoassay (RIA); fluorescence immunoassay (FIA)	Measures levels of substances based on ability to bind specific antibodies; binding is detected using enzyme, radioactive isotope, or fluorescence labeling
Flow cytometry	Detects surface, cytoplasmic, and nuclear antigens; measures calcium flux, DNA synthesis, and viability

biomarker research. Several of these techniques hold particular promise for this type of research. In two-dimensional polyacrylamide gel electrophoresis (2D-PAGE), proteins are separated twice, first in one direction according to their isoelectric point (pI) and then in a direction perpendicular to the first according to their molecular weights, using PAGE under reducing conditions. This technique greatly increases the number of proteins that can be visualized in a sample and yields two pieces of information on each instead of one. The difficult issues of producing reproducible results and analyzing the very large amount of data in 2D gels have been lessened by mechanized systems for casting the gels and computer-based densitometry systems for scanning gels, indexing the protein spots, and comparing different gels. These high resolution and simultaneous analysis attributes make 2D-PAGE well suited for looking for protein biomarkers in complex solutions such as serum. The array of immunochemistry techniques allows specific detection or measurement of biomolecules and provides great flexibility in developing methods.

Use of Animal Models to Investigate Musculoskeletal Biomarkers

One important area of biomarker research is the investigation of potential biomarkers in animal models, in which factors that could confound human studies, for example, medical history, can be controlled. These biomarkers then, of course, must be validated in human studies. A research project that illustrates this point is currently underway at the National Institute for Occupational Safety and Health (NIOSH) facility in Cincinnati, Ohio. In preliminary studies, progressive experimental joint disease was induced in rabbits of approximately 4–5 kg by immobilizing one leg in extension with a thermoplast splint for up to 12 weeks. Blood samples were taken before application of the splints, at weekly intervals, and at necropsy, and were analyzed. The lesions produced were predominately inflammatory in nature, with some cartilage degeneration as well. Grossly, the capsules were thickened and the cartilage surfaces were roughened. Microscopically, the lesions were characterized by fibrillation of the cartilage, synovitis, villinodular proliferation, and some chondrophyte formation. Analysis by nondenaturing PAGE (ND-PAGE) yielded bands that were more intense in treated rabbits than in controls (Figure 21.3). These bands, which appeared as early as 4 weeks, were shown by Western blotting to be haptoglobin. A 25-kDa lectin-binding protein also was identified to be present in higher concentration in treated animals (Figure 21.4). Cartilage matrix glycoprotein was detected in 3 of 4 treated animals, but only 1 of 4 controls. Finally, analysis by enzyme-labeled immunosorbent assay (ELISA) showed β_2-microglobulin to be lower in all control rabbits, except one, than in the treated rabbits. These results show three serum proteins that might be useful as biomarkers for degenerative joint disease. Studies are currently underway to assess changes in these, as well as other, serum proteins during the early stages of immobilization-induced joint disease. A field study of workers at risk for occupational joint disease (carpet layers) will be conducted to determine if the biomarkers identified in these animal studies are applicable in this human population, that is to validate these potential biomarkers of degenerative joint disease. The aim is ultimately to characterize biomarkers of OA at the earliest preclinical and reversible stages to permit successful intervention or preventative measures.

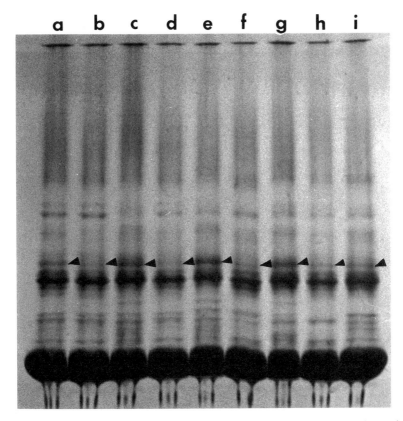

FIGURE 21.3   Native electrophoresis gel of rabbit serum. Lanes a, c, e, g, and i are from rabbits treated by immobilization of one leg for 9 to 16 weeks. Lanes b, d, f, and h are from controls. Arrowheads indicate bands representing haptoglobin. Note that these bands are more intense in the sera from treated rabbits. The samples were run on 8–25% gradient gels and stained with Coomassie Blue.

Conclusions

Research Needs

Research on musculoskeletal biomarkers should focus on further development of biomarkers in animal models and their validation in human populations. An example of a research project to develop better biomarkers of traumatic osteoarthritis is described here. Emerging technologies should be employed to develop better methods that will provide the necessary sensitivity and specificity to evaluate musculoskeletal biomarkers properly. Also, biomarkers are needed that reflect more subtle preclinical (reversible) disorders, especially for the less debilitating but more common and troublesome

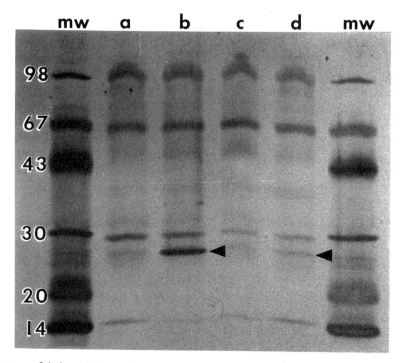

FIGURE 21.4   SDS electrophoresis gel of fucose-containing glycoproteins purified from rabbit serum using lectin-affinity purification. Lanes b and d are serum samples from rabbits treated by immobilization for 9 and 12 weeks, respectively. Lanes a and c are from control rabbits. Arrowheads indicate protein bands that are more intense in sera from treated rabbits (especially lane b) than from controls.

musculoskeletal conditions that commonly become progressively worse over time. Also, reference values for these biomarkers are needed.

Potential Applications for Field Studies: Examples and Considerations

The potential application for biomarkers is great because markers based on noninvasive measures of exposure, and changes of structure or function, will substantially improve future field studies. The greatest potential for biomarkers in field studies will be to clarify the most important mechanisms for musculoskeletal disorders. Determining the relative importance of acute and chronic inflammation, the role of intermittent ischemia, and the role of mediators of chronic pain potentially could resolve some of the greatest debates about the nature of these disorders. Although it is likely that biomarkers will need extensive validation in laboratory and pilot clinical studies before they can be used effectively in large scale epidemiologic studies, their development holds great promise.

References

Arad-Dann, H., Isenberg, D., Ovadia, E., Shoenfeld, Y., Sperling, J., and Sperling, R. (1989). Autoantibodies against a nuclear 56 kDa protein: A marker for inflammatory muscle disease. *J. Autoimmun.* **2**(6), 877–888.

Austin, A. K., Hobbs, R. N., Anderson, J. C., Butler, R. C., and Ashton, B. A. (1988). Humoral immunity to link protein in patients with inflammatory joint disease, osteoarthritis, and in non-arthritis controls. *Ann. Rheum. Dis.* **47**, 886–892.

Axmacher, B., and Lindberg, H. (1988). Coxarthrosis in farmers as appearing on colon radiographs and urograms. *In* "Progress in Occupational Epidemiology" (C. Hogstedt and C. Reurterwall, ed.), pp. 203–206. Elsevier Science, Amsterdam.

Bjerrum, O. W., Rygaard-Olsen, C., Dahlerup, B., Bang, F. B., Haase, J., Janatzen, E., Overgaard, J., and Sehested, P. C. (1984). The carpal tunnel syndrome and amyloidosis. A clinical and histological study. *Clin. Neurol. Neurosurg.* **86**(1), 29–32.

Brandt, K. D. (1989a). A pessimistic view of serologic markers for diagnosis and management of osteoarthritis. Biochemical, immunologic and clinicopathologic barriers. *J. Rheumatol. (Suppl. 18)* **16**, 39–42.

Brandt, K. D. (1989b). Lack of association between serum keratan sulfate concentrations and cartilage changes of osteoarthritis after transection of the anterior cruciate ligament in the dog. *Arthr. Rheum.* **32**(5), 647–651.

Campion, G. V., McCrae, F., Schnitzer, T. J., Lenz, M. E., Dieppe, P. A., and Thonar, E. J.-M. A. (1991). Levels of keratan sulfate in the serum and synovial fluid of patients with osteoarthritis of the knee. *Arthr. Rheum.* **34**(10), 1254–1259.

Dean, J. H., and Murray, M. J. (1991). Toxic responses of the immune system. *In* "Caserett and Doull's Toxicology" (M. O, Amdur, J. Doull, and C. D. Klaassen, ed.), 4th Ed., pp. 282–333. Pergamon Press, New York.

Dean, J. H., Cornacoff, J. B., Rosenthal, G. J., and Luster, M. I. (1989). Immune system: Evaluation of injury. *In* "Principles and Methods of Toxicology" (A. W. Hayes, ed.), 2d Ed., pp. 745–746. Raven Press, New York.

Denko, C. W., and Gabriel, P. (1979). Serum proteins—Transferrin, ceruloplasmin, albumin, α_1-acid-glycoprotein, α_1-antitrypsin—Rheumatic disorders. *J. Rheum.* **6**, 664–672.

Fife, R. S. (1988). Identification of cartilage matrix glycoprotein in synovial fluid in human osteoarthritis. *Arthr. Rheum.* **31**(4), 553–556.

Fife, R. S., and Brandt, K. D. (1989). Cartilage matrix glycoprotein is present in serum in experimental canine osteoarthritis. *J. Clin. Invest.* **84**, 1432–1439.

Fronek, Z., Timmerman, L. A., Alper, C. A., Hahn, B. H., Kalunian, K., Peterlin, B. M., and McDevitt, H. O. (1990). Major histocompatibility complex genes and susceptibility to systemic lupus erythematous. *Arthr. Rheum.* **33**(10), 1542–1553.

Goldberg, R. L., Huff, J. P., Lenz, M. E., Glickman, P., Katz, R., and Thonar, J.-M. A. (1991). Elevated plasma levels of hyaluronate in patients with osteoarthritis and rheumatoid arthritis. *Arthr. Rheum.* **34**(7), 799–807.

Goldberg, V. (1984). The immunology of articular cartilage. *In* "Osteoarthritis Diagnosis and Management" (R. W. Moskowitz, D. S. Howell, V. M. Goldberg, and H. J. Mankin, ed.), pp. 81–92. Saunders, Philadelphia.

Hagberg, M. H., and Wegman, D. H. (1987). Prevalence rates and odd ratios of shoulder-neck diseases in different occupational groups. *Brit. J. Ind. Med.* **44**, 602–610.

Hartmann, D. J., Charriere, G., Ville, G., Vignon, E., and Bejui, J. (1989). Absence of antibodies to native type I and type II collagen in a rabbit model of osteoarthritis. *Arthr. Rheum.* **32**(6), 814–815.

Kelsey, J. L., and Cunningham, L. (1984). Epidemiologic aspects of disability from rheumatic disease. *In* "Epidemiology of the Rheumatic Diseases" (R. C. Lawrence and L. E. Shulman, eds.), pp. 302–311. Gower Medical, New York.

Kyle, R. A., and Greipp, P. R. (1983). Amyloidosis (AL). Clinical and laboratory features in 229 cases. *Mayo Clin. Proc.* **58**(10), 665–683.

Kyle, R. A., Eilers, S. G., Linscheid, R. L., and Gaffey, T. A. (1989). Amyloid localized to teno-synovium at carpal tunnel release. *Am. J. Clin. Path.* **91**(4), 393–397.

McIlwraith, C. W. (1982). Current concepts in equine degenerative joint disease. *J. Am. Vet. Med. Assoc.* **180**(3): 239–250.

Mankin, H., and Brandt, K. (1984). Biochemistry and metabolism of cartilage in osteoarthritis. *In* "Osteoarthritis Diagnosis and Management" (R. W. Moskowitz, D. S. Howell, V. M. Goldberg, and H. J. Mankin, ed.), pp. 43–79. Saunders, Philadelphia.

Meachim, G., and Brooke, G. (1984). The pathology of osteoarthritis. *In* "Osteoarthritis Diagnosis and Management" (R. W. Moskowitz, D. S. Howell, V. M. Goldberg, and H. J. Mankin, ed.), pp. 29–42. Saunders, Philadelphia.

Mehraban, F., and Moskowitz, R. W. (1989). Serum keratan sulfate levels in rabbits with experimentally induced osteoarthritis. *Arthr. Rheum.* **32**(10), 1293–1299.

Mehraban, F., Finegan, C. K., and Moskowitz, R. W. (1991). Serum keratan sulfate: Quantitative and qualitative comparisons in inflammatory versus noninflammatory arthritides. *Arthr. Rheum.* **34**(4), 383–392.

Myllylä, R., Becvar, R., Adam, M., and Kivirikko, K. I. (1989). Markers of collagen metabolism in sera of patients with various rheumatic diseases. *Clin. Chim. Acta* **183**, 243–252.

Nemeth-Csoka, M., Paroczai, C., and Meszaros, Th. (1988). The clinical diagnostic significance of serum antichondrocyte membrane antibodies in osteoarthritis. *Agents Actions* **23**(1/2), 50–51.

Palotie, A., Ott, J., Elima, K., Cheah, K., Väisänen, P., Rhyänen, L., Vikkula, M., and Vuorio, E. (1989). Predisposition to familial osteoarthrosis linked to type II collagen gene. *Lancet* **1**(8644), 924–927.

Parekh, R. B., Dwek, R. A., Sutton, B. J., Fernandes, D. L., Leung, A., Stanworth, D., and Rademacher, T. W. (1985). Association of rheumatoid arthritis and primary osteoarthritis with changes in the glycosylation pattern of total IgG. *Nature (London)* **316**, 452–457.

Parris, W. C. V., Kambam, J. R., Naukam, R. J., and Rama Sastry B. V. (1990). Immunoreactive substance P is decreased in saliva of patients with chronic back pain syndromes. *Anesth. Analg.* **70**, 63–67.

Pujol, J. P., and Loyau, G. (1987). Interleukin-1 and osteoarthritis. *Life Sci.* **41**, 1187–1198.

Puttick, A. H., Briggs, D. C., Welsh, K. I., Vaughn, R., Williamson, E. A., Boyce, M., Jacoby, R. K., and Jones, V. E. (1989). Genes associated with rheumatoid arthritis and mild inflammatory arthritis. I. Major histocompatibility complex class I, II, and III allotypes. *Ann. Rheum. Dis.* **49**(4), 219–224.

Putz-Anderson, V. (1988). "Cumulative Trauma Disorders: A Manual for Musculoskeletal Diseases of the Upper Limbs." Taylor and Francis, London.

Radin, E. L. (1984). Biochemical considerations. *In* "Osteoarthritis Diagnosis and Management" (R. W. Moskowitz, D. S. Howell, V. M. Goldberg, and H. J. Mankin, ed.), pp. 93–107 Saunders, Philadelphia.

Stuart, J. M., Postlethwaite, A. E., Townes, A. S., and Kang, A. H. (1980). Cell-mediated immunity to collagen and collagen α chains in rheumatoid arthritis and other rheumatic diseases. *Am. J. Med.* **69**, 13–18.

Stuart, J. M., Huffstutter, E. H., Townes, A. S., and Kang, A. H. (1983). Incidence and specificity of antibodies to types I, II, III, IV, and V collagen in rheumatoid arthritis and other rheumatic diseases as measured by ^{125}I-radioimmunoassay. *Arthr. Rheum.* **26**, 832–840.

Sukenik, S., Henkin, J., Zimlichman, S., Skibin, A., Neuman, L., Pras, M., and Horowitz, J. (1988). Serum and synovial fluid levels of serum amyloid A protein and C-reactive protein in inflammatory and noninflammatory arthritis. *J. Rheum.* **15**(6), 942–945.

Tanaka, S., Thun, M., Smith, A. B., Halperin, W. E., Lee, S. T., Luggen, M., and Hess, E. V. (1982). "Carpet and Floorlayers." NIOSH HETA 82-065-1664. U.S. Government Printing Office, Washington, D.C.

Tertti, R., and Toivanen, P. (1991). Immune functions and inflammatory reactions in HLA-B27 positive subjects. *Ann. Rheum. Dis.* **50**(10), 731–734.

22

Epilogue

Nathaniel Rothman

Many new and powerful biomarkers are available for epidemiologic research. The purpose of this volume was to present a map of how such markers can be used, as well as a discussion of their strengths and limitations. It has provided an overview of common methodologic issues and specific examples of the use of biomarkers in a wide range of disease categories. This book should guide epidemiologists and laboratory researchers through the promise and potential pitfalls of using these important new tools. In this epilogue, the context of biomarker research will be explored from the vantage point offered by the broad perspective provided in the preceding chapters.

Molecular Epidemiology—A Transitory Nomenclature?

Use of the term "molecular epidemiology" is highly variable across different epidemiologic disciplines. For example, cardiovascular epidemiology uses a wide range of bloodborne lipid markers and DNA-based susceptibility markers, yet has never incorporated molecular epidemiology into its nomenclature. Infectious disease epidemiology has used this term intermittently and inconsistently. Cancer epidemiology, which only recently acquired biomarkers for use in etiologic studies, has adopted the term broadly. Yet all three disciplines use sophisticated biochemical and molecular markers in etiologic research and in assessing the impact of interventions. Further, each field has used or will soon use biomarkers to identify individuals at increased risk of developing disease who will most likely benefit from primary and secondary prevention.

In the future, will all studies that use biomarkers in their research see the need to identify themselves as "molecular"? Or, as biomarkers are "inexorably drawn into the fabric of epidemiologic research" (Hulka, 1990), will the term "epidemiology," without further modification, be adequate to describe the full range of investigations into the etiology and prevention of disease, regardless of the sources of data being used? The answer is unclear, as noted in the preface and Chapter 1. However, for the time being, the term molecular epidemiology does appear to have substantial utility since it heightens awareness about these potentially informative instruments and the methodologic issues unique to their use in epidemiologic investigations.

Some Considerations before Using Biomarkers in Epidemiologic Research

As in all scientific investigations, clarity of thought and purpose at the beginning of the process are critical components to successful research. The following list of questions may provide some assistance to investigators who are beginning to think about using biomarkers in their work.

1. What is the research question being asked?
2. Is there a candidate biomarker, or category of biomarkers, that may be helpful in answering this question?
3. Is enough known about the collection and processing of the biologic sample, the extent of assay measurement error, and the range of the marker in the target population, to design the proposed study effectively?
4. Can the question be answered adequately with alternative, less costly, and perhaps even more effective approaches?
5. Is the research question worth answering?
   a. Will answering the question provide mechanistic insight into a given disease process?
   b. Will answering the question increase our ability to make etiologic inferences about disease causality?
   c. Will answering the question have a positive impact on the public's health?

The commonly heard query "Are there any biomarkers available that can be used to study this disease?" is essentially unanswerable without further clarification. Is the research goal to better assess exposure, genetic susceptibility, or some intermediate outcome? Will the investigation be a cross-sectional, a case-control, or a prospective cohort study? The research question being asked, and the study being envisioned to answer that question, determines the selection of the marker.

The excitement of applying a newly developed biomarker in human populations is a potent driving force for researchers. It is tempting to rush a new marker into the field, without first characterizing the optimal way to process and store the biologic sample, the assay reliability, and the determinants of variability within and between subjects. However, failure to collect this information prior to instituting an etiologic study may result in false-negative or false-positive studies, added inconsistencies to the literature, and wasted time, resources, and energy.

Some epidemiologists would rather wait to cull "validated" biomarkers from the efforts of other investigators. Others do not even consider transitional studies that characterize biomarkers as "true" epidemiologic research, since they are preliminary investigations, often performed on healthy individuals, that are generally incapable of providing direct insight into disease causation. Epidemiologists who are willing to work with laboratory collaborators to determine the optimum way of using a promising new biomarker probably will find that they are able to use the marker sooner, and more effectively, in their population-based studies.

As this book has noted, biomarkers are not always the most appropriate tools for gathering data in epidemiologic studies, particularly for case-control studies of chronic disease, a study design that will continue to dominate much of epidemiologic research because of its efficiency. In this instance, many markers of exposure are limited particularly because of lack of persistence or because of modification by the disease process. For example, biomonitoring workers for current exposure to industrial chemicals is available for many compounds, but retrospective assessment of occupational exposures in case-control studies must rely on the job history, industrial process data, and chemical monitoring data if available, since few chemicals persist in the body or leave a specific impact on some measurable biologic end point. Similarly, urine cotinine levels can help confirm whether someone is currently smoking or not, but a standardized lifetime smoking history obtained by questionnaire is the most valid approach to assessing the pattern of long-term use.

In contrast, biomarkers of genetic susceptibility are particularly valuable for use in case-control studies. Currently available technology can analyze genetic polymorphisms using a small sample of genomic DNA, easily accessible from peripheral blood, a few hair follicles, or a buccal swab. With few exceptions, these assays are not affected by case or control status or previous or current exposures. However, other sources of data, such as questionnaires or record information, will still be necessary to evaluate long-term patterns of exogenous exposures that may modify the expression of, or otherwise interact with, those genetic polymorphisms.

At present, and in the foreseeable future, epidemiology will use questionnaires, medical records, environmental monitoring data, and biomarkers in a complementary fashion to achieve its overall goal: to understand the determinants of human disease and to use that information to control disease through primary, secondary, or tertiary prevention. All sources of epidemiologic data, including biomarkers, are a means to achieving that end.

Looking toward the Future

> By the end of every molecular epidemiologic study that I have conducted, I was told that the assay I had used in analyzing biologic specimens was obsolete . . . by the laboratory collaborator who originally proposed the assay. (Anon.)

Developments in the laboratory that make molecular epidemiology possible are outpacing the traditional time-frame of epidemiologic inquiry. We are in a period of transition to a form of research that will be in a continuous state of flux for the foreseeable future. Studies are being planned with the idea, in fact, with the hope and expectation, that change will occur. Researchers are looking into their "crystal balls" to predict which markers will be most critical to their work in the future, so appropriate biologic samples can be obtained, processed, and adequately stored in the present.

Laboratory investigators and epidemiologists will need to work more closely and communicate more effectively if molecular epidemiologic studies are to be designed in a flexible and efficient manner in the face of rapidly evolving technology. To facilitate this collaboration, it becomes increasingly important to provide epidemiology students with an overview of relevant laboratory science and "hands-on" laboratory experience. The goal of this training will not be to develop expertise in a given assay, which will inevitably become outdated, but to provide insight into the dynamics of laboratory research and the general characteristics of experimental design, sample analysis, and quality control. Similarly, students of toxicology, biochemistry, and molecular biology who wish to apply their work to human populations will need to develop a basic understanding of epidemiologic design and analysis concepts. As noted by the cardiovascular epidemiologist Geoffrey Rose (1990), "Just as epidemiologists need to look inside the black box, so the molecular biologists need to look outside it." It is an encouraging sign that several schools of public health have, or are putting into place, interdisciplinary training programs in the application of biomarkers to epidemiologic research.

In conclusion, designing, implementing, analyzing, interpreting, and communicating the results of molecular epidemiologic studies is a challenging exercise. It can even be a bit daunting, given the probability that previous findings are likely to become outdated at a faster rate than ever before. However, these cautions should *not* lead to inaction. The following observation, written by Toffler in his introduction to *Future Shock*, provides some perspective:

> In dealing with the future, . . . it is more important to be imaginative and insightful than to be one hundred percent "right." Theories do not have to be "right" to be enormously useful. Even error has its uses. The maps of the world drawn by medieval cartographers were so hopelessly inaccurate, so filled with factual error, that they elicit condescending smiles today when almost the entire surface of the earth has been charted. Yet the great explorers could never have discovered the New World without

them. Nor could the better, more accurate maps of today been drawn until men, working with the limited evidence available to them, set down on paper their bold conceptions of worlds they had never seen. (Toffler, 1971)

References

Hulka, B. S. (1990). Methodologic issues in molecular epidemiology. In "Biological Markers in Epidemiology" (B. S. Hulka, T. C. Wilcosky, and J. D. Griffith, eds.), pp. 214–226. Oxford University Press, New York.

Rose, G. (1990). Preventive cardiology: What lies ahead? *Prev. Med.* **19**, 97–104.

Toffler, A. (1971). "Future Shock." Bantam, New York.

Index